BROOKS/COLE

THOMSON LEARNING

Publisher: *Gary W. Ostedt*	Indexer: *Ted Laux*
Marketing Team: *Karin Sandberg, Samantha Cabaluna*	Interior Design: *Forbes Mill Press/Roy R. Neuhaus*
Assistant Editor: *Carol Ann Benedict*	Cover Design: *Vernon T. Boes*
Editorial Assistant: *Daniel G. Thiem*	Interior Illustration: *Audrey Miller*
Production Coordinator: *Kelsey McGee*	Print Buyer: *Vena M. Dyer*
Production Service: *Forbes Mill Press/Robin Gold*	Composition: *WestWords, Inc.*
Manuscript Editor: *Frank Hubert*	Cover Printing, Printing and Binding: *R.R. Donnelley/*
Permissions Editor: *Sue Ewing*	*Crawfordsville*

For more information about this or any other Brooks/Cole products, contact:
BROOKS/COLE
511 Forest Lodge Road
Pacific Grove, CA 93950 USA
www.brookscole.com
1-800-423-0563 (Thomson Learning Academic Resource Center)

Library of Congress Cataloging-in-Publication Data

Bean, Michael A.
    Probability: The science of uncertainty with applications to investments, insurance, and engineering / Michael A. Bean.
        p. cm.
    Includes index.
    ISBN 0-534-36603-1 (text)
    1. Probabilities. I. Title

QA273 .B36 2001   00-058597
519.2—dc21

Probability: The Science of Uncertainty

With Applications to Investments, Insurance, and Engineering

Michael A. Bean, Ph.D., FSA
University of Western Ontario

BROOKS/COLE

TM

THOMSON LEARNING

Australia • Canada • Mexico • Singapore • Spain • United Kingdom • United States

The Brooks/Cole Series in Advanced Mathematics
Paul J. Sally, Jr., Editor

Probability: The Science of Uncertainty
  with Applications to Investments,
  Insurance, and Engineering
Michael A. Bean
University of Western Ontario
© 2001   ISBN: 0534366031

The Mathematics of Finance:
  Modeling and Hedging
Joseph Stampfli
Victor Goodman
University of Indiana–Bloomington
© 2001   ISBN: 0534377769

Geometry for College Students
I. Martin Isaacs
University of Wisconsin–Madison
© 2001   ISBN: 0534351794

A Course in Approximation Theory
Ward Cheney
The University of Texas–Austin
Will Light
University of Leicester, England
© 2000   ISBN: 0534362249

Introduction to Analysis, Fifth Edition
Edward D. Gaughan
New Mexico State University
© 1998   ISBN: 0534351778

Numerical Analysis, Second Edition
David Kincaid
Ward Cheney
The University of Texas–Austin
© 1996   ISBN: 0534338925

Advanced Calculus,
  A Course in Mathematical Analysis
Patrick M. Fitzpatrick
University of Maryland
© 1996   ISBN: 0534926126

Algebra: A Graduate Course
I. Martin Isaacs
University of Wisconsin–Madison
© 1994   ISBN: 0534190022

Fourier Analysis and Its Applications
Gerald B. Folland
University of Washington
© 1992   ISBN: 0534170943

About the Author

Michael A. Bean, Ph.D., FSA, FCIA, has held teaching and research appointments at universities throughout the United States and Canada, including the University of Michigan at Ann Arbor, the University of Toronto, the University of California at Berkeley, the University of Waterloo, and the University of Western Ontario.

He has also worked as a practicing investment actuary for a major international financial services company based in Toronto, and has written numerous articles in the mathematical sciences that have been published in some of the leading journals of the field.

He has a Ph.D. from the University of Waterloo, a Licentiate from Trinity College of Music in London, England, and is a fully qualified Fellow of the Society of Actuaries and a Fellow of the Canadian Institute of Actuaries.

Preface

The idea to write this book first came to me in the spring of 1995, shortly after I joined the faculty of the University of Michigan. At that time, a major review of the entire undergraduate curriculum was underway, the purpose of which was to ensure that undergraduate education remain relevant in the face of a rapidly changing world. About the same time, the Society of Actuaries, which oversees the education of actuaries in North America through the administration of its professional examinations, embarked on a major redesign of its own curriculum to keep abreast of the extraordinary changes taking place in the financial services industry. This time also saw the emergence of financial engineering as a new profession and the rise of programs in financial engineering and financial mathematics around the world. My goal in writing this book was to update the undergraduate probability curriculum to reflect these changes and to incorporate many of the new and interesting applications of probability arising in the fields of engineering, insurance, and investments.

Key Features of This Text

This book has several features that distinguish it from other probability texts currently on the market:

- Key concepts are introduced through detailed motivating examples.
- Random variables and probability distributions are introduced early in the text.
- The text has a large number of detailed worked-out examples and problems with an emphasis on applications from engineering, insurance, and investments.
- There is a wide range of exercises of varying difficulty, many of which are suitable for student projects or group work.
- The text includes topics not covered or not emphasized in other probability texts, such as the geometric expected value, normal power approximations, mixtures, and portfolio selection models.
- The text is written in a clear, concise, expository style, with extensive graphical illustrations throughout, making it well suited for individual study or self-learning.

How to Use This Book

This book can be used in a variety of probability courses with a variety of teaching styles. There is considerably more material in this book than would normally be covered in a one-semester course. Hence, an instructor will have to be selective in what is covered.

What I consider to be core material for an undergraduate probability course is contained in Chapters 3, 4, 5, and 6. An instructor teaching probability should plan on covering most of the material in these chapters, although discussions of some specialized topics such as the Pareto distribution and the beta distribution can be omitted without loss of continuity.

The material in Chapter 7 and Chapter 8 is also important and should be covered to some extent. Instructors teaching engineering students will probably want to discuss the techniques for determining the distribution of a transformed random variable and the distributions of sums and products (§7.1 and §8.1) quite thoroughly. Instructors teaching other types of students may wish to focus on the law of large numbers (§8.4) instead. The sections labeled as being "optional" may be omitted without loss of continuity.

Chapter 2 is unique in that it uses four extended examples to motivate many of the key concepts covered in the rest of the book. I have found that by discussing these examples at the beginning of the course (i.e., before covering Chapters 3 through 8), students are able to make important conceptual discoveries early on and end up learning a great deal of probability theory in a relatively short period of time. Instructors familiar with the discovery method of learning should be quite comfortable using Chapter 2 in this way. Instructors accustomed to teaching in a more traditional way can begin the course at Chapter 3 (after a brief survey of Chapter 1) and use Chapter 2 selectively or omit it entirely.

The material in Chapters 9 and 10 is supplementary and would not normally be covered in a one-semester course in probability. However, this material is good for student projects.

Chapter summaries are provided in the first four chapters to recap the main ideas and help the reader acquire perspective on the subject. Chapters 5 and 6 are written in a summary style throughout and hence do not require separate summary sections. Chapters 7 through 10 are designed to be covered selectively and do not contain summary sections.

An instructor's manual with solutions to all of the exercises in the book is available with a bound-in CD. This manual contains a wealth of material including detailed descriptions of the *Mathematica* commands for constructing the graphs in this book. It is freely available to instructors who adopt this book as a text for their course. For details on how to obtain a copy, contact your Brooks/Cole representative or visit the Brooks/Cole Web site at www.brookscole.com.

Acknowledgments

Writing a textbook of this magnitude is a major undertaking which requires the assistance of many people. I would like to begin by thanking my publisher, Gary W. Ostedt, for agreeing to take on this project and by thanking Carol Benedict, Kelsey McGee, Karin Sandberg, Dan Thiem, and the rest of the Brooks/Cole team for their part in making this book a reality. Thanks also go to Robin Gold at Forbes Mill Press for keeping production on track under a tight schedule and for being patient with me when other responsibilities demanded my attention.

I would also like to thank the reviewers for their valuable comments and suggestions, many of which have been incorporated into the final manuscript. These reviewers include Phillip Beckwith of Michigan Technological University, John Holcomb of Cleveland State University, Paul Holmes of Clemson University, Ian McKeague of Florida State University, and Harry Panjer of the University of Waterloo, former president of the Canadian Institute of Actuaries.

Special thanks also go to my colleague Jack Goldberg for his helpful advice on the publication process (from a textbook author's perspective) and to John Birge for supporting this project in its early stages. I am also grateful to the National Science Foundation and the Center for Research on Learning and Teaching at the University of Michigan for their support of the curriculum development initiatives that ultimately led to my writing this book. Finally, I would like to thank my parents for instilling in me an appreciation of the importance of education and for supporting me in all my endeavors.

Michael Bean

Contents

1 **Introduction**   1

    1.1   What Is Probability?   1

    1.2   How Is Uncertainty Quantified?   2

    1.3   Probability in Engineering and the Sciences   5

    1.4   What Is Actuarial Science?   6

    1.5   What Is Financial Engineering?   9

    1.6   Interpretations of Probability   11

    1.7   Probability Modeling in Practice   13

    1.8   Outline of This Book   14

    1.9   Chapter Summary   15

    1.10 Further Reading   16

    1.11 Exercises   17

2 **A Survey of Some Basic Concepts Through Examples**   19

    2.1   Payoff in a Simple Game   19

    2.2   Choosing Between Payoffs   25

    2.3   Future Lifetimes   36

    2.4   Simple and Compound Growth   42

    2.5   Chapter Summary   49

    2.6   Exercises   51

3 **Classical Probability** 57

3.1 The Formal Language of Classical Probability 58

3.2 Conditional Probability 64

3.3 The Law of Total Probability 68

3.4 Bayes' Theorem 72

3.5 Chapter Summary 75

3.6 Exercises 76

3.7 Appendix on Sets, Combinatorics, and Basic Probability Rules 85

4 **Random Variables and Probability Distributions** 91

4.1 Definitions and Basic Properties 91

    4.1.1 What Is a Random Variable? 91

    4.1.2 What Is a Probability Distribution? 92

    4.1.3 Types of Distributions 94

    4.1.4 Probability Mass Functions 97

    4.1.5 Probability Density Functions 97

    4.1.6 Mixed Distributions 100

    4.1.7 Equality and Equivalence of Random Variables 102

    4.1.8 Random Vectors and Bivariate Distributions 104

    4.1.9 Dependence and Independence of Random Variables 113

    4.1.10 The Law of Total Probability and Bayes' Theorem (Distributional Forms) 119

    4.1.11 Arithmetic Operations on Random Variables 124

    4.1.12 The Difference Between Sums and Mixtures 125

    4.1.13 Exercises 126

4.2 Statistical Measures of Expectation, Variation, and Risk 130

    4.2.1 Expectation 130

    4.2.2 Deviation from Expectation 143

    4.2.3 Higher Moments 149

    4.2.4 Exercises 153

4.3 Alternative Ways of Specifying Probability Distributions 155

    4.3.1 Moment and Cumulant Generating Functions 155

4.3.2   *Survival and Hazard Functions   167*

4.3.3   *Exercises   170*

4.4   Chapter Summary   173

4.5   Additional Exercises   177

4.6   Appendix on Generalized Density Functions (Optional)   178

5   **Special Discrete Distributions**   **186**

5.1   The Binomial Distribution   187

5.2   The Poisson Distribution   195

5.3   The Negative Binomial Distribution   200

5.4   The Geometric Distribution   206

5.5   Exercises   209

6   **Special Continuous Distributions**   **221**

6.1   Special Continuous Distributions for Modeling
       Uncertain Sizes   221

6.1.1   *The Exponential Distribution   221*

6.1.2   *The Gamma Distribution   226*

6.1.3   *The Pareto Distribution   233*

6.2   Special Continuous Distributions for Modeling Lifetimes   235

6.2.1   *The Weibull Distribution   235*

6.2.2   *The DeMoivre Distribution   241*

6.3   Other Special Distributions   245

6.3.1   *The Normal Distribution   245*

6.3.2   *The Lognormal Distribution   256*

6.3.3   *The Beta Distribution   260*

6.4   Exercises   265

7   **Transformations of Random Variables**   **280**

7.1   Determining the Distribution of a Transformed
       Random Variable   281

7.2   Expectation of a Transformed Random Variable   289

7.3   Insurance Contracts with Caps, Deductibles, and Coinsurance (Optional)   297

7.4   Life Insurance and Annuity Contracts (Optional)   303

7.5   Reliability of Systems with Multiple Components or Processes (Optional)   311

7.6   Trigonometric Transformations (Optional)   317

7.7   Exercises   319

8   **Sums and Products of Random Variables**   **325**

8.1   Techniques for Calculating the Distribution of a Sum   325
  8.1.1   *Using the Joint Density*   *326*
  8.1.2   *Using the Law of Total Probability*   *331*
  8.1.3   *Convolutions*   *336*

8.2   Distributions of Products and Quotients   337

8.3   Expectations of Sums and Products   339
  8.3.1   *Formulas for the Expectation of a Sum or Product*   *339*
  8.3.2   *The Cauchy-Schwarz Inequality*   *340*
  8.3.3   *Covariance and Correlation*   *341*

8.4   The Law of Large Numbers   345
  8.4.1   *Motivating Example: Premium Determination in Insurance*   *346*
  8.4.2   *Statement and Proof of the Law*   *349*
  8.4.3   *Some Misconceptions Surrounding the Law of Large Numbers*   *351*

8.5   The Central Limit Theorem   352

8.6   Normal Power Approximations (Optional)   354

8.7   Exercises   356

9   **Mixtures and Compound Distributions**   **363**

9.1   Definitions and Basic Properties   363

9.2   Some Important Examples of Mixtures Arising in Insurance   366

9.3   Mean and Variance of a Mixture   373

9.4   Moment Generating Function of a Mixture   378

9.5   Compound Distributions   379

  9.5.1   *General Formulas*   *380*

  9.5.2   *Special Compound Distributions*   *382*

9.6   Exercises   384

10   The Markowitz Investment Portfolio Selection Model   396

10.1   Portfolios of Two Securities   397

10.2   Portfolios of Two Risky Securities and a Risk-Free Asset   403

10.3   Portfolio Selection with Many Securities   409

10.4   The Capital Asset Pricing Model   411

10.5   Further Reading   414

10.6   Exercises   415

Appendixes   421

A   The Gamma Function   421

B   The Incomplete Gamma Function   423

C   The Beta Function   428

D   The Incomplete Beta Function   429

E   The Standard Normal Distribution   430

F   *Mathematica* Commands for Generating the
    Graphs of Special Distributions   432

G   Elementary Financial Mathematics   434

Answers to Selected Exercises   437

Index   **441**

1 Introduction

Uncertainty is very much a part of the world in which we live. Indeed, one often hears the well-known cliché that the only certainties in life are death and taxes. However, even these supposed certainties are far from being completely certain, as any actuary or accountant can attest; for although one's eventual death and the requirement that one pay taxes may be facts of life, the timing of one's death and the amount of taxes one must pay are far from certain and are generally beyond one's control.

Uncertainty can make life interesting. Indeed, the world would likely be a very dull place if everything were perfectly predictable. However, uncertainty can also cause grief and suffering. For example, the sudden and premature death of a family breadwinner can cause great financial distress for surviving family members with limited means of support. The age-old fascination of humans with predicting the future, as evidenced by the ever-present popularity of astrology and fortune-telling, and the development of institutions such as insurance to make the effects of an uncertain future less severe are no doubt due in large part to a recognition of the malevolent role that uncertainty can play in one's life.

This book presents the scientific approach to uncertainty, known as probability, which has been developed over the past 350 years and is generally accepted in the scientific community. There are undoubtedly many other approaches, such as mysticism and astrology, which some people use to understand uncertainty. However, these approaches lie beyond the realm of science and will not be considered in this book.

In this introductory chapter, we consider what the nature and scope of probability is and how it arises in engineering and the sciences. We also consider how the notion of a probability should be defined and how it can be interpreted. We then discuss how probability models are constructed in practice. We end this introductory chapter with an outline of the topics covered in the rest of the book.

1.1 What Is Probability?

Probability is the branch of science concerned with the study of mathematical techniques for making quantitative inferences about uncertainty. The key words in this definition are *quantitative* and *inferences*. Indeed, as we will soon see, probability provides a mechanism for making quantitative statements about uncertainty and, more important, allows one to draw quantitative conclusions from such statements using the rules of logic.

Most historians consider the work of Fermat (1601–1665) and Pascal (1623–1662) on games of chance to be the first significant contribution to the study of probability; however, many of Fermat's and Pascal's ideas can be traced to earlier works of Cardan, Kepler, and Galileo. There is also some evidence that the Romans, many centuries before, used mortality tables[1] to predict human lifespans. Since Fermat and Pascal's time, nearly every great mathematician has made some contribution to probability. Among the more famous contributors are the Bernoullis, Laplace, DeMoivre, Poisson, DeMorgan, Venn, Bayes, Markov, and Kolmogorov. A complete and readable account of the history of the subject from the early 17th to the mid-19th century is given in the classic book by Todhunter listed at the end of this chapter. Subsequent developments up to the early 20th century are discussed in the scholarly book of the famous economist John Maynard Keynes, which is also listed at the end of the chapter.

While many scholars have studied probability purely for its intellectual and philosophical appeal, a good deal of the motivation for the subject has come, and continues to come, from practical problems outside of mathematics. Indeed, the development of probability since Fermat's time has been heavily influenced by investigations in gaming, demography, insurance, genetics, and quantum physics, to name just a few. Moreover, the subject itself has had profound implications on everything from economics to engineering and, it could be argued, has played a significant role in the history of the world over the last 200 years. To give a simple example, consider marine insurance, whose issuance can be justified by the well-known law of averages: The availability of marine insurance enabled commercial shipping to develop on a large scale (because it freed maritime shippers from the worry of financial ruin due to a catastrophe at sea), which in turn contributed to the economic and political ascendancy of Britain in the 19th century and to international commerce as we know it.[2]

Today, probability is used in a wide range of fields including engineering, finance, medicine, meteorology, and management. We will encounter numerous applications of probability to these and other fields throughout this book.

1.2 How Is Uncertainty Quantified?

If we agree that probability, from a scientific perspective, is the study of mathematical techniques for making quantitative inferences about uncertainty, then for the subject to have any meaningful content, we must have some precise way of quantifying uncertainty and making inferences about that quantification. That there is considerable controversy over how to precisely formulate such a quantification of uncertainty is an understatement, to say the least. Indeed, some philosophers have gone so far as to argue that the very notion of uncertainty cannot be precisely quantified since to do so would, in effect, make uncertainty certain.

One approach to quantifying uncertainty is to use the concept of *relative frequency*. To describe this concept, consider an experiment with several possible outcomes which

---

[1] A mortality table lists the number of deaths each year for a hypothetical group of individuals assumed to be born at the same time.

[2] We will have more to say about the connection between insurance and probability in §1.4.

can be repeated a large number of times.[3] The **relative frequency** of a particular outcome of such an experiment in a sequence of repetitions of the experiment is the fraction of the total number of repetitions of the experiment that result in the desired outcome. For example, in the sequence of coin tosses resulting in H, T, T, T, H (where H signifies heads and T signifies tails), the relative frequency of heads is 2/5, whereas in the sequence of tosses resulting in T, H, T, H, H, the relative frequency of heads is 3/5. Experience suggests that as the number of repetitions of the experiment increases, the relative frequencies associated with a particular outcome converge to a common value. For example, the relative frequency of heads approaches 1/2 as the number of coin tosses increases, provided that the coin is not biased.[4] This common value to which the relative frequencies converge is called the **probability** of the desired outcome.

This approach to quantifying uncertainty, while intuitively appealing, has some major drawbacks, the most serious of which is the reliance on the ambiguous notion of a *limiting* relative frequency. The early probabilists overcame these logical difficulties by restricting their attention to experiments in which the number of outcomes is *finite* and by assuming that all outcomes of such experiments are *equally likely*, (i.e., have the same probability). While the assumption of equal likelihood of outcomes is admittedly idealized,[5] in the context of the games of chance with which the early probabilists were concerned, it is not unrealistic *provided that one correctly identifies the outcomes of the experiment*. The key to applying the classical principle of equal likelihood correctly is to identify the outcomes in such a way that all information on the underlying experiment is captured and no information is suppressed.

To illustrate the difference between a correct and an incorrect application of the principle of equal likelihood, consider the experiment in which two unbiased coins are tossed. If one identifies the possible outcomes as being head–head, head–tail, tail–head and tail–tail, then one correctly assigns a probability of 1/4 to each of these outcomes and one correctly deduces that the probability of getting exactly one head is 1/2. However, if one fails to distinguish between the coins and identifies the possible outcomes as two heads, one head–one tail, and two tails, then one incorrectly assigns a probability of 1/3 to each of the outcomes head–head, tail–tail, and one incorrectly deduces that the probability of getting exactly one head is 1/3.[6]

The great achievement of the classical probabilists was to initiate a *logical* approach to the study of uncertainty, which to a great extent is still with us today. By avoiding the difficulty inherent in considering probabilities to be limiting relative frequencies and instead assuming that all experimental outcomes are equally likely, they were able to focus their energies on developing a logical system for deducing the probabilities of particular *groups* of observations that were often too difficult to determine accurately by successive repetition of an experiment. This system of logical deduction also enabled them to avoid being misled by potentially faulty intuition.

---

[3] The 'experiment' to which we refer here could be a scientific experiment or some other procedure, such as tossing a coin, which can be repeated and which has several possible outcomes.

[4] A coin is **biased** if it has a tendency to land on one side over the other. A coin is **unbiased** or **fair** if it has no such tendency.

[5] It fails, for example, when considering a coin that is biased.

[6] Interestingly enough, the mathematician D'Alembert is alleged to have believed this incorrect line of reasoning at one point in his career. We will have more to say about this particular example in Chapter 3.

TABLE 1.1   Results from 100 Repetitions
of a "Ten Coin Toss"

Number of Heads	Frequency
0	0
1	1
2	3
3	12
4	21
5	24
6	22
7	13
8	4
9	0
10	0
	100

To give a simple illustration of a situation in which the deductive approach succeeds where intuition might fail, consider the probability of getting exactly five heads in ten tosses of a coin. One might think that if the coin is as likely to land heads as it is to land tails, then this probability should be 1/2 since, according to the relative frequency interpretation of probability, an unbiased coin that is tossed a large number of times will land heads approximately half the time. However, if you thought this, you would be wrong! In fact, under the assumption that the probability of getting heads in a single toss is 50%, one can show deductively (using the methods to be developed in Chapter 3) that the probability of getting exactly five heads in ten tosses of the coin is $63/256 = 0.24609375$. Interestingly enough, the value $63/256$ is in accord with the relative frequency interpretation of probability, as one can confirm by repeatedly tossing a fair coin ten times and computing the corresponding relative frequencies. The results of 100 such repetitions are given in Table 1.1.

The deductive approach to probability taken by the classical probabilists is an example of the *axiomatic approach*. The **axiomatic approach** in mathematics is a deductive technique in which the topic of interest is described by a collection of axioms in the language of sets, and all inferences about the topic are made using only these assumptions and the rules of set theory and formal logic. Mathematicians struggled for many years to find an axiomatic formulation for probability that would encompass all types of experiments, not just the ones considered by the classical probabilists. Finally, in the 1930s, the Russian mathematician A. N. Kolmogorov gave an axiomatic description for the theory of probability that permitted virtually every experiment to be considered. A greatly simplified version of this axiomatization will be discussed in Chapter 3.

In most applications, an intuitive understanding of probability based on relative frequencies is generally sufficient. However, it is nice to know that the subject rests on a firm foundation and that the conclusions we reach have some basis in logic!

Before moving on to the next section, it is instructive to make one more remark about the meaning of probability statements. While one often makes probability statements about individual outcomes of an experiment, it is important to keep in mind that such statements are assertions about *groups* of observations. For example, when a doctor tells a patient who must undergo a lifesaving operation that the probability of survival is, for instance, 80%, the doctor is really making a statement about a group of similarly situated individuals because such probability statements have little or no meaning for individuals in isolation. After all, any given individual either survives the operation or does not survive. We will have more to say about interpretations of probability in §1.6.

1.3 Probability in Engineering and the Sciences

Uncertainty arises in virtually every aspect of engineering and science. However, it is most evident in measurement. Indeed, as any good scientist or engineer will acknowledge, no measurement is ever completely certain. Uncertainty in measurement can arise in two ways:

1. Imprecise instruments and human error can result in recorded observations that are different from the true values.
2. The quantity being measured can be in flux (i.e., continually changing) with the result that the recorded observation need not be descriptive of the quantity in the future. An example of this is the measurement of voltage in an electric circuit that uses alternating current. In an alternating current system, the flow of electrical current changes in a cyclical pattern. Consequently, voltage measurements will fluctuate over time.

Probability can be used to quantify the uncertainty in such measurements.

Probability can also be applied to specific areas of engineering such as reliability, quality control, and the analysis of queues.

Reliability is the branch of engineering concerned with the lifetimes of electrical and mechanical systems. Probability can be used to analyze the uncertain lifetimes of such systems and to assist the engineer in designing systems that are both reliable and cost effective.

Quality control is the branch of engineering concerned with the maintenance of quality in manufacturing processes. The goal of quality control is to minimize the number of defects produced by a manufacturing process without incurring unreasonable costs and to identify defects before they leave the production line. Quality control usually involves testing items in the production line to ensure that appropriate standards are met. Since it is not cost effective, and in some cases not even practical,[7] to test every item on a production line, a *sample* of items is usually tested, and inferences about the number of defectives produced in total are made on the basis of test results. Probability can be used to analyze the uncertainty in these inferences and to determine the size of the sample to be tested.

Queuing is the branch of engineering concerned with the analysis and design of systems involving multiple servers and multiple clients, in which clients may be required to wait for service. The simplest example of a queuing system is the checkout line at the grocery store or the service line at a bank or post office. Examples of queues can also

---

[7] For example, stress tests can destroy an item.

be found in multiuser distributed computing systems (so called client–server systems). Probability can be used to analyze the uncertain waiting and service times in such queues and to assist the engineer or computer scientist in designing a system that meets the demands imposed on it in the most cost-effective way.

We will discuss these and other engineering applications in detail throughout the book.[8]

1.4 What Is Actuarial Science?

Outside of the insurance industry, relatively few people have ever heard of actuarial science. Some people may be aware, from reading or hearing about the *Jobs Rated Almanac*, that the job of an actuary is consistently ranked among the best jobs to have. However, they probably have very little idea what an actuary does or what actuarial science is about.

Actuarial science is the subject whose primary focus is analyzing the financial consequences of future uncertain events. In particular, it is concerned with analyzing the adverse financial consequences of large, unpredictable losses and with designing mechanisms to cushion the harmful financial effects of such losses.

We have already discussed an example of an unpredictable loss—the premature death of a family breadwinner with unfulfilled obligations—that can have a devastating financial impact. However, there are many other large, unpredictable losses that can have equally severe financial effects. Among these are the loss or destruction of property due to fire, theft, or natural disaster; the loss of employment due to an economic contraction or the obsolescence of one's skills; and the loss of health due to accident, sickness, or injury. Insurance systems have evolved to cushion the effects of such large, unpredictable losses.

Insurance is based on the premise that individuals faced with large and unpredictable losses can reduce the financial effects of such losses by forming a group and sharing the losses incurred by the group as a whole.

Consider, for example, a group of homeowners who individually risk having their homes destroyed by a tornado. Clearly, the risk of loss faced by each individual in the group is substantial even if tornadoes are a rare occurrence because the size of a loss, when it occurs, will generally be very large. However, the financial consequences of such losses can be reduced through sharing; indeed, if the members of the group agree to pay an equal portion of the total loss incurred by the group as a whole, then the amount that each individual will be required to pay becomes more certain and the chance that any one person will be financially responsible for a large loss becomes very small.

This important principle of loss sharing, known as the **insurance principle**, forms the foundation of actuarial science. It can be justified mathematically using the law of large numbers[9] from probability theory, which we will do in Chapter 8 after we have developed the required probability concepts.

[8] Section 7.5 is specifically devoted to the reliability of systems with multiple components. The other applications are discussed at various points throughout the book.

[9] In fact, one could consider the insurance principle to be simply a restatement of the law of large numbers in the context of insurance.

For the insurance principle to be valid, essentially four conditions should hold (or very nearly hold):

1. The losses should be **unpredictable.**
2. The risks should be **independent** in the sense that a loss incurred by one member of the group makes additional losses by other members of the group no more or less likely.
3. The risks should be **homogeneous** in the sense that a loss incurred by one member of the group is not expected to be any different in size or likelihood from losses incurred by other members of the group.
4. The group should be **sufficiently large** so that the portion of the total loss that each individual is required to pay becomes relatively certain.

Let's consider each of these conditions in turn.

Losses Should Be Unpredictable   If the time and size of a future loss are known in advance, then apart from finding a charitable benefactor, there is nothing one can do to moderate the financial effects of the loss. For it is unlikely that anyone would be willing to share the expense of such a loss unless they themselves were certain to have a future expense of equal or greater size, in which case there would be no benefit to entering into a loss-sharing arrangement. Hence, when losses are not unpredictable, insurance is ineffective.

Risks Should Be Independent   If the losses incurred by individuals in the group tend to occur in concert with one another (i.e., if they are *dependent*), then the portion of the group's total loss paid by each individual will not be appreciably smaller than the large individual loss that would have to be paid by the individual member in the absence of a loss-sharing arrangement. For instance, if the homeowners in the previous example all live on the same street, then it is quite likely that when a tornado strikes the neighborhood, it will damage many of the houses in the group and the resulting repair cost to each member will still be potentially ruinous even though the total repair cost is spread over the entire group. Hence, when risks are not independent, insurance can be ineffective.

Risks Should Be Homogeneous   If the losses expected to be incurred by each member of the group are not the same, then the members with the lower or the less likely losses will not consider the equal distribution of loss expenses to be fair and will not agree to pay for the losses of the other members of the group. For instance, if some of the homeowners in the previous example had houses that were much more expensive to replace than the rest or were located in an area where tornadoes occurred more frequently, then it is unlikely that the homeowners with the more modest or more protected homes would be willing to share equally in the cost of reconstruction in the event of a tornado. Hence, when risks are not homogeneous, insurance arrangements can break down.

Group Should Be Sufficiently Large   If the number of individuals in the group is too small, then the portion of the total loss that each individual in the group is required to pay will be highly unpredictable and is likely to be prohibitively large. In this case, loss sharing will offer the individual little improvement over bearing a loss alone and could in some cases result in a worse situation because the chance that an individual will have to make a payment is actually greater when the individual is part of a group. Hence, when the group is not sufficiently large, insurance is not very effective.

In practice, risks are not truly independent or homogeneous. Moreover, there will always be situations where the condition of unpredictability is violated.[10] However, as long as the dependence among the risks is weak and the differences among the risks are not too great to cause large inequities in the amounts individuals are required to pay, insurance is generally feasible. In situations where heterogeneous (i.e., nonhomogeneous) risks must be combined to obtain a group that is sufficiently large, an equitable distribution of loss expenses can generally be achieved by appropriately *weighting* the amount payable by each member of the group.

Each of these four conditions has a precise mathematical formulation in the language of probability, as we will see when we formally discuss the law of large numbers in Chapter 8.

It is worth pointing out that insurance arrangements such as the one just illustrated cannot eliminate misfortune; indeed, homes are still going to be destroyed by tornadoes. However, insurance arrangements can make the financial effects of such misfortune less devastating and more certain. This is a common feature of insurance. Many people mistakenly believe that insurance is akin to gambling. However, this is not so because gambling creates a risk where there was none before, whereas insurance manages an existing risk that is unavoidable.

For practical reasons, people seeking insurance do not generally try to form loss-sharing groups by themselves. Instead, they turn to financial intermediaries such as insurance companies to do this for them. In particular, people seeking insurance protection enter into contracts with insurance companies in which they transfer the unwanted risk to the insurance company and agree to pay the insurance company an up-front premium for this service. By entering into a large number of such contracts with different individuals, the insurance company can, in turn, form a group for which the insurance principle holds, thereby ensuring that its aggregate claim expenses are reasonably predictable, and make a profit in the process. It is at this juncture that actuarial science enters the picture.

Actuarial science seeks to address the following three problems associated with any such insurance arrangement:

1. Given the nature of the risk being assumed, what price (i.e., premium) should the insurance company charge?
2. Given the nature of the *overall* risks being assumed, how much of the aggregate premium income should the insurance company set aside in a reserve to meet contractual obligations (i.e., pay insurance claims) as they arise?
3. Given the importance to society and the general economy of having sound financial institutions able to meet all their obligations, how much capital should an insurance company have above and beyond its reserves to absorb losses that are larger than

---

[10] Consider, for example, the well-documented cases of automobile owners who abandon their cars, report them stolen, and then illegally collect compensation from an insurance company in an attempt to obtain more money than the resale value of the car.

expected? Given the actual level of an insurance company's capital, what is the probability of the company remaining solvent?

These are generally referred to as the problems of *pricing, reserving,* and *capital allocation and insolvency.*

Actuaries must also be concerned with two economic problems, *adverse selection* and *moral hazard,* which arise from an insurance company's inability to access perfect information about people purchasing insurance. Economists refer to this more general problem as the problem of *imperfect information.* **Adverse selection** in an insurance context arises from an inability to distinguish completely the good risks from the bad. If an insurer is unable to distinguish between risks and charges a uniform premium for all, then the good risks will find the insurance protection expensive and let their policies lapse, whereas the bad risks will find the insurance protection a bargain and purchase more. The result will be larger claim payments than anticipated and inadequate premium income to cover the payments. **Moral hazard** in an insurance context arises from the behavioral changes that insurance protection induces after it is purchased: Once insured, people are more likely to act in reckless ways than they otherwise would because they know that the insurance company will pay for any losses. These two problems can have a significant impact on pricing, reserving, and solvency. Actuaries generally try to minimize their impact by designing products with features such as deductibles and coinsurance, which serve to align the interests of the policyholder with those of the insurer and encourage policyholders through their choice of policy to reveal information that enables insurers to better determine their risk class.[11]

1.5 What Is Financial Engineering?

Financial engineering is a relatively new discipline that is focused on the analysis of risk in financial markets and the design of products and techniques to manage that risk. This rapidly evolving field, which first came into prominence in the 1990s, relies heavily on the use of sophisticated computers and computational techniques to exploit market anomalies and manage financial risk.

Four factors have contributed to the development of financial engineering: technology, deregulation, globalization, and the increased reliance on capital markets and market forces in public policy.

- Rapid advances in computer and communications technology since the late 1970s combined with the resulting fall in transaction costs have made practical the implementation of sophisticated investment strategies such as program trading, portfolio insurance, dynamic hedging, and portfolio replication.
- Deregulation, such as the repeal in 1999 by the United States Congress of the 1930s era Glass–Steagall Act prohibiting commercial banks, investment banks, and insurance companies from entering each other's businesses, has enabled the development of new financial products combining savings, investment, and insurance features which until now were only theoretically possible.

---

[11] For example, people who anticipate making an insurance claim are more likely to choose a low-deductible policy than people who only desire protection against a catastrophic loss which they consider unlikely.

- Globalization of financial markets and of business has led to the development of global financial institutions and of financial products (e.g., cross-currency options[12]) for managing the risks associated with doing business in different countries with different legal systems and different currencies.
- Finally, the increased reliance on capital markets and market forces as illustrated through the privatization of state-owned enterprises in former socialist countries and the privatization of public pension schemes in Latin America has stimulated the demand for new financial products (e.g., mutual funds with return guarantees) to manage risks previously assumed and obscured by government, but now made explicit by the capital market.

The fundamental principle underlying much of financial engineering is the **principle of no arbitrage.** This principle asserts that two securities that provide the same future cash flow and have the same level of risk must sell for the same price. Equivalently, the principle asserts that a risk-free investment of zero can only have a return of zero. In other words, there is "no free lunch."

Arbitrage opportunities[13] can arise from time to time. However, the systematic exploitation of such opportunities by market players will cause market prices to adjust until a point is reached where the arbitrage opportunity no longer exists or it is no longer profitable to take advantage of the opportunity due to the size of transaction costs. Hence, the principle of no arbitrage implicitly assumes a well-functioning market that is in *equilibrium.*

A related concept that arises frequently in financial engineering is the **principle of optimality.** This principle asserts that investors, when forming investment portfolios, should allocate their funds among the available securities in a way that optimizes the portfolio's *risk-adjusted* return. The classic application of this principle is the Nobel Prize winning portfolio selection model due to Harry Markowitz, which we will discuss in Chapter 10.

Financial engineers combine these two principles with techniques from probability, statistics, mathematical physics, and numerical computation to construct optimal portfolios and manage financial risk.

The reader may have noticed the similarity between the work of financial engineers and the work of actuaries. Indeed, it would appear that both types of professionals are concerned with the financial consequences of future uncertainties, which is our definition of actuarial science. The current differences between the disciplines lie in the *types* of future uncertainties considered (e.g., risk of premature death vs. risk of a market crash) and in the *principles* used (e.g., insurance principle vs. principle of no arbitrage). There is also a historical difference: Actuarial science traditionally was concerned with uncertainty in the *liabilities* of the business, whereas financial engineering developed in

---

[12] An **option** is a financial security that gives the holder the right, but not the obligation, to buy or sell as the case may be a specific item (in this case, a specific amount of currency) for a specific price during a specific time period. Options and other financial contracts will be discussed in greater detail in §7.3.

[13] An **arbitrage opportunity** is an investment opportunity that provides a positive return with no initial investment and no risk. For example, if stock XYZ is selling for $100 in New York and £50 in London, and if the exchange rate is £1 = $1.50, it is possible to make a riskless profit by simultaneously buying shares in London, selling shares in New York, and buying pounds with part of the proceeds to cover the purchase price in London.

response to increasing uncertainty in the *assets* of the business, which had previously been ignored. With the trend by businesses to consider contingencies in their assets and liabilities together, it appears likely that the disciplines of actuarial science and financial engineering will move closer together, and it is possible that they will eventually merge.

1.6 Interpretations of Probability

So far in our discussion, we have considered probabilities to be long-run relative frequencies and we have tacitly assumed that they are fixed even though our *estimate* of them may change from one set of experimental data to another. This perspective, in which probability is considered to be a constant long-run relative frequency, is known as the **frequentist** or **objectivist** interpretation of probability. However, there is another perspective, known as the **Bayesian**[14] or **subjectivist** interpretation, in which probabilities are considered to be measures of personal belief. With the Bayesian perspective, the probability assigned to a particular event[15] can be different for two different people even if it is based on the same set of experimental data and can change over time to reflect new information and evolving opinion.

Both interpretations of probability have coexisted since the time of Fermat. Indeed, the classical probabilists' assumption of equal likelihood of outcomes could be considered an expression of their *belief* in the fairness of the games that they were studying.[16] However, from time to time, one or the other of these perspectives has dominated the thinking of the scholars of the day. For example, during the first half of the 20th century, the frequentist perspective so dominated statistical thinking that anyone suggesting a Bayesian approach to a problem risked becoming an outcast in the scholarly community. Today, both perspectives are generally regarded as valuable. However, this does not mean that the Bayesian interpretation is not subject to debate, sometimes heated!

To fully appreciate the difference between the frequentist and the Bayesian perspective, and to help you form your own opinion about which viewpoint is "more correct," it is instructive to consider a concrete example. Hence, suppose that you are given a coin with two distinguishable sides to be used in some game of chance, and consider how you would interpret the probability of getting heads in each of the two paradigms, both before and after being given historical data on the coin.

Frequentist Perspective Before being told anything about the coin's history, the frequentist observes that there are two possible outcomes, assuming that the coin always lands on one side or the other, but reasons that nothing can be said about the probability of heads since there are no available data on which to make a statement about the long-run relative frequency. Hence, the frequentist concludes only that the probability lies between zero and one, nothing else. If it is then revealed that in 1000 tosses of the

---

[14] Named for Thomas Bayes (1701–1761) who discovered the rule on which much of the subjectivist perspective is based. The rule, known as *Bayes' theorem*, will be discussed in Chapter 3.

[15] The term *event* actually has a precise mathematical meaning in probability, as we will discuss in Chapter 3. However, one can interpret it here in the sense of ordinary English.

[16] The principle of equal likelihood was often referred to as the principle of insufficient reason (see Keynes), suggesting an expression of opinion.

coin 550 heads were observed, the frequentist concludes that an estimate for the probability of heads is 55%, acknowledging that the true probability remains unknown. If it is further revealed that in a different 1000 tosses of the coin 510 heads were observed, the frequentist concludes (noting that a total of 1060 heads have been observed in 2000 tosses) that while the unknown probability of heads is the same as before, the estimate for this probability should now be 53%. For the frequentist, the estimate of the probability of heads can change, but the probability of heads itself remains constant. Hence, according to a frequentist, the probability of heads is an inherent constant that does not change with the arrival of new data.

Bayesian Perspective   Before being told anything about the coin's history, the Bayesian observes that the coin is symmetrically constructed except for the different markings on the two sides of the coin and reasons, by the laws of physics, that there should not be a tendency for the coin to land on one side more than the other. Hence, in the absence of historical data, the Bayesian is led to *believe* that the probability of heads is 50%. If it is then revealed that in 1000 tosses of the coin 550 heads were observed, the Bayesian reasons that the probability of heads is closer to 55% and acknowledges that the original belief about this probability was incorrect. If it is further revealed that in a different 1000 tosses of the coin 510 heads were observed, the Bayesian then reasons that the probability is closer to 53% and acknowledges again that earlier beliefs about this probability were incorrect. For the Bayesian, both the estimate of the probability and the belief about what that probability was in the first place can change over time. Hence, according to a Bayesian, the probability of heads is not an inherent constant; rather, it is a number that can change over time to reflect the experimenter's opinion.

In a more sophisticated Bayesian analysis, one can consider the probability p of getting heads to be a "random" quantity itself, and one can assign "probabilities" to each of the values between zero and one in a way which reflects one's belief that the unknown probability p is a particular value[17]; over time, this assignment of probabilities to the possible values of p changes with the arrival of new information to reflect changes in one's belief about p. For example, prior to seeing any data, one may believe that p is within 1% of .50 "with high probability" (e.g., with 80% probability); however, after learning that 550 heads have been observed in 1000 tosses, one may alter this opinion and instead believe that there is only a 25% chance that p is within 1% of .50. The key point to remember is that the notion of probability as it applies to p is a measure of belief that varies from person to person.

Classical statisticians argue that Bayesian methods suffer from a lack of objectivity because different individuals are free to assign different probabilities to the same event according to their personal opinions. Bayesians counter that the classical methods, based on a frequentist interpretation of probability, have built-in subjectivity (e.g., through the design of an experimental sampling procedure) and that the advantage of the Bayesian approach is that the subjectivity is made explicit.

---

[17] If you are uncomfortable with the notion of p as a random quantity, then imagine instead that the assignment of probabilities to the values in [0, 1] is a quantification of the uncertainty in p.

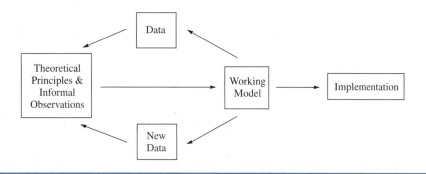

FIGURE 1.1 Schematic Diagram of the Modeling Process

In this book, we will use both the frequentist and the Bayesian interpretations of probability, as dictated by the nature of the application. While we will not explicitly identify which approach is being taken at any given instant, it would be a good exercise for the reader to do so and to try to understand why the interpretation being used is appropriate to the particular problem being discussed. In the exercises at the end of this chapter, you will have an opportunity to explore the two paradigms in greater detail and decide for yourself which of the interpretations you consider more reasonable.

1.7   Probability Modeling in Practice

Most practical problems are addressed using a combination of Bayesian and frequentist methods. Typically, one begins with some simplifying assumptions—such as the assumption that an earthquake in California has no effect on fire insurance claims in Michigan—which one believes to be true or very nearly true; then using theoretical principles, one deduces a working model. This model may contain several unspecified parameters[18]—such as the probability of an earthquake occurring in California in the coming year—which must be estimated using historical data before the model can be used. Once a working model is completely specified, it is then tested on new data for its accuracy, and on the basis of these tests, the underlying assumptions and parameter values are modified. This process is repeated several times until the model is in accord with experience. Finally, the model is implemented to provide a solution to the original problem. The entire modeling process is illustrated in Figure 1.1.

A good model helps one understand why the observed data have the form that they do and does not just give a formula for generating future observations. Indeed, models based solely on historical data should be viewed with some caution since they implicitly assume that the process generating the data, which the model does not describe, will continue to behave in the future as it has in the past. The latter assumption is a dangerous one that can lead to questionable predictions. For example, historically rising housing prices may

---

[18] A **parameter** is an unspecified quantity that is assumed to be constant. For example, in the formula for the velocity of an object thrown vertically into the air, $v(t) = -gt + v_0$, v_0 is a parameter which represents the velocity at time zero. Here, g is the constant acceleration due to gravity and is equal to 9.8 m/s^2.

lead one to predict a 10% annual return on real estate investments for the foreseeable future. However, if personal incomes are not rising with housing prices, then it is unlikely that such a real estate boom can be sustained because a point will be reached where large numbers of people will be unable to afford a house, and the resulting decrease in demand will have a moderating effect on housing price increases.

During the mid-1980s, housing prices in many regions of the country rose by more than 20% per year, leading some "experts" in the industry to predict that the price of an average family home in the year 2000 would be more than $1 million; however, housing prices (outside of New York City or Silicon Valley) in 2000 were nowhere near that predicted level, and it is doubtful that they will reach it for many years to come. Examples such as this have led some cynics to liken statistical forecasting to "driving down the freeway with your eyes focused on the rearview mirror." The point is a valid one. However, more often than not, the problem lies with the person making the predictions rather than with the methodology being used.

1.8 Outline of This Book

We now present a brief outline of the topics to be covered in the rest of the book.

We begin our study of probability in Chapter 2 with a discussion of four concrete examples. These examples are designed to introduce the reader to the important concepts of random variable and probability distribution in a concrete and meaningful way. Examples illustrating arithmetic and geometric means and their uses are also given.

Chapter 3 presents the classical combinatorial probability theory based on sets and counting, and the assumption of equally likely outcomes, in the context of insurance risk classification. Classical probability theory is an effective tool for addressing risk classification and related problems where decisions must be made on the basis of imperfect information. Our discussion is relatively brief so as not to stray too far from our main theme—the random variable—and includes statements and applications of the important law of total probability and Bayes' theorem.

Chapter 4 gives a detailed treatment of random variables and probability distributions, including discussions of distribution functions, density functions, survival functions, hazard functions, expectation, variance, and moment generating functions. This chapter also considers random vectors and bivariate distributions, and it introduces the reader to sums and products of random variables. The meaning of equality for random variables is also considered.

Chapters 5 and 6 present detailed discussions of special distributions that are useful in probability modeling. These chapters also present numerous applications of the special distributions to engineering, insurance, and investment problems.

Chapter 7 addresses the important topic of transformations of random variables. This chapter presents several detailed examples of how to calculate the distribution function for a transformed random variable (i.e., the distribution function for a function of a random variable). Transformations of random variables arise frequently in applications, and it is important to be at ease with them. This chapter also contains optional discussions of insurance contracts and the reliability of systems with multiple components.

Chapter 8 considers two special transformations of random variables: sums and products. Methods for calculating the distribution functions of sums and products are developed, and several detailed examples are given. The chapter also includes a discussion of two important theorems: the law of large numbers and the central limit theorem. Applications of the theorems are given.

Chapter 9 presents a detailed discussion of mixtures and compound distributions. Mixtures arise frequently in actuarial science, particularly in situations where multiple risk classes are present and the risk class of an individual is uncertain. Important formulas for calculating the unconditional expectation and variance are developed through concrete examples.

Chapter 10 concludes our presentation with a discussion of the Markowitz portfolio selection model mentioned in §1.5. This material does not appear in other probability texts of this level. However, it is a nice application of the basic theory and is very accessible.

1.9 Chapter Summary

Probability is the branch of science concerned with the study of mathematical techniques for making quantitative inferences about uncertainty. The term **probability** also refers to the number between zero and one that quantitatively measures the uncertainty in any particular event, with a probability of one indicating that the event is certain to occur and a probability of zero indicating that the event is certain not to occur.

Probability arises naturally in engineering and scientific problems where the measurement of quantities is uncertain. It also plays a principal role in reliability engineering, quality control, and the analysis of queues.

The theory of probability can be formulated using the **axiomatic approach**. This is a deductive technique, common in mathematical study, in which the subject of interest is described by a collection of axioms in the language of sets and all inferences about the subject are made using only these assumptions and the rules of set theory and formal logic.

In concrete applications, probabilities are usually interpreted in one of two ways. The **frequentist** or **objectivist** interpretation of probability is a perspective in which probabilities are considered to be constant long-run relative frequencies. The **Bayesian** or **subjectivist** interpretation of probability is a perspective in which probabilities are considered to be measures of belief that can change over time to reflect new information.

Classical statisticians argue that Bayesian methods suffer from a lack of objectivity because different individuals are free to assign different probabilities to the same event according to their own personal opinions. Bayesians counter that the classical methods, based on a frequentist interpretation of probability, have built-in subjectivity (e.g., through the choice and design of a sampling procedure) and that the advantage of the Bayesian approach is that the subjectivity is made explicit.

Most practical problems are addressed using a combination of Bayesian and frequentist methods. Typically, a probability model is constructed using simplifying assumptions, theoretical principles, and historical data. Models that enable us to understand why the observations have the form that they do are usually better than models based on historical data alone.

It is important to remember that probability statements only have meaning in the aggregate and do not make sense for individual observations. When a doctor tells a patient who must undergo a lifesaving operation that the probability of survival is, for example, 80%, the doctor is really making a statement about a *group* of similarly situated individuals because any given individual will either survive the operation or will not survive.

Actuarial science is the subject concerned with analyzing the adverse financial consequences of large, unpredictable losses and with designing mechanisms to cushion the harmful financial effects of such losses. **Financial engineering** is the discipline concerned with analyzing risk in financial markets and with designing products and techniques to manage that risk.

Actuarial science is based on applications of the insurance principle, whereas financial engineering is based on applications of the principle of no arbitrage and the principle of optimality. Historically, actuarial science developed to address contingencies in a company's *liabilities*, whereas financial engineering developed to address contingencies in the company's *assets*. With the trend toward a unified analysis of assets and liabilities accelerating, the two professions are likely to move closer together.

1.10 Further Reading

The following classic books on probability will be of interest to the reader who would like to learn more about the history of the subject.

G. BOOLE. *The Laws of Thought*, Macmillan, London 1854. Reprinted by Dover, New York 1951.

A. DeMOIVRE. *Doctrine of Chances*, 1718, 1738, 1756. Reprinted by AMS Chelsea, 1967.

A. DeMORGAN. *An Essay on Probabilities and on Their Application to Life Contingencies and Insurance Offices*, Longman, London 1838.

R. A. FISHER. *Statistical Methods and Scientific Inference*, Third Edition, Macmillan, New York 1973.

N. L. JOHNSON & S. KOTZ, EDS. *Leading Personalities in Statistical Sciences from the Seventeenth Century to the Present*, Wiley, New York 1997.

J. M. KEYNES. *A Treatise on Probability*, Macmillan, London 1921.

A. N. KOLMOGOROV. *Foundations of the Theory of Probability* (English translation of the 1933 German original), Chelsea, New York 1950.

P. S. LAPLACE. *A Philosophical Essay on Probabilities* (English translation of the 1814 French original), Wiley, New York 1917. Reprinted by Dover, New York 1952.

L. J. SAVAGE. *The Foundations of Statistics*, Second Edition, Dover, New York 1972.

I. TODHUNTER. *A History of the Mathematical Theory of Probability*, Macmillan, London 1865. Reprinted by Chelsea, New York 1949.

J. VENN. *The Logic of Chance*, Macmillan, London 1866, 1876, 1888. Reprinted by Chelsea, New York 1962.

1.11 Exercises

1. Without consulting the text, define or otherwise describe the meaning of each of the following terms:

probability	objectivist
reliability	subjectivist
parameter	relative frequency
arbitrage opportunity	axiomatic approach
frequentist	queue
Bayesian	probability model
quality control	no arbitrage
actuarial science	insurance principle
financial engineering	optimality

2. Discuss the differences between each of the following pairs of terms.

   a. Bayesian, frequentist
   b. insurance principle, no-arbitrage principle
   c. moral hazard, adverse selection
   d. actuarial science, financial engineering

3. Toss a standard coin 200 times and record the head–tail outcomes as they occur in succession. From the recorded data, determine the relative frequency of heads in

   a. the first 10 tosses;
   b. the first 30 tosses;
   c. the first 50 tosses;
   d. the first 100 tosses;
   e. the total 200 tosses.

   What conclusions do you reach?

4. What are the drawbacks with using limiting relative frequencies to define probabilities? How did the classical probabilists overcome the logical problems associated with relative frequencies?

5. What types of uncertainty can arise in the measurement of quantities such as velocity, temperature, or electrical current?

6. Give examples of situations in engineering where probability models can be used.

7. What four conditions should be satisfied for a mutual insurance arrangement to be feasible? Under what circumstances might these conditions be violated?

8. What three problems associated with insurance arrangements does actuarial science seek to address?

9. One often reads statements in the press such as, "The probability of the earth being destroyed by a meteor is 1 in 1 million." Are such statements meaningful? Explain. What meaning, if any, could be given to this particular statement?

10. An insurance company is studying the accident experience of a group of newly licensed drivers to determine appropriate insurance rates for people seeking automobile insurance for the first time. One thousand 18-year-olds with a permanent driver's license are selected at random and are observed over a 5-year period. The

number of reported accidents for the group over the 5 years is as follows: 90, 70, 75, 60, 65, respectively.

a. Discuss the frequentist and Bayesian interpretations of probability in the context of this problem.

b. Is it realistic to assume a constant accident frequency over time? What factors might cause the accident frequency to change? How might a frequentist explain a change in accident frequency?

11. Most experienced investors will acknowledge that stock prices are inherently unpredictable, at least over the short term. If it were otherwise, an investor with sufficient means could make unlimited profit without risk by buying a stock just before it moved up and selling it just before it moved down. Evidence suggests that it is virtually impossible to do this on a consistent basis. Nevertheless, there is no shortage of opinion on the future direction of the market.

a. Discuss the frequentist and Bayesian interpretations of probability in the context of stock price movements.

b. In any security transaction, there must be both a buyer and a seller. Explain how the Bayesian interpretation of probability is consistent with this fact. How might a frequentist make sense of this fact?

12. Discuss how one constructs a probability model in practice.

2

A Survey of Some Basic Concepts Through Examples

We begin our study of probability by considering four concrete situations where the theory of probability can be applied. Our objective is to introduce the reader to the important concepts of random variable and probability distribution, which are the primary tools for constructing probability models, in a concrete and meaningful way. These examples also introduce the reader to the concepts of arithmetic and geometric mean in a probabilistic context and illustrate their use in investment decision problems. It is our hope that by introducing these concepts at this point in our presentation, the reader will gain a better appreciation of their significance and have a better understanding of them when they are encountered in the more formal setting of Chapter 4.

2.1 Payoff in a Simple Game

This example introduces the random variable as a monetary payoff in a game of chance. The concepts of probability distribution and arithmetic expectation are also illustrated.

Statement of the Example

Consider the following game:

A single balanced die is rolled. If the result is two, three, or four, we win $1; if it is five, we win $2; but if it is one or six, we lose $3.

Should we play this game?

Initial Observations

Notice that there are two elements of this game to consider: the outcome of rolling the die (one of the numbers 1, 2, 3, 4, 5, 6) and the monetary payoff (which is determined by the observed outcome and the specific formula described in the statement of the game). The outcome of the roll is relatively simple to analyze. Indeed, the assumption that the die is *balanced* means that there is no tendency for the die to land on a particular side, and so according to the relative frequency interpretation of probability, each of the six outcomes has an equal probability of being observed. Hence, the probability of the die landing on any given side is 1/6. The structure of the payoff, however, is more subtle.

A cursory glance at the description of the payoff structure might lead one to believe that we have the advantage since our probability of winning (i.e., of having a positive payoff) is 2/3. The payoff is positive for four of the six possible outcomes of rolling the die; thus, since each outcome is equally likely, the probability of getting one of these four particular outcomes is $4/6 = 2/3$. However, a closer examination of the payoff structure reveals that the potential loss on a single roll is much greater than the potential gain, and so it may not be to our advantage to play the game after all. This observation suggests that it is the payoff rather than the actual outcome of the die roll that should be the focus of our attention.[1]

Formulation of the Problem

It is standard practice in mathematical modeling problems to represent the quantities of interest by letters of the alphabet, and we follow this practice in all the probability models we discuss. Hence, let X denote the payoff amount in the game just described. The payoff X is an example of a *random variable*. It is a *variable* because its value varies in a well-defined way according to the outcome of rolling the die; it is *random* because the underlying "process" (rolling the die) on which it depends is itself random. In fact, X is considered a **discrete random variable** because its possible values belong to the *discrete* set $\{-3, 1, 2\}$.[2]

Generally speaking, a **random variable** is any quantity with real values that depends in a well-defined way on some process whose outcomes are uncertain. By convention, random variables are denoted by uppercase letters to distinguish them from ordinary (nonrandom) variables, which are generally written using lowercase letters. In addition, the notation $X = x$ is used to represent the statement that the random quantity X, which has many possible values, assumes the particular nonrandom value x. Statements such as $X \leq x$ have similar interpretations.

The word *process* in our informal definition of random variable is unfortunate because it is easily confused with the term *stochastic process,* which refers to the evolution of an uncertain quantity over time. To avoid this confusion, the term **experiment** is generally used to refer to a process (in the current sense) whose outcomes are not known in advance. Hence, it is more correct to say that a **random variable** is a real valued quantity that depends in a well-defined way on the outcome of an experiment.

We can make our definition of random variable more precise by using the mathematical concepts of set and function. Indeed, if we let S denote the set of all possible outcomes that can result from a given experiment, then a random variable which depends on the outcome of this experiment is simply a real valued function $X : S \rightarrow \mathbf{R}$. The set S is known as the **sample space** for the experiment. In the current example, $S = \{1, 2, 3, 4, 5, 6\}$ and the payoff X is the function given by

$$X(1) = -3, \quad X(2) = 1, \quad X(3) = 1,$$
$$X(4) = 1, \quad X(5) = 2, \quad X(6) = -3.$$

---

[1] Of course, the payoff still depends on the outcome of the die roll, and so the outcomes cannot be completely ignored. The point is that a consideration of the die outcomes *alone* is insufficient.

[2] A set of numbers is considered a **discrete** set if it is finite or if it is infinite and its elements can be arranged in order using the counting numbers. For example, the set of even numbers is discrete since it can be ordered as $0, 2, -2, 4, 6, -6, \ldots$. However, the set of real numbers is not discrete because it cannot be ordered using the counting numbers.

Note that there can be many different random variables associated with a given experiment because there are many different real valued functions that can be defined on a given sample space. In the context of the current example, this is simply a statement of the fact that there can be many different payoffs associated with the same roll of the die.

Distribution of Probability

We have already noted that the assumption that the die is balanced leads to the conclusion that each outcome of the experiment has probability $1/6$. This gives us a description of the distribution of probability over the elements of the sample space S. However, to determine whether or not we should play the game, we really need to know the distribution of probability over the values of the random *payoff* X.

From our discussion, we know that X has three possible values: 1, 2, and -3. To say that $X = 1$ means that the outcome of the roll was two, three, or four. Hence, the probability that $X = 1$ is just the probability of rolling two, three, or four, which is $1/2$ since the die is balanced. Similarly, the probability that $X = 2$ is $1/6$, and the probability that $X = -3$ is $1/3$.

We can conveniently encode this information on the relative frequencies of X by defining a probability *function* for X. In general, the **probability mass function** p_X associated with a discrete random variable X is defined by letting $p_X(x)$ be the probability that $X = x$. Hence, in our current example,

$$p_X(-3) = \frac{1}{3}, \qquad p_X(1) = \frac{1}{2}, \qquad p_X(2) = \frac{1}{6},$$

or equivalently,

$$p_X(x) = \begin{cases} \frac{1}{3} & \text{if } x = -3, \\ \frac{1}{2} & \text{if } x = 1, \\ \frac{1}{6} & \text{if } x = 2, \\ 0 & \text{otherwise.} \end{cases}$$

Note that $p_X(x)$ is defined for all real numbers x, including numbers which are not possible values of X. Indeed, the statement $p_X(x) = 0$ simply expresses the fact that there is no chance that X can assume the value x.

The probability mass function p_X for a discrete random variable X encodes all the relevant probability information about X in a concise form. In particular, it encodes only the relative frequencies for X and suppresses information on the underlying experiment on which X depends. For example, in the game under consideration, the statement $X = 1$ says only that a payoff of \$1 is obtained and not whether this payoff is the result of rolling two, three, or four. This suppression of information on the underlying experiment is actually a fairly desirable property to have because it enables us to focus on the item of real importance (in this case, the payoff) in a problem. Indeed, by eliminating explicit reference to the underlying experiment, we are in a better position to compare payoffs that arise from different random experiments (e.g., from rolling dice or drawing cards). We will obtain a greater appreciation of this property later when we encounter problems in which we must consider several different random quantities at the same time.

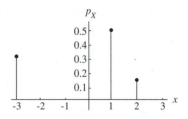

FIGURE 2.1   Mass Plot

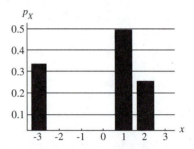

FIGURE 2.2   Bar Chart

The probability mass function of a discrete random variable also has many of the properties of a general probability, which we will discuss in Chapters 3 and 4. For example,

1. $0 \le p_X(x) \le 1$ for all x;
2. $\sum_{\text{all } x} p_X(x) = 1$.

The first property asserts that every function value of p_X is a relative frequency for X. The second asserts that exactly one of the possible payoff amounts will occur. In addition, p_X has the very desirable feature of being defined on the real line, since all the possible values of X are real numbers.[3] Consequently, p_X can be visually represented by a graph.

There are primarily two types of graphs that can be constructed for p_X: a *mass plot* or a *bar chart*. A **mass plot** is a visual representation of a probability mass function in which the precise locations of the probability masses are highlighted using points and lines. A **bar chart** is a visual representation of a probability mass function in which the sizes of probability masses are highlighted using rectangular bars. Mass plots are useful for emphasizing the exact locations of the probability masses. However, bar charts tend to be more appealing visually. The mass plot and bar chart for p_X are given in Figures 2.1 and 2.2.

---

[3] This is not always true of a general probability, as we will see in the next chapter. For example, a random drawing from a standard deck of playing cards will not generally be a real number; of course, it is still possible to assign numerical values to the face cards in this case, but this just amounts to defining a random variable on the underlying random experiment.

Expected Payoff

Let's return to the question of whether or not to play the game in this example. Recall that the probability of winning money on a single roll is 2/3; however, the size of the potential loss is much greater than the size of the potential gain, suggesting that it may not be advantageous to play.

Suppose we were to play this game n times, where n is a large number. Then, according to our naive understanding of probability as a relative frequency, we would expect the following to happen:

- approximately one sixth of the time ($n/6$ games) we would win \$2;
- approximately one half of the time ($n/2$ games) we would win \$1;
- approximately one third of the time ($n/3$ games) we would lose \$3.

Consequently, we would expect our accumulated winnings to be

$$(2)\left(\frac{n}{6}\right) + (1)\left(\frac{n}{2}\right) + (-3)\left(\frac{n}{3}\right)$$
$$= -\frac{1}{6}n.$$

This amounts to an average *loss* of one sixth of a dollar (i.e., $16\frac{2}{3}$ cents) per game.

What does this "loss" actually mean? It does not mean that we stand to lose $16\frac{2}{3}$ cents in one play of the game. It does mean that if we play the game a large number of times, we should expect to lose an average of $16\frac{2}{3}$ cents per game.[4] Thus, while it might be worthwhile to play the game once (since the probability of winning in a single game, i.e., of getting $X > 0$, is 2/3), we should expect to lose money in the long run.

We can express these ideas more succinctly using the notion of probability mass function introduced earlier. Indeed, the expected accumulated winnings over the n plays of the game can be written as

$$(2)\left(\frac{n}{6}\right) + (1)\left(\frac{n}{2}\right) + (-3)\left(\frac{n}{3}\right)$$
$$= n\left(2 \cdot p_X(2) + 1 \cdot p_X(1) + (-3) \cdot p_X(-3)\right)$$
$$= n\left(\sum_{\text{all } x} x\, p_X(x)\right),$$

the latter equality following from the fact that $p_X(x) = 0$ for $x \neq -3, 1, 2$. Hence, the average amount we should expect to win per game when we play a large number of times and arithmetically accumulate our winnings is

$$\sum_{\text{all } x} x\, p_X(x) = -\frac{1}{6}.$$

---

[4] Of course, in any sequence of n repetitions, the actual average loss per game may be very different from $16\frac{2}{3}$ cents. However, we expect it to be close to this average value when n is large.

In general, the **expected value** of a discrete random variable X is defined to be the number $E[X]$ given by

$$E[X] = \sum_{\text{all } x} x \, p_X(x),$$

provided that this sum exists.

The quantity $E[X]$ measures the amount one should expect to gain on average if a game with payoff X is played a large number of times and winnings are accumulated arithmetically. Note that $E[X]$ is the arithmetic average of the values of X weighted by the probability mass function p_X. In §2.4, we will give an example where gains are *compounded* rather than accumulated arithmetically; in such cases, it is often more meaningful to consider *geometric* averages.

There is an interesting graphical interpretation of $E[X]$ which allows us to think of the expected value in more concrete terms:

Suppose that we imagine the real line to be a balance with its fulcrum at the origin, and suppose that at each position x for which $p_X(x) \neq 0$, we place a mass equal to $p_X(x)$ pounds:

In general, the balance will tilt one way or the other. Then the quantity $E[X]$ is the point on the axis to which the fulcrum must be moved to restore balance:

With this interpretation, the expected value is seen to be the *center of mass* or mean of the probability distribution. This explains why the function p_X is called a probability *mass* function.

Summary of Concepts Illustrated in This Example

Before proceeding to the next example, we summarize some of the important probability concepts illustrated in this example.

A **random variable** is any quantity X with real values that depends in a well-defined way on a random experiment. More precisely, it is a real valued function $X : S \to \mathbf{R}$, where S is the set of possible outcomes of the random experiment. The set S is known as the **sample space** of the experiment. The notation $X = x$ is used to represent the statement that the random variable X, which has many possible values, assumes the particular value x. A **discrete random variable** is a random variable X whose set of possible values forms a *discrete* set.

The **probability mass function** p_X associated with a discrete random variable X measures the probability that $X = x$, at each point x, and contains all the relevant information about X. It can be represented visually using a **mass plot** or a **bar chart**. The function p_X is actually defined for all real numbers x, not just the numbers x that are possible values of X; indeed, the statement $p_X(x) = 0$ simply means that there is no chance of having $X = x$.

The **expected value** $E[X]$ of a discrete random variable X is the number $\sum x \, p_X(x)$, where the sum is taken over all possible values of X. It is the arithmetic average amount one should expect to make in a game with payoff X when the game is played a large number of times and winnings are accumulated arithmetically. It is also the *center of mass* of the probability distribution.

2.2   Choosing Between Payoffs

This example illustrates how probability distributions can be effectively used when one must make a choice between two random payoffs. The concepts of independence and equivalence of random variables are also introduced.

Statement of Example

Two coins are tossed and the outcome is observed. Before the coins are tossed, we are given a choice of the following payoffs:

Payoff 1:   Win $1 for each head.
              Lose $3 for getting two tails.

Payoff 2:   Win $1 if the coins are different.
              Win $2 if both coins turn up tails.
              Lose $3 if both coins turn up heads.

Which payoff should we choose?

Initial Observations

Our discussion in §2.1 suggests that the way to approach this problem is to consider the probability distribution of the payoffs. Hence, let Y_1 and Y_2 denote payoff 1 and payoff 2, respectively. Further, let S be the sample space for the underlying experiment. Then, provided that we distinguish between the two coins, the possible outcomes of the experiment are

$$S = \{HH, HT, TH, TT\},$$

where H represents head, T represents tail, and where the order in each of the pairs HH, HT, TH, TT is respected. Note that it is possible to distinguish between the coins by flipping them separately or by using coins of different types (e.g., a nickel and a dime).

With the sample space defined in this way, the payoffs Y_1 and Y_2 are given by

$$Y_1(HH) = 2, \qquad Y_1(HT) = 1, \qquad Y_1(TH) = 1, \qquad Y_1(TT) = -3$$

and

$$Y_2(HH) = -3, \qquad Y_2(HT) = 1, \qquad Y_2(TH) = 1, \qquad Y_2(TT) = 2.$$

The payoffs are clearly different. Hence, after the outcome is revealed, it matters whether we chose payoff 1 or payoff 2 (e.g., if the outcome is two heads, we would win $2 if we had chosen payoff 1, but we would lose $3 if we had chosen payoff 2). However, before we know the outcome, does it matter which payoff we choose? Intuition suggests that it might not.

Distribution of Probability

To determine which payoff we should choose, we need to consider the probability distributions for Y_1 and Y_2. Since the probability distributions for Y_1 and Y_2 are determined by the way probability is distributed over the sample space, we must begin by considering the relative frequencies for the outcomes of the underlying random experiment (tossing two coins).

As we have already noted, the underlying experiment has four possible outcomes: HH, HT, TH, TT, provided that we distinguish between the coins. One way to assign relative frequencies to these outcomes would be to toss the coins together a large number of times and record the frequencies of each outcome. If both coins happened to be fair (i.e., to have no bias favoring one side over the other), we would find that each outcome pair HH, HT, TH, TT occurred approximately one quarter of the time. This suggests that we can assign probabilities by assuming that all outcome pairs are equally likely.

With this assignment of probability, the probability mass functions for Y_1 and Y_2 are given by

$$p_{Y_1}(y) = \begin{cases} \frac{1}{4} & \text{if } y = -3, 2, \\ \frac{1}{2} & \text{if } y = 1, \\ 0 & \text{otherwise,} \end{cases}$$

and

$$p_{Y_2}(y) = \begin{cases} \frac{1}{4} & \text{if } y = -3, 2, \\ \frac{1}{2} & \text{if } y = 1, \\ 0 & \text{otherwise.} \end{cases}$$

For example, the probability that $Y_1 = 1$ is the probability that the experimental outcome is HT or TH, which is $2/4 = 1/2$ since all experimental outcomes are assumed to be equally likely.

Notice that the probability mass functions for Y_1 and Y_2 are exactly the same! This means that before the outcome of the coin toss is known, we are just as likely to win (or lose) any given amount with payoff 1 as with payoff 2. Consequently, before the outcome of the coin toss is known, we should be indifferent to choosing between the given payoffs.

Note that this does not mean that our eventual wealth will be the same regardless of the choice of payoff. Indeed, if the result of the coin toss turns out to be two heads, then the person who had chosen payoff 1 would realize a gain of $2, while the person who had chosen payoff 2 would suffer a loss of $3. Hence, although we should be indifferent to choosing between the two payoffs before the outcome of the coin toss is known, our choice will still affect our ultimate wealth. We can express this fact by saying that the payoffs are *equivalent* but not equal.

In general, two random variables with the same probability distributions are said to be **identically distributed.** We will occasionally use the term **equivalent** in this book to describe such random variables, although this terminology is not as descriptive and is used less frequently in the probability literature. Hence, two discrete random variables Y_1 and Y_2 are identically distributed if and only if

$$p_{Y_1}(y) = p_{Y_2}(y) \qquad \text{for all } y.$$

Choosing Payoffs When Distributions Are Not Identical

We have just seen that when the distributions of two payoffs are the same, we should be indifferent to choosing between the payoffs before the outcome of the experiment is known. However, what if the distributions are not the same? Which payoff should we choose then?

To address these questions, let's consider a third payoff:

Payoff 3:   Win $3 for getting two heads.
               Win $1 for getting one of each.
               Lose $4 for getting two tails.

Let Y_3 denote this payoff. Then the probability mass function of Y_3 is given by

$$p_{Y_3}(y) = \begin{cases} \frac{1}{4} & \text{if } y = -4, 3, \\ \frac{1}{2} & \text{if } y = 1, \\ 0 & \text{otherwise.} \end{cases}$$

The distribution of Y_3 is clearly different from the distributions of Y_1 and Y_2 (Figures 2.3 and 2.4). Note, however, that

$$E[Y_1] = E[Y_2] = E[Y_3] = \frac{1}{4}.$$

That is, all three payoffs have the same expected value. Hence, with each payoff, we should expect to gain the same amount *on average* if we play the game a large number of times (using the chosen payoff each time) and winnings are accumulated arithmetically. Does this mean that we should be indifferent to choosing between payoff 3 and the other two payoffs? Not necessarily.

Notice that the *range* of possible values for Y_3 is wider than the range of possible values for Y_1 and Y_2. In particular, with payoff 3, it is possible both to win more and lose more than with either of the other payoffs in any given play of the game. Hence, if a person is *risk averse* in the sense that they feel the pain of a monetary loss more that they

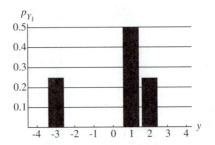

FIGURE 2.3 Bar Chart for p_{Y_1}

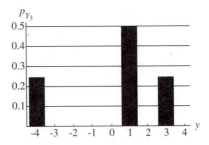

FIGURE 2.4 Bar Chart for p_{Y_3}

experience the joy of a monetary gain, they would probably choose payoff 1 or payoff 2 over payoff 3. On the other hand, if a person is *risk tolerant* in the sense that the joy of a substantial gain outweighs the pain of a substantial loss, they would probably choose payoff 3 over payoff 1 or payoff 2.

Most people are risk averse and thus would probably choose payoff 1 or payoff 2. However, not all people are risk averse all of the time. Hence, when payoff distributions are different, the choice between payoffs is determined to a large extent by personal preference. Nevertheless, if we are given partial information about an individual's preferences, we can often make assertions about what choices are consistent with these preferences. For example, if we are told that an individual prefers payoff 3 to payoff 1, then we can conclude that the individual must also prefer payoff 3 to payoff 2 because payoff 1 and payoff 2 have the same distribution.

Probability Distributions When Coins Are Not Fair

So far, we have made decisions about payoffs under the assumption that the coins in the experiment were both fair. However, not all coins have this property. How do we determine the probability mass functions for the payoffs when the coins are not fair? Note that when the coins are not fair, we can no longer assume that the elements of the sample space $S = \{HH, HT, TH, TT\}$ are equally likely.

Suppose that we label the coins as 1 and 2 and that the probability of heads in a single toss of coin 1 is p_1 while the probability of heads in a single toss of coin 2 is p_2, where at least one of p_1, p_2 is assumed to be different from $1/2$. Our goal is to determine probabilities for each of HH, HT, TH, TT that are consistent with these assumptions.

Let's introduce two new random variables X_1 and X_2 as follows:

$$X_1 = \begin{cases} 1 & \text{if coin 1 lands heads,} \\ 0 & \text{if coin 1 lands tails;} \end{cases}$$

$$X_2 = \begin{cases} 1 & \text{if coin 2 lands heads,} \\ 0 & \text{if coin 2 lands tails.} \end{cases}$$

The quantities X_1 and X_2 are examples of *indicator* random variables. An **indicator random variable** is a random variable whose only values with nonzero probability are 0 and 1; such a random variable "indicates" the occurrence of a specific "event" by assuming the value 1 when the desired event happens and the value 0 otherwise. In the current example, X_1 is an indicator of heads for coin 1 while X_2 is an indicator of tails for coin 2.

With these two random variables, we can recapture the sample space S by introducing the *random vector* $\mathbf{X} = (X_1, X_2)$, where X_1 and X_2 are as defined earlier. A **random vector** is a vector quantity with real components that depends in a well-defined way on the outcome of a random experiment. The distribution of probability for a random vector with two components is referred to as a **bivariate distribution.** The distribution of probability for a random vector with more than two components is referred to as a **multivariate distribution.** When it is desirable to emphasize the interaction of the components of a random vector, the distribution of probability is referred to as the **joint distribution** of the components.

In the current example, the vector $\mathbf{X}$ is such that

$$\mathbf{X}(\text{HH}) = (1, 1), \quad \mathbf{X}(\text{HT}) = (1, 0),$$
$$\mathbf{X}(\text{TH}) = (0, 1), \quad \mathbf{X}(\text{TT}) = (0, 0).$$

Hence, there is a one-to-one correspondence between elements of the sample space S and values of $\mathbf{X}$ with nonzero probability. Consequently, the distribution of probability on the sample space is given by the probability mass function of the random vector $\mathbf{X}$.

Now the probability mass functions of X_1 and X_2 are given, respectively, by

$$p_{X_1}(x_1) = \begin{cases} p_1 & \text{if } x_1 = 1, \\ 1 - p_1 & \text{if } x_1 = 0, \end{cases}$$

and

$$p_{X_2}(x_2) = \begin{cases} p_2 & \text{if } x_2 = 1, \\ 1 - p_2 & \text{if } x_2 = 0. \end{cases}$$

Our goal is to determine the probability mass function of the vector $\mathbf{X} = (X_1, X_2)$. We will appeal to the relative frequency interpretation of probability and the fact that the outcomes of the two coin tosses do not depend on each other. For simplicity, we begin by considering the point $\mathbf{x} = (1, 1)$, which represents the occurrence of two heads.

Suppose that we toss the two coins n times, where n is a sufficiently large number. Then according to the relative frequency interpretation of probability and the assumption that $p_{X_1}(1) = p_1$, approximately $p_1 n$ of the ordered pairs (x_1, x_2) obtained should be of the form $(1, *)$, where $*$ denotes 0 or 1. Suppose that exactly n^* of the ordered pairs obtained are of this form. Then using the fact that the outcome of tossing coin 2 does not depend on the outcome of tossing coin 1, and the assumption that $p_{X_2}(1) = p_2$, together with the relative frequency interpretation of probability, we surmise that approximately $p_2 n^*$ of the n^* ordered pairs $(1, *)$ are of the form $(1, 1)$. Hence, approximately $p_2 n^* \approx p_2 p_1 n$ of the original n ordered pairs (x_1, x_2) should be of the form $(1, 1)$. Since n is a large number, we surmise using the relative frequency interpretation of probability again that

$$p_{\mathbf{X}}(1, 1) = p_1 p_2.$$

In a similar fashion, we can argue that

$$p_{\mathbf{X}}(1, 0) = p_1(1 - p_2),$$
$$p_{\mathbf{X}}(0, 1) = p_2(1 - p_1),$$
$$p_{\mathbf{X}}(0, 0) = (1 - p_1)(1 - p_2).$$

Hence, the probability mass function $p_{\mathbf{X}}$ for $\mathbf{X}$ is

$$p_{\mathbf{X}}(\mathbf{x}) = \begin{cases} p_1 p_2 & \text{if } \mathbf{x} = (1, 1), \\ p_1(1 - p_2) & \text{if } \mathbf{x} = (1, 0), \\ p_2(1 - p_1) & \text{if } \mathbf{x} = (0, 1), \\ (1 - p_1)(1 - p_2) & \text{if } \mathbf{x} = (0, 0), \\ 0 & \text{otherwise.} \end{cases}$$

Consequently, the probability mass functions for payoff 1 and payoff 2, when the probability of heads for coin 1 is p_1 and the probability of heads for coin 2 is p_2, are given by

$$p_{Y_1}(y) = \begin{cases} (1 - p_1)(1 - p_2) & \text{if } y = -3, \\ p_1(1 - p_2) + p_2(1 - p_1) & \text{if } y = 1, \\ p_1 p_2 & \text{if } y = 2, \\ 0 & \text{otherwise,} \end{cases}$$

and

$$p_{Y_2}(y) = \begin{cases} p_1 p_2 & \text{if } y = -3, \\ p_1(1 - p_2) + p_2(1 - p_1) & \text{if } y = 1, \\ (1 - p_1)(1 - p_2) & \text{if } y = 2, \\ 0 & \text{otherwise.} \end{cases}$$

Note that the distributions of Y_1 and Y_2 are generally different because it is usually not the case that $p_1 p_2 = (1 - p_1)(1 - p_2)$. Hence, when the coins are not fair, it is very possible that an individual may prefer payoff 1 over payoff 2, or vice versa.

Numerical Illustration When Coins Are Not Fair

To make these formulas more concrete, let's consider a numerical example. Suppose that the probability of obtaining heads on coin 1 is 1/4 and the probability of obtaining heads on coin 2 is 2/3. Then the probability mass functions of X_1, X_2, $\mathbf{X}$, Y_1, and Y_2 are, respectively,

$$p_{X_1}(x_1) = \begin{cases} \frac{1}{4} & \text{if } x_1 = 1, \\ \frac{3}{4} & \text{if } x_1 = 0; \end{cases} \qquad p_{X_2}(x_2) = \begin{cases} \frac{2}{3} & \text{if } x_2 = 1, \\ \frac{1}{3} & \text{if } x_2 = 0; \end{cases}$$

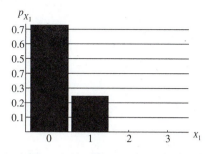

FIGURE 2.5 Bar Chart for p_{X_1}

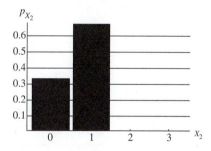

FIGURE 2.6 Bar Chart for p_{X_2}

$$p_{\mathbf{X}}(\mathbf{x}) = \begin{cases} \frac{1}{6} & \text{if } \mathbf{x} = (1,1), \\ \frac{1}{12} & \text{if } \mathbf{x} = (1,0), \\ \frac{1}{2} & \text{if } \mathbf{x} = (0,1), \\ \frac{1}{4} & \text{if } \mathbf{x} = (0,0); \end{cases}$$

$$p_{Y_1}(y) = \begin{cases} \frac{1}{4} & \text{if } y = -3, \\ \frac{7}{12} & \text{if } y = 1, \\ \frac{1}{6} & \text{if } y = 2, \\ 0 & \text{otherwise}; \end{cases} \qquad p_{Y_2}(y) = \begin{cases} \frac{1}{6} & \text{if } y = -3, \\ \frac{7}{12} & \text{if } y = 1, \\ \frac{1}{4} & \text{if } y = 2, \\ 0 & \text{otherwise}. \end{cases}$$

The corresponding bar chart representations are given in Figures 2.5 through 2.9.

Interestingly enough, from the graphs alone, it is not immediately clear what the relationships among the variables X_1, X_2, $\mathbf{X}$, Y_1, and Y_2 are. Hence, while graphs can be very useful for providing us with intuition, we see that they are no substitute for precise analytic formulas.

Independence of Random Variables

You may have noticed that the probability mass function for $\mathbf{X}$ is related to the probability mass functions of its components X_1, X_2 by the formula

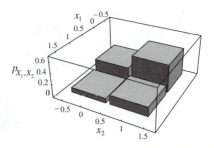

FIGURE 2.7   Bar Chart for $p_{\mathbf{X}}$

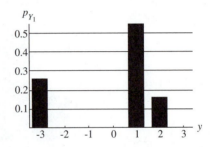

FIGURE 2.8   Bar Chart for p_{Y_1}

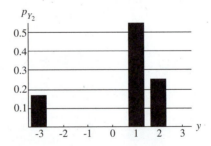

FIGURE 2.9   Bar Chart for p_{Y_2}

$$p_{X_1,X_2}(x_1, x_2) = p_{X_1}(x_1) \cdot p_{X_2}(x_2)$$

for all x_1, x_2.[5] In fact, without a relationship such as this between the probability mass functions p_{X_1}, p_{X_2} and the probability mass function $p_{\mathbf{X}}$, we would not have been able to deduce values for the $p_{\mathbf{X}}(\mathbf{x})$ regardless of how we had assigned values to the $p_{X_1}(x_1)$ and the $p_{X_2}(x_2)$. It is natural to wonder where this formula comes from and whether it is universally true for bivariate distributions.

---

[5] When we wish to emphasize the form of the components of $\mathbf{X}$, we write $p_{X_1,X_2}(x_1, x_2)$ in place of $p_{\mathbf{X}}(\mathbf{x})$.

TABLE 2.1 Contingency Table for Two Indicator Variables

		X_2		$p_{X_1}(x_1)$
		0	1	
X_1	0	$p_{\mathbf{X}}(0, 0)$	$p_{\mathbf{X}}(0, 1)$	$p_{X_1}(0)$
	1	$p_{\mathbf{X}}(1, 0)$	$p_{\mathbf{X}}(1, 1)$	$p_{X_1}(1)$
$p_{X_2}(x_2)$		$p_{X_2}(0)$	$p_{X_2}(1)$	1

A little reflection reveals that the foregoing product formula for $p_{\mathbf{X}}$ is a consequence of the *independence* of X_1 and X_2. Indeed, recall that in our earlier heuristic derivation of the value $p_{\mathbf{X}}(1, 1)$, we argued that in a large number of random tosses of the two coins, the fraction of the outcomes of the type $(1, *)$ that are actually of the type $(1, 1)$ is approximately p_2 $(= p_{X_2}(1))$; however, this "argument" was only valid under the assumption that knowledge of the first component (i.e., knowledge that $X_1 = 1$) had no effect on the probability distribution of the second component (i.e., no effect on the probability that $X_2 = 1$).

In general, we say that two random variables X_1 and X_2 are **independent** if knowledge of one of the variables has no effect on the probabilities assigned to the values of the other. If, in addition, X_1 and X_2 are both discrete, then the relative frequency interpretation of probability suggests that the probability mass function of the joint distribution $\mathbf{X} = (X_1, X_2)$ is the product of the probability mass functions of the scalar components,

$$p_{X_1, X_2}(x_1, x_2) = p_{X_1}(x_1) \cdot p_{X_2}(x_2) \qquad \text{for all } x_1, x_2.$$

One can show in a more formal setting that the validity of this product formula for $p_{\mathbf{X}}$ is actually *equivalent* to the independence of X_1 and X_2 (i.e., it is both a consequence and a requirement of independence). Hence, we should not expect this product formula to hold universally. This suggests that, in the absence of an independence condition on X_1 and X_2, we should not expect to deduce values for $p_{\mathbf{X}}$ directly from the probability mass functions of X_1 and X_2 without understanding how knowledge of one of the quantities X_1, X_2 affects the probability distribution of the other.

Contingency Tables

When a random vector $\mathbf{X} = (X_1, X_2)$ with two components has only a small number of possible values, the probability mass function of $\mathbf{X}$ can be displayed in tabular form in a way that better reveals the relationships between $\mathbf{X}$ and its components. For example, if the only possible values of X_1 and X_2 are 0 and 1, as in the current example, then the probability mass functions of $\mathbf{X}$, X_1, X_2 can be displayed as in Table 2.1. Such a representation is known as a **contingency table** for X_1 and X_2.

Notice that the values of the joint probability mass function $p_{\mathbf{X}}(\mathbf{x})$ are given in the interior square while the values of the probability mass functions for the components X_1, X_2 are given in the "margins." For this reason, the distributions of X_1 and X_2 are often referred to as the **marginal distributions** of the vector **X**. The 1 in the lower right-hand corner of the table simply expresses the fact that

$$\sum_{\text{all } \mathbf{x}} p_{\mathbf{X}}(\mathbf{x}) = \sum_{\text{all } x_1} p_{X_1}(x_1) = \sum_{\text{all } x_2} p_{X_2}(x_2) = 1.$$

That is, $p_{\mathbf{X}}$, p_{X_1}, p_{X_2} are all probability mass functions.

The marginal values of any contingency table can be obtained from the interior values by summing the appropriate row or column. To be precise,

$$p_{X_1}(x_1) = \sum_{\text{all } x_2} p_{X_1, X_2}(x_1, x_2),$$

$$p_{X_2}(x_2) = \sum_{\text{all } x_1} p_{X_1, X_2}(x_1, x_2).$$

However, the interior values can only be obtained from the marginal values when the marginal distributions are *independent*. In that case, and only in that case, the interior values are *products* of marginal values:

$$p_{X_1, X_2}(x_1, x_2) = p_{X_1}(x_1) \cdot p_{X_2}(x_2) \qquad \text{for all } x_1, x_2.$$

Summary of Concepts Illustrated in This Example

We now summarize some of the important probability concepts illustrated in this example.

Two discrete random variables X_1 and X_2 are **identically distributed** if they have the same probability mass functions, that is, if

$$p_{X_1}(x) = p_{X_2}(x) \quad \text{for all } x.$$

This condition is often expressed by writing $X_1 \sim X_2$.

Being identically distributed does not mean that $X_1 = X_2$. Rather, it means that X_1 and X_2 are equivalent in the sense that everyone in the world regardless of risk preference should be indifferent to choosing between them when they represent payoffs in a game of chance.

An **indicator random variable** is a random variable X whose only possible values are 0 and 1. Such a random variable indicates the occurrence of an event by assuming the value 1 when the desired event happens and the value 0 otherwise.

A **random vector** is any vector quantity $\mathbf{X} = (X_1, \ldots, X_n)$ with real components that depends in a well-defined way on the outcome of a random experiment. The notation $\mathbf{X} = \mathbf{x}$ means that the vector quantity $\mathbf{X}$, which has many possible vector values, assumes the particular vector value $\mathbf{x}$; in component form, this notation becomes $(X_1, \ldots, X_n) = (x_1, \ldots, x_n)$ and means that the random variables $X_1, \ldots, X_n$ assume the particular real values $x_1, \ldots, x_n$, respectively.

The statement that $\mathbf{X}$ is a random vector is often expressed by saying that $\mathbf{X}$ has a **multivariate distribution** or, when it is desirable to emphasize the interaction of the components, a **joint distribution**. In particular, when $\mathbf{X}$ has exactly two components, it is said to have a **bivariate distribution**.

A **discrete random vector** is a random vector whose components are all discrete random variables. The **probability mass function** $p_{\mathbf{X}}$ associated with a discrete vector random variable $\mathbf{X}$ measures the probability that $\mathbf{X} = \mathbf{x}$ at each vector $\mathbf{x}$ and is defined for all real vectors $\mathbf{x}$, not just the vectors $\mathbf{x}$ which are possible values of $\mathbf{X}$; indeed, the statement $p_{\mathbf{X}}(\mathbf{x}) = 0$ simply expresses the fact that $\mathbf{x}$ is not a possible value of $\mathbf{X}$.

When $\mathbf{X}$ has exactly two components, X_1 and X_2, the probability mass function is usually written as p_{X_1, X_2} to emphasize the components X_1 and X_2 and can be represented visually using a *three-dimensional* mass plot or bar chart. When the number of nonzero values of p_{X_1, X_2} is finite and small, the probability mass function can also be represented using a **contingency table.**

Two random variables X_1 and X_2 are **independent** if knowledge of one of the variables has no effect on the probabilities assigned to the values of the other. If X_1 and X_2 are both discrete, then the relative frequency interpretation of probability suggests that the joint probability mass function is the product of the marginal probability mass functions,

$$p_{X_1, X_2}(x_1, x_2) = p_{X_1}(x_1) \cdot p_{X_2}(x_2) \qquad \text{for all } x_1, x_2.$$

In words, this equation states that the probability of having $X_1 = x_1$ and $X_2 = x_2$ at the same time is the product of the probabilities of having $X_1 = x_1$ and $X_2 = x_2$ individually.

2.3   Future Lifetimes

This example illustrates how the concept of random variable applies to continuous quantities such as time and how a probability distribution can be specified for such continuous quantities. The notion of expected value for continuous random variables is also developed in the context of life expectancy.

The Basic Problem

There are many insurance and investment products in which the obligations of the insurer are contingent on the death or continued survival of a given person. For example, **life insurance** is a financial security product in which an insurance company agrees to pay a predetermined amount of money on the death of a particular individual in exchange for a regular stream of payments while the insured person is alive. **Annuities,** on the other hand, are financial security products in which an insurance company agrees to make regular payments to a particular individual while that individual is alive in exchange for a fixed amount of money at the commencement of the annuity contract.[6] Life insurance provides protection to families with children against the adverse financial consequences of the premature death of a family provider with unfulfilled obligations (e.g., raising and educating children), whereas annuities provide protection to the elderly against the adverse financial consequences of outliving the savings they have accumulated for retirement.

The future lifetime of any given individual is uncertain. However, the survival pattern for large groups of individuals is generally quite predictable. Consequently, the insurance mechanism is usually feasible for large groups of similarly situated individuals. Nevertheless, to determine the appropriate rates to charge for insurance protection, the insurer needs to have a pretty good idea what the survival pattern for any given group will be. How should the insurer do this?

Initial Observations

The survival pattern for a group of 20 year olds is very different from the survival pattern for a group of octogenarians. Hence, when modeling future lifetimes, we need to keep track of the ages of the members of the group. Since our objective in this example is to extend the concept of random variable and probability distribution to *continuous* quantities such as time, we are going to make the simplifying assumption that all members of the group under consideration are newborns. Hence, let T be the future lifetime of a randomly selected newborn.

One way of determining a probability model for T would be to conduct a study on a large group of newborns. The study would follow all of the subjects until their deaths and at its conclusion would provide a list of the dates of death for the subjects in the study. Unfortunately, the researchers who initiate such a study would very likely all be

---

[6] It is more correct to refer to this type of product as a *life* annuity to distinguish it from an annuity *certain* whose payments continue for a fixed time period and are not contingent on the survival of the insured. However, to avoid unnecessary complications at this point in the book, we conform to popular usage of the term.

dead before the study's conclusion! Even if we arranged for successive generations of researchers to continue the study until its completion, we would probably have to wait over 100 years to get the results, and in the meantime, mortality may have improved to such an extent that the results would be of little value.

A more practical approach would be to analyze the survival pattern for a *cross-section* of the entire population (i.e., a group consisting of people of all ages) for a specific period of time (e.g., 1 year) and then infer a probability model for T on the basis of this information. This type of study design is referred to as a **cross-sectional study design,** whereas the type of study design described in the previous paragraph is referred to as a **longitudinal study design.** The respective terms are generally quite descriptive of the study design type.

In the interest of simplicity, we are going to assume that a longitudinal study has been completed (despite the practical difficulties associated with carrying out such a study on humans). Hence, suppose that we are given the lifespans in days for a large group of individuals who had been under observation since birth. Based on our discussion in §2.1 and §2.2, it is tempting to construct a probability model by considering the relative frequencies for a "days of life" random variable D. However, such an approach would not lead to a very meaningful description of future lifetime since it is very likely that every subject in the study will live for a different number of days. Consequently, if we considered the relative frequencies for the days of life, we would find that on most days the relative frequency is zero, while on days where there is a death, the relative frequency is $1/n$, n being the number of subjects at the commencement of the study; since n is assumed to be large, this means that for all values of the "days of life" random variable, the relative frequency is *essentially zero!*

An alternative approach is to consider one of the following quantities:

1. For each possible lifespan t, the fraction of the original population that dies before reaching that lifespan or has that lifespan precisely;
2. For each possible lifespan t, the fraction of the original population whose lifespans exceed that lifespan.

Let T be the future lifetime random variable (i.e., the lifetime of a randomly chosen individual in the given population). Then the first of these two "quantities" is the function $\hat{F}_T$ given by

$$\hat{F}_T(t) = \text{fraction of population for which } T \leq t,$$

while the second of these quantities is the function $\hat{S}_T$ given by

$$\hat{S}_T(t) = \text{fraction of population for which } T > t.$$

Here, each t represents a possible lifespan, and T represents the actual lifespan for a randomly chosen person in the population. Note that $\hat{F}_T(t) + \hat{S}_T(t) = 1$ for all t. Hence, it suffices to consider only one of these functions to develop a probability model. Let's focus our attention on $\hat{F}_T$.

The function $\hat{F}_T$ measures *cumulative* relative frequencies of death. Note that $\hat{F}_T$ is a step function with a jump at each point where a death occurs. (See Figure 2.10 for a possible graph when $n = 4$.) If the size of the original population is large (e.g., if

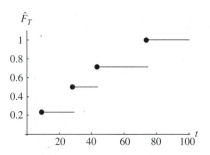

FIGURE 2.10 Possible Graph of $\hat{F}_T$

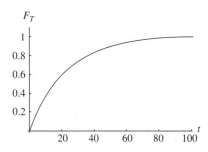

FIGURE 2.11 Possible Graph of F_T

$n \geq 10,000$), then the size of the jumps will be very small (e.g., $1/10,000$ if $n = 10,000$). Consequently, as the population size increases, the graph of $\hat{F}_T$ will approach the graph of a continuous (increasing) function F_T (Figure 2.11). This continuous "limiting" function F_T can be used to define probabilities for the random future lifetime T.

Distribution of Probability

The preceding observations suggest that the distribution of probability for T is best described by the function F_T defined by

$$F_T(t) = \Pr(T \leq t) \qquad \text{for all } t,$$

where the notation $\Pr(T \leq t)$ can be interpreted for the moment as shorthand for the English statement "the probability that $T \leq t$," or by the function S_T defined by

$$S_T(t) = \Pr(T > t) \qquad \text{for all } t,$$

where $\Pr(T > t)$ is shorthand for "the probability that $T > t$."

For any random variable[7] X, the function F_X defined by

$$F_X(x) = \Pr(X \leq x) \qquad \text{for all } x$$

---

[7] We use the notation X here to denote a generic random variable that could represent time or length or some other quantity.

is known as the **cumulative distribution function** or simply the **distribution function** for X, and the function S_X defined by

$$S_X(x) = \Pr(X > x) \qquad \text{for all } x$$

is known as the **survival function** for X.

Using either of these functions, it is a straightforward matter to compute probabilities for intervals of values of T. For example, the probability that $a < T \le b$ is simply $F_T(b) - F_T(a)$. Notice that $a < T \le b$ represents the "event" that a randomly selected newborn survives to age a but dies at or before age b. Hence, $F_T(b) - F_T(a)$ is the probability that a randomly selected newborn survives to age a but dies at or before age b.

If F_T is continuous, as we are assuming for the future lifetime random variable T, then $F_T(b) - F_T(a) \to 0$ as $a \to b$. Consequently, the probability that T assumes any single value is zero; that is, $\Pr(T = t) = 0$ for all t. This observation may at first seem surprising. However, it is merely a statement of the intuitive observation that the chance of a given person dying at a given instant is virtually zero.

Since $\Pr(T = t) = 0$ for all t, it is not meaningful to consider a probability mass function for T, as we did for the discrete random variables in §2.1 and §2.2. However, one can consider a related function based on probability *densities* (i.e., probabilities per unit length).

From remarks in the preceding paragraphs, we know that $F_T(b) - F_T(a)$ is the probability that $a < T \le b$ and, hence, can be interpreted as the relative frequency of deaths in the *interval* $(a, b]$.[8] Consequently, $\{F_T(b) - F_T(a)\}/(b - a)$ is the relative frequency density for this interval. From calculus, we know that

$$F_T'(b) = \lim_{a \to b} \frac{F_T(b) - F_T(a)}{b - a}$$

provided that F_T is differentiable. Thus, $F_T'(b)$ is the *instantaneous* relative frequency density for T at the point b.

Consequently, we can define the **probability density function** for T to be the function f_T given by

$$f_T(t) = F_T'(t) \qquad \text{for all } t,$$

provided that this makes sense. Using the fundamental theorem of calculus, the probability that T lies between a and b is then

$$\Pr(a < T \le b) = F_T(b) - F_T(a)$$
$$= \int_a^b f_T(t)\,dt.$$

Life Expectancy

In §2.1 and §2.2, we found the concept of expected value to be useful in analyzing payoffs. One can introduce an analogous concept for *continuous* random variables such

---

[8] The notation $(a, b]$ is shorthand for the collection of real numbers x defined by $a < x \le b$. The notations $[a, b)$, $[a, b]$, and (a, b) have similar meanings.

as T. When T represents a future lifetime, the expected value of T is usually referred to as the *life expectancy*. Intuitively, the life expectancy (at birth) is the average length of life for subjects in the given group.

The average length of life for members in a group under study is the total number of years lived by all members of the group divided by the number of members initially in the group. For example, if a group consists of three members who die at exact ages 4, 6, and 7, then the average length of life for members in this group is

$$\frac{4+6+7}{3} = 5\frac{2}{3}.$$

If the group is large and we are not given death times for individuals, but instead are given the fractions of the original population still alive at the end of each year—that is, we are given $\hat{S}_T(1), \hat{S}_T(2), \ldots$—then an alternative way to determine the average number of years lived is to use the formula

$$\hat{S}_T(1) + \hat{S}_T(2) + \cdots.$$

For example, for the three-member group considered earlier, the average number of years lived is

$$\frac{1+1+1}{3} + \frac{1+1+1}{3} + \frac{1+1+1}{3} + \frac{1+1+1}{3}$$
$$+ \frac{0+1+1}{3} + \frac{0+1+1}{3} + \frac{0+0+1}{3} + 0 + \cdots$$
$$= \frac{(1+1+1+1+0+0+0+\cdots)}{3}$$
$$+ \frac{(1+1+1+1+1+1+0+\cdots)}{3}$$
$$+ \frac{(1+1+1+1+1+1+1+0+\cdots)}{3}$$
$$= \frac{4}{3} + \frac{6}{3} + \frac{7}{3}$$
$$= 5\frac{2}{3}.$$

The first equality in this series is obtained by adding the first numbers in each of the numerators, then the middle numbers in each of the numerators, and finally the last numbers in each of the numerators. This agrees with the earlier calculation, as it should since the members are assumed to die on their birthdays.

This suggests that the life expectancy when the distribution of T is given by a continuous survival function S_T is

$$E[T] = \int_0^\infty S_T(t)dt.$$

Using integration by parts, we can write this in a form that more closely resembles the definition of expectation for discrete random variables given in §2.1. Indeed,

$$\int_0^\infty S_T(t)dt = \int_0^\infty 1 \cdot S_T(t)dt$$

$$= t S_T(t)\big|_0^\infty - \int_0^\infty t S_T'(t)dt$$

$$= 0 + \int_0^\infty t f_T(t)dt.$$

The final equality follows from the fact that $S_T'(t) = \frac{d}{dt}(1 - F_T(t)) = -F_T'(t) = -f_T(t)$ and the assumption that $S_T(t) = 0$ for all t greater than some specific t^* (i.e., no one lives longer than t^*). This suggests the following definition of the expected value for a continuous random variable: For any continuous random variable X with density function f_X, the **expected value** of X is the number $E[X]$ given by

$$E[X] = \int_{-\infty}^\infty x \, f_X(x)dx,$$

provided that this exists.

Note that for lifetime random variables T, $f_T(t) = 0$ for all $t < 0$ since lifespans cannot be negative, and so $E[T] = \int_0^\infty t f_T(t)dt$, as given before.

Summary of Concepts Illustrated in This Example

Let's summarize some of the important probability concepts illustrated in this example.

The probability distribution for a random variable can be described by a cumulative distribution function or a survival function. For any random variable X, the **cumulative distribution function** or simply the **distribution function** of X is the function F_X that assigns to each value x the probability accumulated up to x,

$$F_X(x) = \Pr(X \le x) \qquad \text{for all } x.$$

For any random variable X, the **survival function** of X is the function S_X that assigns to each value x the probability that $X > x$,

$$S_X(x) = \Pr(X > x) \qquad \text{for all } x.$$

A **continuous random variable** is a random variable X with the property that $\Pr(X = x) = 0$ for all x. Equivalently, it is a random variable for which the distribution function F_X is continuous.

The **probability density function** for a continuous random variable X is the function f_X that assigns to each point x the instantaneous relative frequency density at that point,

$$f_X(x) = \lim_{\varepsilon \to 0} \frac{\Pr(x - \frac{\varepsilon}{2} \leq X \leq x + \frac{\varepsilon}{2})}{\varepsilon}.$$

Probabilities associated with a continuous random variable X can be calculated by integrating the probability density function over the appropriate region. In particular,

$$\Pr(a \leq X \leq b) = \int_a^b f_X(x)dx.$$

Hence,

$$F_X(x) = \int_{-\infty}^x f_X(s)ds$$

and also

$$f_X(x) = F_X'(x),$$

provided that f_X is continuous.

The **expected value** of a continuous random variable X is the number

$$E[X] = \int_{-\infty}^{\infty} x\, f_X(x)dx.$$

It is the arithmetic average value of the function f_X and can also be interpreted as the *center of mass* of the probability distribution.

2.4   Simple and Compound Growth

In the final example of this chapter, we consider the effect of compounding on uncertainty. This is particularly important in investment or population analysis. However, it also arises whenever the uncertain outcome of one random experiment is an input of some other random experiment.

Statement of the Example

Consider the following investment opportunity:

A security, whose value fluctuates from day to day, may be purchased today for $1. The daily return on the security in any given day is determined by the toss of a fair coin. If the coin lands heads, the value of the investment increases 50%. If the coin lands tails, the value decreases 40%.

Is this a good investment?

Initial Observations

At first glance, it would appear that this is indeed a good investment because the expected return on any given day is positive. However, we must be careful with how we interpret this positive expected return. Although it is true that the given opportunity can be a good investment under certain circumstances, surprisingly enough, it is also true that it can be a very bad investment under other circumstances. In particular, it turns out to be an extremely bad investment for the person who plans to buy the security today and hold it for an extended period of time.[9]

Problems such as this routinely arise in investment decision making, and yet a surprising number of people still fail to solve them correctly. Since we think that you would rather be in the company of the people making the profits on such opportunities than in the company of the people suffering the losses, we're going to analyze this particular problem very carefully. In the process, we will discover a new type of expected value—the *geometric* expected value to which we alluded at the end of §2.1—which turns out to be extremely useful in the analysis of investments over multiple time periods.

Investment Over a Single Period

Let's begin by considering the return and the accumulation over a single day. Hence, let X be the simple rate of return over a particular day and let Y be the accumulation of $1 over that day (i.e., Y is the value of $1 after the coin has been flipped). Then, from the description of the given investment opportunity, we have

$$X = \begin{cases} 50\% & \text{with probability } 1/2, \\ -40\% & \text{with probability } 1/2 \end{cases}$$

and

$$Y = \begin{cases} 1.5 & \text{with probability } 1/2, \\ 0.6 & \text{with probability } 1/2. \end{cases}$$

Hence, $E[X] = 5\%$, $E[Y] = 1.05$, and so it appears as though this is a good investment opportunity over a single day.

However, before we jump to any conclusions, let's consider what the statement $E[Y] = 1.05$ actually means. Recall from §2.1 that the quantity $E[Y]$ can be interpreted to be the average per game winnings one can expect to make in a game with payoff Y when the game is played a large number of times and winnings are added together. Hence, in the current example, the quantity $E[Y]$ represents the average per investment accumulation one can expect to obtain when a large number of independent investments,

---

[9] To hold a security means to maintain ownership of the security throughout a given period of time. This does not mean that the value of the security stays the same since other investors will be buying and selling identical securities with the result that the security you own will fluctuate in value.

each with accumulation Y, are made on the same day and accumulations are added together. Consequently, the statement $E[Y] = 1.05$ means that if we place a large number of independent bets on the same day and add our accumulations together, then our arithmetic average accumulation per investment is 1.05, (i.e., our average return per investment is 5%).

Therefore, if we can make a large number of independent investments of the type described for a single day, then the opportunity presented is a good one because we can expect to make 5% return on average per investment. However, what if we must hold our investment for several days? Is the opportunity presented still a good one under this scenario?

Investment Over Several Periods

To answer this question, let's now suppose that we purchase one unit of the security for $1 and hold it for n days. Then our accumulation after n days is

$$(1 + x_1)(1 + x_2) \cdots (1 + x_n),$$

where $x_1, x_2, \ldots, x_n$ are the returns on days $1, 2, \ldots, n$, respectively; in particular, our gains and losses are *compounded* from one day to the next. Now, if n is sufficiently large, then approximately half the returns x_j will be 50% gains and approximately half will be 40% losses. Hence, we should expect the accumulation after n days to be approximately

$$(1 + .5)^{n/2}(1 - .4)^{n/2} = \left(\sqrt{(1.5)(.6)}\right)^n$$
$$= (\sqrt{.9})^n.$$

Since $(\sqrt{.9})^n \to 0$ as $n \to \infty$, this means that we should expect our $1 investment to be essentially worthless after a certain length of time. In fact, we should expect the value of our investment to be less than the initial $1 amount after just 2 days! This shows that if we must hold the investment for more than a single day, then the opportunity presented is not a good one. Hence, we see that the profitability of an investment such as this can be dramatically affected by the time period over which it is considered.

Geometric and Arithmetic Expectations

The preceding discussion clearly illustrates that when gains and losses are compounded over several time periods, we cannot determine whether an investment is good just by looking at its arithmetic expected accumulation $E[Y]$ over a single period. It is natural to wonder whether there is a number, analogous to $E[Y]$, which we can use to judge the profitability of an investment compounded over several time periods. Fortunately, there is such a number, called the *geometric* expectation, which we now proceed to describe.

In the given investment prospect, there are two possible accumulations per period: 1.5 and 0.6. However, for the present discussion, we will find it more convenient to assume that there are k possible accumulations per period. Hence, let Y be the per period accumulation of $1 and suppose that the possible values of Y are $y_1, \ldots, y_k$ with respective probabilities $p_1, \ldots, p_k$. Then, over n periods where n is a large number, we expect:

approximately $p_1 n$ of the accumulation factors to be y_1;
approximately $p_2 n$ of the accumulation factors to be y_2;
$$\vdots$$
approximately $p_k n$ of the accumulation factors to be y_k.

Hence, over n periods, we expect \$1 to accumulate to approximately

$$y_1^{p_1 n} y_2^{p_2 n} \cdots y_k^{p_k n} = (y_1^{p_1} \cdots y_k^{p_k})^n$$

and so, we expect the "average" accumulation per period to be

$$y_1^{p_1} \cdots y_k^{p_k}.$$

Note that the case $k = 2$ is exactly as described earlier for the investment prospect with two possible accumulation factors.

The number $y_1^{p_1} \cdots y_k^{p_k}$ is known as the *geometric mean* of the numbers $y_1, \ldots, y_k$ with respective weights $p_1, \ldots, p_k$. Note that the *arithmetic* mean of these numbers with the given weights is $p_1 y_1 + \cdots + p_k y_k$ and that for the given random variable Y, the *arithmetic expected value* $E[Y]$ is $\sum_{j=1}^{k} p_j y_j$. Hence, it makes sense to define the *geometric expected value* for Y to be $\prod_{j=1}^{k} y_j^{p_j}$.

In general, the **geometric expected value** for a positive discrete random variable Y is the number

$$E_g[Y] = \prod_{\text{all } y} y^{p_Y(y)},$$

provided that this product exists. It is the geometric average accumulation *per period* when a single investment with per period accumulation random variable Y is *compounded* over a large number of independent periods. For example, in the given investment situation with equally likely accumulations 1.5 and 0.6, the geometric expectation is

$$E_g[Y] = (1.5)^{1/2}(.6)^{1/2} = \sqrt{.9}.$$

Confusion can sometimes arise when considering geometric and arithmetic expectations in the same problem. Hence, when we wish to emphasize that the expectation under consideration is arithmetic, we will write $E_a[Y]$. However, keep in mind that the notation $E[Y]$ (without the subscript) always refers to the arithmetic expectation even if a particular discussion may seem to suggest otherwise.

Connections Between Arithmetic and Geometric Expectation

From the definition of $E_g[Y]$ just given and using properties of exponentials, we have

$$E_g[Y] = \prod_{\text{all } y} y^{p_Y(y)}$$

$$= \prod_{\text{all } y} e^{(\log y) \cdot p_Y(y)}$$

$$= \exp\left(\sum_{\text{all } y} (\log y) \cdot p_Y(y)\right).$$

Note that the quantity $\sum_{\text{all } y}(\log y) \cdot p_Y(y)$ is a weighted average of the values of $\log Y$ weighted by the probabilities $p_Y(y)$. From the definition of arithmetic expectation, it is tempting to identify this quantity as $E_a[\log Y]$. However, one must be careful since $E_a[\log Y]$ can be interpreted as $E_a[W]$ where $W = \log Y$, that is, as the sum $\sum w \cdot p_W(w)$, where $W = \log Y$, and it is not obvious that $\sum w \cdot p_W(w) = \sum_{\text{all } y}(\log y) \cdot p_Y(y)$. In fact, one can show that these two quantities are the same.[10]

Consequently, we have the following important relationship between arithmetic and geometric expectation:

$$E_g[Y] = e^{E_a[\log Y]}.$$

We could actually define the geometric expectation in terms of the arithmetic expectation using this formula. This would allow us to define the notion of geometric expectation for all positive random variables Y, not just discrete ones, since $E_a[\log Y]$ can be calculated for all such Y for which the average exists.

Because of the close connection between arithmetic and geometric expectations, and the fact that geometric expectations are only defined on *positive* random variables, most books in probability ignore the geometric expectation altogether. However, this is unfortunate because, as you can see, the geometric expectation is just the tool we need to determine the profitability of an investment that is compounded over many periods. Even if you don't plan on being an investment analyst, you will still find the geometric expectation to be a handy tool. Indeed, this discussion suggests that it will arise naturally in any situation, such as population growth or the evolution of a system over time, where compounding effects occur.

Arithmetic-Geometric Means Inequality

There is a fundamental inequality between arithmetic and geometric means that will help explain why average per period accumulations are depressed when compounding occurs. In one of its simplest forms, the arithmetic-geometric means inequality (AGM inequality, for short) states that

$$\text{GM}(a_1, \ldots, a_k) \leq \text{AM}(a_1, \ldots, a_k),$$

where $\text{AM}(a_1, \ldots, a_k)$, $\text{GM}(a_1, \ldots, a_k)$ denote the arithmetic and geometric means of the numbers $a_1, \ldots, a_k$, respectively, with equality holding precisely when all the a_is are the same. Hence, in terms of arithmetic and geometric expectations, we have the inequality

$$E_g[Y] \leq E_a[Y]$$

with equality holding precisely when Y is a point mass.

Now $E_g[Y] = e^{E_a[(\log Y)]}$. Hence, using the substitution[11] $X = \log Y$ and the monotonicity of the logarithmic function, the AGM inequality for random variables can be written in the following two forms:

---

[10] In the interest of simplicity, we postpone the demonstration of this fact until later in the book.

[11] This X should not be confused with the *simple* return variable X defined at the beginning of the section. Here, X is a *continuously compounded* return.

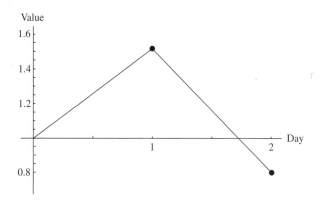

FIGURE 2.12 Effect of a 50% Increase Followed by a 50% Decrease on a $1 Investment

$$e^{E_a[X]} \le E_a[e^X],$$
$$E_a[\log Y] \le \log(E_a[Y]).$$

These inequalities are actually special cases of a more general inequality relating the expectation of a convex or concave function of X to the function of the expectation, that is, relating $E[g(X)]$ to $g(E[X])$ when g is convex or concave. This more general inequality is known as *Jensen's inequality* and will be discussed later when we consider how to estimate the arithmetic expectation of a function of a random variable X.[12]

Why Does Compounding Depress Returns?

The arithmetic-geometric means inequality gives us a *mathematical* reason for compounding to depress returns. However, we can give a more concrete *financial* reason. To keep things simple, let's return to our original investment problem, but assume that daily gains and losses are both 50%. Suppose further that we invest $1 for 2 days and that our investment increases 50% on the first day and then decreases 50% on the second day. Then notice what happens: Our increase in value on the first day is 50 cents (50% of $1), but our decrease in value on the second day is 75 cents (50% of $1.50); in particular, our loss in dollar value on the second day is greater than our gain in dollar value on the first day, even though the size of the percentage gain/loss on both days is the same! The reason is that the percentage loss on the second day is taken with respect to the larger principal value of $1.50, as illustrated in Figure 2.12.

Now we can understand why some people make poor judgments in investment situations where there is compounding. Indeed, such people focus on the *percentage returns* when they should instead focus on the *dollar accumulations*. In particular, they fail to

---

[12] Our earlier comments about the expectation of log Y apply even more so to the expectation of $g(X)$ for any random variable X. The quantity $E[g(X)]$ has two interpretations: It can be interpreted as

$$E[g(X)] = \begin{cases} \sum g(x) \cdot p_X(x) & \text{if } X \text{ is discrete,} \\ \int g(x) \cdot f_X(x)\, dx & \text{if } X \text{ is continuous,} \end{cases}$$

and it can also be interpreted as $E[W]$, where $W = g(X)$. Fortunately, these two interpretations agree.

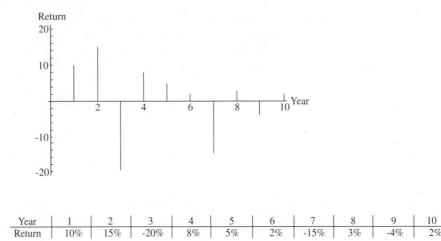

Year	1	2	3	4	5	6	7	8	9	10
Return	10%	15%	-20%	8%	5%	2%	-15%	3%	-4%	2%

FIGURE 2.13   10-Year Performance of a Hypothetical Fund

realize that a gain of 50% in a given time period does not offset a loss of 50% in a prior period!

This discussion suggests that unscrupulous mutual fund managers who want to make their investment performance look better than it really is can do so by publishing arithmetic average returns. Hence, *the buyer should beware!* Indeed, even managers who are not so crass as to quote arithmetic averages will often publish graphs of the annual percentage returns over a 10-year period, thereby giving customers a visually distorted view of dollar value performance. Such a graph is illustrated in Figure 2.13.

Believe it or not, if you had invested $1 at the beginning of this 10-year period, it would still be worth only $1 today, in spite of what the graph suggests! Indeed, from the given data, we see that a $1 investment compounded at the given rates over the 10-year period has value at the end of the 10 years equal to

$$(1.00)\{(1.10)(1.15)(0.80)(1.08)(1.05)(1.02)(0.85)(1.03)(0.96)(1.02)\}$$
$$\approx 1.00.$$

Now aren't you glad that you learned about the difference between arithmetic and geometric expectations? Just think of the effect this new knowledge could have on your future financial health! If nothing else, you've learned how to get rid of fast-talking mutual fund salespeople: Just ask them whether the averages they're throwing at you are arithmetic or geometric!

Summary of Concepts Illustrated in This Example

We conclude this section by reviewing the important differences between arithmetic and geometric expectations.

The **arithmetic expectation** $E_a[Y]$ of a random variable Y is the number

$$E_a[Y] = \begin{cases} \sum_{\text{all } y} y \cdot p_Y(y) & \text{if } Y \text{ is discrete,} \\ \int_{-\infty}^{\infty} y \cdot f_Y(y) & \text{if } Y \text{ is continuous.} \end{cases}$$

The arithmetic expectation $E_a[Y]$ represents the average accumulation *per invest-ment* when a large number of independent investments, each with accumulation random variable Y, are placed over a single time period.

The **geometric expectation** $E_g[Y]$ of a positive random variable Y is the number

$$E_g[Y] = e^{E_a[\log Y]};$$

in particular, if Y is discrete, it is the number

$$E_g[Y] = \prod_{\text{all } y} y^{p_Y(y)}.$$

The geometric expectation $E_g[Y]$ represents the average accumulation *per period* when a single investment, with per period accumulation random variable Y, is *compounded* over a large number of independent periods.

2.5   Chapter Summary

In this chapter, we have developed some of the basic concepts of probability by considering four concrete situations where probability models are used in practice. In particular, we have introduced the important concepts of random variable, probability distribution, and arithmetic and geometric expectation. Our hope is that by introducing these concepts through these concrete examples, the reader has gained a better appreciation of their significance in probability modeling. We close this chapter by reviewing some of the major concepts we have illustrated.

A **random variable** is any quantity X with real values that depends in a well-defined way on the outcome of a random experiment. More precisely, it is a real valued function $X : S \to \mathbf{R}$, where S is the set of all possible outcomes of the random experiment. The set S is known as the **sample space** of the experiment. A **probability distribution** for a random variable X is an assignment of relative frequencies to the values and intervals of values of X.

Probability distributions can be specified by a *distribution function* or a *survival function*. For any random variable X, the **distribution function** of X is the function F_X defined by

$$F_X(x) = \Pr(X \leq x) \qquad \text{for all } x;$$

the **survival function** of X is the function S_X defined by

$$S_X(x) = \Pr(X > x) \qquad \text{for all } x.$$

For discrete random variables X, the **mass function** is the function p_X that assigns to each point x the probability that X assumes the value x:

$$p_X(x) = \Pr(X = x).$$

For continuous random variables X, the **density function** is the function f_X that assigns to each point x the instantaneous relative frequency density of X at x, that is,

$$f_X(x) = \lim_{\varepsilon \to 0} \frac{\Pr(x - \varepsilon/2 \le X \le x + \varepsilon/2)}{\varepsilon}$$

and is the derivative of the distribution function:

$$F'_X(x) = f_X(x).$$

Two important concepts are the notions of *independence* and *identical distribution*. Random variables X and Y are said to be **independent** if knowledge of one of the variables has no effect on the distribution of probability for the other. They are said to be **identically distributed** if they have the same distribution functions; that is, if

$$F_X(t) = F_Y(t) \qquad \text{for all } t.$$

Being identically distributed does not mean that $X = Y$. Rather, it means that X and Y are equivalent in the sense that everyone in the world, regardless of risk preference, should be indifferent to choosing between them when they represent payoffs in a game of chance.

Compounding can have a significant effect on an uncertain quantity, and it is important to take this into consideration when making statements about "average values." The **arithmetic expected value** of X, denoted $E_a[X]$, is the arithmetic average of the values of X weighted by the probability masses and densities:

$$E_a[X] = \begin{cases} \sum_{\text{all } x} x \cdot p_X(x) & \text{if } X \text{ is discrete,} \\ \int_{-\infty}^{\infty} x \cdot f_X(x)\, dx & \text{if } X \text{ is continuous.} \end{cases}$$

It represents the average accumulation *per investment* when a large number of independent investments, each with accumulation random variable X, are placed over a single time period.

On the other hand, the **geometric expected value** of X, denoted $E_g[X]$, is the geometric average of the values of X weighted by the probability masses and densities:

$$E_g[X] = \prod_{\text{all } x} x^{p_X(x)}$$

$$= e^{E_a[\log X]}.$$

It represents the average accumulation *per period* when a single investment with per-period accumulation random variable X is *compounded* over a large number of independent periods.

2.6   Exercises

1. On a particular television game show, contestants spin a wheel to obtain cash prizes. The wheel has 12 equal sections and is perfectly balanced:

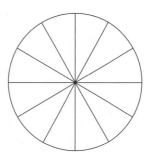

   In 11 sections, there is a number that indicates the dollar amount of the prize to be won. Six of the prizes are for $500, three are for $1000, one is for $2000, and one is for $5000. The remaining section of the wheel has the word BANKRUPT printed on it; a contestant whose spin stops on this section loses all the money accumulated on the show up to that point.

   A contestant whose accumulated winnings are $15,000 is about to spin the wheel. Let X denote the value of this contestant's accumulated winnings after the spin.

   a. Determine the probability mass function of X.
   b. Sketch a mass plot of p_X.
   c. Determine $E[X]$.
   d. If you were this contestant and had the option of not spinning the wheel, what would you do?

2. A computer generates a random number between 1 and 100. The computer is programmed so that all 100 numbers are equally likely. If the number generated is of the form $2^x \cdot 3^y$, then we win $x + y$ dollars; otherwise, we win nothing. Each time we play this game, we must pay $1 up-front. Let G be our net gain.

   a. Determine the probability mass function for G.
   b. Sketch a graph of p_G.
   c. What is our expected gain?
   d. What is the probability of having a positive net gain?
   e. Would you play this game? Explain.

3. Three standard fair coins are tossed. You are given a choice of the following two payoffs:

Payoff 1:	Win $1 for each head.
	Lose $1 for each tail.
Payoff 2:	Win $1 for each tail.
	Lose $6 for three heads.

   Let Y_1, Y_2 be the respective payoffs.

   a. Write out the sample space for this experiment.

   b. Determine the probability mass functions for Y_1 and Y_2.

   c. Discuss the pros and cons of choosing one payoff over the other.

4. Three standard fair coins are tossed. You are given a choice of the following two payoffs:

Payoff 1:	Win $1 for each tail.
	Lose $2 whenever two or more heads turn up.
	Lose $1 for two heads and one tail.
Payoff 2:	Win $3 for all heads.
	Lose $2 for all tails.
	Win $2 for exactly two tails.
	Lose $1 for exactly two heads.

A friend advises you to choose the first payoff. How should you respond to this advice?

5. The probability mass function for a discrete random variable X is given by

$$p_X(x) = \begin{cases} \frac{1}{4}x + \frac{1}{2} & \text{for } x = -2, -1, 0, \\ -\frac{1}{4}x + \frac{1}{2} & \text{for } x = 1, 2, \\ 0 & \text{otherwise.} \end{cases}$$

   a. Specify the distribution function F_X.

   b. Determine the expected value of X.

   c. Sketch the graphs of p_X and F_X.

6. The distribution function for a discrete random variable X is given by

$$F_X(x) = \begin{cases} 0 & \text{if } x < -1, \\ \frac{1}{3} & \text{if } -1 \le x < 2/3, \\ 1 & \text{if } x \ge 2/3. \end{cases}$$

   a. Specify the probability mass function p_X. *Hint:* Consider the graph of F_X and observe the location and size of the jumps.

   b. Calculate the expected value of X.

   c. Determine the probability that $X < 2/3$.

   d. Sketch the graphs of p_X and F_X.

7. The probability mass function for a particular discrete random variable X having nonnegative integer values is defined by the recurrence relation

$$p_X(x) = \frac{1}{2} p_X(x-1), \qquad x = 1, 2, \ldots$$

Determine the probability that X is less than 10. What is the probability that X is less than the integer n?

8. The probability mass function for a particular discrete random variable X having nonnegative integer values is defined by the relation

$$p_X(x) = \frac{1}{x} p_X(x-1), \qquad x = 1, 2, \ldots$$

Determine the value of $p_X(0)$.

9. A nickel and a dime are tossed together. Let X_1 and X_2 be the following indicator random variables:

$$X_1 = \begin{cases} 1 & \text{if the nickel turns up heads,} \\ 0 & \text{if the nickel turns up tails;} \end{cases}$$

$$X_2 = \begin{cases} 1 & \text{if the dime turns up heads,} \\ 0 & \text{if the dime turns up tails.} \end{cases}$$

   a. Give interpretations for each of the following random variables: $X_1 X_2$, $X_1(1 - X_2)$, $(1 - X_1)X_2$, $(1 - X_1)(1 - X_2)$.
   b. Suppose that we win \$1 for each head and lose \$3 for getting two tails. Write this payoff as a linear combination of $X_1 X_2$, $X_1(1 - X_2)$, $(1 - X_1)X_2$, $(1 - X_1)(1 - X_2)$.
   c. Show that every payoff for this game can be written in the form $aX_1 X_2 + bX_1 + cX_2 + d$.

10. Two fair coins are tossed. Let X_1 be an indicator of heads on coin 1 and let X_2 be an indicator of heads on coin 2.

   a. Determine the probability mass function for $X_1 + X_2$. What does $X_1 + X_2$ represent?
   b. Determine the probability mass function for $X_1 X_2$. What does $X_1 X_2$ represent?
   c. Show that $X_1 + X_2$ and $2X_1$ are not identically distributed even though $X_1 \sim X_2$.
   d. Show that $X_1 X_2$ and X_1^2 are not identically distributed even though $X_1 \sim X_2$.
   e. What general conclusion do you reach from parts c and d with regard to substitution of equivalent random variables? Would you reach a different conclusion if it were the case that $X_1 = X_2$?

11. In each of the following cases, determine whether X and Y are independent:

   a. X is the number rolled on a balanced die. Y is the number of heads in two tosses of a fair coin.
   b. X is the height of a randomly chosen student. Y is the height of the student's father.
   c. X is the number on the first card drawn from a deck of cards whose face cards have been removed. Y is the number on the second card drawn.
   d. X is the number rolled on a die. Y is the number of prime factors of X.

12. Two cards are drawn from a special deck consisting of two hearts and two diamonds, and they are placed face down in front of us. The two cards are then turned over and their suits are observed.
   Let X_1 and X_2 be defined as follows:

$$X_1 = \begin{cases} 1 & \text{if the card on our left is a heart,} \\ 0 & \text{if the card on our left is a diamond;} \end{cases}$$

$$X_2 = \begin{cases} 1 & \text{if the card on our right is a heart,} \\ 0 & \text{if the card on our right is a diamond.} \end{cases}$$

   a. What is the sample space for this experiment? *Hint:* Should we distinguish between two cards of the same suit or not?
   b. Are X_1 and X_2 independent? Are they identically distributed?

c. Determine the probability mass function for the random vector $\mathbf{X} = (X_1, X_2)$.

d. Write out a contingency table for X_1 and X_2.

13. The future lifetime for a particular subject can be modeled using the survival function

$$S(x) = \exp\left\{-\left(\frac{x}{5}\right)^3\right\}, \qquad x \geq 0,$$

where x is measured in years. Determine the probability that this subject lives another 5 years. What is the probability that this subject lives another 10 years?

14. The future lifetime for a randomly chosen subject in a particular group can be modeled using the survival function

$$S(x) = e^{-x/4} + \frac{x}{4}e^{-x/4}, \qquad x \geq 0,$$

where x is measured in years. Determine the life expectancy for a member of this group.

15. The probability density function for a continuous random variable X is given by

$$f_X(x) = \begin{cases} \frac{2}{x^3} & \text{for } x \geq 1, \\ 0 & \text{otherwise.} \end{cases}$$

a. Determine a formula for the distribution function F_X.

b. Calculate the expected value of X.

c. Determine the probability that $X > 4$.

d. Sketch the graphs of f_X and F_X.

16. The distribution function for a continuous random variable X is given by

$$F_X(x) = \begin{cases} 1 - \frac{1}{x^3} & \text{if } x \geq 1, \\ 0 & \text{otherwise.} \end{cases}$$

a. Determine a formula for the probability density function of X.

b. Calculate the expected value of X.

c. Determine the probability that $2 < X < 4$.

d. Sketch the graphs of f_X and F_X.

17. The distribution function for a continuous random variable X is given by

$$F_X(x) = \begin{cases} 0 & \text{if } x < 0, \\ \frac{x^2}{8} & \text{if } 0 \leq x < 2, \\ x - \frac{x^2}{8} - 1 & \text{if } 2 \leq x < 4, \\ 1 & \text{if } x \geq 4. \end{cases}$$

a. Determine a formula for the probability density function of X.

b. Calculate the expected value of X.

c. Determine the probability that $1 < X \leq 3$.

d. Sketch the graphs of f_X and F_X.

18. Identify which of the following functions can be distribution functions and which cannot.

a. $F(x) = e^{-x}, \; x \geq 0$

b. $F(x) = 1 - \dfrac{1}{x}, \; x \geq 1$

c. $F(x) = x^2, \; x \geq 0$

d. $F(x) = \sin x, \; x \in \mathbf{R}$

e. $F(x) = \dfrac{1}{x^2}, \; x > 1$

19. The probability mass function for the random vector $\mathbf{X} = (X_1, X_2)$ is given by

$$p_{X_1,X_2}(x_1, x_2) = \begin{cases} \frac{x_1 x_2 + 1}{13} & \text{for } x_1 = 1, 2, \; x_2 = 1, 2, \\ 0 & \text{otherwise.} \end{cases}$$

a. Construct a contingency table for the joint probability mass function $p_{\mathbf{X}}$.

b. Specify the marginal probability mass functions p_{X_1}, p_{X_2}.

c. Are X_1 and X_2 independent? Are they identically distributed? Explain.

d. Calculate the probability that $X_1 + 2X_2 \geq 3$. What is the probability that $X_1 X_2 > 1$?

e. Sketch graphs of $p_{\mathbf{X}}, p_{X_1}$ and p_{X_2}. What is the relationship of the graphs of p_{X_1} and p_{X_2} to the graph of $p_{\mathbf{X}}$?

20. Suppose that X and Y are discrete independent random variables. Complete the following contingency table for the joint probability mass function $p_{X,Y}$:

		1	2	3	4	
	1	.24			.12	
X	2					.4
			.3			

(above the columns: Y)

21. A security, whose value fluctuates from day to day, may be purchased today for $1. The daily return on the security in any given day is determined by the toss of a coin. On days when the coin lands heads, the value of the investment increases 50%, whereas on days when the coin lands tails, the value decreases 40%. The probability that the coin lands heads in any given toss is p.

a. Suppose that we can make a large number of such investments, each independent from the rest, for a single day. For what values of p is the average accumulation per investment greater than $1?

b. Suppose that we may only purchase one such security and we must hold it for at least a year. For what values of p is the average accumulation per day greater than $1?

22. A particular investment, which is to be held for a long time, gains $100g\%$ or loses $100l\%$ with equal probability on any given day. Gains and losses on distinct days are independent.

a. Show that, for this to be a reasonable investment, g and l should satisfy the following inequalities:

$$l \le \frac{g}{1+g},$$

$$g \ge \frac{l}{1-l}.$$

b. If the daily gain (when there is one) is 25%, what should be true of the daily loss? What if the daily gain is $33\frac{1}{3}$%? 50%? 100%?

23. A particular investment gains 50% or loses 40% with equal probability on any given day. Gains and losses on distinct days are independent. The investment is to be held for 2 consecutive days and then sold. Let V be the value of a $1 investment in this security 2 days hence.

   a. Based on the discussion of this chapter, do you think that this security should be held for 2 days? Explain.
   b. Specify the probability mass function for V.
   c. Calculate $\Pr(V > 1)$ and $E[V]$. Does it surprise you that $\Pr(V > 1) < 1/2$ but $E[V] > 1$? Explain.
   d. Under what circumstances could this investment be considered a good one? *Hint:* Consider the meaning of the statement $E[V] > 1$.

24. A particular investment gains 50%, loses 40%, or remains unchanged each day. On any given day, the probability of a gain is 40%, the probability of a loss is 35%, and the probability of no change is 25%, independent of the returns on the other days. Is this a good investment? Explain.

25. In a particular market, there are a large number of securities with independent and identically distributed returns. For each such security, the percentage returns from one day to the next are themselves independent. A portfolio manager is considering two possible strategies:

   Strategy 1:   Invest everything in one security and hold it for many days.
   Strategy 2:   Invest in a large number of securities whose returns are independent and identically distributed for a single day and then sell.

   a. Suppose that strategy 1 is deemed profitable in the sense that the average accumulation per day is greater than the amount invested. Should we expect strategy 2 to be profitable as well?
   b. Suppose that strategy 2 is deemed profitable in the sense that the average accumulation per investment is greater than the amount invested. Should we expect strategy 1 to be profitable as well?
   c. Suppose that strategy 1 is deemed unprofitable. Should we expect strategy 2 to be unprofitable as well?
   d. Suppose that strategy 2 is deemed unprofitable. Should we expect strategy 1 to be unprofitable as well?

   It is important to realize that in assessing our expectation of profitability in this question, we are not considering the riskiness or variability of accumulations. We will return to this important aspect in later chapters when we discuss the variance of a distribution and the law of large numbers.

3 Classical Probability

One of the major difficulties faced by an insurance company is that people purchasing insurance generally know more about their own risk profiles than the insurance company does. For example, drivers seeking automobile insurance generally know more about their personal driving habits such as how often they drive over the speed limit or how frequently they change lanes than does the insurer. This informational asymmetry can lead to adverse selection and a breakdown of the insurance mechanism, as discussed in §1.4.

Insurers try to overcome their lack of complete information about policyholders by developing risk-classification schemes based on readily observable characteristics of policyholders such as the number of speeding convictions or the number of driver's license demerit points acquired. The difficulty with such classification schemes is that they are subject to error and misinterpretation.

Consider an auto insurer who observes that 70% of its policyholders who have an at-fault accident during the current year had a prior speeding conviction. On the basis of this information, it is tempting to conclude that policyholders with a speeding conviction should be classified as high risk. However, one must be careful in drawing such conclusions. It may also be true that 70% of the policyholders who have an at-fault accident happened to own life insurance. Does this mean that people who own life insurance have a greater risk of causing an automobile accident? Not likely.[1]

To determine whether a prior speeding conviction should be used in the classification of drivers, one needs to consider the percentage of drivers with speeding convictions who are involved in an at-fault accident, not the percentage of drivers involved in an at-fault accident who have a prior speeding conviction, as we did in the previous paragraph. Note that these two percentages are conceptually quite different. An important result, known as **Bayes' theorem**, allows one to determine each of these percentages from the other. The goal of this chapter is to develop an understanding of Bayes' theorem and its application in risk classification.

We begin this chapter with a brief introduction to the formal language of classical probability based on sets and functions. This formal language is particularly useful in risk-classification problems, and it enables us to precisely define the notion of a

---

[1] In fact, a case could be made that people who purchase life insurance are more risk averse than average and, hence, potentially more careful drivers. Moreover, people who purchase life insurance tend to be older, which means that they are likely to be more experienced drivers.

conditional probability, the understanding of which is critical to any classification scheme. We then derive the important **law of total probability** and give examples of its application. Finally, we derive Bayes' theorem for "inverting" conditional probabilities and discuss its application in risk classification.

There are many situations besides insurance risk classification in which decisions must be made on the basis of incomplete information. Examples include disease detection and treatment in medical science, screening of applicants for college admission or employment, and oil exploration in the petroleum industry. The methods of classical probability developed in this chapter can be easily adapted to address such problems. In the exercises accompanying this chapter, the reader will have the opportunity to apply the methods of classical probability to a wide variety of decision problems involving imperfect information.

3.1 The Formal Language of Classical Probability

Classical probability theory is best formulated using sets and functions. To the practically oriented person, the use of sets may at first appear to be an unnecessary abstraction that makes the study of probability more complicated than it need be. However, the formalism of sets actually makes the solution of complex risk-classification problems simpler because it removes the ambiguity that can arise in the ordinary English language. The advantages of this formalism will become clearer as we proceed through this chapter.[2]

Classical probability theory is based on two key assumptions:

1. The number of possible outcomes of the random experiment[3] is finite.
2. All outcomes are equally likely; that is, each outcome has the same probability.

These assumptions are quite strong. Indeed, apart from experiments involving standard card games or dice, one might be hard pressed to give a concrete example of an experiment for which the number of outcomes is finite and all outcomes are equally likely. For instance, it is certainly not the case that people of all different ages have the same chance of dying over the coming year. However, the assumption of equal likelihood does apply in a very important situation: survey sampling. Indeed, when selecting a random sample of individuals from a target population, it is generally assumed that each member of the target population has the same chance of being selected. Hence, the assumptions of classical probability theory are not as restrictive as might first appear.

If classical probability were only concerned with the probabilities associated with individual observations, it would not be very interesting since, by assumption, all such probabilities are equal to $1/n$, where n is the number of different possible outcomes. What makes the subject interesting is determining probabilities associated with *groups* of observations. For example, one may be interested in determining the probability that two dice land with the same number facing up or the probability that the sum of the numbers on the two dice is 7. Since there are many experimental outcomes for which the

---

[2] A review of sets and their properties is contained in the appendix to this chapter.

[3] Recall that a **random experiment** is any procedure with several possible outcomes that can be repeated a large number of times and whose outcome is not known in advance.

numbers on the dice are the same or for which the sum of the numbers is 7, determining such probabilities is, in general, a nontrivial problem.

We are now ready to give some formal definitions. The **sample space** of an experiment is the set S of all possible outcomes of the experiment. Each subset E of the sample space S represents a particular group of possible observations. The subsets of S are referred to as **events**. To say that the event E occurred means that the observed outcome of the random experiment is contained in the restricted subset E. The collection of all events for a given experiment is referred to as the **event space.**

The following two examples should clarify the meaning of these definitions.

EXAMPLE 1:   Suppose that a coin with two distinguishable sides is tossed once. Let H denote the outcome heads and let T denote the outcome tails. The sample space for this experiment is

$$S = \{H, T\}.$$

The event space for this experiment is

$$\mathcal{E} = \{\emptyset, \{H\}, \{T\}, \{H, T\}\}.$$

The subset $\{H\}$ represents the event of obtaining heads. The subset $\{H, T\}$ represents the event of obtaining either heads or tails; since no outcomes other than H or T are possible, the event $\{H, T\}$ is certain to occur. The subset $\emptyset$ represents the event of observing none of the possible outcomes; this event can never occur. ∎

EXAMPLE 2:   A standard die with six distinguishable sides is rolled and the outcome is observed. The sample space for this experiment is

$$S = \{1, 2, 3, 4, 5, 6\}.$$

The event space for this experiment is

$$
\begin{aligned}
\mathcal{E} = \{ & \emptyset, \{1\}, \{2\}, \{3\}, \{4\}, \{5\}, \{6\}, \\
& \{1, 2\}, \{1, 3\}, \{1, 4\}, \{1, 5\}, \{1, 6\}, \{2, 3\}, \{2, 4\}, \\
& \{2, 5\}, \{2, 6\}, \{3, 4\}, \{3, 5\}, \{3, 6\}, \{4, 5\}, \{4, 6\}, \{5, 6\}, \\
& \{1, 2, 3\}, \{1, 2, 4\}, \{1, 2, 5\}, \{1, 2, 6\}, \{1, 3, 4\}, \\
& \{1, 3, 5\}, \{1, 3, 6\}, \{1, 4, 5\}, \{1, 4, 6\}, \{1, 5, 6\}, \\
& \{2, 3, 4\}, \{2, 3, 5\}, \{2, 3, 6\}, \{2, 4, 5\}, \{2, 4, 6\}, \\
& \{2, 5, 6\}, \{3, 4, 5\}, \{3, 4, 6\}, \{3, 5, 6\}, \{4, 5, 6\}, \\
& \{1, 2, 3, 4\}, \{1, 2, 3, 5\}, \{1, 2, 3, 6\}, \{1, 2, 4, 5\}, \{1, 2, 4, 6\}, \\
& \{1, 2, 5, 6\}, \{1, 3, 4, 5\}, \{1, 3, 4, 6\}, \{1, 3, 5, 6\}, \{1, 4, 5, 6\}, \\
& \{2, 3, 4, 5\}, \{2, 3, 4, 6\}, \{2, 3, 5, 6\}, \{2, 4, 5, 6\}, \{3, 4, 5, 6\}, \\
& \{1, 2, 3, 4, 5\}, \{1, 2, 3, 4, 6\}, \{1, 2, 3, 5, 6\}, \{1, 2, 4, 5, 6\}, \\
& \{1, 3, 4, 5, 6\}, \{2, 3, 4, 5, 6\}, \{1, 2, 3, 4, 5, 6\}\}.
\end{aligned}
$$

Note that S has $2^6 = 64$ subsets!

The subset $\{1, 3\}$ represents the event that either 1 or 3 is observed. The subset $\{2, 4, 5\}$ represents the event that either 2 or 4 or 5 is observed. The subset $\{5\}$ represents the event that 5 (and only 5) is observed. Other members of the event space $\mathcal{E}$ have similar interpretations. ∎

Now let's consider the probabilities associated with events. Let $\Pr(E)$ denote the probability that the observed outcome of the random experiment is one of the elements of E. It is customary to refer to $\Pr(E)$ as simply the "probability of event E." Then under the classical assumption that all *individual* outcomes of the random experiment are equally likely, the probability of event E is given by

$$\Pr(E) = \frac{|E|}{|S|},$$

where $|E|$ is the number of elements in the event set E and $|S|$ is the number of elements in the sample space S. Consequently, calculating probabilities under the classical assumption reduces to a problem in combinatorics.[4]

EXAMPLE 3:   A standard die is rolled. Determine the probability that the outcome is a multiple of 3.

The desired event is $E = \{3, 6\}$. Since $|E| = 2$ and $|S| = 6$, the desired probability is $1/3$. ∎

EXAMPLE 4:   Two standard distinguishable coins are tossed and the outcome is observed. Determine the probability that both coins land on the same side.

The sample space for this experiment is

$$S = \{(H, H), (H, T), (T, H), (T, T)\}.$$

Note the use of *ordered pairs* to distinguish between the coins. The event space is

$$\begin{aligned}
\mathcal{E} = \{ &\emptyset, \{(H, H)\}, \{(H, T)\}, \{(T, H)\}, \{(T, T)\}, \\
&\{(H, H), (H, T)\}, \{(H, H), (T, H)\}, \{(H, H), (T, T)\}, \\
&\{(H, T), (T, H)\}, \{(H, T), (T, T)\}, \{(T, H), (T, T)\}, \\
&\{(H, H), (H, T), (T, H)\}, \{(H, H), (H, T), (T, T)\}, \\
&\{(H, H), (T, H), (T, T)\}, \{(H, T), (T, H), (T, T)\}, \\
&\{(H, H), (H, T), (T, H), (T, T)\}\}.
\end{aligned}$$

The desired event is $E = \{(H, H), (T, T)\}$. Thus, under the classical assumption of equally likely outcomes,

$$\Pr(E) = \frac{|E|}{|S|} = \frac{2}{4} = \frac{1}{2}. \qquad ∎$$

An Important Comment About Sample Spaces   In applying the classical assumption of equally likely outcomes, it is critically important that we use the "correct" sample space. Consider, for example, an experiment in which two standard coins are tossed.

If we do not distinguish between the coins in this experiment, perhaps because the coins are identical and we toss them together, then we would consider the possible outcomes of the experiment to be

• two heads;
• one head and one tail;
• two tails.

---

[4] Basic counting techniques are discussed in the appendix to this chapter.

However, if we do distinguish between the coins, perhaps by using coins of different denominations or keeping the coins separate, then we would consider the possible outcomes of the experiment to be

- head, head;
- head, tail;
- tail, head;
- tail, tail.

Clearly, the numerical probability assigned to the event of getting two heads, for instance, will be very different in these two formulations (1/3 vs. 1/4), suggesting that these formulations cannot both be right (at least not under the classical assumption of equally likely experimental outcomes). Intuition and experimental evidence based on the relative frequency interpretation of probability suggest that the probability of getting two heads is 1/4, not 1/3. Consequently, the correct sample space under the classical assumptions is

$$S = \{(H, H), (H, T), (T, H), (T, T)\},$$

which is obtained by distinguishing the two coins from one another.

The reason that the set

{two heads, one head and one tail, two tails}

is not an appropriate sample space under the classical assumptions is that it suppresses information about the random experiment by not distinguishing between the coins when it is possible to do so. This suggests that to apply the classical assumption of equally likely outcomes correctly, one must choose the sample space in such a way that all information about the random experiment is captured.

In all of the examples presented so far in this chapter, the number of elements in the sample space has been explicitly given, and probabilities of events have been relatively easy to determine. However, there are many situations in practice where the size of the sample space will not be given and it will be necessary to determine the probability of a desired event from information given about other events. The following example illustrates such a situation.

EXAMPLE 5:   In a recent study on teenage drug and alcohol use, researchers found that one in ten teenagers admitted to using marijuana at least once a month, one in five admitted to drinking alcohol at least once a week, and one in four admitted to smoking cigarettes on a daily basis. The researchers also found that 10% of respondents admitted to both smoking cigarettes daily and drinking alcohol at least once a week; 5% of respondents admitted to both smoking cigarettes daily and using marijuana at least once a month; 3% of respondents admitted to both drinking alcohol at least once a week and using marijuana at least once a month; and 1% of respondents admitted to doing all three.

A rapidly graying middle-aged parent reading these statistics becomes very concerned that his teenager, who took part in the study and whom he considers to be as likely as any of the other teens in the study to engage in such behaviors, may be at serious risk. Is this concerned justified?

Assuming that the teenager is as likely as the others in the study to smoke, drink, or take drugs, what is the probability that the teenager engages in one of these three poor behaviors? What is the probability that the teenager does not engage in any of these behaviors? ∎

In this example, one does not know the size of the sample space. However, one can still determine the desired probabilities from the given information using some basic properties of probabilities.

The three basic properties of probabilities, from which all other properties can be deduced, are as follows:

Basic Properties of Probabilities

1. $0 \leq \Pr(E) \leq 1$ for all events E;
2. $\Pr(S) = 1$;
3. For any sequence of events $E_1, E_2, E_3, \ldots$ with the property that $E_i \cap E_j = \emptyset$ whenever $i \neq j$,

$$\Pr\left(\bigcup E_i\right) = \sum \Pr(E_i).$$

These properties are so fundamental that one could take them as basic axioms of probabilities. With this axiomatic approach, one postulates the existence of a function $\Pr(\cdot)$, known as a **probability measure**,[5] which assigns to each event E a numerical probability $\Pr(E)$ such that the given three axioms are satisfied. One can then determine the probability of any desired event using the axioms and the assumed probability measure.

When the number of events under consideration is small, as in the preceding example, an efficient way to calculate the probability of a desired event is to use a Venn diagram. A **Venn diagram**[6] is a visual representation of selected events and their interaction with one another and the underlying sample space. An illustration of a Venn diagram with three events highlighted is given in Figure 3.1.

The advantage of a Venn diagram is that it allows us to concentrate on the events of interest and to disregard the rest. To illustrate how Venn diagrams can be used to simplify the calculation of probabilities, let's return to the example on teenage drug, alcohol, and tobacco use.

TEENAGE DRUG, ALCOHOL, AND TOBACCO USE STUDY REVISITED: Let M, A, and C represent the following events:

---

[5] We use the term *probability measure* here rather than *probability function* to avoid confusion with the probability mass function p_X, which we initially referred to as a probability function in §2.1. Often, we will just refer to $\Pr(\cdot)$ as a *probability*.

[6] Named after J. Venn who popularized their use. See the reference in §1.10.

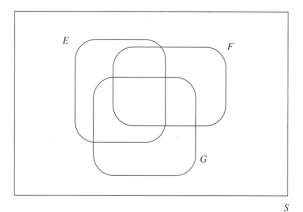

FIGURE 3.1   A Venn Diagram with Three Events Highlighted

M: Teenager admits to using marijuana at least once a month.
A: Teenager admits to consuming alcohol at least once a week.
C: Teenager admits to smoking cigarettes on a daily basis.

The information given in the study can be formulated using probability statements as follows:

$$\Pr(M) = .10, \qquad \Pr(A) = .20, \qquad \Pr(C) = .25,$$

$$\Pr(C \cap A) = .10, \qquad \Pr(C \cap M) = .05, \qquad \Pr(A \cap M) = .03,$$

$$\Pr(M \cap A \cap C) = .01.$$

Using the basic properties of probabilities, one can easily construct a Venn diagram for these events which includes all of the relevant probability information. Generally, the best procedure to follow in completing such a Venn diagram is to begin with the event at the "center" of the diagram (in this example, at the event $M \cap A \cap C$) and work out from that point until probabilities have been assigned to all of the events in the diagram. The relevant Venn diagram for this example is given in Figure 3.2.

From this diagram, one can easily see that

$$\Pr(M \cup A \cup C) = .38$$

and that[7]

$$\Pr((M \cup A \cup C)^c) = 1 - .38 = .62.$$

Hence, according to the survey, 38% of teenagers admitted to engaging in at least one of the undesirable behaviors, whereas 62% claimed that they do not engage in any of these behaviors. ∎

When the number of events under consideration is bigger than four or five, Venn diagrams can become quite complicated and even visually confusing. In such situations, it is generally better to use pure algebra to calculate probabilities.

---

[7]Throughout the text, we will write E^c for the **complement** of the event E; that is, $E^c = \{s \in S : s \notin E\}$.

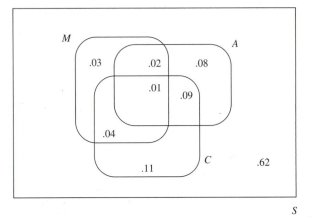

FIGURE 3.2   Venn Diagram for the Drug, Alcohol, and Tobacco Use Example

The following properties of probabilities are very useful in calculating the probabilities for desired events:

Additional Properties of Probabilities

1. $\Pr(E \cup F) = \Pr(E) + \Pr(F) - \Pr(E \cap F)$
2. $\Pr(E^c) = 1 - \Pr(E)$
3. $\Pr\left(\left(\bigcup_{i=1}^{n} E_i\right)^c\right) = \bigcap_{i=1}^{n} \Pr(E_i^c)$
4. $\Pr\left(\left(\bigcap_{i=1}^{n} E_i\right)^c\right) = \bigcup_{i=1}^{n} \Pr(E_i^c)$
5. $\Pr(E) \leq \Pr(F)$ whenever $E \subset F$

These properties are direct consequences of the three probability axioms and can be proved using properties of sets. For details, the reader should consult the appendix to this chapter.

3.2   Conditional Probability

When an auto insurer sells a collision insurance policy to a newly licensed driver, it must determine the premium on the basis of limited information because the new driver has no driving record on which to base a rate. The rate charged in such situations will typically be based on the historical experience of a previous group of similarly situated drivers (e.g., on the observation that 5% of newly licensed drivers in a particular geographic area have historically been involved in an at-fault accident within their first 3 years of being fully licensed) and may take into account factors such as the completion of a formal driver training course or the school grades of the driver.

As time passes, the insurance company acquires new information about the policyholder that may affect insurance rates at policy renewal. Such information includes the number of at-fault accidents, the number of traffic violations, or the diagnosis of some severe health problem such as epilepsy. This new information will cause the insurer to alter its view of the policyholder's chance of being involved in a future at-fault accident and to adjust the insurance premium charged accordingly.

Probabilities that are adjusted to account for new information are referred to as **conditional probabilities.** In this section, we define precisely the notion of a conditional probability and give a formula for calculating conditional probabilities in practice. We begin with a simple example.

EXAMPLE 1:   A standard six-sided die is rolled. Consider the following events:

E: Outcome is an even number.
F: Outcome is 1, 2, 3, or 5.

Without any information other than the fact that $S = \{1, 2, 3, 4, 5, 6\}$, we conclude that

$$\Pr(E) = \frac{1}{2} \text{ and } \Pr(F) = \frac{2}{3}.$$

However, suppose that after the die is rolled, but before we have a chance to observe the actual outcome, we are informed that event F occurred (i.e., the outcome was 1, 2, 3, or 5). What now is the probability that the outcome is an even number?

Note that the only way the outcome can be even under these circumstances is if it is 2. Moreover, the only outcomes now possible are 1, 2, 3, or 5. Hence, the question is this: What is the probability of observing 2 under the assumption that the only possible observations are 1, 2, 3, or 5?

Under the classical assumption of equally likely outcomes, this probability is clearly 1/4. The probability is also 1/4 using the relative frequency interpretation of probability, as the reader is invited to verify.

In terms of sets, the desired probability is

$$\frac{|E \cap F|}{|F|}.$$

Notice that

$$\frac{|E \cap F|}{|F|} = \frac{|E \cap F|/|S|}{|F|/|S|} = \frac{\Pr(E \cap F)}{\Pr(F)}.$$

Hence, to determine the conditional probability of E given F, we need only consider the probabilities $\Pr(E \cap F)$ and $\Pr(F)$.

Using this observation, it is straightforward to determine the probability of event F given the information that E has occurred (i.e., the probability that the outcome of the die roll is 1, 2, 3, or 5 when informed that the outcome is even). This probability is simply

$$\frac{\Pr(F \cap E)}{\Pr(E)} = \frac{1/6}{1/2} = \frac{1}{3}.$$

Note that the probability of E given F is not the same as the probability of F given E. This is generally the case for any pair of events E and F. ■

The preceding example suggests the following formal definition for conditional probabilities:

Definition of Conditional Probability

Consider a random experiment with sample space S and events E and F. The **conditional probability** of event E given the information that event F occurred is the number $\Pr(E|F)$ given by

$$\Pr(E|F) = \frac{\Pr(E \cap F)}{\Pr(F)}.$$

The notation $E|F$ is used to represent the event E *under the condition* that F has occurred.

It would appear from the definition that conditional probabilities are relatively simple to calculate. However, depending on the nature of the information given in a problem, calculating them in practice can sometimes be tricky. One must also be careful to calculate the correct conditional probability in a problem and not to confuse $\Pr(E|F)$ with $\Pr(F|E)$. The following example illustrates a typical conditional probability calculation.

EXAMPLE 2: A large auto insurance company is studying the claim experience of teenage drivers and its possible relationship to the teenager's school grades and to whether or not the teenager took a formal driver education program. The insurer bases the study on its own data accumulated over a 10-year period.

The insurer finds that 6% of teenage drivers are involved in an at-fault accident in their first year; 70% of teenage drivers have taken driver education; and 10% of teenage drivers are A students. The insurer also finds that 1 in 14 teenage drivers who have taken driver education are A students; 1 in 3 teenage drivers involved in an at-fault accident in their first year of driving took driver education; and 1 in 100 teenage drivers were both A students and had an at-fault accident in their first year. Of teenage drivers in the study, 22% were neither A students nor in an at-fault accident the first year nor did they take driver education.

Determine:

1. the probability that a teenage driver who has taken driver education has an at-fault accident in the first year of driving;
2. the probability that a teenage driver who is an A student has an at-fault accident in the first year of driving;
3. the probability that a teenage driver who is both an A student and has taken driver education has an at-fault accident in the first year of driving.

This example illustrates once again how the use of Venn diagrams and the formal language of probability can greatly simplify the calculation of probabilities and ensure that the answer we get is correct! Let A, B, and C represent the following events:

A: Driver is an A student.

B: Driver has taken formal driver education.

C: Driver has an at-fault accident in the first year.

Then, from the given information, we have the following:

$$\Pr(A) = .10, \qquad \Pr(B) = .70, \qquad \Pr(C) = .06,$$
$$\Pr(A|B) = 1/14, \qquad \Pr(B|C) = 1/3,$$
$$\Pr(A \cap C) = .01,$$
$$\Pr((A \cup B \cup C)^c) = .22.$$

In probability notation, the probabilities we must determine are, respectively: (1) $\Pr(C|B)$, (2) $\Pr(C|A)$, (3) $\Pr(C|A \cap B)$. To calculate these probabilities directly from the definition of conditional probability, we need values for $\Pr(B \cap C)$, $\Pr(B)$, $\Pr(A \cap C)$, $\Pr(A)$, $\Pr(A \cap B \cap C)$, and $\Pr(A \cap B)$. Three of these six quantitites are given explicitly in the statement of the problem; the other three—$\Pr(B \cap C)$, $\Pr(A \cap B)$, and $\Pr(A \cap B \cap C)$—must be determined from the given information.

We now have a choice: We can try to construct a Venn diagram or we can use the algebraic properties of probability measures to determine the desired probabilities. The approach we will actually take is a hybrid of the two.

From the definition of conditional probability and the given information,

$$\Pr(B \cap C) = \Pr(B|C) \Pr(C)$$
$$= \left(\frac{1}{3}\right)(.06)$$
$$= .02$$

and

$$\Pr(A \cap B) = \Pr(A|B) \Pr(B)$$
$$= \left(\frac{1}{14}\right)(.70)$$
$$= .05.$$

To determine $\Pr(A \cap B \cap C)$, we will construct a Venn diagram with $\Pr(A \cap B \cap C)$ unspecified and then use the fact that $\Pr(S) = 1$.

Put $x = \Pr(A \cap B \cap C)$. A Venn diagram with x unspecified is given in Figure 3.3. Using the fact that $\Pr(S) = 1$, we have

$$(.04 + x) + (.05 - x) + (.01 - x) + x + (.63 + x)$$
$$+ (.02 - x) + (.03 + x) + .22 = 1$$

from which we conclude that $x = 0$. Hence, $\Pr(A \cap B \cap C) = 0$.

Returning to the original question, we have

$$\Pr(C|B) = \frac{\Pr(C \cap B)}{\Pr(B)} = \frac{.02}{.70} = \frac{1}{35},$$
$$\Pr(C|A) = \frac{\Pr(C \cap A)}{\Pr(A)} = \frac{.01}{.10} = \frac{1}{10},$$
$$\Pr(C|A \cap B) = \frac{\Pr(C \cap A \cap B)}{\Pr(A \cap B)} = \frac{0}{.05} = 0.$$

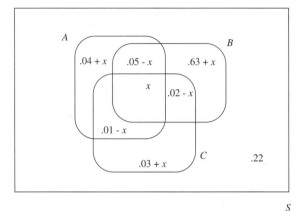

FIGURE 3.3   Venn Diagram for Teenage Driver Example

It is important to interpret these probability statements correctly. The statement $\Pr(C|A \cap B) = 0$, for example, does not mean that a new teenage driver who is both an A student and has taken driver education can never have an at-fault accident in the first year of driving. What it does mean is that none of the teenage drivers *in this study* who were both A students and had taken driver education had an at-fault accident in their first year. ■

3.3   The Law of Total Probability

In the previous section, we considered how probabilities of specific events can change with the arrival of new information, and we gave a formula for calculating such conditional probabilities using their unconditional analogs. In this section, we discuss an important result, known as the law of total probability, which allows us to recapture unconditional probabilities from conditional probabilities.

It may at first appear strange that we would want to calculate unconditional probabilities, which incorporate no new information, if we are already given conditional probabilities, which do incorporate new information. However, as we will soon see, the significance of the law of total probability is that it allows us to decompose an unconditional probability into conditional probability components that may be easier to determine in practice.

We begin with a formal statement of the result:

Law of Total Probability

Suppose that $F_1, F_2, \ldots, F_n$ are events such that $F_i \cap F_j = \emptyset$ whenever $i \neq j$ and $F_1 \cup \cdots \cup F_n = S$. Then for any event E,

$$\Pr(E) = \Pr(E|F_1)\Pr(F_1) + \cdots + \Pr(E|F_n)\Pr(F_n).$$

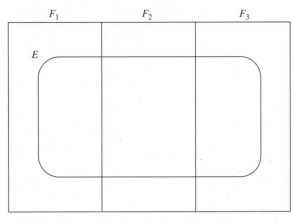

FIGURE 3.4   Venn Diagram Illustrating the Law of Total Probability

Before deriving this result, let's consider a simple application of it.

EXAMPLE 1:   An auto insurer classifies its policyholders as either average or substandard risks according to some classification procedure. Seventy percent are classified as average risks. During the year, 1% of the average risks are involved in an at-fault accident, and 5% of the substandard risks are involved in an at-fault. What fraction of policyholders are involved in an at-fault?

Let A, B, and C represent the following events:

A: Policyholder classifed as average risk.
B: Policyholder classified as substandard risk.
C: Policyholder involved in an at-fault accident.

Then, from the given information,

$$\Pr(A) = .70, \quad \Pr(B) = .30,$$
$$\Pr(C|A) = .01, \quad \Pr(C|B) = .05.$$

By the law of total probability, the desired fraction is

$$\Pr(C) = \Pr(C|A)\,\Pr(A) + \Pr(C|B)\,\Pr(B)$$
$$= (.01)(.70) + (.05)(.30)$$
$$= .022.$$ ∎

The law of total probability is relatively easy to derive from the properties of probabilities discussed in §3.1 and the definition of conditional probability given in §3.2. A Venn diagram illustrating the basic ideas when the sample space is partitioned into three events is given in Figure 3.4.

DERIVATION OF THE LAW: Suppose that S is a disjoint union of the events $F_1, \ldots,$ F_n (i.e., $S = F_1 \cup \cdots \cup F_n$ and $F_i \cap F_j = \emptyset$ whenever $i \neq j$). Let E be any event. Then since the F_j are mutually exclusive (i.e., $F_i \cap F_j = \emptyset$ whenever $i \neq j$), E is a disjoint union of $E \cap F_1, \ldots, E \cap F_n$. Consequently,

$$\Pr(E) = \Pr(E \cap F_1) + \cdots + \Pr(E \cap F_n).$$

From the definition of conditional probability, we have

$$\Pr(E \cap F_j) = \Pr(E|F_j) \Pr(F_j).$$

Hence,

$$\Pr(E) = \Pr(E|F_1) \Pr(F_1) + \cdots + \Pr(E|F_n) \Pr(F_n)$$

as claimed. ∎

After seeing how simple the law of total probability is to derive, one may wonder why this proposition is considered so important. It is, after all, referred to as a *law*. The reason is that it allows us to decompose a desired unconditional probability into conditional probability components that may be easier to determine and more meaningful in practice. We illustrate this use of the law in the following example.

EXAMPLE 2:   Consider an insurer that provides group health insurance to a number of different employers. Each group consists of both healthy and impaired lives in varying proportions. The insurer is concerned with the incidence of catastrophic claims (e.g., claims exceeding \$100,000). Such claims clearly will have a material effect on the insurer's profit and, if numerous enough, could result in the insurer becoming insolvent.

Intuition suggests that impaired lives are more likely than healthy lives to file catastrophic claims. However, the probability of an impaired life filing a catastrophic claim will very likely be the same for each group. Likewise, the probability of a healthy life filing a catastrophic claim will be the same for each group. Consequently, to determine the catastrophic claim incidence probability for any specific group, the insurer need only weight the catastrophic claim probabilities for impaired lives and for healthy lives by the proportions of impaired and healthy lives in the specific group in accordance with the law of total probability. The alternative approach—determining separate catastrophic claim incidence probabilities for each group from scratch—is generally not very practical.

Apart from simplifying calculation, the law of total probability also allows one to break the underlying uncertainty into its component parts. In this case, there are two sources of uncertainty: the catastrophic claim incidence for people known to be impaired or healthy (as the case may be) and the number of impaireds in the group. In practice, claim incidence probabilities must be estimated using some statistical procedure and, hence, are subject to error. By breaking the determination of the overall catastrophic claim incidence probability into probabilities that are easier to estimate, we increase the accuracy of the estimate for the catastrophic claim incidence probability. ∎

We conclude this section with two additional numerical examples illustrating various uses of the law of total probability.

EXAMPLE 3:   A bank is considering extending credit to a new customer and is interested in the probability that the client will default on the loan. Based on data that the bank has accumulated over many years of business, there is a 5% chance that a customer who has overdrawn an account will default on the loan, whereas there is only a 0.5% chance that a customer who has never overdrawn an account will default on the loan. Unfortunately, the bank does not know for sure what type of risk the new customer will turn out to be. Based on background checks and credit reports, the bank believes that there is a 30% chance that the new customer will overdraw the account. Determine the probability that the customer will default on the loan if credit is extended.

Let D and O represent the following events:

D: Customer defaults on loan.
O: Customer overdraws account.

Then, from the given information,

$$\Pr(D|O) = .05, \qquad \Pr(D|O^c) = .005, \qquad \Pr(O) = .30.$$

Hence, by the law of total probability

$$\begin{aligned} \Pr(D) &= \Pr(D|O)\Pr(O) + \Pr(D|O^c)\Pr(O^c) \\ &= (.05)(.30) + (.005)(.70) \\ &= .0185. \end{aligned}$$

Notice how application of the law of total probability enables the bank to determine the desired default probability from data that it is more likely to have access to in practice.

■

EXAMPLE 4:   A health maintenance organization (HMO) concerned with the adverse financial effects of catastrophic medical expenses for its members enters into a reinsurance arrangement in which the reinsurer (i.e., the insurer providing insurance to the HMO) agrees to pay the amount of each plan member's annual hospital expenses that exceed $100,000. Thus, for example, if the HMO had exactly two members and these two members had annual hospital expenses of $130,000 and $70,000, respectively, then the reinsurer would pay the HMO $30,000. Under the terms of this agreement, the HMO will be required to pay an additional reinsurance premium above and beyond the initial reinsurance premium if the fraction of the plan membership for which a reinsurance claim is filed exceeds 5%. This condition helps minimize any incentive the HMO may have to increase its enrollment of substandard risks after entering into the reinsurance arrangement.

The probability of a catastrophic claim for any given member clearly depends on the age and family status of the member. For example, members who are expecting children are at a higher risk of incurring large neonatal expenses than members without families. On the basis of historical data, the plan estimates that 10% of members over the age of 65 will incur hospital expenses exceeding $100,000, 4% of members between the ages of 40 and 65 will incur hospital expenses exceeding $100,000, and 3% of members under the age of 40 will incur hospital expenses exceeding $100,000.

Twenty percent of the HMO's members are currently over the age of 65. If the HMO wishes to maintain this level of members over the age of 65 (perhaps for competitive or regulatory reasons) and wishes to avoid exceeding the 5% trigger point in the reinsurance agreement, how large must the HMO's under-40 population be?

Let C, F_1, F_2, and F_3 represent the following events:

C: Member has hospital expenses exceeding \$100,000.
F_1:   Member is under the age of 40.
F_2:   Member is between the ages of 40 and 65.
F_3:   Member is over 65.

From the given information, we have

$$\Pr(C|F_1) = .03, \qquad \Pr(C|F_2) = .04, \qquad \Pr(C|F_3) = .10,$$
$$\Pr(F_3) = .20.$$

To avoid paying an additional reinsurance premium, the HMO requires that

$$\Pr(C) \le .05.$$

By the law of total probability,

$$\Pr(C) = \Pr(C|F_1)\Pr(F_1) + \Pr(C|F_2)\Pr(F_2) + \Pr(C|F_3)\Pr(F_3)$$
$$= (.03)\Pr(F_1) + (.04)\Pr(F_2) + (.10)(.20)$$
$$= (.03)\Pr(F_1) + (.04)\Pr(F_2) + .02.$$

Hence, the requirement $\Pr(C) \le .05$ is equivalent to

$$(.03)\Pr(F_1) + (.04)\Pr(F_2) \le .03;$$

that is,

$$\Pr(F_1) + \frac{4}{3}\Pr(F_2) \le 1.$$

Put $x = \Pr(F_1)$. Then $\Pr(F_2) = 1 - \Pr(F_1) - \Pr(F_3) = 1 - x - .2 = .8 - x$. Substituting these values into the previous inequality, we have

$$x + \frac{4}{3}(.8 - x) \le 1$$

from which we deduce that

$$x \ge .2.$$

Hence, to avoid exceeding the 5% trigger point in the reinsurance agreement, at least 20% of the HMO's members should be under the age of 40. ∎

3.4   Bayes' Theorem

At the beginning of this chapter, we considered the informational asymmetry problem faced by insurers with respect to their policyholders and we discussed the importance of risk-classification schemes in overcoming this problem. In this section, we develop a formula for "inverting" conditional probabilities, known as **Bayes' theorem.** This result provides us with an important tool to develop accurate risk-classification schemes and to analyze the error associated with any such scheme.

We begin with a formal statement of the theorem:

Bayes' Theorem

For any events E and F, the conditional probabilities $\Pr(E|F)$ and $\Pr(F|E)$ are connected by the following formula:

$$\Pr(E|F) = \frac{\Pr(F|E)\,\Pr(E)}{\Pr(F)}.$$

Bayes' theorem is straightforward to derive using the definition of conditional probability twice. Its derivation is left as an exercise for the reader.

In the remainder of this section, we give examples that illustrate some of the many possible applications of Bayes' theorem. Other applications are developed in the exercises at the end of the chapter.

EXAMPLE 1:   An auto insurer observes from its claim files that 70% of policyholders who have an at-fault accident during the year had a prior speeding conviction. This prompts the insurer to analyze the driving records of all its policyholders to determine if prior speeding convictions are a good predictor of accident occurrence. The insurer determines that 25% of policyholders who did not have an at-fault accident during the year had a prior speeding conviction. Of policyholders in total, 10% had an at-fault accident during the year. Based on these observations, is a prior speeding conviction a good predictor of a future at-fault accident?

Let A and B represent the following events:

A: Policyholder has an at-fault accident during the year.
B: Policyholder has a prior speeding conviction.

We are interested in determining $\Pr(A|B)$. We are given that

$$\Pr(B|A) = .70, \qquad \Pr(B|A^c) = .25, \qquad \Pr(A) = .10.$$

By Bayes' theorem,

$$\begin{aligned}
\Pr(A|B) &= \frac{\Pr(B|A)\,\Pr(A)}{\Pr(B)} \\
&= \frac{(.70)(.10)}{\Pr(B)} \\
&= \frac{.07}{\Pr(B)}.
\end{aligned}$$

To determine $\Pr(B)$, we must use the law of total probability:

$$\begin{aligned}
\Pr(B) &= \Pr(B|A)\,\Pr(A) + \Pr(B|A^c)\,\Pr(A^c) \\
&= (.70)(.10) + (.25)(.90) \\
&= .295.
\end{aligned}$$

Thus, the desired probability is

$$\Pr(A|B) = \frac{.07}{.295} = \frac{14}{59} \approx 24\%.$$

Hence, we see that the predictive power of a prior speeding conviction is not nearly as great as one may have expected! ■

EXAMPLE 2:   A bank classifies customers as either good or bad credit risks. On the basis of extensive historical data, the bank has observed that 1% of good credit risks and 10% of bad credit risks overdraw their account in any given month. A new customer opens a checking account at this bank. On the basis of a check with a credit bureau, the bank believes that there is a 70% chance the customer will turn out to be a good credit risk.

Suppose that this customer's account is overdrawn in the first month. How does this alter the bank's opinion of this customer's creditworthiness? What if the customer overdraws the account in the second month?

Let G and O represent the following events:

G: Customer considered a good risk.
O: Customer overdraws checking account.

From the bank's historical data, we have

$$\Pr(O|G) = .01, \qquad \Pr(O|G^c) = .10.$$

On the other hand, the bank's initial opinion about the customer's creditworthiness is given by

$$\Pr(G) = .70.$$

In this application, the probabilities $\Pr(O|G)$, $\Pr(O|G^c)$ are based on historical data and remain constant for all future months. However, the probability $\Pr(G)$ is a representation of the bank's belief about the customer's creditworthiness and will change to reflect new information (e.g., overdraft) and the bank's evolving opinion of the customer.

We are told that the customer has an overdraft in the first month. Hence, at the end of the first month, the bank's revised opinion about the customer's creditworthiness is given by the probability $\Pr(G|O)$. Using Bayes' theorem and the law of total probability, this probability is

$$
\begin{aligned}
\Pr(G|O) &= \frac{\Pr(O|G)\,\Pr(G)}{\Pr(O|G)\,\Pr(G) + \Pr(O|G^c)\,\Pr(G^c)} \\
&= \frac{(.01)(.70)}{(.01)(.70) + (.10)(.30)} \\
&= \frac{7}{37}.
\end{aligned}
$$

The probability $\Pr(G)$ (before the arrival of new information) is often referred to as the **prior probability,** and the probability $\Pr(G|O)$ (after the arrival of new information) is referred to as the **posterior probability.** The rationale for this terminology should be clear.

Now let's consider what happens in the second month. At the beginning of the month, the bank believes that there is a 7/37 chance that the customer will turn out to be a good credit risk. Hence, at the beginning of the second month, $\Pr(G) = 7/37$.[8] That is, the prior probability of the customer being a good risk is $7/37$. Since the customer has an overdraft in the second month, the bank's opinion about the customer's creditworthiness at the end of the month is given by

$$\Pr(G|O) = \frac{\Pr(O|G)\Pr(G)}{\Pr(O|G)\Pr(G) + \Pr(O|G^c)\Pr(G^c)}$$

$$= \frac{(.01)(7/37)}{(.01)(7/37) + (.10)(30/37)}$$

$$= \frac{7}{307}.$$

Clearly, the bank no longer believes that this customer is a good credit risk.

What would be the bank's opinion of the customer's creditworthiness at the end of the second month if there had not been an overdraft in the second month?   ∎

3.5   Chapter Summary

Classical probability theory is an effective tool for addressing problems such as insurance risk classification where decisions must be made on the basis of imperfect information. Classical probability is generally formulated using the language of sets and functions. The advantage of this formal approach is that it removes the ambiguities of ordinary English language and ensures that our probability calculations are correct.

The **sample space** for a given random experiment is the set S of all possible outcomes of the experiment. Subsets of the sample space are called **events.** The collection of all events is known as the **event space.** A **probability measure** is a function that assigns to each event E a number $\Pr(E)$, known as the probability of event E, with the following three properties:

1. $0 \leq \Pr(E) \leq 1$ for all events E;
2. $\Pr(S) = 1$;
3. For any sequence of events $E_1, E_2, E_3, \ldots$ with the property that $E_i \cap E_j = \emptyset$ whenever $i \neq j$,

$$\Pr\left(\bigcup E_i\right) = \sum \Pr(E_i).$$

Under the strict classical assumption that all outcomes of the experiment are equally likely,

$$\Pr(E) = \frac{|E|}{|S|}$$

for all events E.

When using the classical assumption of equally likely outcomes, it is critical that one correctly identify the sample space S. The correct sample space in this context is the one

---

[8] Note the two different uses of the notation $\Pr(G)$. We could distinguish between these two uses by introducing new notation for the probability of being a good risk at the end of the first month, but that might confuse the reader. Besides, our use of notation is consistent with common practice.

that captures all information about the experiment and does not suppress information necessary to distinguish between different outcomes.

Probabilities for specific events can change with the arrival of new information. Such changes are formalized using the concept of conditional probability. The **conditional probability** of event E given the information that event F has occurred is the number $\Pr(E|F)$ given by

$$\Pr(E|F) = \frac{\Pr(E \cap F)}{\Pr(F)}.$$

In general, the conditional probability of E given F—$\Pr(E|F)$—is not the same as the conditional probability of F given E—$\Pr(F|E)$.

The **law of total probability** is an important result that allows us to decompose an unconditional probability into conditional probability components. Its formal statement is as follows: If $F_1, \ldots, F_n$ are mutually exclusive events whose union is the entire sample space, then for any event E

$$\Pr(E) = \Pr(E|F_1)\Pr(F_1) + \cdots + \Pr(E|F_n)\Pr(F_n).$$

Bayes' theorem is an important result that relates the conditional probability $\Pr(E|F)$ to its "inverse" $\Pr(F|E)$ and allows us to calculate either of these conditional probabilities from the other. For any events E, F, Bayes' theorem asserts that

$$\Pr(E|F)\Pr(F) = \Pr(F|E)\Pr(E).$$

Bayes' theorem has important applications in insurance risk classification and related problems where decisions must be made on the basis of incomplete information. Some of these applications are developed in the exercises.

3.6   Exercises

1. Suppose that E and F are events such that $\Pr(E) = .4$, $\Pr(F) = .6$, and $\Pr(E \cup F) = .8$. Calculate $\Pr(E|F)$ and $\Pr(F|E)$.
2. Suppose that E and F are events such that $\Pr(E) = .3$, $\Pr(F) = .5$, and $\Pr(E|F) = .4$. Calculate $\Pr(E \cap F)$, $\Pr(E \cup F)$, and $\Pr(F|E)$.
3. Suppose that E and F are events. Determine which of the following statements are true and which are false. If the statement is true, explain why; if the statement is false, give a counterexample.
   a. $\Pr(E \cap F) \leq \Pr(E|F)$
   b. $\Pr(E|F) = \Pr(F|E)$
   c. $\Pr(E) + \Pr(F) \leq \Pr(E \cup F) + \Pr(F|E)$
   d. $\Pr(E) \leq \Pr(E|F)$
   e. $\Pr(E \cap F|F) = \Pr(E|F)$
4. Use Venn diagrams to convince yourself that each of the following properties of probabilities is true. Then use the axioms for probability to prove them directly.
   a. $\Pr(E \cup F) = \Pr(E) + \Pr(F) - \Pr(E \cap F)$
   b. $\Pr(E^c) = 1 - \Pr(E)$
   c. If $E \subset F$, then $\Pr(E) \leq \Pr(F)$

5. Use Venn diagrams to find formulas for $\Pr(E \cup F \cup G)$ and $\Pr(E \cup F \cup G \cup H)$. What do you think the general formula for

$$\Pr(E_1 \cup E_2 \cup \cdots \cup E_n)$$

should be?

6. Boole's inequality states that

$$\Pr\left(\bigcup_{i=1}^{n} E_i\right) \leq \sum_{i=1}^{n} \Pr(E_i).$$

Use Venn diagrams to convince yourself that Boole's inequality is true in the cases $n = 1, 2, 3$. Then use induction to prove it for all n. Under what circumstances do you think that Boole's inequality is useful in practice?

7. Bonferroni's inequality states that

$$\Pr(E \cap F) \geq \Pr(E) + \Pr(F) - 1.$$

Use a Venn diagram to convince yourself that Bonferroni's inequality is true. Then use the properties of probabilities to prove it directly. Under what circumstances do you think that Bonferroni's inequality is useful in practice?

8. Suppose that we conduct an experiment and use relative frequencies to assign probabilities to two events E and F.

a. If we only know that $\Pr(E) = .7$ and $\Pr(F) = .4$, what is the strongest statement we can make about $\Pr(E \cup F)$ and $\Pr(E \cap F)$?

b. If we only know that $\Pr(E) = .6$ and $\Pr(F) = .2$, what is the strongest statement we can make about $\Pr(E \cup F)$ and $\Pr(E \cap F)$?

c. Under what conditions does Boole's inequality provide nontrivial information?

d. Under what conditions does Bonferroni's inequality provide nontrivial information?

e. Is it possible for Boole's inequality and Bonferroni's inequality to simultaneously provide nontrivial information?

9. A computer generates a random whole number between 0 and 3 inclusive. Write out the sample space and event space for this experiment.

10. I have four cards in my hand: a heart, a diamond, a spade and a club. My opponent selects two of the four cards at random. Write out the sample space and event space for this experiment.

11. In a particular defined contribution pension plan, participants may direct their contributions to an equity fund, a bond fund, a money-market fund, or some combination of the three. The equity fund invests primarily in blue-chip stocks; the bond fund invests primarily in long-term government securities and high-grade corporate bonds; and the money-market fund invests solely in short-term government securities.

The pension plan administrator is analyzing the participation of eligible employees in each of the funds to determine if greater employee education is required. The plan administrator has determined that 15% of eligible employees own shares in the equity fund, 28% own shares in the bond fund, and 30% own shares in the money-market fund. Moreover, 8% of eligible employees own shares in both the equity fund and the bond fund, 10% own shares in both the equity fund and the money-market fund, 15% own shares in both the bond fund and the money-market fund, and 5% own shares in all three.

a. What percentage of eligible employees are currently participating in the pension plan?

b. What percentage of eligible employees are not currently participating in the pension plan?

c. What percentage of participating employees direct their contributions to a single fund?

d. What fraction of participating employees direct their contibutions to at least two different funds?

e. What fraction of participants with bond shares also own stock shares?

f. What fraction of participants with money-market shares also own stock shares?

12. A particular casualty insurance company specializes in homeowner's insurance, automobile insurance, and professional liability insurance (e.g., medical malpractice insurance). Of its customers, 60% have a homeowner's policy issued by the company, 50% have an automobile policy issued by the company, and 35% have a professional liability policy issued by the company. Of its customers, 30% have both homeowner's and auto insurance, 25% have both homeowner's and professional liability insurance, and 25% have both auto and professional liability insurance. There are 15% of its customers that do not have any of these types of insurance.

a. What fraction of this company's customers have all three types of insurance?

b. What fraction of the company's customers have only professional liability insurance?

c. What fraction of customers with professional liability insurance also have homeowner's insurance?

d. What fraction of customers with auto insurance also have homeowner's insurance?

13. An employee benefits consulting firm has just completed a survey of the health plans offered by employers in a particular state. Among the findings are the following: 40% of employers offer no health insurance to their employees at all; 10% offer only an HMO; 20% offer only a traditional reimbursement plan; 25% offer a point-of-service (POS) plan; 10% offer both an HMO and a traditional reimbursement plan; 10% offer both a POS plan and a traditional plan; and 5% offer all three types of plans. Moreover, of employers who offer an HMO to their employees, 60% offer another type of plan as well.

a. What fraction of employers offer only a POS plan?

b. What fraction of employers with a traditional reimbursement plan also offer a POS plan?

c. What fraction of employers with a POS plan offer some other type of plan as well?

d. What fraction of employers offer their employees a choice of health plans?

14. A pension consulting firm has just completed a survey on the retirement savings of the working-age population. Among the findings are the following: 25% of the work force belong to a company-sponsored pension plan; 20% of the work force have a tax-favored individual retirement account (IRA); 30% have private savings

other than a company pension or a tax-favored IRA in excess of $5000; 5% have all three types of retirement savings; 55% have no retirement savings of any kind other than social security. The survey also revealed that 60% of people who belong to a company pension plan have an IRA, and one third of people with private savings in excess of $5000 have an IRA.

a. What fraction of people with an IRA also have private retirement savings in excess of $5000?

b. What fraction of people with an IRA also belong to a company pension plan?

c. What fraction of people who belong to a company pension plan have no other retirement savings besides social security?

d. What fraction of people with private savings in excess of $5000 do not participate in a company-sponsored pension plan?

15. A test preparation agency offers classes to people who are planning to take the Law School Admissions Test. Of the agency's clients, 65% are undergraduate students, 25% are graduate students, and 10% are people not currently attending school. The agency claims that 80% of its clients improve their scores by taking its preparatory classes. It bases this claim on sample tests which each client takes before and after completion of its classes.

a. In a recent survey of students who had enrolled in the agency's course, 80% of the undergraduates and 50% of the graduates said that their scores improved by taking the course. If these numbers are accurate, what does this suggest about the agency's claim? Explain.

b. In a different survey of students who had enrolled in the course, 95% of the undergrads and 75% of the grads said that their scores improved. What does this suggest about the agency's claim? Explain.

c. In yet another survey of students who had enrolled in the course, 70% of the undergrads and 50% of the grads said that their scores improved by taking the course. Based on these numbers, what is the highest success rate that the agency can possibly claim? What is the smallest possible value for the probability that a client's score improved?

16. A broker handles futures contracts on oil, barley, and orange juice. Of the broker's orders, 60% are for oil futures, 30% are for barley futures, and 10% are for orange juice futures. On a given day, 40% of the orders for oil futures are buys, 55% of the orders for barley futures are sells, and 35% of the orders for orange juice futures are sells. At the end of the day, a clerk notices that one of the sell orders has the name of the commodity future sold omitted. What is the probability that this order was for a future on oil? On barley? On orange juice?

17. An auto insurance company classifies its policyholders as good, bad, or average risks: 30% are deemed good risks, 20% are deemed bad risks, and 50% are deemed average risks. Historical data suggest that 5% of the good risks, 40% of the bad risks, and 10% of the average risks will be involved in an accident in the coming year.

a. What is the probability that a randomly chosen customer files an accident claim in the coming year?

b. An accident claim has just been filed with the company. What is the probability that this customer was classified as a good risk? A bad risk? An average risk?

c. The company would like to have accident claims on at most 10% of its policies. Consequently, it decides to cancel the policies of some bad risk customers and replace these policies with average risks. Of the company's customers, 30% will remain classified as good risks. What is the smallest percentage of the company's customers who must be classified as average risks for the fraction of customers filing accident claims to be at most 10%?

18. An oil company executive is trying to decide whether his company should drill for oil on a newly acquired site. The executive believes that there is a 50% chance of striking oil on this site. To assist him in making this decision, the executive hires a geological consulting agency to perform a seismic survey. The consulting agency will either recommend or not recommend drilling at the site based on the results it obtains. Historical records indicate that this agency has recommended drilling on 90% of known oil producing fields and has recommended against drilling on 80% of known dry holes.

   a. Why is it reasonable for the executive to believe (before obtaining the agency's recommendation) that there is a 50% chance of striking oil? What implicit assumption is he making?
   b. What is the probability that the consulting agency will recommend drilling at this site?
   c. Suppose that the agency recommends drilling at this site. What is the probability of striking oil?
   d. Suppose that the agency recommends against drilling at this site. What is the probability of striking oil?
   e. What is the chance that the agency recommends against drilling but the site actually contains oil?
   f. What is the chance that the agency recommends drilling but the site is dry?

19. The personnel department of a particular company has observed that 30% of the people the company hires are dismissed within a year because they are unable to perform adequately. To reduce the amount of turnover, the company decides to administer a test to all applicants. Data collected over several years suggest that 80% of new hires who remain with the company pass the test, and 90% of new hires who are dismissed fail the test.

   a. What fraction of new hires pass the test?
   b. What fraction of new hires who pass the test will be dismissed within a year?
   c. What fraction of new hires who fail the test will be dismissed within a year? Is this the same as the fraction of dismissed new hires that failed the test? Explain.
   d. You are interviewing a candidate who has failed the test, but you decide to hire the candidate anyway. What is the probability that this person will be with the company 1 year from now?

20. A pharmaceutical company has developed a new test for detecting a particular disease. Historical data suggest that 20% of the general population either have the disease or are at risk of developing it. In recent clinical trials, 95% of people known to have the disease tested positive, and 10% of people known to be healthy tested positive. Assuming that these numbers are also accurate for the general population, determine

   a. the fraction of the population that will test positive;

b. the probability of having the disease if the test is positive;

c. the probability of having the disease if the test is negative.

21. Consider the following experiment:

Two fair coins are tossed and the outcome is observed.

Hector is trying to formulate this experiment using the language of probability but is having some difficulty. He chooses the sample space to be

$$S = \{(H, H), (H, T), (T, H), (T, T)\}$$

and the event space $\mathcal{E}$ to be the collection of subsets of S. Hector observes that the assumption that the first coin is fair is equivalent to the statement

$$\Pr(\{(H, H), (H, T)\}) = \frac{1}{2},$$

and likewise, the assumption that the second coin is fair is equivalent to the statement

$$\Pr(\{(H, H), (T, H)\}) = \frac{1}{2}.$$

He convinces himself that this assignment of probability should be sufficient to completely define a probability measure but has difficulty calculating $\Pr(E)$ for some of the events $E \in \mathcal{E}$.

a. Using appropriate properties of probability measures, show that

$$\Pr(\{(T, H), (T, T)\}) = \frac{1}{2}$$

and

$$\Pr(\{(H, T), (T, T)\}) = \frac{1}{2}.$$

Interpret these statements.

b. Show that Hector's formulation does not uniquely determine the probability measure. *Hint:* Put $\alpha = \Pr(\{(H, H)\})$, $\beta = \Pr(\{(H, T)\})$, $\gamma = \Pr(\{(T, H)\})$, $\delta = \Pr(\{(T, T)\})$, and express the four preceding equations as a linear system in $\alpha, \beta, \gamma, \delta$; then show that this system has an infinite number of solutions.

c. Under Hector's formulation, what are the possible values of $\Pr(\{(H, H)\})$? Does it surprise you that $\Pr(\{(H, H)\})$ need not be $1/4$?

d. What condition has Hector neglected to include in his formulation? What additional condition on the probability measure is sufficient to obtain the correct model?

e. Can you think of a different experiment whose formulation satisfies all of Hector's conditions but for which $\Pr(\{(H, H)\}) \neq 1/4$?

22. Consider the following experiment:

Two cards are drawn from a well-shuffled deck of playing cards containing only hearts and diamonds, and they are placed face down on a table. The cards are turned up one at a time and the suit of each card is observed.

Suppose that we formulate this experiment in the following way. Let

$$S = \{(H, H), (H, D), (D, H), (D, D)\},$$

where, for example, (H,D) denotes the observation that the first card was a heart and the second card was a diamond. Let $\mathcal{E}$ be the collection of subsets of S and put

$$\Pr\left(\{(H, H)\}\right) = \frac{6}{25},$$

$$\Pr\left(\{(H, D)\}\right) = \frac{13}{50},$$

$$\Pr\left(\{(D, H)\}\right) = \frac{13}{50},$$

$$\Pr\left(\{(D, D)\}\right) = \frac{6}{25}.$$

a. Without performing any calculations, explain why it is reasonable to assign different values to $\Pr\left(\{(H, H)\}\right)$ and $\Pr\left(\{(H, D)\}\right)$.

b. Using the relative frequency interpretation of probability, show that the preceding assignment of $\Pr\left(\{(H, H)\}\right)$ is consistent with the given experiment. *Hint:* Imagine repeating the experiment a large number of times. Argue that approximately half of the pairs should have a heart as the first card. Then argue that of the pairs whose first card is a heart, approximately 12/25 should have a heart in the second location as well.

c. By arguing as in part b, show that the assignments of $\Pr(\{(H, D)\})$, $\Pr(\{(D, H)\})$, and $\Pr(\{(D, D)\})$ are also consistent with the given experiment. Explain why we should expect the probability measure $\Pr(\cdot)$ in this experiment to be invariant with respect to interchanging the roles of H and D.

d. From the assignment of probability, deduce that

$$\Pr\left(\{(H, H), (H, D)\}\right) = \frac{1}{2},$$

$$\Pr\left(\{(H, H), (D, H)\}\right) = \frac{1}{2},$$

$$\Pr\left(\{(D, H), (D, D)\}\right) = \frac{1}{2},$$

$$\Pr\left(\{(H, D), (D, D)\}\right) = \frac{1}{2}.$$

Interpret these statements.

e. Suppose that we define random variables X_1 and X_2 as follows:

$$X_1 = \begin{cases} 1 & \text{if the first card is a heart,} \\ 0 & \text{if the first card is not a heart;} \end{cases}$$

$$X_2 = \begin{cases} 1 & \text{if the second card is a heart,} \\ 0 & \text{if the second card is not a heart.} \end{cases}$$

By referring to part d, specify the individual probability mass functions for X_1 and X_2. Are X_1 and X_2 independent? Are they identically distributed?

f. What is surprising about the statement

$$\Pr(\{(H, H), (D, H)\}) = \frac{1}{2}?$$

How do you reconcile this statement with the fact that the first card is not replaced in the deck before the second card is drawn?

g. How does this experiment differ from the experiment in question 21 in which two fair coins are tossed?

23. A certain college has observed that 20% of its incoming freshmen are unqualified and drop out within the first 6 months. To better predict a student's success, the college has decided to administer a test to all freshmen when they first enroll. The college observes over a period of many years that 85% of qualified students pass the test, and 80% of unqualified students fail the test. Let Q represent the event that a student is qualified and let P represent the event that a student passes the test.

a. Describe in words what the quantities $\Pr(Q^c|P)$, $\Pr(Q|P^c)$ measure. What do you think the college would like to be true of these quantities? Explain.

b. If the college is primarily concerned with screening unqualified applicants (i.e., reducing the fraction of unqualified students who gain admission), which of the quantities $\Pr(Q^c|P)$, $\Pr(Q|P^c)$ is more important?

c. If the college is primarily concerned with admitting as many qualified applicants as possible (i.e., reducing the fraction of qualified students who are denied admission), which of the quantities $\Pr(Q^c|P)$, $\Pr(Q|P^c)$ is more important?

d. Calculate $\Pr(Q^c|P)$ and $\Pr(Q|P^c)$. Do you think this is a good test? Explain.

e. What practical considerations make it more likely that a college administering such a test would observe the values $\Pr(P|Q)$ and $\Pr(P^c|Q^c)$, as was done by this college, rather than the values $\Pr(Q^c|P)$ and $\Pr(Q|P^c)$?

24. The college in question 23 is considering an alternative admissions test. It has observed that on this alternative test, 84% of qualified students pass, and 90% of unqualified students fail. Let Q be the event that a student is qualified and let P_i be the event that a student passes test i.

a. Without performing any calculations, which of these tests do you think is better for screening unqualified applicants? Which do you think is better for ensuring that qualified applicants are not refused admission?

b. What is the chance that a student who passed the second test is unqualified? What is the chance that a student who failed the second test is qualified?

c. Considering your answers to part b, which test do you now think is better for screening unqualified applicants? Which do you think is better for ensuring that qualified applicants are not rejected? Are you surprised at your answers?

25. A certain college has observed that 25% of its incoming freshmen are unqualified and drop out within the first 4 months. To enable it to better predict a student's success, the college has administered numerous tests over a period of many years to freshmen at the time of enrollment. The college is now considering using one of these tests as the sole basis for admission.

For each of the following pairs of tests, determine which test should be chosen if the college wishes to minimize the chance that an unqualified student is admitted and which should be chosen if the college wishes to minimize the chance that a qualified student is rejected. Are you surprised by any of your answers?

a. Test 1: 90% of qualified freshmen pass.
80% of unqualified freshmen fail.

Test 2: 85% of qualified freshmen pass.
90% of unqualified freshmen fail.

b. Test 1: 95% of qualified freshmen pass.
85% of unqualified freshmen fail.

Test 2: 75% of qualified freshmen pass.
88% of unqualified freshmen fail.

c. Test 1: 85% of qualified freshmen pass.
80% of unqualified freshmen fail.

Test 2: 84% of qualified freshmen pass.
90% of unqualified freshmen fail.

d. Test 1: 50% of qualified freshmen pass.
49% of unqualified freshmen fail.

Test 2: 48% of qualified freshmen pass.
51% of unqualified freshmen fail.

26. A college administers two tests to each incoming student to better predict student success. Let P_1 be the event that a student passes the first test and let P_2 be the event that a student passes the second test. Let Q be the event that a student is qualified. Put

$$\alpha_1 = \Pr(P_1|Q), \qquad \alpha_2 = \Pr(P_2|Q),$$
$$\beta_1 = \Pr(P_1^c|Q^c), \qquad \beta_2 = \Pr(P_2^c|Q^c),$$
$$q = \Pr(Q).$$

a. Describe in words the meaning of the quantities α_1, β_1, α_2, β_2, q. What is the meaning of the quantities $\Pr(Q^c|P_1)$, $\Pr(Q|P_1^c)$, $\Pr(Q^c|P_2)$, $\Pr(Q|P_2^c)$?

b. Show that

$$\Pr(Q^c|P_i) = \frac{(1 - \beta_i)(1 - q)}{\alpha_i q + (1 - \beta_i)(1 - q)}$$

and

$$\Pr(Q|P_i^c) = \frac{(1 - \alpha_i)q}{(1 - \alpha_i)q + \beta_i(1 - q)}.$$

c. What must be true of $\Pr(Q^c|P_1)$ and $\Pr(Q^c|P_2)$ if the first test is considered better for screening unqualified applicants?

d. What must be true of $\Pr(Q|P_1^c)$ and $\Pr(Q|P_2^c)$ if the first test is considered less likely to eliminate qualified candidates?

e. Using the formulas derived in part b, show that

$$\Pr(Q^c|P_1) < \Pr(Q^c|P_2) \Longleftrightarrow (1 - \beta_1)\alpha_2 < (1 - \beta_2)\alpha_1$$

and

$$\Pr(Q|P_1^c) < \Pr(Q|P_2^c) \Longleftrightarrow (1 - \alpha_1)\beta_2 < (1 - \alpha_2)\beta_1.$$

Interpret these statements. Do you find it surprising that the conditions on the right sides of these statements are independent of q?

f. From the criteria established in part e, deduce that if $\alpha_1 < \alpha_2$ and $\beta_1 < \beta_2$, then the second test is better both for screening unqualified applicants and for ensuring that qualified applicants are not eliminated.

g. Give examples to illustrate that if $\alpha_1 > \alpha_2$ and $\beta_1 < \beta_2$, then each of the following situations could happen:

   i. Test 1 could be better both for screening unqualified applicants and for ensuring that qualified applicants are not eliminated;

   ii. Test 2 could be better both for screening unqualified applicants and for ensuring that qualified applicants are not eliminated;

   iii. Test 1 could be better for screening unqualified applicants but worse for ensuring that qualified applicants are not eliminated;

   iv. Test 2 could be better for screening unqualified applicants but worse for ensuring that qualified applicants are not eliminated.

   What conclusion do you reach?

h. Repeat part g under the assumption that $\alpha_1 < \alpha_2$ and $\beta_1 > \beta_2$.

3.7   Appendix on Sets, Combinatorics, and Basic Probability Rules

In this appendix, we review some basic properties of sets and some basic counting techniques. We also give proofs for some of the probability rules stated in §3.1.

Sets

A **set** is a collection of objects of a particular kind. Members of a set are called **elements.** The notation $e \in S$ is used to denote the statement that e is an element (i.e., member) of the set S.

A **subset** of a given set F is a set E whose elements are also members of F. Hence, E is a subset of F if and only if $e \in F$ whenever $e \in E$. The notation $E \subset F$ is used to denote the statement that E is a subset of F.[9] When E is a subset of F, it is common to say that E is *contained in F* or, alternatively, that *F contains E*.

The **complement** of a set E with respect to some universe S is the set $E^c = \{s \in S : s \notin E\}$. A **universe** is a set that contains all sets of current interest. In probability problems, the term *universe* is synonymous with *sample space*.

The **union** of two sets E and F is the set whose elements belong to either E or F or possibly to both. The union of E and F is denoted $E \cup F$ and can be formally defined in set notation as

$$E \cup F = \{x \in S : x \in E \text{ or } x \in F\}.$$

Here, S represents the universe or sample space. The union of two sets E and F is said to be **disjoint** if E and F have no elements in common.

---

[9] Some authors use the notation $E \subseteq F$ instead of $E \subset F$ and reserve the notation $E \subset F$ for the situation where E is a proper subset of F (i.e., E is a subset of F but E is not equal to F). However, we will not follow this convention in this book.

The **intersection** of two sets E and F is the set whose elements belong to both E and F. The intersection of E and F is denoted $E \cap F$ and can be formally defined in set notation as

$$E \cap F = \{x \in S : x \in E \text{ and } x \in F\}.$$

The sets E and F are said to be **disjoint** if $E \cap F = \emptyset$. Here, $\emptyset$ denotes the **empty set** (i.e., the set with no elements).

The operations of union and intersection for sets are like the operations of addition and multiplication for ordinary arithmetic, and they satisfy associative, commutative, and distributive laws:

Associative Laws:
$$E \cup (F \cup G) = (E \cup F) \cup G$$
$$E \cap (F \cap G) = (E \cap F) \cap G$$

Commutative Laws:
$$E \cup F = F \cup E$$
$$E \cap F = F \cap E$$

Distributive Laws:
$$(E \cup F) \cap G = (E \cap G) \cup (F \cap G)$$
$$(E \cap F) \cup G = (E \cup G) \cap (F \cup G)$$

The associative laws allow us to write unions and intersections unambiguously in the form $\cup_{i=1}^{n} E_i$ and $\cap_{i=1}^{n} E_i$ (i.e., without including brackets). The commutative laws allow us to disregard order in such unions and intersections.

Two important results in basic set theory are **DeMorgan's laws**:

$$\left(\bigcup_{i=1}^{n} E_i \right)^c = \bigcap_{i=1}^{n} E_i^c,$$

$$\left(\bigcap_{i=1}^{n} E_i \right)^c = \bigcup_{i=1}^{n} E_i^c.$$

Set equalities of the type $E = F$ can be proved by showing that $E \subset F$ and $F \subset E$. We use this approach to demonstrate the truth of the first form of DeMorgan's law.

DEMONSTRATION OF DEMORGAN'S LAW: Suppose that $x \in (\cup E_i)^c$. Then $x \notin \cup E_i$. Hence, $x \notin E_i$ for all i (i.e., $x \in E_i^c$ for all i). Consequently, $x \in \cap E_i^c$. Thus, $(\cup E_i)^c \subset \cap E_i^c$. Now suppose that $x \in \cap E_i^c$. Then $x \in E_i^c$ for all i. Hence, $x \notin E_i$ for all i and so $x \notin \cup E_i$. Consequently, $x \in (\cup E_i)^c$. Thus, $\cap E_i^c \subset (\cup E_i)^c$. Since $(\cup E_i)^c \subset \cap E_i^c$ and $\cap E_i^c \subset (\cup E_i)^c$, it follows that $(\cup E_i)^c = \cap E_i^c$, which is the required result. The demonstration of the other form of DeMorgan's law is similar and is left as an exercise for the reader. ■

A more intuitive way to convince oneself of the validity of DeMorgan's laws is to consider Venn diagrams. We invite the reader to use this approach to give an alternative proof of DeMorgan's laws.

Combinatorics

In classical probability, the calculation of probabilities reduces to a problem in combinatorics since, under the classical assumption that all outcomes of a random experiment are equally likely, the probability of event E is simply

$$\Pr(E) = \frac{|E|}{|S|},$$

where $|E|$ is the number of elements in the event set E and $|S|$ is the number of elements in the sample space S. Hence, it is important to have efficient methods of counting when the sample space becomes large. We now review some basic combinatorial techniques.

The most basic combinatorial principle is the **multiplication rule.** This rule asserts that the number of ways of selecting one element from each of two sets is simply the product of the number of elements in the sets. In algebraic terms, the rule asserts that if set S_1 has n_1 elements and set S_2 has n_2 elements, then the number of ways of selecting one element from S_1 and one element from S_2 is $n_1 n_2$. More generally, if we are given r sets $S_1, S_2, \ldots, S_r$ with respective sizes $n_1, n_2, \ldots, n_r$, then the number of ways of selecting one element from each set is $n_1 n_2 \cdots n_r$.

Using the multiplication rule, we can solve the following basic combinatorial problems:

1. Given n distinct objects (e.g., n different books), how many ways are there to arrange these objects in a row?
2. Given n objects of two different types, r of Type I and $n - r$ of Type II (e.g., r letter A's and $n - r$ letter B's), how many ways are there to arrange these objects in a row?
3. Given n distinct objects, how many ways are there to arrange r of these objects in a row, where $r < n$?
4. Given n identical objects, how many ways are there to choose r of them?

We will solve each of these problems in turn.

SOLUTION TO COMBINATORIAL PROBLEM 1:   Consider first the arrangement of n distinct objects in a row. For concreteness, suppose that the objects are books. There are n ways to choose the first book. After the first book has been chosen, there are $(n - 1)$ ways to choose the second book from the remaining $(n - 1)$ books. After the jth book has been chosen, there are $(n - j)$ ways to choose the $(j + 1)$th book from the remaining $(n - j)$ books. Hence, by the multiplication rule, the number of ways of arranging n distinct books in a row is

$$n(n - 1)(n - 2) \cdots (3)(2)(1).$$

This product is usually abbreviated as $n!$ (pronounced "n factorial").   ∎

SOLUTION TO COMBINATORIAL PROBLEM 2:   Next consider the arrangement of n books of two different types in a row. Suppose that r of the books are copies of *Wuthering Heights* (Type I) and the remaining $n - r$ are copies of *The Grapes of Wrath* (Type II). Suppose further that the copies of *Wuthering Heights* are indistinguishable from each other and the copies of *The Grapes of Wrath* are indistinguishable from each other. If all n books were distinguishable, then by the previous paragraph, the number of arrangements would be $n!$. However, the r copies of *Wuthering Heights* are not distinguishable from one another, and the $n - r$ copies of *The Grapes of Wrath* are not distinguishable from one another. The r copies of *Wuthering Heights* can be rearranged in $r!$ ways on the shelf without altering the appearance of the overall arrangement, and similarly the $n - r$ copies of *The Grapes of Wrath* can be rearranged in $(n - r)!$ ways on the shelf without altering the appearance of the overall arrangement. For example, if two copies of *Wuthering*

Heights are switched, the appearance on the shelf will be the same, even though the books are not in exactly the same locations as they were originally. Hence, the number of different-looking arrangements of the r copies of *Wuthering Heights* and the $n - r$ copies of *The Grapes of Wrath* is

$$\frac{n!}{r!(n-r)!}.$$ ■

SOLUTION TO COMBINATORIAL PROBLEM 3:   Next consider the arrangement of n distinguishable books on a shelf that only has room for r books, where $r < n$. There are n choices for the first book, $n - 1$ choices for the second book, and more generally, $(n - j + 1)$ choices for the jth book $(1 \leq j \leq r)$. Hence, by the multiplication rule, the number of arrangements is

$$n(n-1)(n-2)\cdots(n-(r-1)) = \frac{n!}{(n-r)!}.$$ ■

SOLUTION TO COMBINATORIAL PROBLEM 4:   Finally, consider the number of ways to choose r books from n indistinguishable books. Suppose that we place a removable mark on each book that distinguishes it from the rest. Then, by the previous paragraph, the number of ways to choose r books from the n would be $n!/(n-r)!$. For any particular set of r books, there are $r!$ ways of arranging these r books on the shelf. Thus, to determine the number of ways to choose r books from n when the marks distinguishing the books from one another are removed, we simply divide $n!/(n-r)!$ by $r!$. Hence, the required number of ways is

$$\frac{n!}{(n-r)!r!}.$$ ■

It is common to use the notation $\binom{n}{r}$ for the latter quantity:

$$\binom{n}{r} = \frac{n!}{(n-r)!r!}.$$

The notation $\binom{n}{r}$ is read "n choose r." Note that $\binom{n}{r}$ has two interpretations: It is the number of ways of choosing r objects from n identical objects. It is also the number of ways of arranging n objects in a row when r of these objects are of one type and the remaining $n - r$ are of a different type (see the solution to the second combinatorial problem). The interpretations are equivalent because an arrangement of n objects of two different types is completely determined by selecting the locations for objects of one particular type.

The preceding discussion considered two types of arrangements: ordered and unordered. An *ordered* arrangement is a **permutation.** On the other hand, an *unordered* arrangement is a **combination.** In classical probability problems, it is often possible to enumerate the elements of events and sample spaces using either permutations or combinations. However, it is important to be consistent. For example, if the sample space is enumerated using combinations, then the events should also be enumerated using combinations. Similar comments apply to the use of permutations.

The following examples illustrate the use of these combinatorial techniques in classical probability.

EXAMPLE 1: An eight-bit computer word is a sequence of eight 0s and 1s. The number of possible eight-bit words is $2^8 = 256$. (Use the multiplication rule.) ■

EXAMPLE 2: The number of eight-bit computer words with exactly three 1s and five 0s is $\binom{8}{3} = \binom{8}{5} = 56$. (Consider the solution to combinatorial problem 4.) ■

EXAMPLE 3: The Birthday Problem. Determine the probability that in a group of 30 people, at least 2 have the same birthday.

Suppose that no one in the group was born during a leap year. Then the number of possible birthdays for each person is 365. The sample space for this experiment consists of 30-tuples of birthdays with the jth coordinate representing the birthday for the jth person. Hence, the number of elements in the sample space is 365^{30}. To determine the desired probability, we first consider the probability that none of the people in the group of 30 has the same birthday. The number of elements of the sample space for which there are no matching birthdays is the number of permutations of 30 unique birthdays selected from 365, which is simply $(365)(364) \cdots (336) = 365!/335!$. Hence, under the classical assumption of equal likelihood of outcomes, the probability of no matching birthdays is

$$\frac{365!/335!}{365^{30}} \approx .2937.$$

Consequently, the probability that at least 2 of the 30 have the same birthday is $.7063 \approx 71\%$!

Note that the classical assumption of equal likelihood means that each day of the year is as likely to be one's birthday as any other. However, births tend to follow seasonal patterns and cycles, at least in the aggregate. Consequently, this result should be interpreted in that light. ■

Probability Rules

Every rule in probability can be derived from the rules of set theory and the following three axioms:

1. $0 \leq \Pr(E) \leq 1$ for all events E;
2. $\Pr(S) = 1$;
3. For any sequence of events $E_1, E_2, E_3, \ldots$ with the property that $E_i \cap E_j = \emptyset$ whenever $i \neq j$,

$$\Pr(\cup E_i) = \sum \Pr(E_i).$$

In the examples that follow, we derive some of the basic probability rules stated in §3.1.

EXAMPLE 4: Show that $\Pr(E) \leq \Pr(F)$ whenever $E \subset F$.

Suppose that $E \subset F$. Using properties of sets, we can write F as the disjoint union

$$F = (F \cap E) \cup (F \cap E^c).$$

Indeed, if S is the sample space, then

$$F = F \cap S = F \cap (E \cup E^c) = (F \cap E) \cup (F \cap E^c).$$

Since the union is disjoint, by probability axiom 3, we have

$$\Pr(F) = \Pr(F \cap E) + \Pr(F \cap E^c).$$

However, since $E \subset F$, we have $F \cap E = E$ and so $\Pr(F \cap E) = \Pr(E)$. Further, by probability axiom 1, $\Pr(F \cap E^c) \geq 0$. Consequently,

$$\Pr(F) = \Pr(E) + \Pr(F \cap E^c)$$
$$\geq \Pr(E).$$

Therefore, if $E \subset F$, $\Pr(E) \leq \Pr(F)$ as required.   ■

EXAMPLE 5:   Show that $\Pr(E^c) = 1 - \Pr(E)$ for every event E.

Let E be an event. Then, from the definition of E^c and the definition of set union, we see that S is the disjoint union of E and E^c (i.e., $S = E \cup E^c$). Hence, by probability axiom 3,

$$\Pr(S) = \Pr(E) + \Pr(E^c).$$

However, by probability axiom 2, $\Pr(S) = 1$. Consequently,

$$1 = \Pr(E) + \Pr(E^c),$$

and thus,

$$\Pr(E^c) = 1 - \Pr(E)$$

as required.   ■

EXAMPLE 6:   Show that $\Pr(E \cup F) = \Pr(E) + \Pr(F) - \Pr(E \cap F)$.

Let E and F be events. Note that $E \cup F$ can be written as a disjoint union in the following way:

$$E \cup F = (E \cap F^c) \cup (E \cap F) \cup (F \cap E^c).$$

Hence, again by probability axiom 3,

$$\Pr(E \cup F) = \Pr(E \cap F^c) + \Pr(E \cap F) + \Pr(F \cap E^c).$$

Similarly, E can be written as the disjoint union

$$E = (E \cap F) \cup (E \cap F^c)$$

and F can be written as the disjoint union

$$F = (F \cap E) \cup (F \cap E^c).$$

Hence, again by probability axiom 3,

$$\Pr(E) = \Pr(E \cap F) + \Pr(E \cap F^c),$$
$$\Pr(F) = \Pr(E \cap F) + \Pr(F \cap E^c).$$

Combining the three probability equations, we have

$$\Pr(E \cup F) = \Pr(E) + \Pr(F) - \Pr(E \cap F)$$

as required. This result can also be proved using Venn diagrams. We leave this as an exercise for the reader.   ■

4 Random Variables and Probability Distributions

In Chapter 2, we introduced the concepts of random variable and probability distribution through concrete examples. In this chapter, we develop these important concepts more completely.

We begin by giving a definition of random variable that makes use of the language of probability introduced in Chapter 3. We then proceed to discuss seven different ways of specifying the probability distribution for a random variable.

In the course of our discussion, we consider what it means for two different random variables to have the same distribution, how one should define and interpret addition and multiplication of random variables, and what it means for random variables to be independent. We also present distributional analogs of the law of total probability and Bayes' theorem, which were introduced in Chapter 3. Finally, we discuss appropriate numerical statistics for summarizing the information on expectation and risk embedded in probability distributions.

4.1 Definitions and Basic Properties

4.1.1 What Is a Random Variable?

In Chapter 2, we said that a **random variable** is any quantity X with real values that depends in a well-defined way on the outcome of a random experiment. However, we did not describe with any precision what we meant by *random experiment*. With the benefit of Chapter 3, we are now in a position to give a more precise description of the term random experiment and hence a more precise definition of a random variable.

Recall from Chapter 3 that classical probability theory can be formalized using the concepts of sample space, event space, and probability measure. In the classical paradigm, the sample space is the set S of all possible outcomes of an experiment, the event space is a collection $\mathcal{E}$ of subsets of S, and the probability measure is a function $\Pr(\cdot)$ defined on the event space with the properties that

1. $0 \leq \Pr(E) \leq 1$ for all $E \in \mathcal{E}$;
2. $\Pr(S) = 1$;

3. $\Pr\left(\bigcup\limits_{j=1}^{\infty} E_j\right) = \sum\limits_{j=1}^{\infty} \Pr(E_j)$ for any sequence of events $E_1, E_2, \ldots$ with the property that $E_i \cap E_j = \emptyset$ whenever $i \neq j$.

This suggests that a **random experiment** can be more formally described as a triple $\mathcal{P} = < S, \mathcal{E}, \Pr(\cdot) >$ consisting of a set S (the sample space), a collection $\mathcal{E}$ of subsets of S (the event space), and a real valued function $\Pr(\cdot)$ defined on $\mathcal{E}$ (the probability measure) with the three properties just listed.

Taking this to be the mathematical definition of the term random experiment, the concept of random variable can then be defined as follows: A **random variable** for the experiment described by the triple $\mathcal{P} = < S, \mathcal{E}, \Pr(\cdot) >$ is a real valued function X defined on the sample space S of the experiment (i.e., $X : S \to \mathbf{R}$).[1]

The precise mathematical definitions of random experiment and random variable are actually a little more delicate than the preceding discussion suggests. However, these descriptions of the concepts are sufficient for our purposes.

4.1.2   What Is a Probability Distribution?

In most applications, one only needs to determine probabilities associated with random variables, and one does not need to completely understand the underlying experiment on which a random variable depends. That is, one does not need to have an explicit description of the random experiment $\mathcal{P} = < S, \mathcal{E}, \Pr(\cdot) >$. Probability distributions are mechanisms for summarizing the relevant probability information about a particular random quantity X.

A **probability distribution** for a random variable X is an assignment of relative frequencies to the values and intervals of values of X. This assignment is completely determined by the underlying experiment $\mathcal{P} = < S, \mathcal{E}, \Pr(\cdot) >$ and the definition of X.

In §2.3, we introduced the concept of *distribution function* as a means of summarizing the relevant probability information about a random variable X. Recall that the **distribution function** for a random variable X is the function F_X defined by

$$F_X(x) = \Pr(X \leq x) \qquad \text{for all } x.$$

In §2.3, we interpreted the notation $\Pr(X \leq x)$ as shorthand for the English statement "the probability that $X \leq x$." However, with the benefit of the formalism of §4.1.1, we can now interpret this notation in a precise mathematical way. Indeed, $\Pr(X \leq x)$ is simply $\Pr(E)$, where E is the event given by $E = \{s \in S : X(s) \leq x\}$. Hence, we see that the distribution function F_X is completely determined by the function $\Pr(\cdot)$ and the definition of X.

EXAMPLE 1:   Consider the payoff random variable X from §2.1. Determine a formula for the distribution function F_X and describe the graphical properties of F_X.

---

[1] Note that $\Pr(\cdot)$ is a real valued function defined on the *event* space $\mathcal{E}$, whereas X is a real valued function defined on the *sample* space S. Moreover, the values $\Pr(E)$ (for $E \in \mathcal{E}$) can be interpreted as relative frequencies, whereas the values $X(s)$ (for $s \in S$) have no such interpretation.

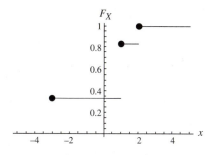

FIGURE 4.1 Distribution Function F_X for the Payoff Random Variable X of §2.1

Recall that the distribution of probability for the payoff random variable X discussed in §2.1 is given by

$$X = \begin{cases} -3 & \text{with probability } 1/3, \\ 1 & \text{with probability } 1/2, \\ 2 & \text{with probability } 1/6. \end{cases}$$

We can determine the distribution function for this X by considering separately each of the following cases:

$$x < -3, \quad x = -3, \quad -3 < x < 1, \quad x = 1, \quad 1 < x < 2, \quad x = 2, \quad x > 2.$$

When we do this, we find that

$$F_X(x) = \Pr(X \le x)$$
$$= \begin{cases} 0 & \text{for } x < -3, \\ \frac{1}{3} & \text{for } -3 \le x < 1, \\ \frac{5}{6} & \text{for } 1 \le x < 2, \\ 1 & \text{for } x \ge 2. \end{cases}$$

For example, for the x in the interval $1 < x < 2$, the statement $X \le x$ means that either $X = -3$ or $X = 1$, and so $\Pr(X \le x) = \Pr(X = -3 \text{ or } X = 1) = \Pr(X = -3) + \Pr(X = 1) = \frac{1}{3} + \frac{1}{2} = \frac{5}{6}$. The values of $F_X(x)$ for other x can be verified in a similar manner.

The graph of F_X is given in Figure 4.1. From this graph, it is clear that F_X is an increasing step function that is continuous *from the right* at every point and has the property that $F_X(x) \to 1$ as $x \to +\infty$ and $F_X(x) \to 0$ as $x \to -\infty$. ∎

The distribution function of a general random variable has properties that are very similar to those just illustrated for the payoff random variable of §2.1:

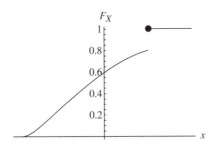

FIGURE 4.2   Distribution Function F_X for Some X

Properties of Distribution Functions

1. $F_X(x_1) \leq F_X(x_2)$ whenever $x_1 \leq x_2$; that is, F_X is increasing.
2. $\lim_{\varepsilon \to 0^+} F_X(x + \varepsilon) = F_X(x)$ for all x; that is, F_X is continuous from the right at every point.
3. $\lim_{x \to -\infty} F_X(x) = 0$ and $\lim_{x \to +\infty} F_X(x) = 1$; that is, $F_X(x) \to 0$ as $x \to -\infty$ and $F_X(x) \to 1$ as $x \to +\infty$.

These properties can be proved using properties of $\Pr(\cdot)$.

Consider, for example, property 1. Suppose that $x_1 \leq x_2$ and let E, F be the events given by $E = \{s \in S : X(s) \leq x_1\}$, $F = \{s \in S : X(s) \leq x_2\}$. Then $F_X(x_1) = \Pr(X \leq x_1) = \Pr(E)$, $F_X(x_2) = \Pr(X \leq x_2) = \Pr(F)$, and so it suffices to show that $\Pr(E) \leq \Pr(F)$. However, since $x_1 \leq x_2$, it is clear that $E \subset F$ (i.e., $s \in F$ whenever $s \in E$). Moreover, it is a general property of $\Pr(\cdot)$ that $\Pr(E) \leq \Pr(F)$ whenever $E \subset F$ (see §3.1). Hence, $\Pr(E) \leq \Pr(F)$ and so $F_X(x_1) \leq F_X(x_2)$ as required.

Property 2 and property 3 can also be proved using properties of $\Pr(\cdot)$. However, their proofs require a deeper understanding of $\Pr(\cdot)$ than we have discussed in this book and, hence, will be omitted.

The three properties of distribution functions that have been given actually *characterize* distribution functions in the sense that under reasonable conditions every function F with the three stated properties is the distribution function for some random variable X (i.e., $F = F_X$ for some random variable X). Hence, the function illustrated in Figure 4.2 is the distribution function of some random variable. The proof of the fact that distribution functions are characterized by the three given properties lies beyond the scope of this book. However, it is a useful fact to know, which is why we state it here without proof.

4.1.3   Types of Distributions

There are three types of random variables and probability distributions: continuous, discrete, and mixed. In Chapter 2, we said that a random variable is continuous if its distribution function is continuous at every point, and a random variable is discrete if

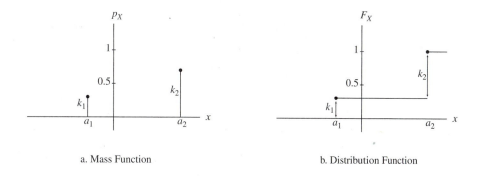

a. Mass Function   b. Distribution Function

FIGURE 4.3 Mass Function and Corresponding Distribution Function for Discrete Random Variable with Two Probability Masses

its probability is distributed over a discrete set (i.e., a set that is finite or can be ordered using the counting numbers). We did not encounter mixed random variables in Chapter 2; however, as a preliminary definition, we can say that a random variable is mixed if it is neither discrete nor continuous.

It is possible to describe the three types of distributions in terms of the properties of their associated distribution functions. Indeed, we have already noted that the distribution function of a continuous random variable is continuous. We now show that the distribution function of a discrete random variable is a step function. It then follows that the distribution function of a mixed random variable is neither continuous nor a step function.

Hence, suppose that X is a discrete random variable. For simplicity, assume that X has only two[2] possible values a_1, a_2 (with $a_1 < a_2$) and let $k_1 = \Pr(X = a_1), k_2 = \Pr(X = a_2)$. Then the probability mass function for X is given by

$$p_X(x) = \begin{cases} k_1 & \text{if } x = a_1, \\ k_2 & \text{if } x = a_2, \\ 0 & \text{otherwise,} \end{cases}$$

where $k_1, k_2 \in (0, 1)$, and $k_1 + k_2 = 1$. This is shown in Figure 4.3a.[3] The distribution function of X can be determined by considering separately the cases $x < a_1$, $x = a_1$, $a_1 < x < a_2$, $x = a_2$, $x > a_2$, as we did in Example 1 of §4.1.2. When we do this, we find that

$$F_X(x) = \Pr(X \leq x)$$
$$= \begin{cases} 0 & \text{for } x < a_1, \\ k_1 & \text{for } a_1 \leq x < a_2, \\ 1 & \text{for } x \geq a_2. \end{cases}$$

Hence, F_X is a step function as claimed. Moreover, the jumps in the step function occur at the locations a_1, a_2 and have respective sizes k_1, k_2, as in Figure 4.3b. This suggests

---

[2] The demonstration for more general discrete random variables is similar.

[3] Probability mass functions were introduced in §2.1 and will be reviewed in §4.1.4.

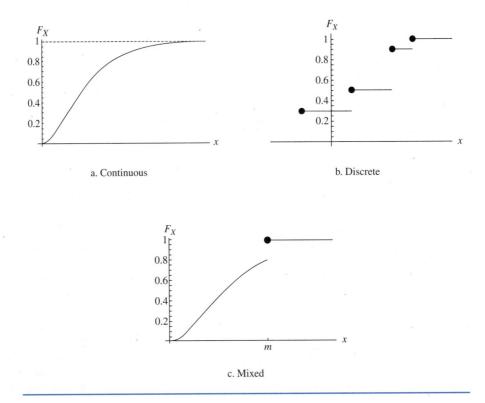

FIGURE 4.4 Types of Distributions

that we can define a discrete random variable to be a random variable whose distribution function is a step function.

Consequently, we see that random variables and probability distributions can be classified in the following way: A random variable is **continuous** if its distribution function is continuous at every point; it is **discrete** if its distribution function is a step function; and it is **mixed** if its distribution function is not a step function but is discontinuous at some point. Figure 4.4 illustrates possible graphs of F_X for each type of distribution.

Discrete distributions are generally associated with discrete quantities such as the number of claims an insurer receives on a given block of business, whereas continuous distributions are generally associated with continuous quantities such as the time a claim occurs. Occasionally, a discrete quantity such as a monetary payoff for which the number of possible values is large and the probability of the payoff assuming any particular value is small will be modeled using a continuous distribution. The justification for this is the fact that continuous random variables have the property that $\Pr(X = x) = 0$ for all x.[4] In fact, the property $\Pr(X = x) = 0$ for all x can be taken to be an alternative criterion

---

[4] $\Pr(X = x) = \Pr(X \leq x) - \Pr(X < x) = F_X(x) - F_X(x^-)$. If F_X is continuous, then $F_X(x^-) = F_X(x)$ and so $\Pr(X = x) = 0$.

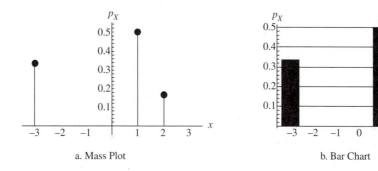

a. Mass Plot   b. Bar Chart

FIGURE 4.5 Graphical Representations of a Probability Mass Function

for a distribution to be continuous. In insurance applications, the size of a claim, known as the claim *severity*, is often modeled as a continuous quantity for these reasons.

Mixed distributions are hybrids of continuous and discrete distributions and arise quite frequently in insurance applications, particularly in connection with contract caps and deductibles. For example, the payout to the customer on an insurance contract with a cap at m but no deductible could be modeled as a mixed random variable with a continuous distribution of probability on the interval $(-\infty, m)$ and a jump at m (Figure 4.4c). Mixed distributions also arise frequently in electrical engineering (see §7.1).

4.1.4   Probability Mass Functions

As noted in §2.1 and the previous section, the distribution of probability for a discrete random variable X can be defined using a *probability mass function* as well as the distribution function F_X. The **probability mass function** for a discrete random variable X is the function p_X defined by

$$p_X(x) = \Pr(X = x) \qquad \text{for all } x.$$

This function can be represented visually using a mass plot or a bar chart, as illustrated in Figure 4.5. Note that the probability masses are simply the sizes of the respective jumps on the graph of the distribution function F_X. Figure 4.6 compares p_X and F_X. Hence, p_X could be defined alternatively as

$$
\begin{aligned}
p_X(x) &= F_X(x) - F_X(x^-) \\
&= F_X(x) - \lim_{\varepsilon \to 0^+} F_X(x - \varepsilon).
\end{aligned}
$$

4.1.5   Probability Density Functions

It is not possible to specify the probability distribution of a continuous random variable using a probability *mass* function, since for any continuous random variable X, $\Pr(X = x) = 0$ for all x. However, we can specify the distribution using a probability *density* function, as discussed in §2.3.

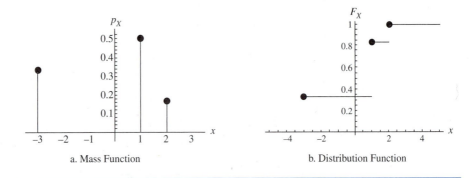

a. Mass Function   b. Distribution Function

FIGURE 4.6 Comparison of the Mass Function and the Distribution Function for a Particular Discrete Random Variable

The **probability density function** for a continuous random variable X is the function f_X defined by

$$f_X(x) = \lim_{\varepsilon \to 0} \frac{\Pr(x - \varepsilon/2 \le X \le x + \varepsilon/2)}{\varepsilon} \quad \text{for all } x.$$

This function measures at each point x the instantaneous relative frequency density at that point.

Since $\Pr(x - \varepsilon/2 \le X \le x + \varepsilon/2) = F_X(x + \varepsilon/2) - F_X(x - \varepsilon/2),$[5] we have

$$f_X(x) = \lim_{\varepsilon \to 0} \frac{F_X(x + \varepsilon/2) - F_X(x - \varepsilon/2)}{\varepsilon}$$

$$= F_X'(x),$$

provided that F_X is differentiable. Hence, by the fundamental theorem of calculus,

$$F_X(x) = \int_{-\infty}^{x} f_X(t)dt$$

and more generally,

$$\Pr(a \le X \le b) = \int_{a}^{b} f_X(x)dx.$$

Consequently, for continuous random variables X, probabilities have the interpretation of being areas under the curve defined by the density function (Figure 4.7).

Note the similarity of the formula $f_X = F_X'$ for continuous X to the earlier relationship $p_X(x) = F_X(x) - F_X(x^-)$ for discrete X. The graphs of a possible density function f_X and its corresponding distribution function F_X are illustrated in Figure 4.8.

In most applications, the distribution functions of continuous random variables are differentiable. However, one may occasionally encounter a distribution function that is not, in which case the density function, as we have described it, will not be defined at every point. If the set of points where the distribution function is not differentiable is

---

[5] More precisely, $\Pr(x - \varepsilon/2 \le X \le x + \varepsilon/2) = F_X(x + \varepsilon/2) - F_X(x - \varepsilon/2) + \Pr(X = x - \varepsilon/2) = F_X(x + \varepsilon/2) - F_X(x - \varepsilon/2)$ since $\Pr(X = t) = 0$ for all t when X has a continuous distribution.

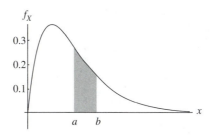

FIGURE 4.7 Probability as an Area

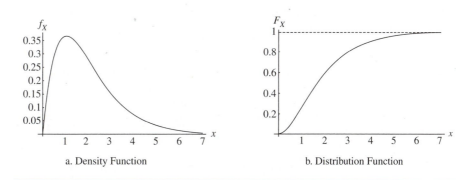

a. Density Function

b. Distribution Function

FIGURE 4.8 Density Function and Corresponding Distribution Function for a Particular Continuous Random Variable

a discrete set, then there will be no effect on the calculation of probabilities since the integral of any bounded function over a discrete set is zero. Provided that this is the case, the density function can be defined arbitrarily at such points.

EXAMPLE 1: The distribution function for a continuous random variable X is given by

$$F_X(x) = \begin{cases} 1 - e^{-x^2} & \text{for } x \geq 0, \\ 0 & \text{for } x < 0. \end{cases}$$

Determine the density function.

Since F_X is differentiable, the density function is given by

$$f_X(x) = F_X'(x)$$
$$= \begin{cases} 2xe^{-x^2} & \text{for } x \geq 0, \\ 0 & \text{for } x < 0. \end{cases}$$ ∎

EXAMPLE 2: The density function for a continuous random variable X is given by

$$f_X(x) = \begin{cases} xe^{-x} & \text{for } x \geq 0, \\ 0 & \text{for } x < 0. \end{cases}$$

Determine the probability that $X > 2$.

Using integration by parts,

$$
\begin{aligned}
\Pr(X > 2) &= \int_2^\infty f_X(x)dx \\
&= \int_2^\infty xe^{-x}dx \\
&= -xe^{-x}\big|_2^\infty - \int_2^\infty 1 \cdot (-e^{-x})\,dx \\
&= 2e^{-2} + e^{-2} \\
&= 3e^{-2}.
\end{aligned}
$$

■

4.1.6 Mixed Distributions

Mixed distributions arise frequently in applications, and consequently, it is important to be comfortable with them. They can at times be quite tricky to work with due to their hybrid nature (being neither discrete nor continuous). In this section, we introduce a hybrid of the mass and density functions discussed in §4.1.4 and §4.1.5 that is useful for calculating probabilities and expectations[6] for mixed distributions. We then illustrate the use of this hybrid mass-density function with some examples.

The simplest kind of mixed distribution is one whose distribution function resembles the function shown in Figure 4.4c. Distributions of this kind are continuous on an interval of the form $(0, m)$ and have a single jump discontinuity at $x = m$. Such distributions arise in connection with insurance contracts with caps or electrical circuits with limiters. For example, consider an insurance contract that indemnifies (i.e., compensates) an individual for the full amount of a random loss L up to a maximum amount m. Then the amount of the insurer's payment to the individual is $X = \min(L, m)$, and so if the distribution of L is continuous, as we will generally assume, the distribution of X will be mixed with a jump at $x = m$ of size $\Pr(L \geq m)$ representing the probability of a loss equal to or exceeding the maximum allowable payment on the policy.

We have already seen in our discussion of continuous distributions that $\Pr(X = x) = 0$ at all points x where F_X is continuous. Indeed, $\Pr(X = x) = \Pr(X \leq x) - \Pr(X < x) = F_X(x) - F_X(x^-)$ and $F_X(x) = F_X(x^-)$ at any point x where F_X is continuous. Hence, we should not expect to be able to describe the probability distribution of a mixed random variable in a meaningful way using mass functions. However, we can describe the distribution of a mixed random variable using a generalization of the density function. Recall from §4.1.5 that the density function is given by

$$
f_X(x) = \lim_{\varepsilon \to 0} \frac{F_X(x + \varepsilon/2) - F_X(x - \varepsilon/2)}{\varepsilon}
$$

for all x where this limit exits. Suppose that F_X is differentiable except at the points where F_X has a jump discontinuity. This is clearly the case for the function illustrated in Figure 4.4c and is true in most applications involving mixed distributions. Then we can define a density function on the continuous part of the distribution in the usual way. However, at points of discontinuity, the density becomes infinite because the numerator

---

[6] The expectation for a random variable was introduced in Chapter 2 and will be discussed in detail in §4.2.

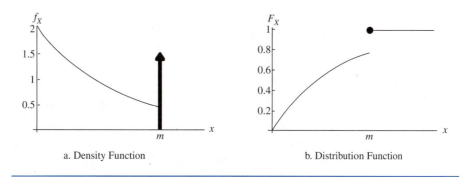

a. Density Function   b. Distribution Function

FIGURE 4.9 Density Function and Distribution Function for a Particular Mixed Random Variable

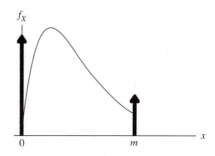

FIGURE 4.10 Mixed Density with Two Probability Masses

of the ratio $\left(F_X(x + \varepsilon/2) - F_X(x - \varepsilon/2)\right)/\varepsilon$ is bounded below by the size of the jump discontinuity and the denominator tends to zero (Figure 4.4c). Hence, the traditional density function is only meaningful at the points where F_X is continuous.

Nevertheless, we can define a generalized density function by superimposing the probability masses representing the jump sizes at the points of discontinuity on a density function of the type just described. An illustration of such a generalized density along with the corresponding distribution function is given in Figure 4.9. Note that the area under the continuous part of this density is not equal to 1. Rather, it equals $1 - k$, where k is the size of the discrete probability mass at the point where F_X has a jump discontinuity.

The arrowhead on the probability mass in Figure 4.9a reminds us that, from the perspective of probability densities (which is what the vertical scale of this graph is measuring), this mass is really an infinity point on the graph of f_X. That is, $f_X(x) = \infty$ at points x for which F_X has a jump discontinuity. Note that the heights of such probability masses on a generalized density plot will not always represent the true sizes of the probability masses, particularly if the scale of the density plot would make the masses appear to be otherwise invisible, as is the case with the distribution in Figure 4.9. However, the relative sizes of the probability masses should always be correct. This is shown in Figure 4.10. From this figure, it is clear that the size of the probability mass at the origin is about twice the size of the probability mass at the point m.

Distributions of the type illustrated in Figure 4.10 arise in connection with insurance contracts where payments are capped but a loss need not occur. Indeed, for such a contract, the insurer's payment to the policyholder is given by

$$X = \begin{cases} 0 & \text{if no loss occurs,} \\ L & \text{if a loss occurs and it is at most } m, \\ m & \text{if a loss occurs and it is greater than } m. \end{cases}$$

These kinds of mixed distributions will be discussed in detail in Chapter 9.

The concept of a generalized density can be developed more formally using delta functions. For the reader familiar with delta functions, we have included a discussion of this formalization in the appendix to this chapter.

We conclude this section with two worked examples.

EXAMPLE 1:   A nonnegative random variable X has a continuous distribution of probability along the positive x axis and a point mass at $x = 0$. On the continuous part, the density is given by

$$f_X(x) = \frac{1}{2}e^{-2x}, \qquad x > 0.$$

Determine the size of the point mass.

The area under the continuous density is given by

$$\Pr(X > 0) = \int_0^\infty \frac{1}{2}e^{-2x}dx = \frac{1}{4}.$$

Hence, the probability mass at $X = 0$ is

$$\Pr(X = 0) = 1 - \Pr(X > 0) = \frac{3}{4}.$$ ■

EXAMPLE 2:   The distribution function for a mixed random variable X is given by

$$F_X(x) = \begin{cases} 0 & \text{for } x < 0, \\ 1 - \frac{3}{4}e^{-x} & \text{for } 0 \le x < 1, \\ 1 & \text{for } x \ge 1. \end{cases}$$

Determine the probability that $0 \le X < 1$.

A sketch of the graph of F_X reveals jumps at $x = 0$ and $x = 1$ of size $\frac{1}{4}$ and $\frac{3}{4}e^{-1}$, respectively. Hence,

$$\Pr(0 \le X < 1) = \Pr(X \le 1) - \Pr(X = 1) - \Pr(X \le 0) + \Pr(X = 0)$$
$$= F_X(1) - F_X(0) - \Pr(X = 1) + \Pr(X = 0)$$
$$= 1 - \frac{1}{4} - \frac{3}{4}e^{-1} + \frac{1}{4}$$
$$= 1 - \frac{3}{4}e^{-1}.$$ ■

4.1.7   Equality and Equivalence of Random Variables

The concept of a probability distribution is useful primarily for three reasons:

1. It enables us to summarize in a meaningful way the *relevant* probability information about a given random variable.
2. It enables us to give a *visual* representation of the uncertainty in a given random variable.
3. It enables us to recognize *equivalence* between two seemingly different random prospects.

We have already discussed how distributions allow one to summarize the relevant probability information in a visually appealing way. In this section, we discuss *equivalence* of random variables and how this concept differs from *equality*.

To introduce the notion of equivalence, consider the following situation:

In a simple game, two fair (i.e., unbiased) coins are to be tossed and the payoff is to be determined from the outcome. Before the coins are tossed, we may choose one of the following payoffs:

Payoff 1:	Win \$1 for each head.
	Lose \$2 for two tails.
Payoff 2:	Win \$1 if the coins are different.
	Win \$2 if the coins are both tails.
	Lose \$2 if the coins are both heads.

Which payoff should we choose?

After some reflection, it becomes clear that although the payoffs are different, it does not matter which one we choose since, if the coins are truly fair, we are just as likely to win any given amount with one payoff as the other.

To see this formally, let X_1, X_2 denote the respective payoffs and suppose that the two coins are distinguishable in some way so that the possible outcomes of the coin toss are HH, HT, TH, TT. Then

$$X_1 = \begin{cases} 2 & \text{if the outcome is HH,} \\ 1 & \text{if the outcome is HT or TH,} \\ -2 & \text{if the outcome is TT,} \end{cases}$$

and

$$X_2 = \begin{cases} -2 & \text{if the outcome is HH,} \\ 1 & \text{if the outcome is HT or TH,} \\ 2 & \text{if the outcome is TT.} \end{cases}$$

Further, since each of the outcomes HH, HT, TH, TT is equally likely,

$$p_{X_1}(x) = \begin{cases} \frac{1}{4} & \text{for } x = -2, 2, \\ \frac{1}{2} & \text{for } x = 1, \\ 0 & \text{otherwise,} \end{cases}$$

$$= p_{X_2}(x).$$

Consequently, before the outcome of the coin toss is known, we are just as likely to win (or lose) any given amount with payoff 1 as with payoff 2, and so we should be indifferent to choosing between them.

Note that this does not mean that our eventual wealth will be the same regardless of the choice of payoff. Indeed, if the result of the coin toss turns out to be two heads, then the person who had chosen payoff 1 would realize a gain of \$2, while the person who had chosen payoff 2 would suffer a loss of \$2. Hence, although we should be indifferent to choosing between the two payoffs before the outcome of the coin toss is known, our choice will still affect our ultimate wealth. We can express this fact by saying that the payoffs are *equivalent* but not equal.

In general, two random variables with the same probability distribution are said to be **identically distributed.** We will also use the term **equivalent** in this book to describe such random variables, although this terminology is not as descriptive and is used less frequently in the probability literature. Hence, X_1 and X_2 are identically distributed if and only if

$$F_{X_1}(x) = F_{X_2}(x) \qquad \text{for all } x.$$

Note that it is possible for two random variables with the same probability distribution to have different *density* functions because changing the density function at a point on the continuous part of the distribution does not alter the distribution of probability; however, for the distributions to be the same, the set of points at which the density functions differ cannot be too large. Hence, we can say that X_1 and X_2 are identically distributed if and only if

$$f_{X_1}(x) = f_{X_2}(x)$$

for almost all x.

Throughout this book, we will often find it convenient to work with density functions rather than distribution functions. To avoid being pedantic when stating identities in terms of density functions, we will suppress the qualifying phrase *almost all* from hereon in, unless we feel that its inclusion is particularly significant. However, the reader should remember that most identities involving densities are technically only correct when this qualifying phrase is included. If we restrict our attention to density functions that are continuous, then the qualifying *almost* can be dropped entirely.

Equivalence and equality of random variables are clearly not the same thing. Indeed, for two random variables X_1 and X_2 to be equal, they must be identical as functions defined on the sample space; that is, $X_1(s) = X_2(s)$ for all $s \in S$. However, for them to be equivalent, they need only have identical probability distributions; that is, $F_{X_1}(x) = F_{X_2}(x)$ for all $x \in \mathbf{R}$. Clearly, two random variables that are equal must have identical distributions; however, two random variables with identical distributions need not be equal.

To distinguish between equivalence and equality, we will use the notation $X_1 = X_2$ to indicate equality and $X_1 \sim X_2$ to indicate equivalence. Note that in the simple game illustrated earlier, $X_1 \sim X_2$ but $X_1 \neq X_2$.

4.1.8 Random Vectors and Bivariate Distributions

In §2.2, we introduced the concepts of random vector and bivariate distribution in connection with an experiment in which two biased coins with potentially different biases are tossed. In applications, random vectors arise naturally when two or more uncertain quantities associated with the same subject are considered at the same time. For example, when pricing an automobile insurance policy, a casualty actuary must consider how frequently accident claims will be filed (since repeat claims are indeed possible) and how severe accident claims will be when they are filed.

In this section, we consider how the notions of distribution function and density function can be extended to random vectors. For simplicity, we restrict our attention to vectors with two random components. However, most of our observations apply, in a suitably modified form, to vectors with more than two components as well.

Bivariate Distribution Functions

A **random vector** is a vector quantity with real components that depends in a well-defined way on the outcome of a random experiment. More precisely, a **random vector** with two components for the experiment $\mathcal{P} = <S, \mathcal{E}, \Pr(\cdot)>$ is a vector valued function $\mathbf{X} : S \rightarrow \mathbf{R}^2$ with the property that $\{s \in S : \mathbf{X}(s) \leq \mathbf{x}\} \in \mathcal{E}$ for all $\mathbf{x}$. Here, the notation $\mathbf{X} \leq \mathbf{x}$ is shorthand for $X_1 \leq x_1$, and $X_2 \leq x_2$, where $\mathbf{X} = (X_1, X_2)$ and $\mathbf{x} = (x_1, x_2)$.

A **bivariate distribution** for a random vector $\mathbf{X} = (X_1, X_2)$ with two components is an assignment of relative frequencies to the values and collections of values of $\mathbf{X}$. This assignment is completely determined by the underlying experiment and the definition of $\mathbf{X}$, and it can be summarized by a **bivariate distribution function** $F_{\mathbf{X}}$, which is given by

$$F_{\mathbf{X}}(\mathbf{x}) = \Pr(\mathbf{X} \leq \mathbf{x}) \qquad \text{for all } \mathbf{x}.$$

That is,

$$F_{X_1, X_2}(x_1, x_2) = \Pr(X_1 \leq x_1, X_2 \leq x_2) \qquad \text{for all } x_1, x_2.$$

Figure 4.11 illustrates a possible bivariate distribution function.

Bivariate distribution functions have properties analogous to the properties of univariate distribution functions stated in §4.1.2:

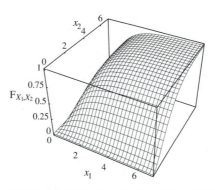

FIGURE 4.11 A Bivariate Distribution Function

Properties of Bivariate Distribution Functions

1. $F_{X_1,X_2}(a_1, a_2) \leq F_{X_1,X_2}(b_1, b_2)$ whenever $a_1 \leq b_1$ and $a_2 \leq b_2$.
2. $\lim_{\substack{\varepsilon_1 \to 0^+ \\ \varepsilon_2 \to 0^+}} F_{X_1,X_2}(x_1 + \varepsilon_1, x_2 + \varepsilon_2) = F_{X_1,X_2}(x_1, x_2)$ for all x_1 and x_2; that is F_{X_1,X_2} is continuous from the right and from above.
3. $F_{X_1,X_2}(-\infty, x_2) = F_{X_1,X_2}(x_1, -\infty) = F_{X_1,X_2}(-\infty, -\infty) = 0$ and $F_{X_1,X_2}(\infty, \infty) = 1$.

Each of these properties can be demonstrated by appealing to similar properties of $\Pr(\cdot)$. Note that, unlike in the univariate case, a function of two variables $F(x_1, x_2)$ with the three stated properties need not be the bivariate distribution function of any random vector $\mathbf{X} = (X_1, X_2)$.

Bivariate Density Functions

The probability distribution for a random vector with two components can also be specified using a bivariate *density* function. For any random vector $\mathbf{X} = (X_1, X_2)$ with two components, the **bivariate density function** for $\mathbf{X}$ is the function $f_{\mathbf{X}}$ given by

$$f_{\mathbf{X}}(\mathbf{x}) = \lim_{\varepsilon \to 0} \frac{\Pr(\mathbf{x} - \varepsilon/2 \leq \mathbf{X} \leq \mathbf{x} + \varepsilon/2)}{\varepsilon_1 \varepsilon_2},$$

where $\varepsilon = (\varepsilon_1, \varepsilon_2)$, provided that this makes sense. Equivalently, the bivariate density is the function given by

$$f_{X_1,X_2}(x_1, x_2) = \frac{\partial^2}{\partial x_1 \partial x_2} F_{X_1,X_2}(x_1, x_2),$$

provided that this makes sense.

EXAMPLE 1: The distribution function of $\mathbf{X} = (X_1, X_2)$ is given by

$$F_{X_1,X_2}(x_1, x_2) = \begin{cases} (1 - e^{-x_1})(1 - (1 + 2x_2)e^{-2x_2}) & \text{for } x_1 \geq 0, x_2 \geq 0, \\ 0 & \text{otherwise.} \end{cases}$$

Determine the bivariate density function.

The bivariate density on the region $x_1 > 0$, $x_2 > 0$ is determined by calculating the appropriate mixed partial derivative:

$$\begin{aligned} f_{X_1,X_2}(x_1, x_2) &= \frac{\partial^2}{\partial x_1 \partial x_2} F_{X_1,X_2}(x_1, x_2) \\ &= \frac{\partial}{\partial x_1}(1 - e^{-x_1}) \cdot \frac{\partial}{\partial x_2}(1 - (1 + 2x_2)e^{-2x_2}) \\ &= e^{-x_1} \cdot 4x_2 e^{-2x_2} \\ &= 4x_2 e^{-x_1 - 2x_2}. \end{aligned}$$

For all other x_1, x_2, the bivariate density is zero. ∎

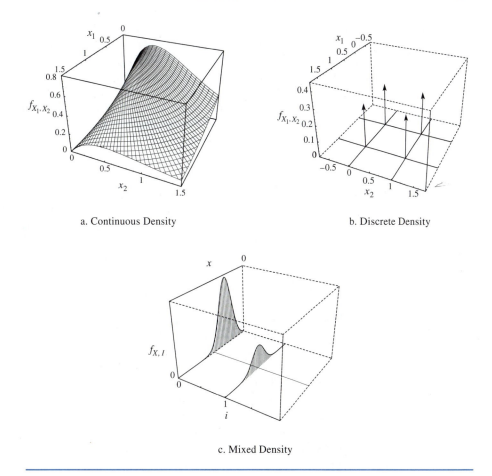

a. Continuous Density

b. Discrete Density

c. Mixed Density

FIGURE 4.12 Types of Bivariate Densities

Figure 4.12 illustrates three different possible bivariate density functions. Note that the variety in the densities of mixed random *vectors* is much greater than it is for mixed random *variables*.

One may at first think that mixed random vectors (i.e., vectors with both continuous and discrete parts) are a bit artificial. However, these types of vectors actually arise quite frequently in applications. The following examples illustrate some situations where the consideration of a mixed vector is required.

EXAMPLE 2:   Glaucoma is an eye disease, primarily affecting people over the age of 40, in which fluid accumulates near the retina. The resulting pressure on the optic nerve kills optical cells and, if left untreated, can lead to complete blindness.

One of the ways of detecting glaucoma in its early stages is to consider eye pressure measurements. Indeed, since a buildup of fluid on the retina increases eye pressure, a higher-than-normal pressure reading tends to be associated with a greater-than-average risk of having the disease. However, it is still possible, given the naturally occurring variability in individual eye pressures, for a healthy eye to have higher-than-normal pressure and for a diseased eye to have lower-than-normal pressure. Hence, an eye pressure reading alone does not allow one to diagnose with certainty the presence or absence of the disease.

To analyze the uncertainty in using eye pressure measurements to detect the presence of glaucoma, one needs to consider the vector quantity $\mathbf{V} = (X, I)$, where X is the eye pressure of a randomly chosen subject and I is an indicator of glaucoma for this subject. Since eye pressure measurements are best modeled as a continuous quantity and I is clearly discrete, the random vector $\mathbf{V}$ has a mixed density, as illustrated in Figure 4.12c. ∎

EXAMPLE 3: Defined benefit pension plans are pension plans in which participants receive fixed benefit payments at retirement (possibly indexed for inflation). The benefit payments depend only on a participant's length of service and/or wages during the participant's working life and do not depend on the investment performance of the underlying pension fund or on the length of time that the participant happens to live. Defined benefit pension plans in effect guarantee a level of benefit in retirement to each retiree until the retiree's death even though the plan does not know in advance how long each retiree will live.[7]

Not every worker who participates in a company's pension plan stays with the company until retirement. Some leave for better employment opportunities, others are fired, and some die before reaching retirement. The obligations of the pension plan to the participant who terminates prematurely depend on the nature of the termination.

For the participant who dies while employed, the pension plan generally pays a quite substantial survivor pension and lump sum death benefit to the participant's spouse and dependent children. On the other hand, for the participant who terminates to accept other employment or as a result of being fired, the monetary benefits are considerably less generous and usually consist of the terminated employee acquiring a right to a partial pension at retirement based on service and earnings to date but not adjusted for inflation prior to retirement.

The obligations of the pension plan to a plan member clearly depend on the amount of time the member contributes to the plan (years of service) and the cause of termination. However, when an employee first joins the plan, the pension plan administrator has no idea when the employee will terminate and what type of obligation the plan will have on the employee's termination.

To address this problem, pension plans must consider the vector $\mathbf{V} = (T, J)$, where T is the future time contributing to the plan and J is a discrete random variable that specifies the cause of termination. This vector clearly has a mixed density. ∎

Probabilities for events defined by random vectors can generally be calculated by integrating the density function over the desired region. For example, the probability

---

[7] Another type of pension plan, known as a *defined contribution* plan, provides no guarantee on the level of benefit at retirement, but does allow a plan member to participate in the gains and losses of the underlying pension fund. With this type of plan, the employer's contribution, rather than the retiree's benefit, is defined and participants are free to direct the contributions to any of a number of approved investments. Generally speaking, defined benefit plans tend to favor long-service career employees, whereas defined contribution plans tend to favor employees who change employers frequently. An example of a defined contribution plan is a 401K plan.

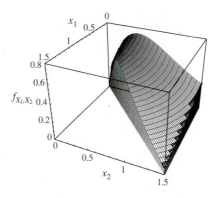

FIGURE 4.13 Probability as a Volume

that $\mathbf{X}$ assumes a value in the region $\mathcal{R}$ is given by

$$\Pr(\mathbf{X} \in \mathcal{R}) = \iint_{(x_1, x_2) \in \mathcal{R}} f_{X_1, X_2}(x_1, x_2) dx_1 dx_2.$$

From calculus, we recognize this as simply the volume under the surface defined by the density and lying above the region $\mathcal{R}$.

As a special case, when $\mathcal{R}$ is the region $\mathcal{R} = \{(s, t) \in \mathbf{R}^2 : s \leq x_1, t \leq x_2\}$, we obtain

$$F_{X_1, X_2}(x_1, x_2) = \int_{-\infty}^{x_1} \int_{-\infty}^{x_2} f_{X_1, X_2}(t_1, t_2) dt_2 dt_1.$$

EXAMPLE 4: The density function for the vector $\mathbf{T} = (T_1, T_2)$ whose components are positive random variables is given by

$$f_{T_1, T_2}(t_1, t_2) = t_2 e^{-t_1} e^{-t_2}, \qquad t_1 > 0, t_2 > 0.$$

Determine the probability that $T_2 > T_1$.

The required probability is the volume shown in Figure 4.13. Using iterated integration and integration by parts, this probability is given by

$$\begin{aligned}
\Pr(T_2 > T_1) &= \int_0^{\infty} \int_0^{t_2} t_2 e^{-(t_1 + t_2)} \, dt_1 \, dt_2 \\
&= \int_0^{\infty} t_2 e^{-t_2} (1 - e^{-t_2}) \, dt_2 \\
&= \int_0^{\infty} t_2 e^{-t_2} \, dt - \int_0^{\infty} t_2 e^{-2t_2} \, dt_2 \\
&= 1 - \frac{1}{4} \\
&= \frac{3}{4}
\end{aligned}$$

Details are left to the reader. ∎

TABLE 4.1 Contingency Table for $\mathbf{X} = (X_1, X_2)$

		X_2		
		0	1	$p_{X_1}(x_1)$
X_1	0	$\dfrac{1}{6}$	$\dfrac{1}{12}$	$\dfrac{1}{4}$
	1	$\dfrac{1}{3}$	$\dfrac{5}{12}$	$\dfrac{3}{4}$
$p_{X_2}(x_2)$		$\dfrac{1}{2}$	$\dfrac{1}{2}$	1

Marginal Distributions

From a bivariate density function, it is possible to recapture the univariate distributions for a random vector's scalar components. Indeed, if $\mathbf{X} = (X_1, X_2)$, then

$$F_{X_1}(x_1) = \Pr(X_1 \le x_1)$$
$$= \int_{-\infty}^{x_1} \int_{-\infty}^{\infty} f_{X_1,X_2}(t_1, t_2) dt_2 dt_1,$$

and similarly,

$$F_{X_2}(x_2) = \int_{-\infty}^{\infty} \int_{-\infty}^{x_2} f_{X_1,X_2}(t_1, t_2) dt_2 dt_1.$$

Consequently, by the fundamental theorem of calculus,

$$f_{X_1}(x_1) = \frac{d}{dx_1} F_{X_1}(x_1)$$
$$= \int_{-\infty}^{\infty} f_{X_1,X_2}(x_1, t) dt,$$

and similarly,

$$f_{X_2}(x_2) = \int_{-\infty}^{\infty} f_{X_1,X_2}(t, x_2) dt.$$

When X_1 and X_2 are discrete, the latter two formulas reduce to

$$p_{X_1}(x_1) = \sum_{\text{all } x_2} p_{X_1,X_2}(x_1, x_2),$$

$$p_{X_2}(x_2) = \sum_{\text{all } x_1} p_{X_1,X_2}(x_1, x_2),$$

which the alert reader will recognize from §2.2 as formulas for particular row sums and column sums in the contingency table for $\mathbf{X}$.

To see this more clearly, consider the random vector $\mathbf{X} = (X_1, X_2)$ whose probability mass function is given by the contingency table in Table 4.1. Note that the entries in the right and bottom margins of this table correspond to the values of the probability

TABLE 4.2 Contingency Table for $\mathbf{Y} = (Y_1, Y_2)$

		\(Y_2\) 0	1	\(p_{Y_1}(y_1)\)
Y_1	0	$\dfrac{1}{10}$	$\dfrac{3}{20}$	$\dfrac{1}{4}$
	1	$\dfrac{2}{5}$	$\dfrac{7}{20}$	$\dfrac{3}{4}$
$p_{Y_2}(y_2)$		$\dfrac{1}{2}$	$\dfrac{1}{2}$	1

mass functions for the scalar components X_1, X_2, respectively, and that these marginal values can be obtained from the interior values by calculating an appropriate row sum or column sum. In light of this fact, the distributions of X_1 and X_2 when considered individually are often called **marginal distributions** and their corresponding densities are called **marginal densities.** These terms are also used even when it isn't practical to construct a contingency table for $\mathbf{X}$.

While it is always possible to determine the distributions of a random vector's components (i.e., the marginal distributions) from the (bivariate) distribution of the vector itself, it is not possible to determine the distribution of the vector from the distributions of the individual components without understanding how the components interact. Consider, for example, the random vector $\mathbf{Y} = (Y_1, Y_2)$ whose probability mass function is given by the contingency table in Table 4.2. Comparing this table to the contingency table in Table 4.1, it is clear that the marginal distributions of $\mathbf{X}$ and $\mathbf{Y}$ are the same, but their bivariate distributions are different.

To emphasize the importance of the *interaction* of the components of a random vector $\mathbf{X} = (X_1, X_2)$, the distribution of $\mathbf{X}$ is sometimes referred to as the **joint distribution** of X_1 and X_2, and the functions $F_{\mathbf{X}}$, $f_{\mathbf{X}}$ are called, respectively, the **joint distribution function** and the **joint density function** of the random variables X_1 and X_2.

EXAMPLE 5: The joint density of X and Y is given by

$$f_{X,Y}(x, y) = xe^{-x(y+1)}, \qquad x > 0, y > 0.$$

Determine the marginal densities.

From the formulas for marginal densities, we have

$$
\begin{aligned}
f_X(x) &= \int_{-\infty}^{\infty} f_{X,Y}(x, y)\,dy \\
&= \int_0^{\infty} xe^{-x(y+1)}\,dy \\
&= -e^{-x(y+1)}\Big|_{y=0}^{\infty} \\
&= e^{-x}
\end{aligned}
$$

and

$$f_Y(y) = \int_{-\infty}^{\infty} f_{X,Y}(x, y)dx$$

$$= \int_0^{\infty} xe^{-x(y+1)}dx$$

$$= \frac{xe^{-x(y+1)}}{-(y+1)}\Bigg|_{x=0}^{\infty} - \int_0^{\infty} \frac{e^{-x(y+1)}}{-(y+1)}dx$$

$$= \frac{1}{(y+1)^2}.$$

Alternatively, we can find the joint distribution function first:

$$F_{X,Y}(x, y) = \int_0^x \int_0^y se^{-s(t+1)}dt\, ds$$

$$= \int_0^x -e^{-s(t+1)}\Bigg|_{t=0}^{y} ds$$

$$= \int_0^x (e^{-s} - e^{-s(y+1)})ds$$

$$= \frac{y}{y+1} - e^{-x}\left(1 - \frac{e^{-xy}}{y+1}\right).$$

Then

$$F_X(x) = F_{X,Y}(x, \infty) = 1 - e^{-x}$$

and

$$F_Y(y) = F_{X,Y}(\infty, y) = 1 - \frac{1}{y+1}.$$

Hence,

$$f_X(x) = e^{-x}$$

and

$$f_Y(y) = \frac{1}{(y+1)^2}$$

as before. ■

EXAMPLE 6:   The joint density for a mixed random vector $\mathbf{Y} = (N, X)$ is given by

$$f_{N,X}(n, x) = \frac{x^n e^{-2x}}{n!}, \qquad x \geq 0; n = 0, 1, 2, \ldots.$$

Determine the probability that $N < 2$ and the probability that $X > 4$.

The marginal probability mass function for N is given by

$$p_N(n) = \int_0^{\infty} \frac{x^n e^{-2x}}{n!}dx, \qquad n = 0, 1, 2, \ldots.$$

Thus, the probability that $N < 2$ is

$$p_N(0) + p_N(1) = \int_0^\infty e^{-2x}dx + \int_0^\infty xe^{-2x}dx$$

$$= \frac{1}{2} + \frac{1}{4}$$

$$= \frac{3}{4}.$$

The marginal density function for X is given by

$$f_X(x) = e^{-2x}\sum_{n=0}^{\infty}\frac{x^n}{n!}$$

$$= e^{-2x}\cdot e^x$$

$$= e^{-x}.$$

Thus, the probability that $X > 4$ is

$$\Pr(X > 4) = \int_4^\infty e^{-x}dx = e^{-4}.\qquad\blacksquare$$

4.1.9   Dependence and Independence of Random Variables

In §2.2, we introduced the notion of independence for random variables. In this section, we consider what it means for two random variables to be *dependent,* and we develop techniques for determining probabilities associated with such dependent random variables.

Speaking informally, two random variables are **independent** if knowledge of the value assumed by one of the variables has no effect on the probability distribution of the other. On the other hand, two random variables are **dependent** if knowledge of the value assumed by one of the variables can change the probability distribution of the other.

Consider, for example, the random variables X_1, X_2, which indicate heads on the toss of a nickel and dime, respectively:

$$X_1 = \begin{cases} 1 & \text{if the nickel lands heads,} \\ 0 & \text{if the nickel lands tails;} \end{cases}$$

$$X_2 = \begin{cases} 1 & \text{if the dime lands heads,} \\ 0 & \text{if the dime lands tails.} \end{cases}$$

Clearly, knowledge of the outcome of tossing the nickel has no effect on how the dime will land, and vice versa. Hence, X_1 and X_2 in this example are independent.

On the other hand, consider the random variables C, N, which represent, respectively aggregate claim payouts and total number of claims on a given block of some insurer's business during a particular time period. Clearly, knowledge of the actual number of claims paid during the underwriting period without knowledge of the corresponding claim sizes can greatly affect the distribution of aggregate claims. Indeed, if N is known to be large, then C is very likely to be large as well. Hence, the random variables C, N in this example are dependent.

Note that the meaning of dependence in the context of random variables is different from its meaning with respect to deterministic (i.e., nonrandom) variables. Indeed, two deterministic quantities x, y are considered dependent if they are connected by some deterministic relationship such as $y = 2x + 1$. However, two random variables can be dependent without satisfying such a deterministic relationship.

Conditional Distributions

To analyze the dependence of two or more random variables, we require some new notation. For simplicity, we only consider the analysis of two dependent random variables. However, the analysis of more than two random variables is similar.

Let's write $X_1|X_2 = x_2$ to represent the quantity X_1 under the condition that X_2 is known to assume the value x_2 and write $X_2|X_1 = x_1$ to represent the quantity X_2 under the condition that X_1 is known to assume the value x_1. Note that for each x_2, the quantity $X_1|X_2 = x_2$ is a random variable, and for each x_1, the quantity $X_2|X_1 = x_1$ is a random variable. However, $X_1|X_2 = x_2$ need not have the same distribution as X_1, and $X_2|X_1 = x_1$ need not have the same distribution as X_2. Indeed, if X_1 and X_2 are dependent, then the distributions of X_1 and $X_1|X_2 = x_2$ will differ for some x_2, and the distributions of X_2 and $X_2|X_1 = x_1$ will differ for some x_1. The reason is that $X_1|X_2 = x_2$ represents the quantity X_1 under the condition that $X_2 = x_2$, whereas X_1 alone represents the quantity unconditionally. A similar comment applies to $X_2|X_1 = x_1$ and X_2.

To emphasize that the distribution of $X_1|X_2 = x_2$ for a given x_2 is conditional on X_2 assuming the value x_2 and to distinguish the distribution of $X_1|X_2 = x_2$ from the distribution of the unconditional quantity X_1, the distribution of $X_1|X_2 = x_2$ is usually referred to as a **conditional distribution.** When X_1 is discrete, the conditional distribution of $X_1|X_2 = x_2$ can be specified by a **conditional probability mass function** $p_{X_1|X_2=x_2}$. More generally, when X_1 is continuous or mixed, the conditional distribution of $X_1|X_2 = x_2$ can be specified by a **conditional probability density function** $f_{X_1|X_2=x_2}$, where density is understood in the sense of "generalized density" discussed in §4.1.6. Similar comments apply to the variables $X_2|X_1 = x_1$.

Formulas for Conditional Densities

To determine a formula for the conditional density, let's fix x_2 and let's assume for the moment that X_1 and X_2 are both discrete. Then $X_1|X_2 = x_2$ is a discrete random variable, and its distribution can be specified by the probability mass function $p_{X_1|X_2=x_2}$.

For each x_1, $p_{X_1|X_2=x_2}(x_1)$ is the probability that $X_1 = x_1$ given that $X_2 = x_2$:

$$p_{X_1|X_2=x_2}(x_1) = \Pr(E|F),$$

where E is the event $X_1 = x_1$ and F is the event $X_2 = x_2$. From the definition of conditional probability given in §3.2,

$$\Pr(E|F) = \frac{\Pr(E \cap F)}{\Pr(F)}.$$

However, $\Pr(E \cap F) = \Pr(X_1 = x_1 \text{ and } X_2 = x_2) = p_{X_1,X_2}(x_1, x_2)$ and $\Pr(F) = \Pr(X_2 = x_2) = p_{X_2}(x_2)$. Hence, the probability mass function for $X_1|X_2 = x_2$ is given by

$$p_{X_1|X_2=x_2}(x_1) = \frac{p_{X_1,X_2}(x_1, x_2)}{p_{X_2}(x_2)} \qquad \text{for all } x_1.$$

Consequently, the probability mass function for $X_1|X_2 = x_2$ can be obtained from the joint probability mass function of X_1 and X_2 by scaling the appropriate values of $p_{X_1,X_2}(x_1, x_2)$ by the marginal probability mass $p_{X_2}(x_2)$.

This suggests that the conditional density for $X_1|X_2 = x_2$ is, in general, given by

$$f_{X_1|X_2=x_2}(x_1) = \frac{f_{X_1,X_2}(x_1, x_2)}{f_{X_2}(x_2)} \qquad \text{for all } x_1.$$

Using appropriate techniques from calculus, one can show that this is indeed the correct formula for $f_{X_1|X_2=x_2}$. Note that the units on both sides of this formula are, in general, length^{-1} since the bivariate density f_{X_1,X_2} measures probability per unit area. The conditional density $f_{X_2|X_1=x_1}$ for the random variable $X_2|X_1 = x_1$ for fixed x_1 has a similar formula:

$$f_{X_2|X_1=x_1}(x_2) = \frac{f_{X_1,X_2}(x_1, x_2)}{f_{X_1}(x_1)} \qquad \text{for all } x_2.$$

Before proceeding further, let's highlight these formulas for future reference:

Formulas for Conditional Densities

$$f_{X_1|X_2=x_2}(x_1) = \frac{f_{X_1,X_2}(x_1, x_2)}{f_{X_2}(x_2)}$$

$$f_{X_2|X_1=x_1}(x_2) = \frac{f_{X_1,X_2}(x_1, x_2)}{f_{X_1}(x_1)}$$

The densities $f_{X_1|X_2=x_2}$, $f_{X_2|X_1=x_1}$ have an interesting graphical interpretation. Indeed, from the graphs in Figure 4.14, one can see that the density $f_{X_2|X_1=x_1}$ is the cross-section of the bivariate density f_{X_1,X_2}, which is taken parallel to the coordinate plane $x_1 = 0$ and scaled by the marginal density value $f_{X_1}(x_1)$. Similarly, the density $f_{X_1|X_2=x_2}$ is the cross-section of f_{X_1,X_2} taken parallel to $x_2 = 0$ and scaled by the factor $f_{X_2}(x_2)$.

This graphical interpretation of conditional densities should be compared to the graphical interpretation of marginal densities, which were considered in §4.1.8. From the formulas $f_{X_1}(x_1) = \int_{-\infty}^{\infty} f_{X_1,X_2}(x_1, t)dt$ and $f_{X_2}(x_2) = \int_{-\infty}^{\infty} f_{X_1,X_2}(s, x_2)ds$ developed in §4.1.8, it is clear that the marginal densities are simply the *projections* of the bivariate density onto the respective coordinate axes. Hence, marginal densities are projections; conditional densities are scaled cross-sections.

EXAMPLE 1: Consider the contingency table given in Table 4.1 (§4.1.8). From this table, the probability mass functions for $X_1|X_2 = 0$, $X_1|X_2 = 1$, $X_2|X_1 = 0$, $X_2|X_1 = 1$ are, respectively,

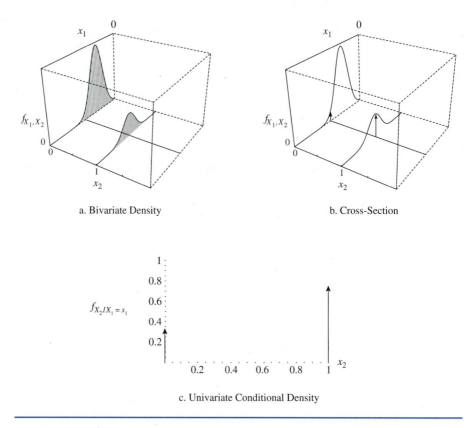

a. Bivariate Density

b. Cross-Section

c. Univariate Conditional Density

FIGURE 4.14 Conditional Densities as Scaled Cross-Sections

$$p_{X_1|X_2=0}(x_1) = \begin{cases} \frac{1}{3} & \text{for } x_1 = 0, \\ \frac{2}{3} & \text{for } x_1 = 1; \end{cases}$$

$$p_{X_1|X_2=1}(x_1) = \begin{cases} \frac{1}{6} & \text{for } x_1 = 0, \\ \frac{5}{6} & \text{for } x_1 = 1; \end{cases}$$

$$p_{X_2|X_1=0}(x_2) = \begin{cases} \frac{2}{3} & \text{for } x_2 = 0, \\ \frac{1}{3} & \text{for } x_2 = 1; \end{cases}$$

$$p_{X_2|X_1=0}(x_2) = \begin{cases} \frac{4}{9} & \text{for } x_2 = 0, \\ \frac{5}{9} & \text{for } x_2 = 1. \end{cases}$$

Note that the probability mass function of $X_2|X_1 = 0$, for example, is obtained from the first row of the contingency table by scaling the respective elements of the first row by the row sum. ∎

The formulas for the conditional densities $f_{X_1|X_2=x_2}$ and $f_{X_2|X_1=x_1}$ can be rearranged to give formulas for the density f_{X_1,X_2} of the joint distribution. Indeed,

$$f_{X_1,X_2}(x_1, x_2) = f_{X_1|X_2=x_2}(x_1) \cdot f_{X_2}(x_2)$$
$$= f_{X_2|X_1=x_1}(x_2) \cdot f_{X_1}(x_1).$$

These formulas have three important applications, and we discuss one of these applications now. The other two will be considered in the next section where we develop distributional analogs of the law of total probability and Bayes' theorem.

To see the first application, suppose that X_1 and X_2 are independent in the sense defined at the beginning of the section. Then for each x_2, $X_1|X_2 = x_2$ must have the same distribution as X_1, and so

$$f_{X_1|X_2=x_2}(x_1) = f_{X_1}(x_1)$$

for all x_1 and x_2. Substituting this into the preceding formula for the joint density f_{X_1,X_2}, we find that

$$f_{X_1,X_2}(x_1, x_2) = f_{X_1}(x_1) \cdot f_{X_2}(x_2)$$

and also

$$f_{X_2|X_1=x_1}(x_2) = f_{X_2}(x_2).$$

Therefore, for independent random variables X_1, X_2, we must have

$$f_{X_1,X_2}(x_1, x_2) = f_{X_1}(x_1) \cdot f_{X_2}(x_2)$$

for all x_1, x_2, except possibly for a small number of values x_1, x_2.[8]

We can use this observation to give a formal definition of independence for random variables:

Definition of Independence

Two random variables X_1, X_2 are **independent** if their joint density is the product of the marginal densities. That is,

$$f_{X_1,X_2}(x_1, x_2) = f_{X_1}(x_1) \cdot f_{X_2}(x_2).$$

Otherwise, X_1 and X_2 are **dependent.**

We conclude this section with some illustrative calculations for two joint density functions that were introduced in the examples of §4.1.8.

EXAMPLE 2: The joint density of X and Y is given by

$$f_{X,Y}(x, y) = xe^{-x(y+1)}, \qquad x > 0, y > 0.$$

Determine the conditional densities.

---

[8] This allows for the fact that densities can differ at a point without changing the probability distribution. See comments in §4.1.7 on our conventions for stating identities involving densities.

In Example 5 of §4.1.8, we showed that the marginal densities for the given X and Y are

$$f_X(x) = e^{-x}, \qquad x > 0$$

and

$$f_Y(y) = \frac{1}{(y+1)^2}, \qquad y > 0.$$

Hence, from the formulas for conditional densities, we have

$$f_{X|Y=y}(x) = \frac{xe^{-x(y+1)}}{(y+1)^{-2}}$$
$$= x(y+1)^2 e^{-x(y+1)}$$

and

$$f_{Y|X=x}(y) = \frac{xe^{-x(y+1)}}{e^{-x}}$$
$$= xe^{-xy}.$$

It is clear from these formulas that X and Y are not independent. The dependence of X and Y is also clear from the observation that $f_X(x) \cdot f_Y(y) \neq f_{X,Y}(x, y)$.   ■

EXAMPLE 3:   The joint density of N and X is given by

$$f_{N,X}(n, x) = \frac{x^n e^{-2x}}{n!}, \qquad x \geq 0; n = 0, 1, 2, \ldots .$$

Determine the conditional densities.

From Example 6 of §4.1.8, the marginal probability mass function for N is given by

$$p_N(n) = \int_0^\infty \frac{x^n e^{-2x}}{n!} dx, \qquad n = 0, 1, 2, \ldots ,$$

and the marginal probability density function for X is given by

$$f_X(x) = e^{-x}, \qquad x > 0.$$

By recursively calculating the integral in the formula for p_N, we find that

$$p_N(n) = \left(\frac{1}{2}\right)^{n+1}, \qquad n = 0, 1, 2, \ldots .$$

Consequently, the conditional probability mass function for $N|X = x$ is

$$p_{N|X=x}(n) = \frac{f_{N,X}(n, x)}{f_X(x)}$$
$$= \frac{x^n e^{-x}}{n!} \qquad n = 0, 1, 2, \ldots$$

and the conditional probability density function for $X|N = n$ is

$$f_{X|N=n}(x) = \frac{f_{N,X}(n, x)}{p_N(n)}$$

$$= \frac{2^{n+1}x^n e^{-2x}}{n!} \qquad x > 0.$$

Once again, it should be clear that X and N are not independent. ∎

4.1.10 The Law of Total Probability and Bayes' Theorem (Distributional Forms)

The formulas connecting the joint, marginal, and conditional densities for X_1 and X_2 presented in the preceding section can be used to derive distributional forms of the law of total probability and Bayes' theorem. In this section, we develop these distributional analogs of the law of total probability and Bayes' theorem, and we illustrate their use in practice.

For clarity, we begin by restating the formulas connecting joint, marginal, and conditional densities:

$$f_{X_1|X_2=x_2}(x_1) = \frac{f_{X_1,X_2}(x_1, x_2)}{f_{X_2}(x_2)},$$

$$f_{X_2|X_1=x_1}(x_2) = \frac{f_{X_1,X_2}(x_1, x_2)}{f_{X_1}(x_1)}.$$

We also restate the formulas for calculating marginal densities from joint densities developed in §4.1.8:

$$f_{X_1}(x_1) = \int_{-\infty}^{\infty} f_{X_1,X_2}(x_1, t)dt,$$

$$f_{X_2}(x_2) = \int_{-\infty}^{\infty} f_{X_1,X_2}(s, x_2)ds.$$

Law of Total Probability

Recall from Chapter 3 that the law of total probability is a result that allows us to decompose an unconditional probability into conditional probability components. In §4.1.9, we introduced the concept of conditional probability for random variables. We noted that marginal distributions encode unconditional probabilities about a random variable, whereas conditional distributions encode probabilities for a random variable under the condition that some related random variable assumes a particular value. The distributional form of the law of total probability is a result that allows one to decompose a marginal (i.e., unconditional) density into conditional density components, which may be easier to specify in practice.

Law of Total Probability (Distributional Form)

$$f_{X_1}(x_1) = \int_{-\infty}^{\infty} f_{X_1|X_2=t}(x_1)f_{X_2}(t)dt$$

$$f_{X_2}(x_2) = \int_{-\infty}^{\infty} f_{X_2|X_1=s}(x_2)f_{X_1}(s)ds$$

These formulas follow directly from the formulas connecting joint, marginal, and conditional densities developed in the previous two sections. Indeed,

$$f_{X_1}(x_1) = \int_{-\infty}^{\infty} f_{X_1,X_2}(x_1, t)dt$$

$$= \int_{-\infty}^{\infty} f_{X_1|X_2=t}(x_1)f_{X_2}(t)dt$$

and

$$f_{X_2}(x_2) = \int_{-\infty}^{\infty} f_{X_1,X_2}(s, x_2)ds$$

$$= \int_{-\infty}^{\infty} f_{X_2|X_1=s}(x_2)f_{X_1}(s)ds.$$

Note the similarity of these formulas to the original statement of the law of total probability given in Chapter 3.

By performing appropriate integrations, we can obtain analogous formulas for the marginal distribution functions F_{X_1} and F_{X_2}. Indeed,

$$F_{X_1}(x_1) = \int_{-\infty}^{x_1} f_{X_1}(s)ds$$

$$= \int_{-\infty}^{x_1}\int_{-\infty}^{\infty} f_{X_1|X_2=t}(s)f_{X_2}(t)\,dt\,ds$$

$$= \int_{-\infty}^{\infty}\int_{-\infty}^{x_1} f_{X_1|X_2=t}(s)f_{X_2}(t)\,ds\,dt$$

$$= \int_{-\infty}^{\infty} F_{X_1|X_2=t}(x_1)f_{X_2}(t)\,dt.$$

Similarly,

$$F_{X_2}(x_2) = \int_{-\infty}^{\infty} F_{X_2|X_1=s}(x_2)f_{X_1}(s)\,ds.$$

These formulas are also referred to as the law of total probability.

In many important applications, one of the variables X_1, X_2 will be continuous and the other will be discrete. In such situations, the law of total probability has a slightly simpler form. Indeed, if X_1 is continuous and X_2 is discrete, then

$$f_{X_1}(x_1) = \sum_{\text{all } x_2} f_{X_1|X_2=x_2}(x_1)p_{X_2}(x_2),$$

whereas if X_1 is discrete and X_2 is continuous, then

$$p_{X_1}(x_1) = \int_{-\infty}^{\infty} p_{X_1|X_2=t}(x_1)f_{X_2}(t)\,dt.$$

We now illustrate the use of these formulas by considering some examples from insurance.

EXAMPLE 1: An auto insurer is trying to develop a model for claim size in an attempt to better price its products. On the basis of historical data, it has determined that the claim size distribution for policyholders classified as good risks has density

$$f_X(x) = 2e^{-2x}, \qquad x > 0$$

and the claim size distribution for policyholders classified as bad risks has density

$$f_X(x) = \frac{1}{3}e^{-x/3}, \qquad x > 0,$$

where claims are measured in thousands of dollars. There is a 30% chance that a given policyholder is a bad risk. What is the probability that this policyholder's claim exceeds $1000?

Let X be the claim size and let I be an indicator for being a bad risk. Then, from the given information,

$$I = \begin{cases} 1 & \text{with probability .30,} \\ 0 & \text{with probability .70;} \end{cases}$$

and further,

$$f_{X|I=0}(x) = 2e^{-2x}, \qquad x > 0,$$

$$f_{X|I=1}(x) = \frac{1}{3}e^{-x/3}, \qquad x > 0.$$

Hence, by the law of total probability, the density function for X is given by

$$f_X(x) = f_{X|I=0}(x)p_I(0) + f_{X|I=1}(x)p_I(1)$$

$$= (2e^{-2x})(.70) + \frac{1}{3}e^{-x/3}(.30)$$

$$= 1.4e^{-2x} + 0.1e^{-x/3}.$$

Thus, the desired probability is

$$\Pr(X > 1) = \int_1^{\infty} f_X(x)\,dx$$

$$= (1.4)\int_1^{\infty} e^{-2x}\,dx + (0.1)\int_1^{\infty} e^{-x/3}\,dx$$

$$= 1.4e^{-2} + 0.1e^{-1/3}. \qquad \blacksquare$$

COMMENT: In this example, the claim size distribution function is a weighted sum of some other distribution functions. To be precise,

$$F_X(x) = (.70)F_{X_1}(x) + (.30)F_{X_2}(x),$$

where X_1 is the random variable with density $f_{X_1}(x) = 2e^{-2x}$ and X_2 is the random variable with density $f_{X_2}(x) = \frac{1}{3}e^{-x/3}$. A random variable with this property is referred to as a *mixture* or, more precisely, a *discrete mixture*.

A **discrete mixture** is a random variable X whose distribution function has the form

$$F_X(x) = p_1 F_{X_1}(x) + \cdots + p_n F_{X_n}(x),$$

where $0 < p_i < 1$ for all i and $\sum p_i = 1$. Mixtures arise in connection with the law of total probability and are extremely important in insurance applications where the risk class of a subject is uncertain. We will discuss mixtures in detail in Chapter 9.

EXAMPLE 2:   An insurance company has just sold a group health insurance plan to a new employer. Experience with similar employers suggests that the number of hospitalization claims per month can be modeled using the mass function

$$p_N(n) = \frac{\lambda^n e^{-\lambda}}{n!}, \qquad n = 0, 1, 2, \ldots,$$

where λ is a parameter that describes the group's expected utilization. The insurer is uncertain what the true value of λ for this group is (i.e., the insurer is uncertain about how many claims to expect) and decides to model this uncertainty using the density function

$$f_\Lambda(\lambda) = e^{-\lambda}, \qquad \lambda > 0.$$

Determine the probability that the group has two or more hospitalization claims in the next month.

Let N be the number of hospitalization claims in the next month. We are given that

$$p_{N|\Lambda=\lambda}(n) = \frac{\lambda^n e^{-\lambda}}{n!}, \qquad n = 0, 1, 2, \ldots,$$

and

$$f_\Lambda(\lambda) = e^{-\lambda}, \qquad \lambda > 0.$$

To determine the desired probability, we require the unconditional probability mass function of N.

By the law of total probability,

$$p_N(n) = \int_{-\infty}^{\infty} p_{N|\Lambda=\lambda}(n) f_\Lambda(\lambda) d\lambda$$

$$= \int_0^\infty \left(\frac{\lambda^n e^{-\lambda}}{n!} \right) e^{-\lambda} d\lambda$$

$$= \frac{1}{n!} \int_0^\infty \lambda^n e^{-2\lambda} d\lambda.$$

Hence, the desired probability is

$$\Pr(N \geq 2) = 1 - p_N(0) - p_N(1)$$
$$= 1 - \int_0^\infty e^{-2\lambda} d\lambda - \int_0^\infty \lambda e^{-2\lambda} d\lambda.$$

Using integration by parts,

$$\int_0^\infty \lambda e^{-2\lambda} d\lambda = \frac{1}{4}.$$

Consequently, the desired probability is

$$\Pr(N \geq 2) = 1 - \frac{1}{2} - \frac{1}{4} = \frac{1}{4}.$$

Hence, there is a 25% chance that the group will have two or more hospitalization claims next month. ■

Bayes' Theorem

Recall from Chapter 3 that Bayes' theorem is a result for "inverting" conditional probabilities. In distributional form, Bayes' theorem can be stated as follows:

Bayes' Theorem (Distributional Form)

$$f_{X_1 | X_2 = x_2}(x_1) = \frac{f_{X_2 | X_1 = x_1}(x_2) f_{X_1}(x_1)}{f_{X_2}(x_2)}$$

This result is an immediate consequence of the formulas connecting joint, marginal, and conditional densities.

EXAMPLE 3: Consider the previous example where the insurer models the number of hospitalization claims in a month using the probability mass function

$$p_{N | \Lambda = \lambda}(n) = \frac{\lambda^n e^{-\lambda}}{n!}, \qquad n = 0, 1, 2, \ldots$$

and initially models the uncertainty in the utilization parameter λ using the density

$$f_\Lambda(\lambda) = e^{-\lambda}, \qquad \lambda > 0.$$

Suppose that there is one hospitalization in the first month. How does this information alter the insurer's belief about the true value of λ?

At the end of the first month, the insurer's belief about the parameter λ is captured by the conditional density $f_{\Lambda | N = 1}$. Using Bayes' theorem, we find that

$$f_{\Lambda|N=1}(\lambda) = \frac{f_{N|\Lambda=\lambda}(1)f_{\Lambda}(\lambda)}{f_N(1)}$$

$$= \frac{p_{N|\Lambda=\lambda}(1)f_{\Lambda}(\lambda)}{p_N(1)}$$

$$= \frac{(\lambda e^{-\lambda})(e^{-\lambda})}{p_N(1)}.$$

By the law of total probability,

$$p_N(1) = \int_{-\infty}^{\infty} p_{N|\Lambda=\lambda}(1)f_{\Lambda}(\lambda)d\lambda$$

$$= \int_0^{\infty} (\lambda e^{-\lambda})(e^{-\lambda})d\lambda$$

$$= \frac{1}{4}.$$

Hence,

$$f_{\Lambda|N=1}(\lambda) = 4\lambda e^{-2\lambda}, \qquad \lambda > 0.$$

As a particular illustration, consider the insurer's belief about the parameter λ exceeding two. At the beginning of the month, the insurer believes that the chance of λ exceeding two is

$$\Pr(\Lambda > 2) = \int_2^{\infty} e^{-\lambda}d\lambda = e^{-2}.$$

However, at the end of the month, after learning that one hospitalization claim was received during the month, the insurer believes that the chance of λ exceeding two is now

$$\Pr(\Lambda > 2|N = 1) = \int_2^{\infty} 4\lambda e^{-2\lambda}d\lambda = 5e^{-4}.$$

Calculation details are left to the reader. ∎

4.1.11   Arithmetic Operations on Random Variables

There are many situations in probability modeling where sums and products of uncertain quantities arise. For example, the aggregate loss S on a group of insurance policies is the sum of the losses on individual policies,

$$S = X_1 + \cdots + X_n,$$

where $X_1, \ldots, X_n$ are the uncertain losses on the individual policies. On the other hand, the accumulated value P of \$1 over multiple time periods is the product of the accumulation factors for each time period,

$$P = (1 + R_1) \cdots (1 + R_n),$$

where $R_1, \ldots, R_n$ are the uncertain rates of return over individual time periods. To address such problems properly, we need a mechanism for performing addition and multiplication of random variables.

Since random variables are actually real valued functions, addition and multiplication are easy to define. Indeed, if X and Y are random variables on a given sample space S and α is a real constant, then $X + Y$, $X \cdot Y$, and αX are the random variables given by

$$(X + Y)(s) = X(s) + Y(s),$$
$$(X \cdot Y)(s) = X(s) \cdot Y(s),$$
$$(\alpha X)(s) = \alpha X(s),$$

respectively. The real difficulty is in determining the *distributions* of $X + Y$, $X \cdot Y$, and αX.

In Chapter 8, we will develop techniques for systematically determining the distributions of sums and products. In the remainder of this section, we consider the effect of substitution on identities involving sums and products.

Simply put, substitution of *equal* random variables is valid, but substitution of *equivalent* (i.e., identically distributed) random variables is not. In particular, if $X_1 = X_2$ then, from the properties of real valued functions,

$$X_1 + X_2 = 2X_1 = 2X_2$$

and

$$X_1 X_2 = X_1^2.$$

However, if it is only known that $X_1 \sim X_2$ (i.e., X_1 and X_2 are identically distributed), then it need not be true that $X_1 + X_2 \sim 2X_1$ or $X_1 X_2 \sim X_1^2$.

For example, suppose that two fair coins are tossed and X_1 and X_2 are indicators of heads on the respective coins,

$$X_j = \begin{cases} 1 & \text{if coin } j \text{ lands heads,} \\ 0 & \text{if coin } j \text{ lands tails.} \end{cases}$$

Then one can show that $X_1 \sim X_2$, but $X_1 + X_2 \not\sim 2X_1$ and $X_1 X_2 \not\sim X_1^2$ (see question 10 of the exercises from Chapter 2).

In many insurance applications, one is interested in the sum of a collection of identically distributed losses (i.e., the sum $S = X_1 + \cdots + X_n$, where $X_j \sim X$ for each j). If one blindly follows the rules of ordinary arithmetic, it is easy to make the faulty deduction that $S \sim nX$ (i.e., the sum of the losses is equivalent to a multiple of one of the losses). However, as the preceding example illustrates, this line of reasoning is wrong since substitution of equivalent random variables is not valid! After we have discussed the concept of variance in §4.2.2, we will see in another way why the statement $X_1 + \cdots + X_n \sim nX$ cannot hold.

4.1.12 The Difference Between Sums and Mixtures

In §4.1.10, we introduced the notion of a mixture. Recall that a (discrete) mixture is a random variable X whose distribution function has the form

$$F_X(x) = p_1 F_{X_1}(x) + \cdots + p_n F_{X_n}(x)$$

for some random variables $X_1, \ldots, X_n$, where $0 < p_i < 1$ for all i and $\sum p_i = 1$ (i.e., the distribution function is a weighted sum of distribution functions). Mixtures arise naturally when considering a random quantity whose distribution depends on some uncertain risk parameter.

Because of the way in which mixtures are defined, it is easy to confuse them with *sums*. However, it is important to keep in mind that mixtures are not the same as sums. Indeed, for any random variables X_1, X_2 and any number $\alpha \in (0, 1)$, the weighted sum $\alpha X_1 + (1 - \alpha) X_2$ is generally not the same as the mixture of X_1 and X_2 with mixing weights α, $1 - \alpha$. That is, it is generally not the case that $F_{\alpha X_1 + (1-\alpha)X_2}(x) = \alpha F_{X_1}(x) + (1 - \alpha) F_{X_2}(x)$ for all x. We can better understand why this is so by considering a concrete example that emphasizes the differences in the meanings of these two concepts.

Hence, let X_1 and X_2 represent simple returns for the securities S_1 and S_2 over a given period and consider the following two portfolios:

- Portfolio $\mathcal{P}_1$ consists of precisely one of the securities S_1, S_2, but it is not known which one; the probability of the security being S_1 is α.
- Portfolio $\mathcal{P}_2$ consists of both securities, with the fraction invested in S_1 being α.

Further, let R_1, R_2 be the returns on the portfolios $\mathcal{P}_1$, $\mathcal{P}_2$, respectively. Then R_1 is the *mixture* of the returns X_1, X_2 with *mixing* weights α, $1 - \alpha$, and R_2 is the *sum* of the returns X_1, X_2 with *portfolio* weights α, $1 - \alpha$. Clearly, the returns on these two portfolios will generally be different.

This example illustrates that the distribution of a weighted sum is not the same as a weighted sum of distributions.

4.1.13   Exercises

1. Indicate which of the following statements are true and which are false. Give reasons for your choice.

   a. The distribution function of a random variable is continuous at every point.
   b. The density function of a continuous random variable is unique.
   c. The distribution function of a random variable is continuous from the right at every point.
   d. The probability mass function of a discrete random variable is unique.
   e. $\Pr(X = a)$ is zero at every point a of a continuous distribution.
   f. $\Pr(X = a)$ is zero at every point a of a mixed distribution.
   g. $\Pr(X = a)$ is zero at all but a finite number of points of a discrete distribution.
   h. Discrete random variables have only a finite number of possible values.
   i. The density function of a continuous random variable is continuous.
   j. Every nondecreasing step function is the distribution function of some random variable.
   k. Every nonnegative continuous function is the density of some continuous random variable.
   l. If the density function of a continuous random variable is continuous, then the median is unique. (A *median* for the random variable X is a number m with the property that $\Pr(X \le m) \ge \frac{1}{2}$ and $\Pr(X \ge m) \ge \frac{1}{2}$. See the discussion in §4.2.2 for more details).
   m. If the distribution function is discontinuous and strictly increasing on some interval, then the distribution is mixed.

n. If the density function of a continuous distribution is discontinuous at $X = a$, then $\Pr(X = a) > 0$.

2. The function

$$F(x) = k\left(1 - \left(\frac{1}{2}\right)^{[x]}\right), \qquad x > 0$$

is the distribution function for a discrete random variable X. Here, $[x]$ denotes the *integer part* of x (i.e., the greatest integer less than or equal to x).

a. Determine the value of k.
b. Specify the probability mass function of X.

3. The function

$$F(x) = k\left(1 - \frac{1}{x^2}\right), \qquad 1 \le x < 2$$

is the distribution function for a continuous random variable X. Determine the value of k and specify the probability density function for X.

4. The function

$$F(x) = \begin{cases} 0 & \text{for } x < 0, \\ \frac{1}{4}x + \frac{1}{6} & \text{for } 0 \le x < 2, \\ 1 & \text{for } x \ge 2 \end{cases}$$

is the distribution function of a mixed random variable X. Determine the generalized density for X. Write F_X as a weighted sum of a continuous distribution function and a discrete distribution function.

5. A biased coin is tossed ten times. Suppose that the probability of getting heads on a single toss is p. Let X be the number of heads obtained.

a. Give an algebraic formula for the probability mass function of X.
b. Construct a table of values for p_X in the cases $p = 1/10, 1/5, 4/5$. Use your table to graph p_X in each of these cases.
c. Give a qualitative description of p_X using the graphs constructed in part b. Describe how the graph of p_X changes as p varies. What is the relationship between the distributions with $p = 1/5$ and $p = 4/5$?
d. What do you think $E[X]$ should be? (The quantity $E[X]$ was introduced in §2.1 and will be discussed in greater detail in §4.2.)

6. Suppose that X is a random variable and $Y = aX + b$, where a and b are fixed constants.

a. If X is discrete, what is the relationship between the graphs of p_X and p_Y?
b. If X is continuous, what is the relationship between the graphs of f_X and f_Y?
c. Are the relationships in parts a and b the same or different? Explain.

7. The distribution function for a discrete random vector $\mathbf{X} = (X_1, X_2)$ is given by

$$F_{X_1,X_2}(x_1, x_2) = \begin{cases} 0 & \text{if } x_1 < 0 \text{ or } x_2 < 0, \\ \frac{1}{8} & \text{if } 0 \le x_1 < 1 \text{ and } 0 \le x_2 < 1, \\ \frac{1}{4} & \text{if } x_1 \ge 1 \text{ and } 0 \le x_2 < 1, \\ \frac{3}{8} & \text{if } 0 \le x_1 < 1 \text{ and } x_2 \ge 1, \\ 1 & \text{if } x_1 \ge 1 \text{ and } x_2 \ge 1. \end{cases}$$

a. Using five different degrees of shading or color, indicate the regions in the x_1-x_2 plane where the function F_{X_1,X_2} assumes each of its values.
b. Using your answer to part a, sketch a three-dimensional graph of F_{X_1,X_2} in the x_1-x_2-z coordinate system.
c. Specify the probability mass function $p_\mathbf{X}$ for $\mathbf{X}$.
d. Sketch a graph of $p_\mathbf{X}$.
e. By referring to the graphs in parts a and b, discuss how a bivariate distribution function is similar to and different from a univariate distribution function.

8. The probability density function for the continuous random vector $\mathbf{X} = (X_1, X_2)$ is given by

$$f_{X_1,X_2}(x_1, x_2) = \begin{cases} 2x_1 & \text{for } 0 < x_1 < 1 \text{ and } 0 < x_2 < 1, \\ 0 & \text{otherwise.} \end{cases}$$

a. Sketch a graph of $f_\mathbf{X}$ in three-dimensional space.
b. Determine formulas for the marginal and conditional densities by considering only graphical interpretations. Verify that your formulas are correct by using algebraic definitions and performing appropriate integrations.
c. Determine whether X_1 and X_2 are independent or dependent, first by considering the conditional densities and then by considering the marginal densities. Do you reach the same conclusion?
d. Determine the distribution function $F_\mathbf{X}$.

9. The probability density function for the continuous random vector $\mathbf{X} = (X_1, X_2)$ is given by

$$f_{X_1,X_2}(x_1, x_2) = \begin{cases} \frac{1}{2} & \text{if } x_1 + x_2 \le 2, x_1 \ge 0, x_2 \ge 0, \\ 0 & \text{otherwise.} \end{cases}$$

a. Sketch the region of nonzero probability in the x_1-x_2 plane. Then sketch the graph of $f_\mathbf{X}$ in three-dimensional space.
b. Using only the graphical interpretation of conditional densities, determine formulas for $f_{X_1|X_2=x_2}$ and $f_{X_2|X_1=x_1}$. Are X_1 and X_2 independent?
c. Using the graphical interpretation of marginal densities and the geometry of the graph of $f_\mathbf{X}$, determine formulas for f_{X_1} and f_{X_2}. Verify that your formulas are correct by performing an appropriate integration.
d. Using the algebraic definition of a conditional density, i.e., $f_{X_1|X_2=x_2}(x_1) = f_{X_1,X_2}(x_1, x_2)/f_{X_2}(x_2)$, determine formulas for $f_{X_1|X_2=x_2}$ and $f_{X_2|X_1=x_1}$. Do these formulas agree with the ones surmised in part b?

    e. Sketch graphs of the marginal and conditional densities. Are these graphs consistent with the general graphical interpretation of marginal and conditional densities as projections and scaled cross-sections?

    f. Calculate the probability that $X_1 > 2X_2$.

10. For each of the following functions F, determine k such that F is a distribution function. Assume that all the probability lies in the indicated region.

    a. $F(x, y) = kxy(2x + 3y)$, $0 \le x \le 1$, $0 \le y \le 1$.

    b. $F(x, y) = kxy(2x + 3y)$, $x + y \le 1$, $x > 0$, $y > 0$.

    c. $F(x, y) = kx^2y(x + y)$, $0 \le x \le 2$, $0 \le y \le 1$.

    d. $F(x, y) = kxy(x^2 + y)$, $x + 2y \le 4$, $x > 0$, $y > 0$.

11. On a particular type of insurance policy, claim size can be modeled using the density

$$f_X(x) = \lambda e^{-\lambda x}, \qquad x > 0,$$

    where λ is a parameter that is related to the expected size of a claim. In fact, $\lambda = 1/E[X]$. The insurer is uncertain what the true value of λ for a given group of policyholders is and decides to model this uncertainty using the density function

$$f_\Lambda(\lambda) = 4\lambda e^{-2\lambda}, \qquad \lambda > 0.$$

    a. Determine the unconditional probability density function for claim size. What is the probability that the claim size exceeds two?

    b. Suppose that a claim of size two is received. How does this affect the insurer's belief about the true value of λ?

12. The joint density for N and X is given by

$$f_{N,X}(n, x) = \frac{2x^n e^{-3x}}{n!}, \qquad x > 0; \; n = 0, 1, 2, 3, \ldots.$$

    Determine the marginal and conditional distributions. Are N and X independent?

13. Consider the random vector $\mathbf{V} = (C, N)$, where N is the random number of claims filed by a particular group in the next month and C is the total dollar amount of the claims. Sketch a possible graph for the joint density of C and N.

14. The random variables X_1 and X_2 are independent and identically distributed with common density

$$f_X(x) = e^{-x}, \qquad x > 0.$$

    Determine the distribution function for the random variable Y given by $Y = X_1 + X_2$.

15. In each of the following situations, determine whether the desired model is a sum or mixture of random variables:

    a. An insurer is interested in the total claim payments to be made on a group of 100 policies.

    b. An investment analyst is interested in the return on the stock of a particular company. The company is involved in several different businesses with the result that its stock can sometimes behave like a "growth stock" and at other times behave like a "blue-chip stock."

    c. An insurer is interested in modeling the loss on a group of contracts for which the risk class of the insureds is uncertain.

    d. A portfolio manager is interested in the return on a portfolio of securities and in selecting the security weights that maximize expected return while minimizing risk.

4.2   Statistical Measures of Expectation, Variation, and Risk

Distribution functions and density functions provide complete descriptions of the distribution of probability for a given random variable. However, they do not allow us to easily make comparisons between two different distributions. *Descriptive statistics* enable us to summarize particular information about a probability distribution in a numerical form suitable for making such comparisons. In this section, we discuss statistical measures of expectation, variation, and risk that arise frequently in applications.

4.2.1   Expectation

Consider a group of n independent, homogeneous risks[9] and let $X_1, \ldots, X_n$ be the respective losses incurred by members of the group. For initial clarity, assume that the only possible losses are $a_1, \ldots, a_k$ and let $p_X(a_1), \ldots, p_X(a_k)$ be the respective relative frequencies with which these losses occur. Then, if the group is sufficiently large, we expect:

    approximately $np_X(a_1)$ of the members to incur a loss of a_1;

    approximately $np_X(a_2)$ of the members to incur a loss of a_2;

$$\vdots$$

    approximately $np_X(a_k)$ of the members to incur a loss of a_k.

Consequently, when the group is large, we expect the total loss incurred by the group to be

$$np_X(a_1) \cdot a_1 + \cdots + np_X(a_k) \cdot a_k$$

and the average loss per member to be

$$\frac{1}{n} \cdot \sum_{j=1}^{k} np_X(a_j)\, a_j = \sum_{j=1}^{k} a_j\, p_X(a_j).$$

If the distribution of the individual losses is continuous rather than discrete, as assumed, with relative frequency density function f_X, then for a sufficiently large group of size n, we expect the total loss to be

$$n \cdot \int_{-\infty}^{\infty} x f_X(x)\, dx$$

---

[9] See §1.4 for a discussion of homogeneous risks in insurance.

and the average loss per member to be

$$\int_{-\infty}^{\infty} x f_X(x) dx.$$

This example suggests the following definition:

Definition of Arithmetic Expectation

For any continuous random variable X, the **expected value** of X is the number $E[X]$ defined by

$$E[X] = \int_{-\infty}^{\infty} x f_X(x) dx,$$

provided that this number exists. For any discrete random variable X, the **expected value** of X is the number $E[X]$ defined by

$$E[X] = \sum_{\text{all } x} x p_X(x),$$

provided that this number exists. The quantity $E[X]$ is also referred to as the **expectation** of X or the **mean** of the distribution for X.

The formula for the expectation of a mixed random variable will be given later.

Note that $E[X]$ is an *arithmetic* average as opposed to a *geometric* average. When we wish to emphasize this fact, we will refer to $E[X]$ as the **arithmetic expected value** or the **arithmetic expectation** of X and write $E_a[X]$ in place of $E[X]$. From a graphical perspective, $E[X]$ is the **center of mass** of the distribution of X. That is, it is the point at which a fulcrum should be placed for a corresponding distribution of masses to be perfectly balanced (see §2.1).

Two comments about $E[X]$ are worth making. First, $E[X]$ need not be a possible value of X (i.e., a value with nonzero probability). For example, in a game in which we win \$1 if a particular unbiased coin lands heads and lose \$1 if it lands tails, the expected gain is \$0, a payoff which is not attainable on a single toss. Second, $E[X]$ need not be the actual average loss per member observed. Indeed, it is very likely that the actual average loss will be different from $E[X]$ most of the time. However, the law of large numbers (to be discussed in Chapter 8) asserts that the actual average loss will be close to $E[X]$ with high probability.

The arithmetic expectation need not exist or be finite. An example of a distribution without a mean is given in §7.6.

There is another type of expectation—the *geometric* expectation—which arises frequently in investment problems. This expectation is a particular geometric average that may be defined as follows:

Definition of Geometric Expectation

For any positive random variable X, the **geometric expected value** or **geometric expectation** of X is the number $E_g[X]$ defined by

$$E_g[X] = \exp\left(\int_0^\infty (\log x) f_X(x) dx\right),$$

where $\exp(\cdot)$ denotes the exponential function; that is, $\exp(x) = e^x$.

When X is discrete, this formula for $E_g[X]$ reduces to

$$E_g[X] = \exp\left(\sum_x (\log x) p_X(x)\right)$$
$$= \exp(\log(\prod_x x^{p_X(x)}))$$
$$= \prod_x x^{p_X(x)},$$

which has the more recognizable form of a geometric average.

To see how the geometric expectation arises in practice, consider a \$1 investment in a security that is to be held for n days. Suppose that the returns $R_1, \ldots, R_n$ on the successive days are independent and identically distributed and let $X_1, \ldots, X_n$ be the respective accumulation factors (i.e., $X_i = 1 + R_i$ for $i = 1, \ldots, n$) so that the value of the \$1 investment after n days is $X_1 X_2 \cdots X_n$. For simplicity, assume that the only possible accumulation factors are $a_1, \ldots, a_k$ and let $p_X(a_1), \ldots, p_X(a_k)$ be the respective relative frequencies with which these factors occur. Then, if n is sufficiently large, we expect:

approximately $n p_X(a_1)$ of the accumulation factors to be a_1;

approximately $n p_X(a_2)$ of the accumulation factors to be a_2;

$$\vdots$$

approximately $n p_X(a_k)$ of the accumulation factors to be a_k.

Consequently, when the number of days n is large, we expect the total accumulation of the \$1 investment to be

$$a_1^{n p_X(a_1)} \cdots a_k^{n p_X(a_k)}$$

and the average (compounded) accumulation per day to be

$$(a_1^{n p_X(a_1)} \cdots a_k^{n p_X(a_k)})^{1/n} = \prod_{j=1}^k a_j^{p_X(a_j)}.$$

Since

$$\prod_{j=1}^{k} a_j^{p_X(a_j)} = \prod_{j=1}^{k} (e^{\log a_j})^{p_X(a_j)}$$

$$= \exp\left(\sum_{j=1}^{k} (\log a_j) p_X(a_j)\right),$$

we can extend this argument to the case of a *continuous* distribution of accumulation factors in a natural way. In that case, the expected accumulation per day is

$$\exp\left(\int_0^\infty (\log x) f_X(x) dx\right),$$

where X is the distribution of accumulation factors and is assumed to be positive.

The example just discussed suggests that the geometric expectation may be more appropriate to consider than the arithmetic expectation when analyzing investment problems. However, both types of expectation are widely used. The appropriate one in any given problem really depends on the context. To illustrate this, suppose that X is the accumulation factor for an investment over a particular time period (e.g., 1 day, as earlier). Suppose further that we have a choice of making a large number of independent investments of this type for a single time period or a single investment of this type over a large number of successive and independent time periods. Then the arithmetic expectation should be used in evaluating the first alternative (multiple investments, single period), whereas the geometric expectation should be used in evaluating the second (single investment, multiple periods). Indeed, the arithmetic expectation $E_a[X]$ represents the average accumulation *per investment* when a large number of independent investments, each with accumulation random variable X, are placed over a single time period. On the other hand, the geometric expectation $E_g[X]$ represents the average accumulation *per period* when a single investment, with per period accumulation random variable X, is compounded over a large number of independent periods.

Failure to use the proper expectation can lead to costly mistakes. Indeed, consider an investment whose daily return is either +50% or −40% with equal probability. In this case, the arithmetic expectation of the accumulation factor is 1.05, suggesting an opportunity that should be taken; on the other hand, the geometric expectation is $\sqrt{.9}$, suggesting an investment that should be avoided. Whether the opportunity is a good one or not really depends on how you are able to take advantage of it. If you can invest in a large number of independent securities of this type for a single day, then it is clearly to your advantage to do so. However, if you can only purchase one such security and must hold it for a long period of time, then it is clearly not a good idea to do so.

Notice that the arithmetic expectation in the numerical example just given was greater than the geometric expectation. This is a general result, known as the **arithmetic-geometric means inequality**:

Arithmetic-Geometric Means Inequality

For any positive random variable X,

$$E_g[X] \le E_a[X].$$

Here, as earlier, we use the notation $E_a[X]$ in place of $E[X]$ when we wish to emphasize that $E[X]$ is an arithmetic expectation. The arithmetic-geometric means inequality for random variables is actually a special case of Jensen's inequality, which we will discuss in Chapter 7.

EXAMPLE 1: A discrete random variable X with probability mass function

$$p_X(x) = \frac{e^{-\lambda}\lambda^x}{x!}, \qquad x = 0, 1, 2, \dots$$

is said to have a **Poisson distribution** with parameter λ, where λ is assumed to be a positive real number. (The Poisson distribution arises in many applications and will be discussed in detail in §5.2.) Determine the expected value of X in terms of λ.

First note that, from the power series expansion $e^z = \sum_{n=0}^{\infty} \frac{z^n}{n!}$, it is clear that the given p_X is a probability mass function. Indeed,

$$\sum_{x=0}^{\infty} p_X(x) = e^{-\lambda} \sum_{x=0}^{\infty} \frac{\lambda^x}{x!} = e^{-\lambda} \cdot e^{\lambda} = 1.$$

Hence, from the definition of arithmetic expectation, we have

$$E[X] = \sum_{x=0}^{\infty} x p_X(x)$$

$$= \sum_{x=0}^{\infty} x \cdot \frac{e^{-\lambda}\lambda^x}{x!}$$

$$= \sum_{x=1}^{\infty} x \cdot \frac{e^{-\lambda}\lambda^x}{x!}$$

$$= \lambda \sum_{x=1}^{\infty} \frac{e^{-\lambda}\lambda^{x-1}}{(x-1)!}$$

$$= \lambda \sum_{n=0}^{\infty} \frac{e^{-\lambda}\lambda^n}{n!}$$

$$= \lambda \sum_{n=0}^{\infty} p_X(n)$$

$$= \lambda.$$

Consequently, the expected value of a Poisson random variable with parameter λ is simply λ. ∎

EXAMPLE 2: A continuous random variable X with probability density function

$$f_X(x) = \begin{cases} \lambda e^{-\lambda x} & \text{for } x \geq 0, \\ 0 & \text{for } x < 0 \end{cases}$$

is said to have an **exponential distribution** with parameter λ, where λ is assumed to be a positive real number. (The exponential distribution arises in many applications and will be discussed in detail in §6.1.1.) Show that $E[X] = 1/\lambda$.

First note that the given f_X is a probability density function because

$$\int_{-\infty}^{\infty} f_X(x)dx = \int_{0}^{\infty} \lambda e^{-\lambda x}dx = -e^{-\lambda x}\Big|_{0}^{\infty} = 1.$$

Hence, from the definition of expectation and using integration by parts, we have

$$\begin{aligned} E[X] &= \int_{-\infty}^{\infty} x \cdot f_X(x)dx \\ &= \int_{0}^{\infty} x \cdot \lambda e^{-\lambda x}dx \\ &= x(-e^{-\lambda x})\Big|_{0}^{\infty} - \int_{0}^{\infty}(-e^{-\lambda x})dx \\ &= \int_{0}^{\infty} e^{-\lambda x}dx \\ &= \frac{1}{\lambda} \end{aligned}$$

as claimed. Consequently, the expected value of an exponential random variable with parameter λ is the *reciprocal* of λ. ∎

Expectation for Mixed Random Variables

Mixed random variables arise frequently in applications, particularly in connection with insurance contracts with caps or deductibles and electrical circuits with limiters, and consequently, it is important to be at ease with them. Determining the expectation of a mixed random variable can be tricky, and if one is not careful, it is easy to make mistakes. Hence, we suggest reading this section very carefully.

Recall from §4.1.6 that a mixed distribution has both discrete and continuous parts. If f_X denotes the density for the continuous part and p_X denotes the mass function for the discrete part, then the expectation can be calculated using the following formula:[10]

$$E[X] = \underbrace{\int_{\text{continuous part}} x f_X(x)dx}_{} + \underbrace{\sum_{\text{discrete part}} x p_X(x)}_{}.$$

Note that in this context, $\int f_X(x)dx + \sum p_X(x) = 1$, where the integral is calculated over the continuous part and the sum is calculated over the discrete part.

A concrete example illustrates the use of this formula.

---

[10] This formula is justified in the appendix to this chapter.

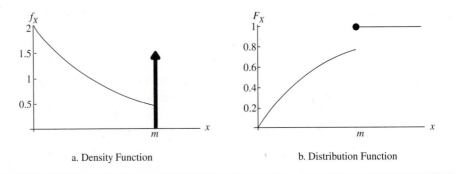

a. Density Function

b. Distribution Function

FIGURE 4.15 The Truncated Exponential with $\lambda = 2$

EXAMPLE 3:   A random variable X with distribution function

$$F_X(x) = \begin{cases} 1 - e^{-\lambda x} & \text{for } 0 \le x < m, \\ 1 & \text{for } x \ge m \end{cases}$$

is said to have a **truncated exponential distribution** with parameters λ and m. The truncated exponential distribution arises frequently in applications, particularly in connection with insurance contracts with a cap (see Chapter 7 for details). Determine the expected value of X.

A sketch of the graph of F_X reveals that the distribution is mixed with a point mass of size $e^{-\lambda m}$ at $x = m$ and a continuous distribution of probability on the interval $0 < x < m$ (Figure 4.15b). The density on the continuous part of the distribution is given by

$$f_X(x) = \lambda e^{-\lambda x}, \qquad 0 < x < m.$$

Hence, the expected value of X is

$$E[X] = \int_{\substack{\text{continuous} \\ \text{part}}} x f_X(x)dx + \sum_{\substack{\text{discrete} \\ \text{part}}} x p_X(x)$$

$$= \int_0^m x \cdot \lambda e^{-\lambda x} \, dx + m \cdot e^{-\lambda m}.$$

Using integration by parts, we have

$$\int_0^m \lambda x e^{-\lambda x} dx = -x e^{-\lambda x}\Big|_0^m + \int_0^m e^{-\lambda x} dx$$

$$= -m e^{-\lambda m} - \frac{1}{\lambda} e^{-\lambda x}\Big|_0^m$$

$$= -m e^{-\lambda m} - \frac{1}{\lambda} e^{-\lambda m} + \frac{1}{\lambda}.$$

Consequently,

$$E[X] = \left(-m e^{-\lambda m} - \frac{1}{\lambda} e^{-\lambda m} + \frac{1}{\lambda} \right) + m e^{-\lambda m}$$

$$= \frac{1}{\lambda}(1 - e^{-\lambda m}).$$

As a check on our calculations, let's determine $\int_0^\infty S_X(x)dx$. Recall from §2.3 that the expected value of any positive random variable can be calculated using the formula $E[X] = \int_0^\infty S_X(x)\,dx$, where S_X is the survival function; that is, $S_X(x) = 1 - F_X(x)$.[11] From the definition of F_X, we see that

$$S_X(x) = \begin{cases} 1 & \text{for } x < 0, \\ e^{-\lambda x} & \text{for } 0 \le x < m, \\ 0 & \text{for } x \ge m. \end{cases}$$

Consequently,

$$\int_0^\infty S_X(x)dx = \int_0^m e^{-\lambda x}dx = \frac{1}{\lambda}(1 - e^{-\lambda m}).$$

This agrees with the value for the expectation obtained earlier.

Note that for large m, $E[X] \approx 1/\lambda$, which from the previous example we recognize as being the expectation of an exponential distribution with parameter λ. This shouldn't be too surprising because the distribution function of a truncated exponential random variable approaches the distribution function of an exponential random variable as $m \to \infty$. In fact, the distribution of a truncated exponential random variable can be determined from the distribution of an exponential random variable by applying the transformation $X = \min(L, m)$. We will discuss this and other transformations in detail in Chapter 7. ∎

Expectation of a Function of a Random Variable

We can write the definition of $E_g[X]$ more compactly, and emphasize the connection between $E_a[X]$ and $E_g[X]$ more clearly by considering the arithmetic expectation of $\log X$ (i.e., $E_a[\log X]$). One might be tempted to immediately conclude that

$$E_a[\log X] = \int_0^\infty (\log x) f_X(x)dx.$$

However, we must be careful because $E_a[\log X]$ actually represents the arithmetic expectation of Y, where $Y = \log X$, and it is not immediately obvious that

$$\int_{-\infty}^\infty y f_Y(y)dy = \int_0^\infty (\log x) f_X(x)dx.$$

In fact, it is not too difficult to show that this must be the case since the transformation $Y = \log X$ is one-to-one. Indeed, using the relationship between density functions and distribution functions, and the fact that $\log X \le y \Longleftrightarrow X \le e^y$, we have

---

[11] It is straightforward to show that $\int_0^\infty S_X(x)dx = \int_0^\infty x f_X(x)dx$ using integration by parts. See §2.3 for details.

$$f_Y(y) = \frac{d}{dy} F_Y(y)$$

$$= \frac{d}{dy} \Pr(Y \leq y)$$

$$= \frac{d}{dy} \Pr(\log X \leq y)$$

$$= \frac{d}{dy} \Pr(X \leq e^y)$$

$$= \frac{d}{dy} F_X(e^y)$$

$$= f_X(e^y) \cdot e^y.$$

Hence,

$$\int_{-\infty}^{\infty} y f_Y(y) dy = \int_{-\infty}^{\infty} y f_X(e^y) \cdot e^y dy.$$

Applying the substitution[12] $y = \log x$ to the latter integral, we obtain

$$\int_{-\infty}^{\infty} y f_Y(y) dy = \int_0^{\infty} (\log x) f_X(x) dx,$$

which is the result claimed.

Consequently, for any positive random variable X,

$$E_a[\log X] = \int_0^{\infty} (\log x) f_X(x) dx$$

and thus,

$$E_g[X] = \exp(E_a[\log X]).$$

Taking the logarithm of both sides, we obtain the following important relationship:

$$\log(E_g[X]) = E_a[\log X].$$

Using similar arguments, one can show that

$$e^{E_a[X]} = E_g[e^X].$$

Relationships Between Arithmetic and Geometric Expectations

$$\log(E_g[X]) = E_a[\log X]$$

$$e^{E_a[X]} = E_g[e^X]$$

---

[12] Note the use of lowercase x and y here since x and y in this context represent real numbers, not random variables.

We will encounter many situations where it will be necessary to calculate the arithmetic expectation of some *function* of the random variable X under consideration. That is, it will be necessary to compute $E_a[h(X)]$ for some real valued function h. From our preceding comments, it should be clear that $E_a[h(X)]$ really represents the expectation of the transformed variable $Y = h(X)$. We will consider transformations of random variables in detail in Chapter 7. For the moment, we will state without any further explanation the following formula for calculating the expectation of a function of X:

Formula for Calculating the Expectation of a Function of a Random Variable

$$E[h(X)] = \int_{-\infty}^{\infty} h(x) f_X(x) dx$$

EXAMPLE 4: Consider the random variable Y given by $Y = X^2 + 1$, where X is an exponential random variable with parameter λ. (The exponential distribution was defined in Example 2.) Determine the expected value of Y.

We could calculate $E[Y]$ directly from the definition of expectation. However, this would require us first to determine the density function for Y (i.e., to determine f_Y), which could be difficult. Alternatively, we can calculate $E[Y]$ using the formula stated in the immediately preceding box. Indeed,

$$
\begin{aligned}
E[Y] &= E[X^2 + 1] \\
&= \int_{-\infty}^{\infty} (x^2 + 1) f_X(x) dx \\
&= \int_{0}^{\infty} (x^2 + 1) \lambda e^{-\lambda x} dx \\
&= \int_{0}^{\infty} x^2 \lambda e^{-\lambda x} dx + \int_{0}^{\infty} \lambda e^{-\lambda x} dx.
\end{aligned}
$$

Now $\int_0^\infty \lambda e^{-\lambda x} dx = 1$ because $\int_{-\infty}^{\infty} f_X(x) dx = 1$ and $\int_0^\infty x^2 \lambda e^{-\lambda x} dx = 2/\lambda^2$. (Use integration by parts twice.) Consequently,

$$E[Y] = \frac{2}{\lambda^2} + 1. \qquad \blacksquare$$

Properties of Arithmetic and Geometric Expectation

The arithmetic and geometric expectations have several important properties that we state here without proof.[13]

---

[13] Proofs will be given in Chapter 8.

Properties of Arithmetic Expectation

1. $E[mX + b] = mE[X] + b$, for all constants m, b.
2. $E[X + Y] = E[X] + E[Y]$.
3. $E[XY] = E[X] \cdot E[Y]$ if X and Y are independent.

The first two properties are quite natural and believable. Just think about what they mean! The product relationship $E[XY] = E[X]E[Y]$ for independent X and Y is somewhat less intuitive and need not hold when X and Y are dependent. Consider, for example, the random variables X, Y given by

$$X = \begin{cases} 1 & \text{with probability } 1/2, \\ -1 & \text{with probability } 1/2, \end{cases}$$

and $Y = X$ (the strongest form of dependence!). Note that $XY = 1$ with certainty so that $E[XY] = 1$, but $E[X] = 0 = E[Y]$.

Although there is no simple equality formula for $E[XY]$ that holds for all X and Y, one can give an inequality for $E[XY]$ that holds for all X and Y:

Cauchy-Schwarz Inequality

$$E[XY]^2 \le E[X^2] \cdot E[Y^2]$$

This inequality has an important implication in risk–reward analysis and will be derived in Chapter 8.

The analogs of the foregoing three properties for geometric expectation are as follows:

Properties of Geometric Expectation

1. $E_g[mX^b] = m \cdot (E_g[X])^b$, for all constants m, b.
2. $E_g[XY] = E_g[X] \cdot E_g[Y]$.
3. $E_g[X^Y] = E_g[X]^{E_a[Y]}$ if Y and $\log X$ are independent.

Each of these properties follows from the analogous property for arithmetic expectation using the relationship $\log(E_g[X]) = E_a[\log X]$ and properties of logarithms.

EXAMPLE 5: The losses X, Y on two distinct lines of business have distributions given by

$$f_X(x) = \frac{2}{(1+x)^3}, \qquad x > 0$$

and

$$f_Y(y) = \frac{3}{(1+y)^4}, \qquad y > 0.$$

Determine the expected size of the combined losses.

The desired quantity is $E[X + Y]$. One way to calculate this expectation is first to determine the distribution of $X + Y$. However, this approach can be tedious, and we don't yet have the tools to calculate the distribution of such a sum! A better way to calculate $E[X + Y]$ is to use the formula $E[X + Y] = E[X] + E[Y]$. Using integration by parts, one can show that

$$E[X] = 1 \text{ and } E[Y] = \frac{1}{2}.$$

Hence, $E[X + Y] = \frac{3}{2}$. Details of the integration are left to the reader.

Notice that the formula $E[X + Y] = E[X] + E[Y]$ simply asserts that the average of a sum is the sum of the averages, a fact that most people would consider fairly intuitive.

■

Markov's Inequality

In many applications, it is important to have good estimates for probabilities of the type $\Pr(X \geq a)$. An event of the form $X \geq a$ represents an "extreme" event, particularly when a is large. For example, if X represents a random loss, then for sufficiently large values of a, the event $X \geq a$ represents a catastrophic loss, and $\Pr(X \geq a)$ represents the probability of such a catastrophic loss. Insurance companies and regulators are generally quite concerned with the risk of catastrophic loss because such a loss could result in the company becoming insolvent.

Markov's inequality gives an upper bound for probabilities of the type $\Pr(X \geq a)$ when X is a nonnegative random variable with finite mean. This bound is particularly useful when there is no simple algebraic formula for $\Pr(X \geq a)$.

Markov's Inequality

Suppose that X is a nonnegative random variable with finite mean. Then for any $a > 0$,

$$\Pr(X \geq a) \leq \frac{E[X]}{a}.$$

PROOF OF MARKOV'S INEQUALITY: Markov's inequality is actually quite simple to derive. Consider for a fixed value of $a > 0$ the indicator random variable I defined by

$$I = \begin{cases} 1 & \text{if } X \geq a, \\ 0 & \text{otherwise.} \end{cases}$$

From the definition of I, it is clearly true that

$$I \leq \frac{X}{a}$$

since $X \geq 0$. Hence,

$$E[I] \leq \frac{E[X]}{a}.$$

However, from the definition of I, it is also true that

$$E[I] = \Pr(I = 1)$$
$$= \Pr(X \geq a).$$

Consequently,

$$\Pr(X \geq a) \leq \frac{E[X]}{a}$$

as claimed.   ∎

We now illustrate the use of Markov's inequality with two examples.

EXAMPLE 6:   The average amount of time that it takes to change a tire at a certain garage is 7 minutes. Give an estimate for the probability that it will take more than 35 minutes to change the next tire.

Let X be the amount of time in minutes to change the next tire. The desired probability is $\Pr(X > 35)$. We are not given the distribution function for X, so we cannot calculate $\Pr(X > 35)$ exactly. However, since we are given the mean of X, we can give an upper bound for $\Pr(X > 35)$ using Markov's inequality. Indeed,

$$\Pr(X > 35) \leq \frac{E[X]}{35} = \frac{7}{35} = \frac{1}{5} = .20.$$

Note that the actual value of $\Pr(X > 35)$ will depend on the form of the distribution of X and may be quite a bit smaller than $1/5$.   ∎

EXAMPLE 7:   Suppose that the time to change the tire in the previous example has an exponential distribution. Determine the exact value of the probability that it will take more than 35 minutes to change the tire.

Recall from Example 2 in §4.2.1 that the density function for an exponential distribution has the form

$$f_X(x) = \begin{cases} \lambda e^{-\lambda x} & \text{for } x \geq 0, \\ 0 & \text{otherwise,} \end{cases}$$

where $\lambda = 1/E[X]$. Since $E[X] = 7$ in this case, the density for the time to change the tire is

$$f_X(x) = \frac{1}{7} e^{-x/7}, \qquad x \geq 0$$

and the desired probability is

$$\Pr(X > 35) = \int_{35}^{\infty} \frac{1}{7} e^{-x/7} dx = e^{-5} \approx .006.$$

Note how much smaller this value is than the estimate provided by Markov's inequality!   ∎

4.2.2 Deviation from Expectation

Whether arithmetic or geometric, the expectation represents a theoretical limiting average value, which can be used as a starting point for setting insurance rates or making predictions about the long-run growth of a stock or a company. However, expectation by itself is insufficient for setting rates or making predictions because it does not incorporate the risk that actual experience will be different from the average. As an extreme example, consider the losses X_1, X_2 given by

$$X_1 = \$5000 \qquad \text{with probability 1;}$$

$$X_2 = \begin{cases} \$0 & \text{with probability } 1/2, \\ \$10,000 & \text{with probability } 1/2. \end{cases}$$

Clearly, the risk in underwriting the loss X_2 is much greater than the risk in underwriting the loss X_1 even though $E[X_1] = E[X_2]$.

In this section, we consider statistics for analyzing the *deviation* from expectation. As might be suspected, the statistics for measuring deviation depend on whether the expectation being considered is arithmetic or geometric.

In ordinary arithmetic, deviation is measured in primarily two ways: as a difference (i.e., using subtraction) or as a ratio (i.e., using division), with differences arising naturally in arithmetic contexts and ratios arising naturally in geometric contexts.[14] This suggests that measures of deviation from arithmetic expectation should be based on *differences*, whereas measures of deviation from geometric expectation should be based on *ratios*.

For simplicity in what follows, we only discuss statistics for measuring deviation from arithmetic expectation. Analogous statistics for measuring deviation from geometric expectation can be developed in a similar way.

Hence, let X be a random variable for which $E[X]$ exists and write μ_X for $E[X]$. Our objective is to develop a suitable statistic for measuring deviation from μ_X. Note that such a quantity is necessarily a nonrandom number depending only on X.[15]

Let's begin by considering the difference $X - \mu_X$. Note that $X - \mu_X$ is a random variable that measures the difference between the observed value of X and the expected value μ_X. Hence, a possible statistic for deviation is $E[X - \mu_X]$ (i.e., the average deviation from the mean). However, after some reflection, it is clear that this statistic is not a suitable measure of deviation since $E[X - \mu_X] = 0$. The reason $E[X - \mu_X] = 0$ is that the positive and negative differences "average out."

As an alternative to $E[X - \mu_X]$, one might consider the statistic $E[|X - \mu_X|]$. This statistic is known as the **mean absolute deviation** of X and is denoted by $\text{MAD}(X)$. Note that the problem of positive and negative differences averaging out does not arise with this statistic because the quantity $|X - \mu_X|$ does not distinguish between positive and negative differences from μ_X of the same magnitude. In fact, $E[|X - \mu_X|] > 0$ unless $X = \mu_X$ with certainty. Hence, $\text{MAD}(X)$ would appear to be a useful measure of deviation from μ_X. However, $\text{MAD}(X)$ is deficient for a subtle, but important, reason. To see why, let's interpret $E[|X - \mu_X|]$ as a *prediction error*.

---

[14] Consider, for example, how change is measured from term to term in arithmetic and geometric progressions.

[15] This is simply the definition of a statistic.

Suppose that the experiment on which X depends is repeated a large number of times, and each time, we predict that the observed value of X will be μ_X. Then the average absolute error in each prediction will be approximately $E[|X - \mu_X|]$.[16] Hence, $E[|X - \mu_X|]$ is the expected absolute error in predicting the value of X to be μ_X. More generally, $E[|X - \alpha|]$ for any given α is the expected absolute error in predicting the value of X to be α.

Now, it can be shown that $E[|X - \alpha|]$ is minimized when α is a *median* of X.[17] Hence, if prediction error is measured using $E[|X - \alpha|]$, then the "best guess" for X (i.e., the value of α for which the error $E[|X - \alpha|]$ is minimized) is a median, not μ_X, unless μ_X happens to be a median. However, μ_X has the interpretation of being the value of X that we *expect* to obtain on average when the experiment on which X depends is repeated a large number of times and values of X are accumulated arithmetically. Hence, if we want μ_X to be the preferred prediction value for X, we will need to measure prediction error in some way other than $E[|X - \alpha|]$. This suggests that the mean absolute deviation $MAD(X)$ is not a suitable measure of deviation from μ_X.

Two measures of deviation from μ_X that overcome these difficulties are $E[(X - \mu_X)^2]$ (the mean square deviation from the mean) and $\sqrt{E[(X - \mu_X)^2]}$ (the root mean square deviation from the mean). Indeed, it is easy to show that when prediction error is measured by $E[(X - \alpha)^2]$ or $\sqrt{E[(X - \alpha)^2]}$, the best guess for X is μ_X; in particular, $\min_\alpha E[(X - \alpha)^2] = E[(X - \mu_X)^2]$. The quantity $E[(X - \mu_X)^2]$ is called the **variance** of X and is denoted by $Var(X)$. Its square root $\sqrt{E[(X - \mu_X)^2]}$ is called the **standard deviation** of X and is denoted by $SD(X)$ or σ_X.

Definition of Variance and Standard Deviation

For any random variable X, the **variance** of X is the number $Var(X)$ given by

$$Var(X) = E[(X - \mu_X)^2]$$

and the **standard deviation** of X is the number σ_X given by

$$\sigma_X = \sqrt{E[(X - \mu_X)^2]}.$$

Here, μ_X is the expected value of X (i.e., $\mu_X = E[X]$).

Note that $SD(X)$ has the same dimensionality as X and μ_X, whereas $Var(X)$ does not. For this reason, $SD(X)$ is sometimes preferred over $Var(X)$ as a measure of deviation from μ_X. There may be other suitable measures of deviation from μ_X; however, $Var(X)$ and $SD(X)$ are the ones commonly used.

From the definition of variance and properties of expectation, one can obtain the following formula for $Var(X)$, which is useful for calculation:

---

[16] We say "approximately" because $E[|X - \mu_X|]$ represents a *limiting* average value.

[17] A **median** of X is a number m with the property that $Pr(X \le m) \ge 1/2$ and $Pr(X \ge m) \ge 1/2$. Note that more than one median is possible.

$$Var(X) = E[X^2] - E[X]^2.$$

Indeed, $Var(X) = E[(X - \mu_X)^2] = E[X^2 - 2\mu_X X + \mu_X^2] = E[X^2] - 2\mu_X E[X] + \mu_X^2 = E[X^2] - 2E[X] \cdot E[X] + E[X]^2 = E[X^2] - E[X]^2$, as claimed. One can also demonstrate the following important properties:

Properties of Variance

1. $Var(c) = 0$ for all constants c.
2. $Var(mX + b) = m^2 Var(X)$.
3. $Var(X + Y) = Var(X) + Var(Y)$ if X and Y are independent.
4. $Var(X + Y) = Var(X) + 2Cov(X, Y) + Var(Y)$.

The quantity $Cov(X, Y)$, which appears in property 4, is known as the **covariance** of X and Y and is defined to be

$$Cov(X, Y) = E[(X - \mu_X)(Y - \mu_Y)].$$

This quantity is a generalization of variance—indeed, $Cov(X, X) = Var(X)$—and has an important interpretation as a measure of association. It also arises when measuring deviation from the center of mass in bivariate distributions.

From the definition of covariance and using properties of expectation, one can show that[18]

$$Cov(X, Y) = E[XY] - E[X]E[Y].$$

Hence, if X and Y are independent, then $Cov(X, Y) = 0$. (See the properties of arithmetic expectation stated earlier in §4.2.1.) However, the converse is not true. Indeed, there are many examples of *dependent* random variables X, Y for which $Cov(X, Y) = 0$. We will discuss these and other properties of covariance in Chapter 8.

Properties of the standard deviation analogous to the ones for variance listed earlier can easily be written. The only property that requires some care is the one for the standard deviation of a multiple:

$$\text{SD}(mX) = |m|\,\text{SD}(X).$$

The reason for the absolute value is to ensure that the expression for $\text{SD}(mX)$ remains nonnegative.

Measures of deviation from the geometric expected value $E_g[X]$ analogous to $Var(X)$ and $\text{SD}(X)$ can be defined as follows:

$$Var_g(X) = e^{Var(\log X)},$$

$$\text{SD}_g(X) = e^{\text{SD}(\log X)}.$$

Note the similarity of these formulas to the definition for geometric expectation $E_g[X]$.

---

[18] The derivation is similar to the derivation of the formula $Var(X) = E[X^2] - E[X]^2$ given earlier.

EXAMPLE 1: A continuous random variable X has a density function of the form

$$f_X(x) = \lambda e^{-\lambda x}, \qquad x > 0.$$

Determine the variance of X in terms of λ.

This is the exponential distribution with parameter λ introduced in Example 2 in §4.2.1. Recall that for such distributions $E[X] = 1/\lambda$. Hence, using the definition of variance, we have

$$Var(X) = E\left[\left(X - \frac{1}{\lambda}\right)^2\right]$$

$$= \int_0^\infty \left(x - \frac{1}{\lambda}\right)^2 \lambda e^{-\lambda x} dx$$

$$= \int_0^\infty \lambda x^2 e^{-\lambda x} dx - \int_0^\infty 2x e^{-\lambda x} dx + \frac{1}{\lambda} \int_0^\infty e^{-\lambda x} dx.$$

Using integration by parts, we have

$$\int_0^\infty x e^{-\lambda x} dx = \frac{1}{\lambda^2}$$

and

$$\int_0^\infty x^2 e^{-\lambda x} dx = \frac{2}{\lambda^3}.$$

Hence,

$$Var(X) = \frac{2}{\lambda^2} - \frac{2}{\lambda^2} + \frac{1}{\lambda^2} = \frac{1}{\lambda^2}.$$

Alternatively, using the formula $Var(X) = E[X^2] - E[X]^2$, we have

$$Var(X) = \int_0^\infty x^2 \lambda e^{-\lambda x} dx - E[X]^2$$

$$= \lambda\left(\frac{2}{\lambda^3}\right) - \frac{1}{\lambda^2}$$

$$= \frac{1}{\lambda^2}.$$

Note that it is generally simpler to calculate the variance using the formula $Var(X) = E[X^2] - E[X]^2$ than to calculate it directly from the definition. ■

EXAMPLE 2: Determine the variance of a truncated exponential distribution with parameters λ and m.

Recall that the truncated exponential distribution has a point mass of size $e^{-\lambda m}$ at $x = m$ and a continuous distribution of probability on the interval $(0, m)$ given by the density function

$$f_X(x) = \lambda e^{-\lambda x}, \qquad 0 < x < m.$$

In Example 3 in §4.2.1, we showed that the expected value for a truncated exponential distribution is

$$E[X] = \frac{1}{\lambda}(1 - e^{-\lambda m}).$$

Hence, to determine the variance, we need only calculate $E[X^2]$. Since X has a mixed distribution,

$$E[X^2] = \int_{\substack{\text{continuous}\\ \text{part}}} x^2 f_X(x)dx + \sum_{\substack{\text{discrete}\\ \text{part}}} x^2 p_X(x)$$

$$= \int_0^m x^2 \lambda e^{-\lambda x}dx + m^2 e^{-\lambda m}.$$

Using integration by parts twice, we have

$$\int_0^m x^2 \cdot \lambda e^{-\lambda x}dx = -m^2 e^{-\lambda m} + \int_0^m 2xe^{-\lambda x}dx$$

$$= -m^2 e^{-\lambda m} - 2\frac{m}{\lambda}e^{-\lambda m} + \frac{2}{\lambda}\int_0^m e^{-\lambda x}dx$$

$$= \frac{2}{\lambda^2}\left(1 - e^{-\lambda m}(1 + \lambda m)\right) - m^2 e^{-\lambda m}.$$

Thus,

$$E[X^2] = \frac{2}{\lambda^2}\left(1 - e^{-\lambda m}(1 + \lambda m)\right).$$

Now,

$$E[X]^2 = \left\{\frac{1}{\lambda}(1 - e^{-\lambda m})\right\}^2.$$

Consequently,

$$Var(X) = \frac{1}{\lambda^2}(1 - 2\lambda m e^{-\lambda m} - e^{-2\lambda m}).$$

Details are left to the reader.

Notice that $Var(X) \to 1/\lambda^2$ as $m \to \infty$. ■

EXAMPLE 3:   Show that the formula

$$Var(X + Y) = Var(X) + 2Cov(X, Y) + Var(Y)$$

for the variance of a sum holds using the formulas for the expectation of a sum and product stated in §4.2.1.

From the definition of variance,

$$Var(X + Y) = E[((X + Y) - E[X + Y])^2].$$

From the formula for the expected value of a sum (i.e., $E[X + Y] = E[X] + E[Y]$), it follows that

$$Var(X + Y) = E[((X - E[X]) + (Y - E[Y]))^2].$$

Now,

$$((X - E[X]) + (Y - E[Y]))^2 = (X - E[X])^2 + 2(X - E[X])(Y - E[Y])$$
$$+ (Y - E[Y])^2.$$

Hence, applying the formula for the expected value of a sum again to this latter equation, we have

$$Var(X + Y) = E[(X - E[X])^2] + E[2(X - E[X])(Y - E[Y])]$$
$$+ E[(Y - E[Y])^2].$$

Consequently, using the fact that $E[mX] = mE[X]$ for constants m and the definition of variance and covariance, we have

$$Var(X + Y) = Var(X) + 2Cov(X, Y) + Var(Y),$$

which is the formula we were required to demonstrate.   ∎

Chebyshev's Inequality

In §4.2.1, we saw that Markov's inequality provides an estimate for probabilities of the type $\Pr(X \geq a)$ when X is a nonnegative random variable. We now describe an inequality for "catastrophic" probabilities of the type $\Pr(|X - \mu_X| \geq k)$ which is suitable for random variables with both positive and negative extreme values.

Chebyshev's Inequality

Suppose that X is a random variable with finite mean and finite variance. Then for any $k > 0$,

$$\Pr(|X - \mu_X| \geq k) \leq \frac{\sigma_X^2}{k^2}.$$

This inequality follows immediately by applying Markov's inequality to the random variable $Y = (X - \mu_X)^2$. Details are left to the reader.

Chebyshev's inequality has important consequences for engineering practice. Indeed, putting $k = 3\sigma_X$, we have

$$\Pr(|X - \mu_X| \geq 3\sigma_X) \leq \frac{1}{9},$$

or equivalently,

$$\Pr(\mu_X - 3\sigma_X < X < \mu_X + 3\sigma_X) \geq \frac{8}{9}.$$

Hence, for any distribution with finite mean and variance, most of the probability is within 3 standard deviations of the mean. This surprising result holds regardless of the amount of symmetry in the distribution!

EXAMPLE 4:   The **Laplacian distribution** is a probability distribution that is used frequently in the computer analysis of speech and images. Its probability density function has the form

$$f_X(x) = \frac{1}{\sqrt{2}\sigma} \exp(-\sqrt{2}|x|/\sigma), \qquad x \in \mathbf{R},$$

where σ is the standard deviation of X. Determine an exact value for $\Pr(|X - \mu_X| \geq 3\sigma_X)$. Compare this value to the estimate given by Chebyshev's inequality.

It is clear from the form of f_X that the distribution of X is symmetric about zero. Hence, assuming that the distribution has a mean, we must have $\mu_X = 0$. Moreover, to determine the desired probability, we need only consider probabilities of the form $\Pr(X \geq a)$ for $a > 0$. From the definition of f_X, we have (for $a > 0$)

$$\begin{aligned}
\Pr(X \geq a) &= \frac{1}{\sqrt{2}\sigma} \int_a^\infty e^{-\sqrt{2}x/\sigma} dx \\
&= -\frac{1}{2} e^{-\sqrt{2}x/\sigma} \Big|_a^\infty \\
&= \frac{1}{2} e^{-a\sqrt{2}/\sigma}.
\end{aligned}$$

Hence, using the symmetry of f_X, we have

$$\begin{aligned}
\Pr(|X - \mu_X| \geq 3\sigma_X) &= \Pr(X \geq 3\sigma) + \Pr(X \leq -3\sigma) \\
&= 2\Pr(X \geq 3\sigma) \\
&= e^{-3\sqrt{2}} \\
&\approx 0.0143696.
\end{aligned}$$

This is quite a bit smaller than the estimate $(1/9 \approx 0.11)$ provided by Chebyshev's inequality.

We leave it as an exercise for the reader to show that the mean of X exists and the standard deviation of X is σ. ∎

4.2.3 Higher Moments

The statistics of variance and standard deviation discussed in the previous section provide measures of deviation from the arithmetic expectation and can be used to measure the risk that actual experience will be different from the historical average. However, are they *sufficient* measures of risk?

To answer this question, consider the random variables X_1, X_2 with respective densities, as illustrated in Figure 4.16. Although it may not be completely obvious from the graphs, the densities are mirror images of each other about the vertical axis,

$$f_{X_1}(x) = f_{X_2}(-x) \qquad \text{for all } x,$$

and the means of both distributions are zero (i.e., $E[X_1] = 0 = E[X_2]$). Hence, by the symmetry in the definition of variance,

$$Var(X_1) = Var(X_2)$$

and also $\mathrm{SD}(X_1) = \mathrm{SD}(X_2)$. That is, the variances and the standard deviations are equal.

However, the risk profiles of X_1 and X_2 are entirely different. Indeed, if X_1 and X_2 are monetary payoffs on particular uncertain prospects, then X_1 represents a prospect with limited downside risk and unlimited upside potential, whereas X_2 represents a

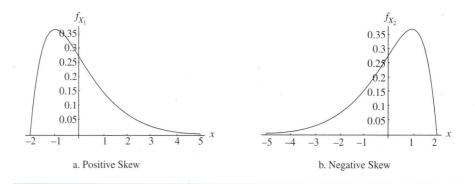

FIGURE 4.16 An Example of Two Distributions with the Same Mean and the Same Variance

prospect with limited upside potential and unlimited downside risk. Clearly, X_2 is much riskier than X_1. However, this is not indicated by the variance or standard deviation. Consequently, the variance and standard deviation are not sufficient measures of risk.

The difficulty with variance and standard deviation is that they don't treat positive and negative deviations from μ_X any differently in the averaging process. However, in many problems in insurance and investments, the sign of the deviation is important. Consider, for example, the difference between an unexpected gain and an unexpected loss in the stock market.

One possible statistic that takes the sign of the deviation into account is $E[(X - \mu_X)^3]$. This statistic gives a measure of the *skewness* of a distribution. Indeed, $E[(X - \mu_X)^3] > 0$ when the distribution is skewed to the right, as in Figure 4.16a, and $E[(X - \mu_X)^3] < 0$ when the distribution is skewed to the left, as in Figure 4.16b. When the distribution is symmetric about the mean—that is when $f_X(x - \mu_X) = f_X(\mu_X - x)$—then $E[(X - \mu_X)^3] = 0$. Distributions that are skewed to the right are often referred to as *positively* skewed, whereas distributions skewed to the left are referred to as *negatively* skewed.

It is fairly clear that the statistic $E[(X - \mu_X)^3]$ can be used to describe the *sign* of the skew. However, it is less effective at describing the *degree* of the skew. The reason for this deficiency is that $E[(X - \mu_X)^3]$ is not invariant under a change of scale. Indeed, for $k > 0$,

$$E[(kX - \mu_{kX})^3] = k^3 \, E[(X - \mu_X)^3]$$
$$\neq E[(X - \mu_X)^3] \quad \text{unless } k = 1.$$

However, changing the scale of X (e.g., by changing monetary units from dollars to pounds) should not affect the size of the skew.[19] This suggests that another measure of skewness should be used.

The measure of skewness that is generally adopted is the statistic γ_X defined by

$$\gamma_X = E\left[\left(\frac{X - \mu_X}{\sigma_X}\right)^3\right].$$

---

[19] The mean and variance are affected by a change of scale; however, unlike the skew, mean and variance should change.

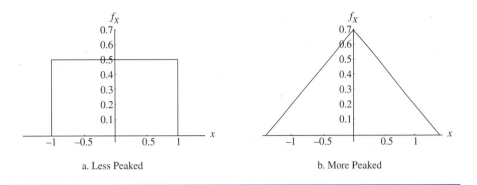

a. Less Peaked   b. More Peaked

FIGURE 4.17 An Example of Two Distributions with the Same Mean, the Same Variance, and the Same Skewness

This statistic, which is known as the **skewness**[20] of X, has the following important properties:

Properties of the Skewness

1. $\gamma_X = 0$ if the distribution of X is symmetric.
2. $\gamma_{mX+b} = sgn(m) \cdot \gamma_X$.
3. $\gamma_{X+Y} = \dfrac{\gamma_X \sigma_X^3 + \gamma_Y \sigma_Y^3}{\sigma_{X+Y}^3}$ if X and Y are independent.

Note that $sgn(m)$ is the *sign function* defined by

$$sgn(m) = \begin{cases} 1 & \text{if } m > 0, \\ -1 & \text{if } m < 0, \\ 0 & \text{if } m = 0. \end{cases}$$

Properties 1 and 2 follow directly from the definition of the skewness. Property 3 follows from an analogous formula for cumulant generating functions to be discussed in §4.3.1.

Are skewness and variance, if taken together, sufficient to describe risk? A consideration of the densities illustrated in Figure 4.17 reveals that they are not sufficient. Indeed, both distributions in Figure 4.17, being symmetric about the origin, have mean and skewness equal to zero. Moreover, one can show that the variances of the two distributions are the same.[21] However, the risk profiles of the two distributions are clearly different. One can express this difference by saying that the distribution in Figure 4.17b is more "peaked" than the one in Figure 4.17a.

---

[20] To avoid confusion with other measures of skewness, such as the one in the previous paragraph, from here on the phrase "the skewness" will refer to the statistic γ_X unless otherwise stated.

[21] To do this, you would need the equations of the densities, which we haven't specified. As an exercise, try to determine the definitions of the densities of the type shown for which the variances of the distributions are equal.

A statistic that measures such differences in peakedness is

$$E\left[\left(\frac{X - \mu_X}{\sigma_X}\right)^4\right].$$

This statistic is known as the **kurtosis** of X.[22]

In light of our earlier comments, one should not expect the mean, variance, skewness, and kurtosis (when taken together) to uniquely determine the probability distribution of a random variable. (We leave it as an exercise for the reader to demonstrate this.) However, distributions with the same mean, variance, skewness, and kurtosis will nevertheless have similar, if not identical, risk profiles.

By considering more statistics of the form $E[(X - \mu_X)^k]$, one can obtain progressively better descriptions of the distribution of X. Intuition suggests that if we specify the entire set of statistics $\{E[(X - \mu_X)^k], k = 2, 3, \ldots\}$ along with the mean μ_X, then the distribution of X will be uniquely determined. Under reasonable conditions, this is true.

Instead of specifying the set $\{E[(X - \mu_X)^k], k = 2, 3, \ldots\} \cup \{\mu_X\}$, it is often more convenient to specify the set $\{E[X^k] : k = 1, 2, \ldots\}$. The two methods are equivalent because

$$E[(X - \mu_X)^k] = \sum_{i=0}^{k} \binom{k}{i} E[X^i] \cdot \mu_X^{k-i}, \quad k = 2, 3, \ldots$$

by the binomial theorem.

The statistic $E[X^k]$ is called the **kth moment** of X and the statistic $E[(X - \mu_X)^k]$ is called the **kth central moment** of X. Collectively, the $E[X^k]$ are referred to as the **moments** of X.

The proof that an arbitrary distribution is uniquely determined by its moments requires techniques in mathematical analysis that lie beyond the scope of this book. However, the proof of this characterization for finite distributions with a known set of possible values is straightforward and will now be given.

PROOF OF CHARACTERIZATION PROPERTY IN A SPECIAL CASE: Suppose that X is a random variable with possible values $a_1, \ldots, a_k$, and let $p_j = \Pr(X = a_j)$. Suppose further that the values $a_1, \ldots, a_k$ are all known. Then we claim that the probabilities $p_1, \ldots, p_k$ are uniquely determined by the moments $E[X], E[X^2], \ldots, E[X^{k-1}]$. To see this, note that

$$E[X^j] = a_1^j p_1 + \cdots + a_k^j p_k, \quad \text{for } j = 1, \ldots, k - 1.$$

Adding the equation $p_1 + \cdots + p_k = 1$ to this collection of $k - 1$ equations, we obtain the linear system

---

[22] Some authors use kurtosis measures that are slightly different from this one.

$$\begin{bmatrix} 1 & 1 & \cdots & 1 \\ a_1 & a_2 & \cdots & a_k \\ a_1^2 & a_2^2 & \cdots & a_k^2 \\ \vdots & \vdots & & \vdots \\ a_1^{k-1} & a_2^{k-1} & \cdots & a_k^{k-1} \end{bmatrix} \begin{bmatrix} p_1 \\ p_2 \\ \vdots \\ p_k \end{bmatrix} = \begin{bmatrix} 1 \\ E[X] \\ \vdots \\ E[X^{k-1}] \end{bmatrix}.$$

Since $a_1, \ldots, a_k$ are distinct, the coefficient matrix is invertible. (Its determinant is a Vandermonde determinant.) Consequently, $p_1 \ldots, p_k$ are uniquely determined by $E[X], E[X^2], \ldots, E[X^{k-1}]$ as claimed. It follows that a distribution with a finite number of possible values, all of which are known, is uniquely determined by its moments. ∎

While the complete set of moments is required in theory to characterize an arbitrary distribution, in practice, knowledge of the mean, variance, and skewness is generally adequate to obtain a good approximation of the distribution. Consequently, most practitioners focus attention on the statistics μ_X, σ_X, and γ_X.

In Chapter 8, we will consider some important distributional approximations that are based on knowledge of the mean, variance, and skewness.

4.2.4 Exercises

1. The probability mass function of a discrete random variable X is given by

$$p_X(x) = \begin{cases} \frac{x}{15} & \text{for } x = 1, 2, 3, 4, 5, \\ 0 & \text{otherwise.} \end{cases}$$

   Determine the expected value of $X^2 + 1$.

2. The probability density function of a continuous random variable X is given by

$$f_X(x) = \begin{cases} \frac{3}{x^4} & \text{for } x \geq 1, \\ 0 & \text{otherwise.} \end{cases}$$

   Determine the expected value of $\dfrac{X^2 + 1}{X}$.

3. A mixed distribution has probability masses of $1/4$ each at $x = -2$ and $x = 2$ and a continuous distribution of probability on $(-1, 1)$ given by the density

$$f_X(x) = \begin{cases} \frac{1+x}{2} & \text{for } -1 < x < 0, \\ \frac{1-x}{2} & \text{for } 0 \leq x < 1, \\ 0 & \text{otherwise.} \end{cases}$$

   Determine the expected value of $|X + 1|$.

4. On a particular type of insurance policy, an insurance company will indemnify policyholders for losses up to a maximum of m. Thus, for example, if $m = 1000$ and the loss amount is 1500, the insurer would pay 1000. The insurer models claim payouts in thousands of dollars on this policy type using a truncated exponential distribution with parameter λ and cap m. If there were no cap, the expected payout per policy would be $2000. The insurer would like the expected payout to be no more than $1000 per policy. At what level should the insurer set the cap to accomplish this?

5. The probability mass function for a particular discrete random variable N is given by

$$p_N(n) = \frac{\lambda^n e^{-\lambda}}{n!}, \qquad n = 0, 1, 2, \ldots.$$

Determine formulas for $E[N]$ and $Var(N)$ in terms of λ. *Hint:* Use the fact that $e^x = \sum_{n=0}^{\infty} \frac{x^n}{n!}$.

6. Consider the family of discrete random variables X_d, for $d = 1, 2, \ldots$, with probability mass functions given by

$$p_{X_d}(x) = \begin{cases} 1 - \frac{1}{d} & \text{if } x = 0, \\ \frac{1}{d} & \text{if } x = d. \end{cases}$$

a. Graph the mass functions in the cases $d = 1$, $d = 10$, $d = 100$. Qualitatively describe how the distributions change as d increases.

b. Calculate $E[X_d]$ and $Var(X_d)$. What happens as $d \to \infty$?

c. Would you say that the standard deviation is always a good measure of distribution spread? What measure of dispersion would you suggest using here instead?

7. A biased coin is tossed n times. Suppose that the probability of getting heads on a single toss is p. Let X be the number of heads obtained.

a. Give an algebraic formula for the probability mass function of X.

b. What do you think the expected value of X should be? *Hint:* Consider the interpretation of X.

c. Use the algebraic formula for p_X derived in part a to prove that your assertion in part b is correct. *Hint:* Show that

$$x\binom{n}{x} = n\binom{n-1}{x-1}$$

and deduce that

$$\sum_{x=0}^{n} x p_X(x) = np \sum_{x=1}^{n} \binom{n-1}{x-1} p^{x-1}(1-p)^{(n-1)-(x-1)} = np.$$

d. How does the probability mass function p_X change when p is replaced by $(1-p)$? What is the effect of this change on the variance of X? What does this tell you about the variance of X?

e. Calculate $Var(X)$ for $(n, p) = (5, 1/2)$, $(10, 1/2)$, $(10, 1/10)$, $(10, 1/5)$ and examine the value of $Var(X)/E[X]$ in each case. Using these data, give your best guess for a general formula for $Var(X)$. Your formula will necessarily involve the parameters n and p.

8. Suppose that $X_1, \ldots, X_n$ are independent and identically distributed with $X_j \sim X$. Show that

$$Var(X_1 + \cdots + X_n) = n Var(X)$$

and

$$Var(nX) = n^2 Var(X).$$

Conclude that, for $n > 1$,

$$X_1 + \cdots + X_n \nsim nX.$$

9. Suppose that $X_1, \ldots, X_n$ are independent and identically distributed with common skewness γ. Show that

$$\gamma_{X_1 + \cdots + X_n} = \frac{\gamma}{\sqrt{n}} \to 0 \qquad \text{as } n \to \infty.$$

Explain the meaning of this result.

10. The per period return X on a particular security is given by

$$X = \begin{cases} \mu + \sigma & \text{with probability } 1/2, \\ \mu - \sigma & \text{with probability } 1/2. \end{cases}$$

Let $Y = 1 + X$ be the corresponding per period accumulation factor.

a. Show that $E_a[Y] = 1 + \mu$ and $Var_a(Y) = \sigma^2$.

b. Show that $E_g[Y] = (1 + 2\mu + \mu^2 - \sigma^2)^{1/2}$.

c. Using the fact that

$$(1 + z)^\alpha = 1 + \frac{\alpha}{1!}z + \frac{\alpha(\alpha - 1)}{2!}z^2 + \frac{\alpha(\alpha - 1)(\alpha - 2)}{3!}z^3 + \cdots$$

for $|z| < 1$, show that, to second order,

$$E_g[Y] \approx 1 + \mu - \frac{1}{2}\sigma^2.$$

For simplicity, assume that $|2\mu + \mu^2 - \sigma^2| < 1$.

d. Deduce that, to second order,

$$E_g[Y] \approx E_a[Y] - \frac{1}{2}Var_a(Y).$$

4.3 Alternative Ways of Specifying Probability Distributions

So far, we have discussed three ways of describing the distribution of probability for a random variable X: the distribution function F_X, the mass density function f_X, and the set of moments $\{E[X^k] : k = 1, 2, \ldots\}$. In this section, we consider four additional ways of specifying probability distributions. As we will soon see, each of these additional ways is important in its own right.

4.3.1 Moment and Cumulant Generating Functions

As we have stated, the set of moments $\{E[X^k] : k = 1, 2, \ldots\}$ characterizes the distribution of X under reasonable conditions. We can write this information in a more usable form by introducing the formal algebraic expression

$$1 + E[X]t + E[X^2]\frac{t^2}{2!} + E[X^3]\frac{t^3}{3!} + \cdots,$$

where (for the moment) $1, t, t^2, t^3, \ldots$ are placeholders and convergence of the sum in an analytic sense is irrelevant. Such an algebraic expression is referred to in general as a *generating function*.

Using formal algebraic manipulation, we have

$$1 + E[X]t + E[X^2]\frac{t^2}{2!} + \cdots = E[1 + X + \frac{(Xt)^2}{2!} + \cdots]$$
$$= E[e^{tX}].$$

Hence, all of the information about the distribution of X is contained in the quantity $E[e^{tX}]$ considered as a function of t. This suggests the following definition:

Definition of Moment Generating Function

For any random variable X, the **moment generating function** of X is the function M_X defined by

$$M_X(t) = E[e^{tX}]$$

for all values of t for which this makes sense.

Note that $M_X(t)$ is defined without any reference to the earlier infinite sum.

Moment generating functions are extremely useful tools in probability and have the following important properties:

Properties of the Moment Generating Function

1. M_X characterizes the distribution of X under reasonable conditions.
2. $M_X^{(k)}(0) = E[X^k]$ for $k = 0, 1, 2, \ldots$.
3. $M_{aX+b}(t) = M_X(at) \cdot e^{bt}$.
4. $M_{X+Y}(t) = M_X(t) \cdot M_Y(t)$ if X and Y are independent.

Properties 1 and 4 are particularly important. Indeed, by combining them, it may be possible to determine the distribution of an independent sum without having to perform any integration. We illustrate this in the examples that follow.

Properties 2, 3, and 4 are fairly straightfoward to derive using the definition of M_X and properties of expectation. For example, property 2 follows from the fact that

$$M_X^{(k)}(t) = E[X^k e^{tX}] \qquad k = 1, 2, \ldots,$$

which can be obtained by successively differentiating the integral for $E[e^{tX}]$ with respect to t. However, the proof of property 1 requires techniques in mathematical analysis that are beyond the scope of this book. For details, the interested reader should consult a book on advanced probability.

EXAMPLE 1: A random variable X with density function

$$f_X(x) = \frac{1}{\sqrt{2\pi}\,\sigma} e^{-(x-\mu)^2/2\sigma^2}, \qquad x \in \mathbf{R}$$

is said to have a **normal distribution** with parameters μ and $\sigma > 0$. Show that the moment generating function of X is given by

$$M_X(t) = \exp(\mu t + \frac{1}{2}\sigma^2 t^2).$$

Deduce that μ is the mean of X and σ is the standard deviation of X.

From the definition of moment generating function, we have

$$M_X(t) = E[e^{tX}]$$
$$= \int_{-\infty}^{\infty} e^{tx} \cdot \frac{1}{\sqrt{2\pi}\sigma} e^{-(x-\mu)^2/2\sigma^2} dx.$$

Multiplying and dividing the latter integral by $\exp(\mu t + \frac{1}{2}\sigma^2 t^2)$ and then rewriting the integrand with $\mu + \sigma^2 t$ taking the role of μ, we obtain

$$M_X(t) = \exp(\mu t + \frac{1}{2}\sigma^2 t^2) \int_{-\infty}^{\infty} \frac{1}{\sqrt{2\pi}\sigma} e^{-(x-(\mu+\sigma^2 t))^2/2\sigma^2} dx.$$

However,

$$\int_{-\infty}^{\infty} \frac{1}{\sqrt{2\pi}\sigma} e^{-(x-(\mu+\sigma^2 t))^2/2\sigma^2} dx$$
$$= \int_{-\infty}^{\infty} \frac{1}{\sqrt{2\pi}\sigma} e^{-(y-\mu)^2/2\sigma^2} dy$$
$$= \int_{-\infty}^{\infty} f_X(y) dy.$$

(Apply the substitution $y = x - \sigma^2 t$ to the integral.) Moreover, $\int_{-\infty}^{\infty} f_X(y) dy = 1$ because f_X is a probability density.[23] Hence,

$$M_X(t) = \exp(\mu t + \frac{1}{2}\sigma^2 t^2)$$

as claimed.

From the formula $M_X(t) = \exp(\mu t + \frac{1}{2}\sigma^2 t^2)$ and the general formula $E[X^k] = M_X^{(k)}(0)$, it is straightforward to show that μ and σ are the mean and standard deviation, respectively. Indeed, since

$$M_X'(t) = (\mu + \sigma^2 t)\exp(\mu t + \frac{1}{2}\sigma^2 t^2)$$

and

$$M_X''(t) = \sigma^2 \exp(\mu t + \frac{1}{2}\sigma^2 t^2) + (\mu + \sigma^2 t)^2 \exp(\mu t + \frac{1}{2}\sigma^2 t^2),$$

it follows that

$$E[X] = M_X'(0) = \mu$$

---

[23] That f_X is a probability density follows from the nontrivial result that $\int_{-\infty}^{\infty} e^{-x^2} dx = \sqrt{\pi}$. Taking this result as given, it is straightforward to show that f_X is a density. Details are left to the reader.

and

$$E[X^2] = M''_X(0) = \sigma^2 + \mu^2.$$

Hence, from the formula $Var(X) = E[X^2] - E[X]^2$, it follows that $Var(X) = \sigma^2$. Consequently, μ and σ are the mean and standard deviation as claimed.   ∎

EXAMPLE 2:   By considering moment generating functions, show that the sum of two independent normal random variables also has a normal distribution.

Let X_1, X_2 be two independent normal random variables with respective means μ_1, μ_2 and respective standard deviations σ_1, σ_2. From the previous example, the moment generating functions of X_1 and X_2 are, respectively,

$$M_{X_1}(t) = \exp(\mu_1 t + \frac{1}{2}\sigma_1^2 t^2),$$

$$M_{X_2}(t) = \exp(\mu_2 t + \frac{1}{2}\sigma_2^2 t^2).$$

Since X_1 and X_2 are independent, the moment generating function of the sum $X_1 + X_2$ is given by $M_{X_1+X_2}(t) = M_{X_1}(t)M_{X_2}(t)$ (property 4 listed earlier). Hence,

$$M_{X_1+X_2}(t) = \exp(\mu_1 t + \frac{1}{2}\sigma_1^2 t^2) \cdot \exp(\mu_2 t + \frac{1}{2}\sigma_2^2 t^2)$$

$$= \exp((\mu_1 + \mu_2)t + \frac{1}{2}(\sigma_1^2 + \sigma_2^2)t^2).$$

However, the latter function has the form of the moment generating function of a normal random variable with mean $\mu_1 + \mu_2$ and standard deviation $\sqrt{\sigma_1^2 + \sigma_2^2}$. Hence, by the uniqueness property of moment generating functions (property 1), the sum $X_1 + X_2$ must have a normal distribution with mean $\mu_1 + \mu_2$ and standard deviation $\sqrt{\sigma_1^2 + \sigma_2^2}$. Notice how simple it was to conclude that the sum of two independent normal distributions is again normal, using moment generating functions!   ∎

Closely related to the moment generating function, and unfortunately neglected in many probability texts, is the *cumulant generating function*.

Definition of Cumulant Generating Function

For any random variable X, the **cumulant generating function** is the function ψ_X defined by

$$\psi_X(t) = \log M_X(t)$$

for all t for which this makes sense.

Cumulant generating functions have the following properties, which are analogous to the ones stated for moment generating functions.

Properties of the Cumulant Generating Function

1. ψ_X characterizes the distribution of X under reasonable conditions.
2.

$$\psi_X^{(k)}(0) = \begin{cases} 0 & \text{for } k = 0, \\ E[X] & \text{for } k = 1, \\ Var(X) & \text{for } k = 2, \\ E[(X - \mu_X)^3] & \text{for } k = 3; \end{cases}$$

however, $\psi_X^{(k)}(0) \neq E[(X - \mu_X)^k]$ for $k \geq 4$.
3. $\psi_{aX+b}(t) = \psi_X(at) + bt$.
4. $\psi_{X+Y}(t) = \psi_X(t) + \psi_Y(t)$ if X and Y are independent.

Property 2 illustrates that cumulant generating functions are particularly useful for determining the statistics μ_X, σ_X, and γ_X. Indeed,

$$\mu_X = \psi_X'(0),$$
$$\sigma_X = \psi_X^{(2)}(0)^{1/2},$$
$$\gamma_X = \frac{\psi_X^{(3)}(0)}{\psi_X^{(2)}(0)^{3/2}}.$$

Moreover, properties 2 and 4, when combined, provide formulas for the variance and skewness of independent sums. Indeed, if X and Y are independent, then

$$\sigma_{X+Y}^2 = \sigma_X^2 + \sigma_Y^2,$$
$$\gamma_{X+Y} = \frac{\gamma_X \sigma_X^3 + \gamma_Y \sigma_Y^3}{\sigma_{X+Y}^3},$$

which are precisely the formulas for the variance and skewness of independent sums stated in §4.2.2 and §4.2.3. Note that $\psi_X^{(k)}(0) = E[(X - \mu_X)^k]$ for $k = 2$ and 3, but not for $k \geq 4$. To emphasize that the $\psi_X^{(k)}(0)$ are generally not central moments, the number $\psi_X^{(k)}(0)$ is called the **kth cumulant** of X.

There is no simple generic formula for the derivatives $\psi_X^{(k)}$ analogous to the formula for the $M_X^{(k)}$. However, the derivatives $\psi_X^{(k)}$ can be determined iteratively without too much difficulty by introducing the functions

$$m_k(t) = \frac{M_X^{(k)}(t)}{M_X(t)}, \qquad k = 1, 2, \dots.$$

Note that $m_1(t) = \psi_X'(t)$ and $m_k'(t) = m_{k+1}(t) - m_k(t)m_1(t)$. Hence, $\psi_X^{(k)}(t)$ can be written as a polynomial in $m_1(t), \dots, m_k(t)$.

Economic Interpretations of Moment and Cumulant Generating Functions (Optional)

The moment and cumulant generating functions have interesting interpretations when X represents an uncertain continuously compounded[24] rate of return. Indeed, $M_X(t)$ is the arithmetic expected accumulation of a \$1 investment at time t subject to a continuously compounded growth rate of X per unit time. On the other hand, $\psi_X(t)$ is the certain continuously compounded rate of return over the time interval $[0, t]$ that a risk-free investment must have for it to be considered as desirable, in the eyes of a risk-neutral[25] investor, as a risky prospect with uncertain per unit time return X.

Let's consider these interpretations more closely. From elementary financial mathematics, the accumulation of one unit of currency at time t subject to a continuously compounded per unit time rate of return X is

$$e^{Xt}.$$

Hence, if X is uncertain, $E[e^{Xt}]$ represents the arithmetic expected accumulation of one unit of currency at time t. However, $M_X(t) = E[e^{tX}]$ by definition. Consequently, the moment generating function has the interpretation that has been stated.

Now, consider an investor who makes investment decisions solely on the basis of arithmetic expected accumulations, with high expected accumulation preferred over low expected accumulation. Such an investor is called a **risk-neutral** investor.[26] Suppose that this investor can choose between a prospect with uncertain per unit time return X and a risk-free investment with per unit time return r. Suppose further that the investor's time horizon is at time t and is fixed. Then the investor will:

- prefer the risky prospect to the risk-free investment if $E[e^{Xt}] > e^{rt}$;
- prefer the risk-free investment to the risky prospect if $E[e^{Xt}] < e^{rt}$;
- be indifferent between the risky prospect and the risk-free investment if $E[e^{Xt}] = e^{rt}$.

Define r^* to be the number such that $E[e^{Xt}] = e^{r^*t}$. Then a risk-neutral investor will prefer the risky prospect to the risk-free alternative if and only if $r^* > r$ and will be indifferent between the two if and only if $r^* = r$. Consequently, r^* is the certain rate of return that a risk-neutral investor considers equivalent to X.

From the definitions of $\psi_X(t)$ and $M_X(t)$, we have

$$E[e^{Xt}] = e^{\psi_X(t)}.$$

Hence, from the preceding remarks, we see that $\psi_X(t)/t$ is the certain continuously compounded per unit time rate of return that a risk-neutral investor considers equivalent to X. Consequently, $\psi_X(t)$ is the certain continuously compounded rate of return over the

---

[24] If an investment is compounded n times per year at the rate of i per annum, then the value of \$1 at the end of the year is $(1 + \frac{i}{n})^n$. The more frequent the compounding (i.e., the larger the value of n), the greater the accumulation for a given interest rate i. In the limit as $n \to \infty$, the accumulation becomes e^i. This type of compounding is referred to as *continuous* compounding. Hence, if \$1 is invested at a continuously compounded rate of r per annum, then the accumulation of this investment t years later is e^{rt}.

[25] The term *risk-neutral* is defined in a moment.

[26] The reader may wish to compare the term risk-neutral with the terms *risk-averse* and *risk-tolerant* introduced in §2.2. We will have more to say about risk preference when we discuss the Markowitz portfolio selection model in Chapter 10.

time interval $[0, t]$ that a risk-free investment must have for it to be considered equivalent, in the eyes of a risk-neutral investor, to a risky prospect with uncertain per unit time return X. This is precisely the interpretation of $\psi_X(t)$ stated earlier.

Note that the certain rate of return "equivalent" (in the sense just discussed) to a given uncertain rate X depends on the time horizon. We can see this explicitly by considering an investment whose per annum continuously compounded rate of return X has a normal distribution with mean μ and standard deviation σ. For such a normal distribution, $\psi_X(t) = \mu t + \frac{1}{2}\sigma^2 t^2$ (see Example 1). Hence, over the time period $[0, t]$, the per annum rate on an equivalent risk-free investment must be $\psi_X(t)/t = \mu + \frac{1}{2}\sigma^2 t$, which varies with the time horizon t. Since this per annum rate increases with t, we see that the investment with uncertain per annum rate X becomes more attractive in the eyes of a risk-neutral investor as the time horizon increases.

In view of the interpretations of $M_X(t)$ and $\psi_X(t)$ just discussed, the algebraic properties of moment and cumulant generating functions stated earlier become very natural. For example, since $\psi_X(t)$ can be interpreted as a continuously compounded interest rate and since continuously compounded interest rates are additive, the additivity property of cumulant generating functions,

$$\psi_{X+Y}(t) = \psi_X(t) + \psi_Y(t),$$

for independent X, Y should not be surprising. Likewise, since knowledge of $\psi_X(t)$ for each t can be interpreted as knowledge of certain rates equivalent to X at every time t, the uniqueness property of cumulant generating functions (i.e., that ψ_X characterizes the distribution of X) should now seem very natural.

Graphical Properties of Moment and Cumulant Generating Functions (Optional)

In addition to the algebraic properties discussed earlier, moment and cumulant generating functions have important graphical properties, which we now consider.

Graphical Properties of Moment Generating Functions

1. $M_X(t) > 0$ for all t, where $M_X(t)$ is defined.
2. If $X \geq 0$ with certainty, then $M_X'(t) \geq 0$, and M_X is increasing on its domain of definition. If $X \leq 0$ with certainty, then $M_X'(t) \leq 0$, and M_X is decreasing on its domain of definition.
3. $M_X''(t) > 0$ for all t, where $M_X(t)$ is defined, unless X is a point mass at the origin, in which case $M_X''(t) = 0$ for all t. Hence, M_X is concave up on its domain of definition.
4. $M_X(0) = 1$, and the slope of M_X at $t = 0$ is equal to μ_X.

These properties are illustrated in Figure 4.18 for random variables with (a) only positive values, (b) only negative values, and (c) both positive and negative values.

Cumulant generating functions have similar graphical properties:

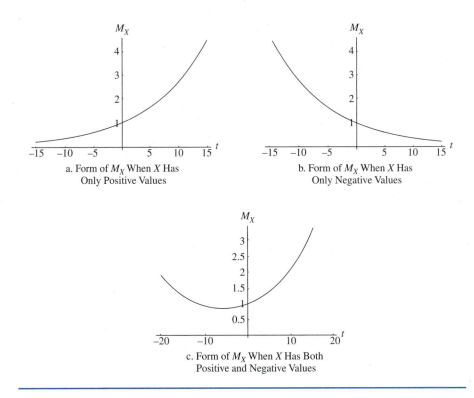

a. Form of M_X When X Has
Only Positive Values

b. Form of M_X When X Has
Only Negative Values

c. Form of M_X When X Has Both
Positive and Negative Values

FIGURE 4.18   Graphical Properties of Moment Generating Functions

Graphical Properties of Cumulant Generating Functions

1. If $X \geq 0$ with certainty, then $\psi_X'(t) \geq 0$, and ψ_X is increasing on its domain of definition. If $X \leq 0$ with certainty, then $\psi_X'(t) \leq 0$, and ψ_X is decreasing on its domain of definition.
2. $\psi_X''(t) \geq 0$ for all t, where $\psi_X(t)$ is defined.
3. $\psi_X(0) = 0$, and the slope of ψ_X at $t = 0$ is equal to μ_X.

Note that, unlike M_X, ψ_X has both positive and negative values because $\psi_X(t) = \log M_X(t)$. Note also that since $\psi_X'(t) = M_X'(t)/M_X(t)$ and $M_X(t) > 0$, the signs of $\psi_X'(t)$ and $M_X'(t)$ are identical for each t; hence, ψ_X and M_X increase and decrease in tandem. The concavity property, $\psi_X''(t) \geq 0$ for all t, is not immediately obvious, since

$$\psi_X''(t) = \frac{M_X''(t)}{M_X(t)} - \left(\frac{M_X'(t)}{M_X(t)}\right)^2 .$$

However, by applying the Cauchy-Schwarz inequality to the random variables $Xe^{tX/2}$ and $e^{tX/2}$, it becomes clear that this property must indeed hold.[27] Note that if X is a point mass at a, then $\psi_X(t) = at$ and $\psi_X''(t) = 0$ for all t.

The Moments of e^X (Optional)

Moment and cumulant generating functions have other important properties that we will develop throughout the text as necessary. However, there is one property relating the moments of e^X to the moment generating function of X that is worth mentioning here.

For any random variable X, the moments of $Y = e^X$ can be determined from the moment generating function of X as follows:

$$E[Y^k] = E[e^{kX}] = M_X(k), \qquad k = 1, 2, \ldots.$$

More generally, the moments of $Y_t = e^{Xt}$ are given by

$$E[Y_t^k] = E[e^{ktX}] = M_X(kt), \qquad k = 1, 2, \ldots.$$

When X has a normal distribution, the variable $Y = e^X$ is said to have a lognormal distribution.[28] Lognormal distributions arise frequently in models of stock prices. From the given formula for $E[Y^k]$ and the fact that $M_X(t) = \exp(\mu t + \frac{1}{2}\sigma^2 t^2)$ for normal random variables X (see Example 1), it follows that the lognormal random variable $Y = e^X$ has mean and variance given by

$$E[Y] = e^{\mu + \frac{1}{2}\sigma^2},$$
$$Var(Y) = E[Y]^2(e^{\sigma^2} - 1).$$

Note that $E[Y]$ which is the arithmetic mean of Y, is not equal to e^μ, as one might naively think. However, e^μ can still be interpreted as an average for Y. Indeed, from the relationship $\log E_g[Y] = E_a[\log Y]$ given in §4.2.1, it is clear that e^μ is the *geometric* mean of Y, i.e. $E_g[Y] = e^\mu$.

Worked Examples

We conclude our discussion of moment and cumulant generating functions with some worked examples.

EXAMPLE 3: The moment generating function for a particular random variable X is given by

$$M_X(t) = \exp\{\lambda(e^t - 1)\}, \qquad t \in \mathbf{R}.$$

Determine the mean, variance, and skewness of X in terms of λ.

The mean, variance, and skewness for any random variable X are given by

---

[27] The Cauchy-Schwarz inequality was stated in §4.2.1 and will be proved in §8.3.2.

[28] A **lognormal** random variable is one whose logarithm is normally distributed. Note that the definition of a normal distribution was given in Example 1 earlier in this section.

$$\mu_X = \psi'_X(0),$$
$$\sigma_X^2 = \psi''_X(0),$$
$$\gamma_X = \frac{\psi_X^{(3)}(0)}{\psi''_X(0)^{3/2}}.$$

Hence, we need only determine $\psi'_X(0)$, $\psi''_X(0)$, and $\psi_X^{(3)}(0)$. From the given information,

$$\psi_X(t) = \log M_X(t) = \lambda(e^t - 1).$$

Hence,

$$\psi'_X(t) = \psi''_X(t) = \psi_X^{(3)}(t) = \lambda e^t.$$

Consequently,

$$\psi'_X(0) = \psi''_X(0) = \psi_X^{(3)}(0) = \lambda,$$

and thus,

$$\mu_X = \lambda,$$
$$\sigma_X^2 = \lambda,$$

and

$$\gamma_X = \frac{1}{\lambda^{1/2}}. \qquad \blacksquare$$

EXAMPLE 4: Suppose that X is a discrete random variable with probability mass function

$$p_X(x) = \begin{cases} \frac{1}{6} & \text{if } x = -1, \\ \frac{1}{3} & \text{if } x = 0, \\ \frac{1}{2} & \text{if } x = 2. \end{cases}$$

Determine the moment generating function of X.

From the definition of moment generating function, we have

$$M_X(t) = E[e^{tX}]$$
$$= \frac{1}{6}e^{-t} + \frac{1}{3}e^0 + \frac{1}{2}e^{2t}$$
$$= \frac{1}{6}e^{-t} + \frac{1}{3} + \frac{1}{2}e^{2t}. \qquad \blacksquare$$

EXAMPLE 5: The moment generating function of a discrete random variable X is given by

$$M_X(t) = \frac{1}{4} + \frac{1}{2}e^t + \frac{1}{4}e^{3t}.$$

Determine the probability that $X > 1$.

This moment generating function has the same form (weighted sum of exponentials) as the moment generating function of the previous example. Since moment generating functions uniquely characterize probability distributions, we surmise that X has a discrete distribution with probability masses of $\frac{1}{4}$, $\frac{1}{2}$, and $\frac{1}{4}$ at $x = 0$, $x = 1$, and $x = 3$, respectively,

$$p_X(x) = \begin{cases} \frac{1}{4} & \text{if } x = 0, 3, \\ \frac{1}{2} & \text{if } x = 1. \end{cases}$$

A quick calculation verifies that the given M_X is the moment generating function associated with the latter probability mass function. Consequently, the required probability is 1/4. ∎

EXAMPLE 6:   Determine the moment generating function for the random variable X with distribution function

$$F_X(x) = \begin{cases} 1 - e^{-\lambda x} & \text{for } 0 \le x < m, \\ 1 & \text{for } x \ge m. \end{cases}$$

Note that this is the **truncated exponential distribution** with parameters λ and m introduced in Example 3 of §4.2.1. From the definitions of moment generating function and expectation for mixed random variables, we have

$$M_X(t) = \int_0^m e^{tx} \cdot \lambda e^{-\lambda x} dx + e^{tm} \cdot e^{-\lambda m}.$$

Hence, for $t \ne \lambda$,

$$M_X(t) = \lambda \left. \frac{e^{(t-\lambda)x}}{t - \lambda} \right|_{x=0}^{m} + e^{-m(\lambda - t)}$$

$$= \frac{\lambda}{\lambda - t}(1 - e^{-m(\lambda - t)}) + e^{-m(\lambda - t)}$$

$$= \frac{\lambda}{\lambda - t} - \frac{t}{\lambda - t} e^{-m(\lambda - t)}$$

while for $t = \lambda$, $M_X(t) = \lambda m + 1$. Note that M_X is actually continuous since

$$\frac{\lambda}{\lambda - t}(1 - e^{-m(\lambda - t)}) + e^{-m(\lambda - t)} \to \lambda m + 1$$

as $t \to \lambda$. (Expand the exponential function in a Taylor series and take limits.)

Notice that if $m = \infty$ (i.e., no truncation), we obtain the formula

$$M_X(t) = \frac{\lambda}{\lambda - t}, \qquad t < \lambda.$$

Here, the restriction $t < \lambda$ is real because $E[e^{tX}]$ does not exist for $t \ge \lambda$. (The integral diverges.) However, using the formula $M_X(t) = \lambda/(\lambda - t)$, it is possible to extend the definition of M_X to all $t \ne \lambda$.

We will often extend the definition of moment generating functions in this way when $M_X(t)$ is not defined for all t, but a closed form for $M_X(t)$ that makes sense for a larger domain of t values is available. One should keep in mind, however, that the graphical properties of the moment generating function stated earlier need not hold for such extended moment generating functions. Indeed, from Figure 4.19, it is clear that

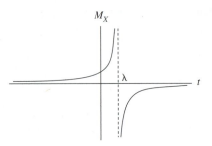

FIGURE 4.19 Extended Moment Generating Function for the Exponential Distribution

the concavity property of moment generating functions does not hold for all values of the extended moment generating function M_X of this example. ∎

EXAMPLE 7:   A random variable X with density function

$$f_X(x) = \begin{cases} \dfrac{\lambda^r x^{r-1} e^{-\lambda x}}{\Gamma(r)} & \text{for } x \geq 0, \\ 0 & \text{for } x < 0, \end{cases}$$

is said to have a **gamma distribution** with parameters r and λ. Here, $\Gamma(\cdot)$ denotes the gamma function.[29] The gamma function is defined such that the given f_X is a density function, i.e., such that $\int_0^\infty f_X(x) = 1$. Show that the moment generating function of X is given by

$$M_X(t) = \left(\frac{\lambda}{\lambda - t}\right)^r, \qquad t < \lambda.$$

From the definition of moment generating function, we have

$$M_X(t) = E[e^{tX}]$$
$$= \int_0^\infty e^{tx} \frac{\lambda^r x^{r-1} e^{-\lambda x}}{\Gamma(r)} dx$$
$$= \left(\frac{\lambda}{\lambda - t}\right)^r \int_0^\infty \frac{(\lambda - t)^r x^{r-1} e^{-(\lambda - t)x}}{\Gamma(r)} dx.$$

However, the integrand of the latter integral has the form of the density for a gamma random variable with parameters r and $\lambda - t$ and, hence, must integrate to 1. Consequently,

$$M_X(t) = \left(\frac{\lambda}{\lambda - t}\right)^r, \qquad t < \lambda$$

as claimed. ∎

EXAMPLE 8:   By considering moment generating functions, show that the sum of independent exponential random variables with the same parameter λ has a gamma distribution.

---

[29] A discussion of the gamma function and its properties is given in Appendix A.

Recall that the exponential random variable with parameter λ has density function $f_X(x) = \lambda e^{-\lambda x}$, $x > 0$ and moment generating function $M_X(t) = \lambda/(\lambda - t)$. The formula for the moment generating function follows from Example 6 on the truncated exponential distribution with $m = \infty$ or from Example 7 on the gamma distribution with $r = 1$. Suppose that $X_1, \ldots, X_r$ are independent exponential random variables with common parameter λ. Then

$$M_{X_1 + \cdots + X_r}(t) = M_{X_1}(t) \cdots M_{X_r}(t)$$

$$= \left(\frac{\lambda}{\lambda - t} \right)^r.$$

However, from the previous example, we recognize this latter formula as being the moment generating function of a gamma distribution with parameters r and λ. Hence, by the uniqueness of moment generating functions, the sum $X_1 + \cdots + X_r$ has a gamma distribution with parameters r and λ as claimed.　■

4.3.2　Survival and Hazard Functions

When the random quantity X under consideration is a lifetime (e.g., life of a person, machine, mortgage, financial contract), one is most often interested in the probabilities of survival to given ages and the instantaneous probabilities of failure at given ages. These probabilities can be formalized by using *survival functions* and *hazard functions,* respectively.

Definition of Survival Function and Hazard Function

For any random variable X, the **survival function** of X is the function S_X defined by

$$S_X(x) = \Pr(X > x) \qquad \text{for all } x.$$

The **hazard function** of X is the function λ_X defined by

$$\lambda_X(x) = \lim_{\Delta x \to 0} \frac{\Pr(x < X < x + \Delta x | X > x)}{\Delta x}$$

for all x where this makes sense.

The hazard function is sometimes referred to as the **failure rate function,** particularly in reliability engineering. In actuarial science, the hazard function is known as the **force of mortality.**[30] In reliability engineering, the survival function is often referred to as the **reliability function.**

Note that the survival function is connected to the distribution function F_X by the relationship

$$S_X(x) = 1 - F_X(x).$$

---

[30] Actuaries have such cheerful terminology!

Hence, the survival function completely determines the distribution of X. The hazard function also determines the distribution of X, as we will soon see.

From the definition of λ_X, it is clear that the hazard function assigns to each x an instantaneous relative frequency density. However, unlike the density function f_X, the values of λ_X are *conditional* relative frequency densities. Indeed, when X is a lifetime, $\lambda_X(x)$ is the instantaneous relative frequency density of failure at age x given survival to age x.

Hazard functions arise naturally in statistical studies because the subjects of such studies are generally known to have attained particular ages at the inception of the study. Hazard functions are also often easier to obtain from statistical data than survival functions. The reason is that most statistical studies out of necessity are unable to follow subjects for the duration of their lifetimes. Indeed, it is quite possible for some subjects to outlive the investigator who initiated the study in the first place!

For any random variable X, the hazard function λ_X and the survival function S_X are related as follows:

$$\lambda_X(x) = \frac{-S_X'(x)}{S_X(x)}.$$

Indeed, using the definition of conditional probability, the fact that $f_X = F_X' = -S_X'$, and the approximation $\Pr(x < X < x + \Delta x) \approx f_X(x)\Delta x$, we have

$$
\begin{aligned}
\lambda_X(x) &= \lim_{\Delta x \to 0} \frac{\Pr(x < X < x + \Delta x \mid X > x)}{\Delta x} \\
&= \lim_{\Delta x \to 0} \frac{\Pr(x < X < x + \Delta x)}{\Pr(X > x) \cdot \Delta x} \\
&= \lim_{\Delta x \to 0} \frac{f_X(x)\Delta x}{S_X(x)\Delta x} \\
&= \frac{f_X(x)}{S_X(x)} \\
&= \frac{-S_X'(x)}{S_X(x)}
\end{aligned}
$$

as claimed.

From this relationship, we can obtain an explicit formula for S_X in terms of λ_X by performing an appropriate integration. Indeed, rewriting the equation $\lambda_X(x) = -S_X'(x)/S_X(x)$ in the form

$$\lambda_X(t) = -\frac{d}{dt}\log S_X(t) \qquad \text{for all } t,$$

integrating this latter equation from $t = -\infty$ to x, and using the fact that $\lim_{x \to -\infty} S_X(x) = 1$, we obtain

$$\int_{-\infty}^{x} \lambda_X(t)dt = -\left.\log S_X(t)\right|_{t=-\infty}^{x},$$

from which it follows that

$$S_X(x) = \exp\left(-\int_{-\infty}^{x} \lambda_X(t)dt\right).$$

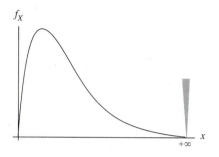

FIGURE 4.20   Possible Density for an Insurer's Lifetime

In particular, when X represents a lifetime, so that $X \geq 0$ with certainty,

$$S_X(x) = \exp\left(-\int_0^x \lambda_X(t)dt\right), \qquad x \geq 0.$$

Let's highlight these relationships for future reference:

Relationship Between Survival Functions and Hazard Functions

$$\lambda_X(x) = \frac{-S_X'(x)}{S_X(x)}$$

$$S_X(x) = \exp\left(-\int_{-\infty}^x \lambda_X(t)dt\right)$$

An important class of lifetime distributions is the one for which λ_X is a power function,

$$\lambda_X(x) = kx^n, \qquad x \geq 0.$$

A distribution with this type of hazard function is known as a **Weibull distribution.** Properties of Weibull distributions will be discussed in §6.2.1.

In actuarial science, survival and hazard functions often arise in connection with the problem of insurer solvency. Indeed, when analyzing the solvency of an insurer, a natural quantity to consider is the time until failure of the company. Unlike the typical lifetime distribution for a person or physical object, the lifetime distribution for an insurer may have a positive probability mass at infinity since the company needn't necessarily fail. (See Figure 4.20 for the graph of a possible density for an insurer's lifetime. Note in this graph, that the inverted shaded triangle represents the probability mass at infinity.) This can complicate the analysis of an insurer's lifetime random variable to some extent.

Properties of Survival Functions

Survival functions have properties that are analogous to the properties of distribution functions:

1. S_X is decreasing.
2. S_X is right continuous.
3. $\lim_{x \to -\infty} S_X(x) = 1$ and $\lim_{x \to +\infty} S_X(x) = 0$.

However, they have an additional property that is useful for calculating the expectation of a lifetime random variable. For any nonnegative random variable X,

$$E[X] = \int_0^\infty S_X(x)dx,$$

provided that $\lim_{x \to \infty} x \cdot S_X(x) = 0$.

The derivation of this fact is straightforward using integration by parts and the fact that $f_X = -S'_X$. Indeed,

$$E[X] = \int_0^\infty x f_X(x)dx$$

$$= x(-S_X(x))\big|_0^\infty - \int_0^\infty (-S_X(x))dx$$

$$= \int_0^\infty S_X(x)dx$$

as claimed.

EXAMPLE 1: Using the formula for $E[X]$ just derived, it is easy to calculate the expected value for the truncated exponential distribution introduced in Example 3 of §4.2.1. Indeed, if X has a truncated exponential distribution with parameters λ and m, then

$$S_X(x) = \begin{cases} e^{-\lambda x} & \text{for } 0 < x < m, \\ 0 & \text{for } x \geq m, \end{cases}$$

and thus,

$$E[X] = \int_0^m e^{-\lambda x}dx = \frac{1}{\lambda}(1 - e^{-\lambda m}),$$

which is precisely the formula derived before. ∎

4.3.3 Exercises

1. Consider a coin for which the probability of getting heads on a single toss is p.

   a. Suppose that the coin is tossed once and let I be an indicator of heads on the toss. Determine the moment generating function for I.
   b. Suppose that the coin is tossed n times in succession and let I_j indicate heads on the jth toss. Determine the moment generating function for $I_1 + \cdots + I_n$. What does this sum represent?
   c. Let X be the number of heads obtained in n tosses of the coin. Determine $E[X]$ and $Var(X)$.

2. The moment generating function of a random variable X is given by

$$M_X(t) = \frac{1}{4} + \frac{1}{8}e^t + \frac{3}{8}e^{2t} + \frac{1}{4}e^{4t}.$$

Determine the probability that X lies between -1 and 3.

3. The moment generating function for a random variable X is given by

$$M_X(t) = \frac{1}{4} + \frac{3}{4} \cdot \frac{1}{1-t}, \qquad t > 1.$$

Determine formulas for the density function f_X and the distribution function F_X. Sketch graphs for f_X and F_X.

4. Two cards are drawn from a well-shuffled deck of playing cards from which the spades and clubs have been removed, and are placed face down on a table. The cards are turned up one at a time and the suit of each card is observed. Let X_1 and X_2 be defined as follows:

$$X_1 = \begin{cases} 1 & \text{if the first card is a heart,} \\ 0 & \text{if the first card is not a heart;} \end{cases}$$

$$X_2 = \begin{cases} 1 & \text{if the second card is a heart,} \\ 0 & \text{if the second card is not a heart.} \end{cases}$$

Further, put $Y = X_1 + X_2$. Determine M_{X_1}, M_{X_2}, and M_Y. Is it true that $M_{X_1+X_2}(t) = M_{X_1}(t)M_{X_2}(t)$?

5. In this question, you will derive two useful formulas for calculating the skewness of a distribution. It is important to know both of these formulas because, when one is difficult to use, the other is often easier to apply.

a. Show that for any random variable X,

$$E[(X - \mu_X)^2] = E[X^2] - E[X]^2,$$
$$E[(X - \mu_X)^3] = E[X^3] - 3E[X^2]E[X] + 2E[X]^3,$$
$$E[(X - \mu_X)^4] = E[X^4] - 4E[X^3]E[X]$$
$$+ 6E[X^2]E[X]^2 - 3E[X]^4.$$

Hint: Expand $(X - \mu_X)^j$ in each case.

b. Put

$$m_k(t) = \frac{M_X^{(k)}(t)}{M_X(t)}.$$

Show that

$$m_k'(t) = m_{k+1}(t) - m_k(t)m_1(t).$$

c. Using the relationship derived in part b, show that

$$\psi_X'(t) = m_1(t),$$
$$\psi_X''(t) = m_2(t) - m_1(t)^2,$$
$$\psi_X^{(3)}(t) = m_3(t) - 3m_2(t)m_1(t) + 2m_1(t)^3,$$
$$\psi_X^{(4)}(t) = m_4(t) - 4m_3(t)m_1(t) - 3m_2(t)^2$$
$$+ 12m_2(t)m_1(t)^2 - 6m_1(t)^4.$$

Deduce that

$$\psi_X'(0) = \mu_X,$$
$$\psi_X''(0) = E[(X - \mu_X)^2],$$
$$\psi_X^{(3)}(0) = E[(X - \mu_X)^3],$$

but that

$$\psi_X^{(4)}(0) \neq E[(X - \mu_X)^4].$$

d. Deduce that the skewness γ_X of a random variable X may be calculated by one of the following formulas:

$$\gamma_X = \frac{E[X^3] - 3E[X^2]E[X] + 2E[X]^3}{\sigma_X^3},$$

$$\gamma_X = \frac{\psi_X^{(3)}(0)}{\psi_X^{(2)}(0)^{3/2}}.$$

6. The moment generating functions $M_X(t)$ for particular random variables X follow. For each of these random variables, determine the mean, variance, and skewness of the associated probability distribution.

a. $M_X(t) = \dfrac{1}{1-t}$

b. $M_X(t) = \dfrac{1}{2}e^t + \dfrac{1}{2}e^{-2t}$

c. $M_X(t) = \left(\dfrac{3}{4}e^t + \dfrac{1}{4}\right)^3$

d. $M_X(t) = e^{t^2/2}$

e. $M_X(t) = \exp(e^t - 1)$

7. The annual continuously compounded rate of return R on a particular investment is to be determined by some random experiment. The distribution of R is as follows:

$$R = \begin{cases} 5\% & \text{with probability } .25, \\ 6\% & \text{with probability } .50, \\ 8\% & \text{with probability } .25. \end{cases}$$

Determine the rate on a risk-free investment that a risk-neutral investor with a time horizon of 1 year would consider equivalent to the given investment opportunity. What would be the rate on an equivalent risk-free investment if the investor's time horizon were only 6 months?

8. The failure rate for a particular device is given by

$$\lambda_X(x) = 3x^2.$$

Determine the probability that $X > 1$.

9. The Pareto distribution (to be discussed in Chapter 6) is used quite frequently in insurance to model claim size because its "heavy tail" captures the risk of unexpected large claims quite well. A special case of this distribution has the survival function

$$S_X(x) = \left(\frac{1}{1+x}\right)^2, \qquad x \geq 0.$$

Determine the expected value of this distribution. What is true of the variance of this particular Pareto distribution?

10. The density function of the random variable X is given by

$$f_X(x) = \begin{cases} \frac{1}{2} & \text{for } -1 < x < 1, \\ 0 & \text{otherwise.} \end{cases}$$

Determine the moment generating function for X.

4.4 Chapter Summary

In this chapter, we have discussed seven different ways of specifying the probability distribution of a random variable X: the distribution function F_X, the mass density function f_X, the set of moments $\{E[X^k] : k = 1, 2, \ldots\}$, the moment generating function M_X, the cumulant generating function ψ_X, the survival function S_X, and the hazard function λ_X. Each of these specifications is important in its own right.

For any random variable X, the **distribution function** is the function F_X defined by

$$F_X(x) = \Pr(X \le x) \qquad \text{for all } x.$$

The distribution function has the following three properties:

1. $F_X(x_1) \le F_X(x_2)$ whenever $x_1 \le x_2$; that is, F_X is increasing.
2. $\lim_{\varepsilon \to 0^+} F_X(x + \varepsilon) = F_X(x)$ for all x; that is, F_X is continuous from the right at every point.
3. $\lim_{x \to -\infty} F_X(x) = 0$ and $\lim_{x \to +\infty} F_X(x) = 1$; that is, $F_X(x) \to 0$ as $x \to -\infty$ and $F_X(x) \to 1$ as $x \to +\infty$.

These properties actually characterize distribution functions in the sense that under reasonable conditions every function F with these properties is the distribution function for some random variable X (i.e., $F = F_X$ for some X).

There are three types of probability distributions: continuous, discrete, and mixed. A random variable is **continuous** if its distribution function is continuous at every point; it is **discrete** if its distribution function is a step function; and it is **mixed** if its distribution function is not a step function but is discontinuous at some point. Mixed distributions are hybrids of continuous and discrete distributions and arise quite frequently in insurance applications, particularly in connection with contract caps and deductibles; they also arise frequently in electrical engineering.

Discrete distributions can be specified by a probability mass function while continuous distributions can be specified by a probability density function. For any discrete random variable X, the **probability mass function** is the function p_X defined by

$$p_X(x) = \Pr(X = x) \qquad \text{for all } x.$$

For any continuous random variable X, the **probability density function** is the function f_X defined by

$$f_X(x) = \lim_{\varepsilon \to 0} \frac{\Pr(x - \varepsilon/2 \le X \le x + \varepsilon/2)}{\varepsilon} \qquad \text{for all } x.$$

The latter function measures at each point x the instantaneous relative frequency density at that point. The probability mass function and the probability density function are related to the distribution function in the following way:

$$p_X(x) = F_X(x) - F_X(x^-),$$
$$f_X(x) = F_X'(x).$$

Mixed distributions can be specified using a hybrid mass density function.

Descriptive statistics summarize particular information about a probability distribution in a numerical form, and provide us with a convenient way to make comparisons between different distributions. Three important statistics are the **mean**, **variance**, and **skewness**. For any random variable X, these statistics are defined as follows:

$$\mu_X = E[X],$$
$$\sigma_X^2 = E[(X - \mu_X)^2],$$
$$\gamma_X = E\left[\left(\frac{X - \mu_X}{\sigma_X}\right)^3\right].$$

Here $E[X]$ denotes the **expected value** of x, and is defined by

$$E[X] = \int_{-\infty}^{\infty} x f_X(x) dx.$$

For any function g, the expected value of $g(X)$ can be calculated using the formula

$$E[g(X)] = \int_{-\infty}^{\infty} g(x) f_X(x)\, dx.$$

The mean is a measure of central tendency for a probability distribution, whereas the variance is a measure of dispersion. When X represents a monetary payoff, σ_X^2 provides a measure of the risk that the actual payoff will be significantly different from the expected payoff μ_X. However, σ_X^2 is not a sufficient measure of risk since it does not distinguish between positive and negative deviations from μ_X in the averaging process. The skewness provides a measure of the extent to which deviations from μ_X are positive or negative.

For any random variable X, the **k-th moment** of X is the statistic $E[X^k]$. An important result in probability is that, under reasonable conditions, the complete set of moments $\{E[X^k] : k = 1, 2, \ldots\}$ uniquely determines the distribution of X.

For any random variable X, the **moment generating function** M_X and the **cumulant generating function** ψ_X are defined as follows:

$$M_X(t) = E[e^{tX}],$$
$$\psi_X(t) = \log M_X(t).$$

Since

$$M_X(t) = 1 + E[X]t + E[X^2]\frac{t^2}{2!} + E[X^3]\frac{t^3}{3!} + \cdots,$$

the moment generating function is completely determined by, and completely determines, the set of moments $\{E[X^k] : k = 1, 2, \ldots\}$. Hence, under reasonable conditions, M_X, and also ψ_X, uniquely determine the distribution of X.

The moment and cumulant generating functions have the following important properties:

$$M_{X+Y}(t) = M_X(t) \cdot M_Y(t) \qquad \text{if } X, Y \text{ are independent;}$$
$$\psi_{X+Y}(t) = \psi_X(t) + \psi_Y(t) \qquad \text{if } X, Y \text{ are independent.}$$

These properties are useful for determining the distribution of an independent sum when the moment or cumulant generating function of the sum has a recognizable form. The cumulant generating function also provides a convenient way to calculate the mean, variance, and skewness of a distribution:

$$\mu_X = \psi'_X(0),$$
$$\sigma_X^2 = \psi''_X(0),$$
$$\gamma_X = \frac{\psi_X^{(3)}(0)}{\psi_X^{(2)}(0)^{3/2}}.$$

The moment and cumulant generating functions have interesting economic interpretations when X represents an uncertain continuously compounded rate of return. Indeed, $M_X(t)$ is the arithmetic expected accumulation of a \$1 investment at time t subject to a continuously compounded growth rate of X per unit time, and $\psi_X(t)$ is the continuously compounded rate of return over the time interval $[0, t]$ that a risk-free investment must have for it to be considered as desirable, in the eyes of a risk-neutral investor, as a risky prospect with uncertain per unit time return X.

For any random variable X, the **survival function** of X is the function S_X defined by

$$S_X(x) = \Pr(X > x) \qquad \text{for all } x.$$

The **hazard function** of X is the function λ_X defined by

$$\lambda_X(x) = \lim_{\Delta x \to 0} \frac{\Pr(x < X < x + \Delta x | X > x)}{\Delta x}$$

for all x where this makes sense. The survival function is connected to the distribution function F_X by the relationship

$$S_X(x) = 1 - F_X(x)$$

and to the hazard function by the relationships

$$\lambda_X(x) = \frac{-S'_X(x)}{S_X(x)},$$
$$S_X(x) = \exp\left(-\int_{-\infty}^{x} \lambda_X(t)\, dt\right).$$

Hence, the survival function and the hazard function uniquely determine the distribution of X.

The survival and hazard functions arise frequently when X represents a lifetime. In this context, $S_X(x)$ is the probability of survival to age x and $\lambda_X(x)$ is the instantaneous relative frequency density of failure at age x given survival to age x. The survival function is also known as the **reliability function,** particularly in systems engineering, and the hazard function is known variously as the **failure rate function** or the **force of mortality.** When X is a nonnegative random variable, the survival function provides the following convenient formula for calculating the expected value of X:

$$E[X] = \int_0^\infty S_X(x)\,dx.$$

In addition to discussing these seven ways of specifying a probability distribution, we also introduced the concepts of *bivariate distribution* and *conditional distribution* in this chapter. These concepts enable one to study the interaction of two or more random variables and provide a mechanism for discussing the distributions of sums and products of random variables.

For any two random variables X, Y, the **bivariate distribution function** for X and Y is the function $F_{X,Y}$ defined by

$$F_{X,Y}(x, y) = \Pr(X \le x \text{ and } Y \le y),$$

and the **bivariate density function** is the function $f_{X,Y}$ given by

$$f_{X,Y}(x, y) = \frac{\partial^2}{\partial x \partial y} F_{X,Y}(x, y)$$

provided that this makes sense.

The **conditional distribution** of X given Y is defined by the density function $f_{X|Y}$ in the following way:

$$f_{X|Y=y}(x) = \frac{f_{X,Y}(x, y)}{f_Y(y)}.$$

This concept extends the concept of conditional probability, which was introduced in Chapter 3, to random variables and probability distributions, and enables us to give distributional analogs for the **law of total probability** and **Bayes' theorem**:

$$f_X(x) = \int_{-\infty}^\infty f_{X|Y=y}(x) f_Y(y)\,dy,$$

$$f_{X|Y=y}(x) = \frac{f_{Y|X=x}(y) f_X(x)}{f_Y(y)}.$$

Two important concepts are the concepts of *independence* and *identical distribution*. Random variables X and Y are **independent** if their joint density is the product of the marginal densities; that is,

$$f_{X,Y}(x, y) = f_X(x) \cdot f_Y(y).$$

They are **identically distributed** if they have the same probability distribution; that is

$$F_X(t) = F_Y(t) \qquad \text{for all } t.$$

Being identically distributed does not mean that $X = Y$. Rather, it means that X and Y are equivalent in the sense that everyone in the world, regardless of risk preference, should be indifferent to choosing between them when they represent payoffs in a game of chance. It is important to keep in mind that in identities involving sums and products, substitution of equal random variables is valid, but substitution of equivalent (i.e., identically distributed) random variables is not.

A **discrete mixture** is a random variable X whose distribution function has the form

$$F_X(x) = p_1 F_{X_1}(x) + \cdots + p_n F_{X_n}(x),$$

for some random variables $X_1, \ldots, X_n$, where $0 < p_i < 1$ for all i and $\sum p_i = 1$, i.e., the distribution function is a weighted sum of distribution functions. Mixtures arise in connection with the law of total probability and are extremely important in insurance applications where the risk class of a subject is uncertain.

Because of the way in which mixtures are defined, it is easy to confuse them with sums. However, mixtures are not the same as sums, since for any random variables X, Y and any number $\alpha \in (0, 1)$ it is generally not the case that $F_{\alpha X + (1-\alpha)Y}(t) = \alpha F_X(t) + (1 - \alpha) F_Y(t)$ for all t.

4.5 Additional Exercises

1. A particular investment gains $100g\%$ or loses $100l\%$ each day. On any given day, the probability of a gain is p, independent of the gains and losses on the other days. Let V_1 be the value of a \$1 investment in this security 1 day from now, let V_2 be the value 2 days from now, and so on.

   a. Construct a tree diagram for the possible paths the investment value could follow over the first 3 days.
   b. Specify the probability mass function for V_k, the value of the \$1 investment after k days.
   c. Calculate $E_a[V_k]$ and $E_g[V_k]$.
   d. Suppose that $p = 1/2$, $g = .05$, $l = .02$. Discuss the different circumstances under which the given investment is a good one.

2. A certain manufacturer wishes to study the reliability of the light bulbs it produces. Experience suggests that, in everyday use, light bulbs fail when they are switched on and not while they are functioning. Hence, the manufacturer decides to study reliability by counting the number of times a light bulb can be switched on.

   Let X be the number of times a bulb can be switched on until it fails for the first time. For example, $X = 10$ means that the light bulb worked the first nine times it was tried, but failed the tenth time. Suppose that we make the following assumptions:

   i. Each time the light bulb is switched on, it either works or it fails, independently of whether it worked on previous trials. However, once a light bulb fails it cannot work on subsequent trials.
   ii. On any trial, the probability that the light bulb fails when it is switched on is p unless the light bulb failed on a previous trial, in which case it is certain to fail again.

   a. Construct a probability model for X (i.e., determine a probability mass function and specify the possible values of X). Verify that your mass function satisfies the condition $\sum_x p_X(x) = 1$. *Hint:* Recall that the sum of the geometric series $\sum_{k=0}^{\infty} r^k$ is $1/(1 - r)$ for $r \in (0, 1)$.
   b. Without referring to the formula for p_X derived in part a, describe the qualitative behavior that you think the graph of p_X should have. Then, using the formula for p_X, construct graphs in the cases $p = 1/4$, $p = 1/2$, $p = 3/4$ and try to show that p_X behaves in the way predicted. Which values of p do you think the manufacturer would prefer?

    c. Determine the probability that a given light bulb functions at least 100 times. For what values of p is this probability at least 95%?

    d. Determine the probability that a given bulb functions at least 10 times but not more than 19 times. For what values of p is this probability at least 5%?

    e. Intuitively, what do you think the mean value of X should be? Prove your conjecture using the mass function p_X derived in part a. *Hint:* Differentiate the geometric series $\sum_{k=0}^{\infty} r^k$ with respect to r.

    f. Specify the location of the mean on each of the graphs constructed in part b. What is the relationship of the mean to the median and the mode in these cases? Can you show this for all values of p? (A *mode* for a distribution with probability mass function p_X is a number x for which $p_X(x)$ is maximized.)

3. The probability model constructed for X in the previous question has some interesting consequences that we now investigate.

    a. Using the definition of conditional probability, calculate the value of $\Pr(X = m + n \mid X > m)$.

    b. How does your answer to part a compare with $\Pr(X = n)$? Give an explanation using the interpretation of X and the assumptions of the model.

    c. Calculate the value of $\Pr(X > m + n \mid X > m)$.

    d. What does your answer to part c suggest about the reliability of a light bulb that has been used for a long period of time? Is this realistic?

4. Suppose that X is a discrete random variable with positive integer values that satisfies the following two properties:

      i. $\Pr(X = 1) = p$

      ii. $\Pr(X > m + n \mid X > m) = \Pr(X > n)$, for all $m, n = 1, 2, 3, \ldots$

    Property ii says that the distribution is *memoryless.*

    By referring to the results of the previous two questions, explain why this gives us an equivalent way to model the reliability of light bulbs. Can you think of any other situation where it is reasonable to assume that X has the memoryless property?

4.6   Appendix on Generalized Density Functions (Optional)

In §4.1.6, we saw that a generalization of the density function could be defined for mixed random variables. In this optional appendix, we give a more formal description of the generalized density function using the *delta function,* with which the reader may be familiar from studies in physics and electricity. We also show that the distribution function of every mixed random variable can be written as a weighted sum of the distribution function for a particular continuous distribution and the distribution function for a particular discrete distribution (i.e., in the language of §4.1.12, every mixed distribution is a mixture of a discrete distribution and a continuous distribution). Finally, we show how the procedure for calculating the expectation of a mixed random variable discussed in §4.2.1 follows from the formal definition of generalized density.

Delta Functions

To introduce the notion of a delta function, let's consider the simplest distribution function possible: the unit step function u defined by

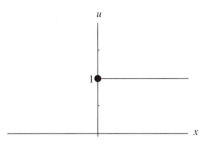

FIGURE 4.21 Unit Step Function at the Origin

$$u(x) = \begin{cases} 1 & \text{if } x \geq 0, \\ 0 & \text{if } x < 0. \end{cases}$$

Note that u is the distribution function for a point mass certain to be at the origin. Its graph is illustrated in Figure 4.21.

The relative frequency density function corresponding to u is the function δ defined by

$$\delta(x) = \lim_{\varepsilon \to 0} \frac{u(x + \varepsilon/2) - u(x - \varepsilon/2)}{\varepsilon}$$
$$= \begin{cases} \infty & \text{if } x = 0, \\ 0 & \text{otherwise.} \end{cases}$$

If the function u were continuous and differentiable at every point, then from the discussion of density functions, given in §4.1.5, we could conclude that

$$\delta(x) = \frac{d}{dx}u(x)$$

and

$$\int_{-\infty}^{\infty} \delta(x)\,dx = 1.$$

However, since u is a step function, these identities need not hold. Indeed, strictly speaking, the function δ cannot even be integrated (in the Riemann sense) because it is not bounded.

Nevertheless, in our quest to define a density function for discrete distributions, it is convenient to introduce a function δ with the following formal properties:

1. $\delta(x) = \begin{cases} \infty & \text{if } x = 0, \\ 0 & \text{otherwise.} \end{cases}$

2. $\displaystyle\int_{-\infty}^{\infty} \delta(x)dx = 1.$

3. $\delta(x) = \dfrac{d}{dx}u(x).$

Such a function is known as a **delta function.**

Densities for Discrete Random Variables

Now consider a discrete random variable X with only two possible values, a_1 and a_2, and let k_1, k_2 be the corresponding probability masses,

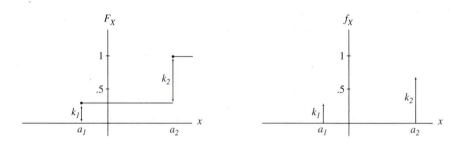

a. Distribution function   b. Density function

FIGURE 4.22 Distribution Function and Corresponding Generalized Density Function for a Discrete Random Variable with Two Probability Masses

$$p_X(x) = \begin{cases} k_1 & \text{if } x = a_1, \\ k_2 & \text{if } x = a_2, \end{cases}$$

where $k_1, k_2 \in (0, 1)$ and $k_1 + k_2 = 1$. The distribution function F_X for this X is clearly a step function with jumps of k_1 and k_2 at a_1 and a_2, respectively (Figure 4.22a).

A little reflection reveals that F_X can be written as a weighted sum of two unit step functions with respective jumps at a_1, a_2 and respective weights k_1, k_2,

$$F_X(x) = k_1 u_{a_1}(x) + k_2 u_{a_2}(x),$$

where for any a, u_a is the unit step function with jump at a defined by

$$u_a(x) = \begin{cases} 1 & \text{if } x \geq a, \\ 0 & \text{if } x < a. \end{cases}$$

In fact, since $u_a(x) = u(x - a)$, F_X can be defined in terms of u as follows:

$$F_X(x) = k_1 u(x - a_1) + k_2 u(x - a_2).$$

Using the properties of the delta function, we can formally differentiate this equation for $F_X(x)$ to obtain

$$\begin{aligned} F_X'(x) &= k_1 \delta(x - a_1) + k_2 \delta(x - a_2) \\ &= k_1 \delta_{a_1}(x) + k_2 \delta_{a_2}(x), \end{aligned}$$

where for any a, $\delta_a(x) = \delta(x - a)$ (i.e., δ_a is the delta function centered at the point a). Consequently, we can formally define the density function f_X for the given random variable X to be the weighted sum of the delta functions $\delta_{a_1}, \delta_{a_2}$ with weights k_1 and k_2,

$$f_X(x) = k_1 \delta_{a_1}(x) + k_2 \delta_{a_2}(x).$$

For such a density function to be meaningful, we must make a distinction between the function δ and the scalar multiple $k\delta$, even though ordinary arithmetic might suggest that $k\delta = \delta$ (since $k \cdot 0 = 0$ and since $k \cdot \infty$ is generally taken to be ∞). Otherwise, the "density" function so obtained will tell us the locations of probability masses, but not their relative sizes. Algebraically, this distinction is made by leaving the formula for f_X as a weighted sum of delta functions intact. Graphically, this distinction is made by displaying the "infinity points" of the density function using vertical lines with upward-pointing arrows and respective heights k_j, as illustrated in Figure 4.22b.

Notice the striking similarity of the density plot in Figure 4.22b to a mass plot. Indeed, the only difference is arrowheads in place of dots on the tops of the vertical lines (see Figure 4.3a in §4.1.3). This suggests that density functions and mass functions may not be so different after all.

Densities for Mixed Random Variables

With the generalized notion of density function just developed, we can now proceed to define a density function for mixed random variables. Recall that a mixed random variable is one whose distribution function is not a step function but is nevertheless discontinuous at some point.

Let X be a mixed random variable and let $a_1, \ldots, a_n$ be the points of discontinuity of F_X.[31] Further, let $k_1, \ldots, k_n$ be the jump sizes at $a_1, \ldots, a_n$, respectively.

Consider the function F given by

$$F(x) = (F_X(x) - \sum_{j=1}^{n} k_j u_{a_j}(x))/(1 - \sum_{j=1}^{n} k_j).$$

Note that F is obtained from F_X by removing the steps in F_X and then scaling the resulting function so that $\lim_{x \to \infty} F(x) = 1$. Hence, F has the properties of a distribution function. Moreover, F is continuous, because the discontinuities have been removed. Consequently, F is the distribution function of some continuous random variable (i.e., $F = F_C$ for some continuous random variable C). Rearranging the formula for $F(x)$ and substituting F_C for F, we obtain the following formula for F_X:

$$F_X(x) = (1 - \sum_{j=1}^{n} k_j) \cdot F_C(x) + \sum_{j=1}^{n} k_j u_{a_j}(x).$$

Now consider the quantity $\sum_{j=1}^{n} k_j u_{a_j}(x)$ which arises in this formula for F_X. Put $k = \sum_{j=1}^{n} k_j$. From our discussion of unit step functions earlier in the section, it is clear that $\sum_{j=1}^{n} (k_j/k) \cdot u_{a_j}$ is the distribution function for some discrete random variable. Let D be a discrete random variable with probability mass function p_D given by

$$p_D(a_j) = \frac{k_j}{k}, \qquad j = 1, \ldots, n.$$

Then

$$F_D(x) = \sum_{j=1}^{n} \frac{k_j}{k} u_{a_j}(x)$$

and thus,

$$F_X(x) = (1 - k) F_C(x) + k F_D(x).$$

Hence, F_X is a weighted sum of particular continuous and discrete distribution functions.

---

[31] For simplicity, we are assuming that F_X is only discontinuous at a finite number of points. However, the discussion that follows applies, with some modification, to more general mixed distributions as well.

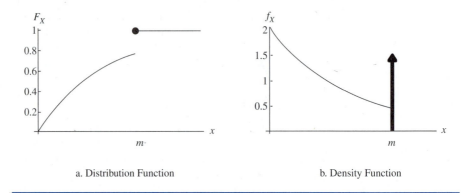

a. Distribution Function   b. Density Function

FIGURE 4.23 Distribution Function and Corresponding Density Function for a Particular Mixed Random Variable

Formally differentiating this expression for F_X, we obtain

$$F_X'(x) = (1 - k)f_C(x) + kf_D(x),$$

where f_C is the ordinary density for C and where f_D is the formal density for D defined earlier,

$$f_D(x) = \sum_{j=1}^{n} \frac{k_j}{k} \delta_{a_j}(x).$$

Consequently, the density function for the mixed random variable X can be defined to be the function f_X given by

$$f_X(x) = (1 - k)f_C(x) + \sum_{j=1}^{n} k_j \delta_{a_j}(x),$$

where C is a continuous random variable whose distribution is obtained from F_X by removing the jumps of F_X and scaling appropriately, and $k = \sum_{j=1}^{n} k_j$.

Figure 4.23 illustrates the graph of a possible distribution function for a mixed random variable and its corresponding generalized density function. Note that the area under the continuous part of the density is not equal to 1. Indeed, it equals $1 - \sum_{j=1}^{n} k_j$, where $\sum_{j=1}^{n} k_j$ is the sum of the probability masses.

An interesting fact that emerges from the preceding discussion is that every mixed random variable X has the property that its distribution function F_X is a weighted sum of the distribution functions F_C and F_D, where C and D are particular continuous and discrete random variables, respectively. A similar statement applies to the density function f_X. Note, however, that this does not mean that X as a random variable is a weighted sum of the random variables C and D (see §4.1.12).

EXAMPLE 1:   The distribution function for a mixed random variable X is given by

$$F_X(x) = \begin{cases} 0 & \text{for } x < 0, \\ 1 - \frac{3}{4}e^{-x} & \text{for } 0 \le x < 1, \\ 1 & \text{for } x \ge 1. \end{cases}$$

Determine the generalized density for X. Write F_X as a weighted sum of a continuous distribution function and a discrete distribution function.

A sketch of the graph of F_X reveals jumps at $x = 0$ and $x = 1$ of size $\frac{1}{4}$ and $\frac{3}{4}e^{-1}$, respectively. Hence,

$$F_X(x) = (1 - \frac{1}{4} - \frac{3}{4}e^{-1})F_C(x) + \frac{1}{4}u_0(x) + \frac{3}{4}e^{-1}u_1(x),$$

where F_C is obtained from F_X by removing the jumps at $x = 0$ and $x = 1$ and then scaling the resulting function so that $\lim_{x\to\infty} F_C(x) = 1$,

$$F_C(x) = \begin{cases} 0 & \text{for } x < 0, \\ \frac{1-e^{-x}}{1-e^{-1}} & \text{for } 0 \le x < 1, \\ 1 & \text{for } x \ge 1. \end{cases}$$

Now,

$$F_C'(x) = \begin{cases} \frac{e^{-x}}{1-e^{-1}} & \text{for } 0 < x < 1, \\ 0 & \text{otherwise.} \end{cases}$$

Consequently, the generalized density function for X is given by

$$f_X(x) = \frac{3}{4}(1 - e^{-1})F_C'(x) + \frac{1}{4}\delta_0(x) + \frac{3}{4}e^{-1}\delta_1(x)$$

$$= \begin{cases} \frac{3}{4}e^{-x} & \text{for } 0 < x < 1, \\ 0 & \text{otherwise;} \end{cases} + \frac{1}{4}\delta_0(x) + \frac{3}{4}e^{-1}\delta_1(x).$$

To express F_X explicitly as a weighted sum of F_C and F_D, put

$$F_D(x) = \frac{u_0(x) + 3e^{-1}u_1(x)}{1 + 3e^{-1}}.$$

This is the distribution function of a discrete random variable D with probability masses of $1/(1 + 3e^{-1})$ and $3e^{-1}/(1 + 3e^{-1})$ at $x = 0$ and $x = 1$, respectively. Then the desired form is

$$F_X(x) = \frac{3}{4}(1 - e^{-1})F_C(x) + \frac{1}{4}(1 + 3e^{-1})F_D(x). \qquad \blacksquare$$

The decomposition of the generalized density allows us to determine the expected value of a mixed random variable in a fairly straightforward manner. Indeed, from the decomposition

$$f_X(x) = (1 - k)f_C(x) + \sum_{j=1}^{n} k_j \delta_{a_j}(x),$$

we have

$$E[X] = \int_{-\infty}^{\infty} x \cdot f_X(x)dx$$

$$= (1-k)\int_{-\infty}^{\infty} x \cdot f_C(x)dx + \sum_{j=1}^{n} k_j \int_{-\infty}^{\infty} x \cdot \delta_{a_j}(x)dx$$

$$= (1-k)E[C] + \sum_{j=1}^{n} k_j \int_{-\infty}^{\infty} x \cdot \delta_{a_j}(x)dx.$$

Now,

$$\int_{-\infty}^{\infty} x \cdot \delta_m(x)dx = m.$$

To see why this is so, note that δ_m represents the density of a point mass certain to be at location m; hence, for the value of the expectation to agree with the value we get using the probability mass function, we must have $\int_{-\infty}^{\infty} x \cdot \delta_m(x)dx = m$. Consequently,

$$E[X] = (1-k)E[C] + \sum_{j=1}^{n} k_j a_j.$$

EXAMPLE 2:   A random variable X has distribution function

$$F_X(x) = \begin{cases} 1 - e^{-\lambda x} & \text{for } 0 \le x < m, \\ 1 & \text{for } x \ge m. \end{cases}$$

Determine the expected value of X in terms of λ and m.

Note that this is the truncated exponential distribution with parameters λ and m introduced in Example 3 of §4.2.1. From our earlier discussions, we know that this distribution is mixed with a point mass of size $e^{-\lambda m}$ at $x = m$ and a continuous distribution of probability on the interval $0 < x < m$ (see Figure 4.15 in §4.2.1). Differentiating the distribution function formally, we find that the density function for X is given by

$$f_X(x) = \begin{cases} \lambda e^{-\lambda x} & \text{for } 0 \le x < m, \\ 0 & \text{otherwise,} \end{cases} + e^{-\lambda m}\delta_m(x),$$

where δ_m is the delta function at $X = m$.

From the definition of expectation,

$$E[X] = \int_{-\infty}^{\infty} xf_X(x)dx$$

$$= \int_{-\infty}^{\infty} x \cdot g(x)dx + \int_{-\infty}^{\infty} x \cdot e^{-\lambda m}\delta_m(x)dx,$$

where

$$g(x) = \begin{cases} \lambda e^{-\lambda x} & \text{for } 0 \le x < m, \\ 0 & \text{otherwise.} \end{cases}$$

Consequently, using the relationship

$$\int_{-\infty}^{\infty} x \cdot \delta_m(x)dx = m,$$

we have

$$E[X] = \int_0^m x \cdot \lambda e^{-\lambda x} dx + e^{-\lambda m} \cdot m.$$

Now, using integration by parts,

$$\int_0^m \lambda x e^{-\lambda x} dx = -x e^{-\lambda x}\Big|_0^m + \int_0^m e^{-\lambda x} dx$$

$$= -m e^{-\lambda m} - \frac{1}{\lambda} e^{-\lambda x}\Big|_0^m$$

$$= -m e^{-\lambda m} - \frac{1}{\lambda} e^{-\lambda m} + \frac{1}{\lambda}.$$

Hence,

$$E[X] = (-m e^{-\lambda m} - \frac{1}{\lambda} e^{-\lambda m} + \frac{1}{\lambda}) + m e^{-\lambda m}$$

$$= \frac{1}{\lambda}(1 - e^{-\lambda m}).$$

This is precisely the formula determined in Example 3 of §4.2.1.

5 Special Discrete Distributions

In applications of probability, certain families of distributions arise quite frequently and it is important to have a thorough understanding of these frequently occurring distributions and their properties. The exponential distribution with parameter λ—that is, the distribution with density function $f(x) = \lambda e^{-\lambda x}, x \geq 0$—which was introduced in the examples of the previous chapter, is one such family of distributions. It represents a *family* of distributions because each distinct value of λ gives rise to a different distribution of probability, and at the same time, the different distributions in the family have common characteristics (e.g., all distributions in the family have density function of the form $f(x) = \lambda e^{-\lambda x}, x \geq 0$ and expected value $1/\lambda$).[1]

In this chapter and the next, we discuss 12 special families of distributions that are important in the study of investments, insurance, and engineering. For each of these special distributions, we consider:

1. the definition of the distribution as given by a mass, density, survival, or distribution function;
2. the interpretation of the distribution and its applications;
3. a derivation of the probability mass or probability density function based on the interpretation given;
4. formulas for calculating the mean, variance, and higher moments, including formulas for the moment and cumulant generating functions where possible;
5. techniques, such as recursive formulas, for calculating probabilities in practice;
6. the effect of arithmetic operations, such as sums and scalar multiples, on the distribution of probability;
7. the relationship of the distribution to other special distributions.

Each special distribution is considered in a separate section to make future reference to these distributions easy.

For the sake of completeness, the discussion in each section on the relationship of a particular distribution to the other special distributions may require reference to a distribution that has not yet been completely discussed.[2] This problem is most acute

---

[1] Several other families of distributions were also introduced in the previous chapter. These included the Poisson family, the gamma family, and the normal family.

[2] However, in most cases, the distribution in question will have been previously introduced in an example in Chapter 4.

in the first few sections of this chapter, where the reader's knowledge of the special distributions may still be limited. To alleviate this problem, we suggest taking one of the following approaches: (a) postpone reading the material on the relationships among the distributions until the end of Chapter 6 or (b) begin by reading over the definitions and interpretations of all 12 special distributions.

To facilitate the reader's understanding of these 12 special distributions and to avoid having this chapter and the next take on the appearance of an encyclopedia, we have organized our discussion of these distributions using the following classification:

1. discrete distributions that are often used to model an uncertain frequency, such as the number of claims received on a given insurance policy or group of insurance policies;
2. continuous distributions that are often used to model an uncertain size, such as the size of a claim on an automobile insurance policy;
3. continuous distributions that are often used to model an uncertain time, such as the future lifetime of a person or machine;
4. other special distributions, which do not easily fit into the three preceding categories, such as the normal and lognormal distributions.

Even though we have chosen to organize our discussion in this way, it is important to keep in mind that each of the 12 distributions can arise in many different contexts. For example, the gamma distribution (to be discussed in the next chapter) can be used to model both the size of an insurance claim and the time until a claim arrival. Hence, our classification of the basic distributions should not be taken too literally.

In this chapter, we discuss four discrete distributions that are commonly used to model an uncertain frequency: the binomial, the Poisson, the negative binomial, and the geometric. As we will soon see, the Poisson is a limiting case of the binomial, the negative binomial is a particular mixture of Poissons, and the geometric is an important special case of the negative binomial. We also discuss how the binomial distribution arises in connection with models of stock prices.

5.1   The Binomial Distribution

Definition and Interpretation

A random variable X is said to have a binomial distribution with parameters n and p if its probability mass function is given by

$$p_X(x) = \binom{n}{x} p^x (1-p)^{n-x}, \qquad x = 0, 1, \ldots, n.$$

Here, n is a positive integer and p is a real number in the interval $[0, 1]$. Possible graphs of p_X are given in Figure 5.1. The notation Binomial(n,p) signifies a binomial distribution with parameters n and p.

A binomial random variable X has the following interpretation: It is the number of successes in n independent trials, where the probability of success on an individual trial is p.

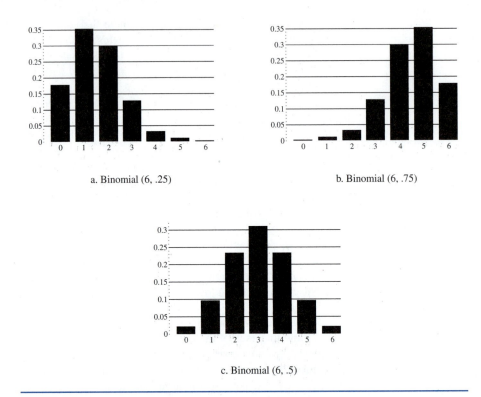

a. Binomial (6, .25)

b. Binomial (6, .75)

c. Binomial (6, .5)

FIGURE 5.1 Particular Binomial Distributions

In insurance, the binomial distribution arises naturally as the number of claims on a group of n policies for which claims are independent, at most one claim per policy is possible, and the probability of a claim on any given policy is p.

In investments, the binomial distribution arises in connection with a simplified model of the evolution of stock prices. To see this connection, consider a security S whose closing price today is S and let $S_1, S_2, \ldots, S_n$ be the successive closing prices for this stock over the next n trading days. Suppose that on each day, the price of the stock either moves up to S^*u or down to S^*d, where S^* is the previous day's closing price and u and d are fixed numbers such that $d = 1/u$. Suppose further that the probability of an upward movement on each day is p, independent of the stock price movements on the previous days. Then the possible values of the stock price n days hence (i.e., the possible values of S_n) are

$$Sd^n, Sd^{n-1}u, Sd^{n-2}u^2, \ldots, Sdu^{n-1}, Su^n,$$

and the respective probabilities for these values are the binomial probabilities

$$q^n, \binom{n}{1}pq^{n-1}, \ldots, \binom{n}{n-1}p^{n-1}q, p^n,$$

where $q = 1 - p$, which is the probability of a downward movement on any given day.

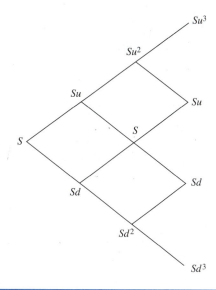

FIGURE 5.2 Binomial Stock Price Model over Three Periods

The simplified model for stock prices just presented is known as the binomial model for stock prices in the finance literature. It is based on the premise that there is no predictive information in past stock prices. That is, past prices and price patterns cannot be used to predict future prices and price movements. This premise is consistent with a weak form of the principle in finance known as the efficient market hypothesis.[3] The binomial model for stock prices is often illustrated schematically using binomial trees. Such a tree for $n = 3$ is displayed in Figure 5.2.

The binomial distribution also arises in connection with random sampling.[4] Indeed, it is a model for the number of subjects with a particular trait in a random sample of size n when the sample is chosen from a fixed population *with replacement* and the fraction of the overall population with the desired trait is p. For example, if 10 balls are chosen at random from a bin containing 20 red balls and 30 green balls, and if each chosen ball is replaced in the bin before the next ball is selected, then the number of red balls in the sample has a Binomial(10, 4) distribution. In the exercises, the reader will develop an exact model for the number of subjects in a sample with a particular trait when the sampling is done *without replacement*. It turns out that this exact model can be approximated by a binomial model when the sample size is small relative to the population size. Hence, for any sampling procedure, the binomial distribution can be

---

[3] The **efficient market hypothesis** asserts that security prices observed in a well-functioning market reflect all available information. This does not necessarily mean that market prices are rational. For further information on the concepts of market efficiency and market rationality, consult Z. Bodie, A. Kane, A. J. Marcus: *Investments,* 4th edition, 1999, Irwin-McGraw-Hill, or E. J. Elton, M. J. Gruber: *Modern Portfolio Theory and Investment Analysis,* 5th edition, 1995, Wiley.

[4] A **random sample** is a collection of items selected at random from a given population. An example of a random sample is a group of people selected at random for a public opinion survey.

used to model the number of subjects in a sample with a desired characteristic, provided that the population is large compared to the sample size.

Random sampling is used in quality control engineering to determine the number of defective items in a large lot of manufactured items without testing every item in the lot. An illustration of this application is given in Example 4 at the end of this section.

Derivation of the Probability Mass Function

The probability mass function for a binomial random variable X can be derived from its interpretation as the number of successes in n independent trials with common success probability p, using basic combinatorial principles. We now proceed to do this in the context of coin tosses for a coin whose probability of heads on a single toss is p.

Suppose that we toss the coin n times and are interested in the probability of getting exactly x heads. There are many ways in which we could obtain exactly x heads in n tosses: The first x tosses could all be heads and the remaining $n - x$ tosses could all be tails; the first $n - x$ tosses could all be tails and the remaining x tosses could all be heads; or the x heads could be interspersed with the $n - x$ tails in some way. The number of ways of getting x heads in n tosses is the same as the number of ways of arranging x Hs and $n - x$ Ts in a row. From basic combinatorics (see §3.7), the number of such arrangements is

$$\binom{n}{x} = \frac{n!}{(n-x)!x!}.$$

Now the probability of obtaining any specific sequence of x heads and $n - x$ tails (e.g., x heads followed by $n - x$ tails) is $p^x(1-p)^{n-x}$, since individual coin tosses are independent. Hence, the probability of getting exactly x heads and $n - x$ tails in any order in the n tosses is

$$\binom{n}{x}p^x(1-p)^{n-x}.$$

Consequently, the formula for the binomial probability mass function follows from the interpretation of X as the number of successes in n independent trials, where the probability of success on each trial is p.

Note that $\sum_{x=0}^{n}\binom{n}{x}p^x(1-p)^{n-x} = 1$ since $\sum_{x=0}^{n}p_X(x) = 1$, p_X being a probability mass function. This algebraic identity also follows directly from the binomial theorem $(a+b)^n = \sum_{x=0}^{n}\binom{n}{x}a^xb^{n-x}$ with $a = p$ and $b = 1 - p$.

Mean, Variance, and Higher Moments

The mean and variance of a binomial random variable X are, respectively,

$$E[X] = np,$$
$$Var(X) = np(1-p).$$

The moment generating function of X is given by

$$M_X(t) = \left(pe^t + (1-p)\right)^n, \qquad t \in \mathbf{R}.$$

The derivation of the moment generating function is fairly straightforward and follows from the identity:

$$X = I_1 + I_2 + \cdots + I_n,$$

where the I_j are defined by

$$I_j = \begin{cases} 1 & \text{if the } j\text{th trial is a success,} \\ 0 & \text{if the } j\text{th trial is not a success.} \end{cases}$$

Indeed, since the trials are independent by assumption, the I_j are independent random variables. Thus, by the formula for the moment generating function of an independent sum, we have

$$M_X(t) = M_I(t)^n,$$

where

$$I = \begin{cases} 1 & \text{with probability } p, \\ 0 & \text{with probability } 1 - p. \end{cases}$$

(See §4.3.1.) Since $M_I(t) = pe^t + (1 - p)$, using the definition of a moment generating function, $M_X(t) = (pe^t + (1 - p))^n$ as claimed.

Formulas for the mean, variance, and also the skewness are then readily obtainable from $M_X(t)$ using the properties of moment generating functions discussed in §4.3.1. Details are left to the reader.

Note that the mean and variance of a binomial random variable X are related by the inequality $Var(X) < E[X]$ because $0 < 1 - p < 1$. The significance of this observation will become clearer after our discussions of the Poisson and negative binomial random variables in §5.2 and §5.3.

Techniques for Calculating Probabilities

The simplest way to compute probabilities is with the recursive formula

$$\frac{p_X(x)}{p_X(x-1)} = \frac{n - x + 1}{x} \cdot \frac{p}{1 - p},$$

which follows directly from the definition of the probability mass function. This recursive formula can also be used to determine the character of the graph of the probability mass function for various values of p. Indeed, using this formula, one can show that for $p \leq 1/(n + 1)$, the function p_X is decreasing; for $p \geq n/(n + 1)$, the function p_X is increasing; and for p such that $1/(n + 1) < p < n/(n + 1)$, the graph of the function p_X first increases and then decreases. One can also use the recursive formula to determine the values of x for which $p_X(x)$ is greatest. Such values x are known as the **mode(s)** of the distribution. Details are left to the reader.

Effect of Arithmetic Operations

An independent sum of binomial random variables with the same parameter p is again binomial. In particular, if $X_1 \sim \text{Binomial}(n_1, p)$, $X_2 \sim \text{Binomial}(n_2, p)$, and X_1, X_2 are independent, then $X_1 + X_2 \sim \text{Binomial}(n_1 + n_2, p)$.[5]

This property follows directly from the formula for the moment generating function of a binomial random variable given previously and the general formula for the moment generating function of an independent sum; that is, $M_{X_1 + X_2}(t) = M_{X_1}(t) \cdot M_{X_2}(t)$. It

---

[5] The notation $X \sim \text{Binomial}(n, p)$ signifies that X has a binomial distribution with parameters n and p.

also makes intuitive sense, given the interpretation of a binomial random variable as the number of successes in independent trials with common success probability. Note that the property only holds when the components of the sum are independent and have common parameter p. Intuition again suggests that this condition is not unreasonable.

Relationship with Other Distributions

If n becomes large and p simultaneously becomes small in such a way that np remains finite, bounded, and nonzero, then

$$p_X(x) \to \frac{\lambda^x e^{-\lambda}}{x!},$$

where $\lambda = np$. This limiting function is the probability mass function for a Poisson random variable with parameter $\lambda = np$. The Poisson distribution was introduced in Example 1 of §4.2.1 and will be considered in the next section, where this limiting property will be demonstrated.

On the other hand, if n becomes large but p does not become correspondingly small, then the distribution of X begins to assume the character of a normal distribution with mean np and variance $np(1 - p)$. The normal distribution was introduced in Example 1 of §4.3.1 and will be considered in detail in §6.3.1. More precisely,

$$\text{Distribution of } \frac{X - np}{\sqrt{np(1 - p)}} \to \text{Normal}(0,1),$$

where Normal(0,1) represents a normal distribution with mean 0 and standard deviation 1, and where convergence here means convergence with respect to distribution functions. That is, $F(x) \to \Phi(x)$, where F represents the distribution function of $(X - np)/\sqrt{np(1 - p)}$ and Φ represents the distribution function of Normal(0,1). The normal distribution and its relationship to the binomial distribution will be discussed in §6.3.1. For now, we simply note that the binomial distribution can be considered the discrete analog of the normal distribution.

Finally, we note that when $n = 1$, the binomial distribution reduces to the distribution for an indicator random variable with success probability p. Such an indicator is often referred to as a **Bernoulli random variable.** Its distribution can be expressed as Bernoulli(p).

Before moving on to a discussion of the Poisson distribution in §5.2, we present a few worked examples.

EXAMPLE 1:   A life insurance company is interested in the number of claims that it will receive on a particular block of life insurance policies in the coming year. All of the policies in the block are written on lives of people who are currently age 60. There are presently 100 active policies in the block. None of the policies in the block is written on more than one life, and no insured person is covered by more than one policy. The company estimates that the probability of any given 60-year-old dying in the next year is .0137604. Determine the probability that the company receives more than two claims on this block of 100 policies in the coming year. Assume that the future lifetimes of the insured persons are independent.

Let N be the number of claims in the coming year. Then since future lifetimes are independent, $N \sim \text{Binomial}(100, p)$ where $p = .0137604$. Using the binomial probability mass function, we have

$$\Pr(N = 0) = (1 - p)^{100} = (.9862396)^{100} \approx .2501749,$$
$$\Pr(N = 1) = 100p(1 - p)^{99} \approx .3490537,$$
$$\Pr(N = 2) = \binom{100}{2} p^2 (1 - p)^{98} \approx .2410716.$$

Hence, the desired probability is

$$\Pr(N > 2) = 1 - \Pr(N = 0) - \Pr(N = 1) - \Pr(N = 2) \approx .1596998. \qquad \blacksquare$$

EXAMPLE 2: An auto insurance company is analyzing the claim frequency on a block of 250 policies. Historical data suggest that 10% of policyholders in this block will file at least one claim in the coming coverage period. What is the probability that more than 12% of policyholders file at least one claim in the coming coverage period? Assume that claim occurrences are independent for distinct policyholders.

Let X be the number of policies in the block for which there is at least one claim in the coming period. Then

$$X = I_1 + \cdots + I_{250},$$

where for each j,

$$I_j = \begin{cases} 1 & \text{if policyholder } j \text{ files at least one claim,} \\ 0 & \text{if policyholder } j \text{ files no claims.} \end{cases}$$

Note that X is not the number of claims because individual policyholders can submit more than one claim during the period. From the given information, $\Pr(I_j = 1) = .10$. Hence, $I_j \sim \text{Binomial}(1, .1)$ for each j, and $X \sim \text{Binomial}(250, .1)$ since the I_j are independent by assumption.

We are interested in the probability that $X > (.12)(250)$; that is, $\Pr(X > 30)$. Using the binomial probability mass function, this is

$$\Pr(X > 30) = \sum_{x=31}^{250} \binom{250}{x} (.10)^x (.90)^{250-x}.$$

Alternatively,

$$\Pr(X > 30) = 1 - \sum_{x=0}^{30} \binom{250}{x} (.10)^x (.90)^{250-x}.$$

Neither of these summation formulas is very appealing. As an alternative, one could use a Poisson distribution approximation, as discussed earlier,

$$p_X(x) \approx \frac{\lambda^x e^{-\lambda}}{x!},$$

where $\lambda = np = (250)(.10) = 25$. Using this approximation,

$$\Pr(X > 30) \approx 1 - \sum_{x=0}^{30} \frac{25^x e^{-25}}{x!}.$$

Evaluation of the latter summation will still require a computer (or a very patient person with a sufficiently accurate hand calculator). We leave this as an exercise for the reader.

∎

EXAMPLE 3:   The current price of a particular stock is $10. On any given day, the stock price will either increase $1 or decrease $1. The probability of an increase on any given day is 60%, independent of the price movements on previous days. Determine the expected value and standard deviation of the stock price 4 days from now.

This simplified model of stock prices is slightly different from the one discussed earlier in the section. With this model, the *dollar* amounts of price changes are the same from day to day, whereas with the earlier model, the *percentage* price changes are the same. The earlier model is actually more realistic because it does not permit the stock price to fall below zero, whereas the model of this example does. Nevertheless, over short time periods, the present model is not too bad and is computationally easier to work with.

Let S_4 be the stock price 4 days hence. Then, from a binomial tree, it is clear that the possible values of S_4 are 6, 8, 10, 12, 14 with respective probabilities q^4, $\binom{4}{1}pq^3$, $\binom{4}{2}p^2q^2$, $\binom{4}{3}p^3q$, p^4, where $p = .60$ and $q = 1 - p = .40$. We could calculate the mean and variance of S_4 directly from this information. However, there is an easier way.

Notice that by shifting the distribution of S_4 to the left by 6 units and dividing the resulting values of S_4 by 2, we obtain a Binomial(4, .6) distribution; that is, the random variable X given by $X = (S_4 - 6)/2$ has the property that $X \sim$ Binomial(4, .6). Hence, $S_4 = 2X + 6$ and thus,

$$E[S_4] = 2\,E[X] + 6,$$
$$Var(S_4) = 4\,Var(X).$$

However, since $X \sim$ Binomial(4, .6), $E[X] = 2.40$ and $Var(X) = 0.96$. Consequently,

$$E[S_4] = 2(2.40) + 6 = 10.8,$$
$$Var(S_4) = 4(0.96) = 3.84.$$

Thus, the mean and standard deviation of the stock price 4 days hence are $10.80 and $1.96, respectively.

∎

EXAMPLE 4:   An automotive parts manufacturer produces transmissions for sport utility vehicles. There is a 2% chance that any given transmission is defective, independent of the other transmissions produced. Determine the probability that more than 2 transmissions in a shipment of 100 are defective.

Let X be the number of defectives in a shipment of 100. Then an exact model for X is Binomial(100, .02) and the desired probability is

$$\Pr(X > 2) = 1 - \Pr(X = 0) - \Pr(X = 1) - \Pr(X = 2)$$

$$= 1 - \binom{100}{0}(.98)^{100} - \binom{100}{1}(.98)^{99}(.02)^1$$

$$- \binom{100}{2}(.98)^{98}(.02)^2$$

$$\approx 1 - .1326196 - .2706522 - .2734139$$

$$\approx .3233.$$

∎

5.2 The Poisson Distribution

Definition and Interpretation

A random variable X is said to have a Poisson distribution with parameter λ if its probability mass function is given by

$$p_X(x) = \frac{e^{-\lambda}\lambda^x}{x!}, \qquad x = 0, 1, 2, \ldots .$$

Here, λ is a positive real number and is in fact the mean of the distribution. Possible graphs of p_X are given in Figure 5.3. The notation Poisson(λ) signifies a Poisson distribution with parameter λ.

A Poisson random variable X has the following interpretation: It is the number of occurrences of a rare event during some fixed time period in which the expected number of occurrences of the event is λ and individual occurrences of the event are independent of each other.

In insurance, the Poisson distribution arises naturally as the number of claims on a large group of policies for which the expected number of claims is known and claims occur independently and infrequently. While the binomial distribution can also be used to model claim frequency, as discussed in the previous section, a Poisson model is often easier to apply in practice. This is particularly true in group insurance applications where the membership of a group frequently changes throughout the period of consideration, making the application of a binomial model difficult, while the expected number of claims remains fairly stable. The Poisson distribution is also useful for modeling claim frequency on individual policies, such as auto insurance, where multiple claims are possible, but rare.

In more general situations, the Poisson distribution arises as the number of arrivals of claims, people, letters, or whatever in a fixed time period in which the expected number of arrivals is known and arrivals are independent of one another.

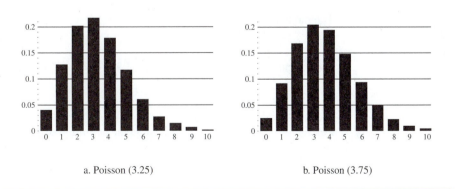

a. Poisson (3.25)   b. Poisson (3.75)

FIGURE 5.3 Particular Poisson Distributions

Derivation of the Probability Mass Function

The Poisson probability mass function can be derived by taking the limit of a binomial probability mass function as $n \to \infty$ under the assumption that np remains constant. Indeed, putting $\lambda = np$ and substituting for p in the formula for the binomial probability mass function, we have

$$\binom{n}{x} p^x (1-p)^{n-x}$$

$$= \frac{n!}{(n-x)!x!} p^x (1-p)^{n-x}$$

$$= \frac{n!}{(n-x)!x!} \left(\frac{\lambda}{n}\right)^x \left(1 - \frac{\lambda}{n}\right)^{n-x}$$

$$= \frac{n(n-1)\cdots(n-x+1)}{n^x} \cdot \frac{\lambda^x}{x!} \left(1 - \frac{\lambda}{n}\right)^n \cdot \left(1 - \frac{\lambda}{n}\right)^{-x}$$

$$\to 1 \cdot \frac{\lambda^x}{x!} \cdot e^{-\lambda} \cdot 1$$

as $n \to \infty$.[6] Hence, the Poisson probability mass function is a particular limit of the binomial probability mass function as claimed. Note that since np is the expected value for Binomial(n, p), the parameter λ is always the expected value of Poisson(λ).

This derivation justifies the approximation Binomial$(n, p) \approx$ Poisson(np) when n is large, p is small, and np is moderate in size. The significance of this approximation for insurance is that the number of claims on a group of n policies for which claim occurrences are independent can be modeled as a Poisson random variable when n is large. This simplifies the calculation of probabilities substantially[7] and provides a means of modeling claim frequency when the group size n is not precisely known.

Mean, Variance, and Higher Moments

The mean, variance, and skewness of a Poisson random variable X are, respectively,

$$E[X] = \lambda,$$
$$Var(X) = \lambda,$$
$$\gamma_X = \frac{1}{\lambda^{1/2}}.$$

The moment generating function of X is given by

$$M_X(t) = \exp\{\lambda(e^t - 1)\}, \qquad t \in \mathbf{R},$$

and the cumulant generating function of X is given by

$$\psi_X(t) = \lambda(e^t - 1), \qquad t \in \mathbf{R}.$$

The moment generating function can be derived directly from the formula for the Poisson probability mass function using the definition of a moment generating function.

---

[6] The well-known fact that $\lim_{n\to\infty}(1 - \lambda/n)^n = e^{-\lambda}$ can be demonstrated by considering the logarithm of $(1 - \lambda/n)^n$ and applying l'Hôpital's rule.

[7] Compare the complexity of calculation for a Poisson probability mass function and a binomial probability mass function when n is large.

Taking this approach, we have

$$M_X(t) = E[e^{tX}]$$

$$= \sum_{x=0}^{\infty} e^{tx} \frac{\lambda^x e^{-\lambda}}{x!}$$

$$= e^{-\lambda} \sum_{x=0}^{\infty} \frac{(\lambda e^t)^x}{x!}$$

$$= e^{-\lambda} \cdot \exp(\lambda e^t)$$

$$= \exp\{\lambda(e^t - 1)\}.$$

Note the use of the relationship $e^x = \sum_{n=0}^{\infty} \frac{x^n}{n!}$.

Formulas for the mean, variance, and skewness can then be obtained from the cumulant generating function, using the general formulas for μ_X, σ_X^2, and γ_X developed in §4.3.1. Note that, for a Poisson random variable, the derivatives of ψ_X are much easier to compute than the derivatives of M_X, making ψ_X the natural choice for determining μ_X, σ_X, and γ_X.

Note that the mean and variance of a Poisson random variable are equal and $\gamma_X \to 0$ as $\lambda \to \infty$ (i.e., the distribution becomes more symmetric as the mean increases). This relationship between $E[X]$ and $Var(X)$ for Poisson random variables can be contrasted to the relationship $Var(X) < E[X]$ for binomial random variables, discussed in §5.1. It shows, in particular, that the variance of a Poisson distribution that is used to approximate a given binomial distribution will be larger than the variance of the given binomial distribution. Hence, tail probabilities—that is, probabilities of the form $\Pr(X > a)$ for large a—will generally be overestimated in the approximation. This is a desirable property for an approximation to have because events of the type $X > a$ generally represent "catastrophes."

Techniques for Calculating Probabilities

Probabilities can be computed using the formula for the probability mass function directly or using the recursive formula

$$\frac{p_X(x)}{p_X(x-1)} = \frac{\lambda}{x},$$

which follows directly from the definition of the probability mass function. Both methods are fairly efficient. However, when a large number of the values $p_X(x)$ are to be summed, the recursive method will be faster. The recursive formula is also useful for determining the character of the graph of p_X and the location of the mode(s). This application of the formula is considered in the exercises.

Effect of Arithmetic Operations

An independent sum of Poisson random variables is again Poisson. In particular, if $X_1 \sim \text{Poisson}(\lambda_1)$, $X_2 \sim \text{Poisson}(\lambda_2)$, and X_1, X_2 are independent, then $X_1 + X_2 \sim \text{Poisson}(\lambda_1 + \lambda_2)$.

This property follows directly from the formula for the moment generating function of a Poisson random variable given earlier and the general formula for the moment generating function of an independent sum. It also makes intuitive sense when one interprets the Poisson random variables as claim frequencies.

Relationship with Other Distributions

We have already seen that the Poisson distribution is a limiting case of the binomial. However, it also related to the exponential and gamma distributions, which were introduced in Example 2 of §4.2.1 and Example 7 of §4.3.1 respectively, in the following way:

Suppose that for each time t, the number of arrivals in the interval $[0, t]$ is Poisson(λt). Then the waiting time until the first arrival is Exponential(λ) and the waiting time until the rth arrival is Gamma(r, λ).

We will consider this property in more detail after we have learned more about the exponential distribution in §6.1.1 and the gamma distribution in §6.1.2. We have stated the property here for completeness and for future reference.

Before moving on to a discussion of the negative binomial distribution in §5.3, we present a few worked examples.

EXAMPLE 1:   Calls to a toll-free telephone hotline service are made randomly and independently at an expected rate of two per minute. The hotline service has five customer service representatives, none of whom is currently busy. Determine the probability that the hotline receives fewer than five calls in the next minute, using a Poisson model.

Let X be the number of calls in the next minute. Since calls are made randomly, independently, and arrive at a constant expected rate, a Poisson model with mean 2 is appropriate. Hence, the desired probability is

$$\Pr(X < 5) = \sum_{x=0}^{4} \frac{2^x e^{-2}}{x!}$$
$$= e^{-2} \left\{ 1 + 2 + 2 + \frac{4}{3} + \frac{2}{3} \right\}$$
$$= 7e^{-2}$$
$$\approx .9473.$$ ■

EXAMPLE 2:   A computer hardware company manufactures a particular type of microchip. There is a 0.1% chance that any given microchip of this type is defective, independent of the other microchips produced. Determine the probability that there are at least 2 defective microchips in a shipment of 1000.

Let X be the number of defective microchips in a shipment of 1000. Then an exact model for X is the binomial model with $n = 1000$ and $p = .001$, since the microchips are assumed to be independent, and the desired probability is

$$\Pr(X \geq 2) = 1 - \Pr(X = 0) - \Pr(X = 1)$$
$$= 1 - \binom{1000}{0}(.999)^{1000}(.001)^0 - \binom{1000}{1}(.999)^{999}(.001)^1$$
$$\approx 1 - .3676954 - .3680635$$
$$= .2642411.$$

Since n is large, p is small, and np is moderate in size, an approximate model for X is the Poisson model with $\lambda = np = 1$. The desired probability, using this approximation, is

$$\Pr(X \geq 2) = 1 - \Pr(X = 0) - \Pr(X = 1)$$
$$\approx 1 - e^{-\lambda} - \frac{\lambda^1 e^{-\lambda}}{1!}$$
$$= 1 - 2e^{-1}$$
$$\approx .2642411. \qquad \blacksquare$$

EXAMPLE 3: The number of visits per minute to a particular Website providing news and information can be modeled using a Poisson distribution with mean 5. The Website can only handle 20 visits per minute and will crash if this number of visits is exceeded. Determine the probability that the site crashes in the next minute.

Let X be the number of visits to the site in the next minute. By assumption, $X \sim$ Poisson(5). Hence, the desired probability is

$$\Pr(X > 20) = 1 - \Pr(X \leq 20)$$
$$= 1 - \sum_{x=0}^{20} \frac{5^x e^{-5}}{x!}$$
$$\approx 1 - 148.41315 e^{-5}$$
$$\approx 0. \qquad \blacksquare$$

EXAMPLE 4: A national fast-food chain provides health insurance to its employees through HMOs (health maintenance organizations) affiliated with a national insurance carrier. The employee turnover is about 40% per year. However, the size and demographic composition of the work force have remained fairly stable over time. On the basis of the experience of employers similar to this one, the insurance company anticipates three hospitalizations per month. Determine the probability that more than four of the fast-food chain's employees are hospitalized in the next month. Assume that hospitalizations are independent for different employees.

Let N be the number of hospitalizations in the next month. If the membership of the insured population did not change and if we were given the probability of a hospitalization for an individual insured, then we could model N using a binomial distribution. However, since the membership of the insured population does change, a binomial model is not appropriate. Fortunately, the size and demographic composition of the insured population remain stable. Consequently, an appropriate model for N is the Poisson distribution with parameter $\lambda = 3$.

Hence, the desired probability is

$$\Pr(N > 4) = 1 - \Pr(N \leq 4)$$
$$= 1 - \sum_{n=0}^{4} \frac{3^n e^{-3}}{n!}$$
$$= 1 - .0497871 - .1493612 - .2240418$$
$$\quad - .2240418 - .1680314$$
$$= .1847367$$
$$\approx 18.5\%. \qquad \blacksquare$$

EXAMPLE 5: The incidence of hospitalization tends to vary with the flu season over the year. Data on a particular group suggests that one should expect one hospitalization in September, one hospitalization in October, two hospitalizations in November, and three hospitalizations in December. Determine the probability that there are fewer than five hospitalizations in the September to December period. Assume that hospitalizations are independent among plan members and from month to month.

Let N_1 be the number of hospitalizations in September, N_2 be the number of hospitalizations in October, N_3 be the number of hospitalizations in November, and N_4 be the number of hospitalizations in December. From the given information,

$$N_1 \sim \text{Poisson}(1), \qquad N_2 \sim \text{Poisson}(1),$$
$$N_3 \sim \text{Poisson}(2), \qquad N_4 \sim \text{Poisson}(3).$$

We are interested in the quantity $X = N_1 + N_2 + N_3 + N_4$. Since the N_j are assumed to be mutually independent, $X \sim \text{Poisson}(1 + 1 + 2 + 3)$ (i.e., $X \sim \text{Poisson}(7)$). Consequently, the desired probability is

$$\Pr(X < 5) = \sum_{x=0}^{4} \frac{7^x e^{-7}}{x!}$$
$$= .0009119 + .0063832 + .0223411 + .0521293$$
$$+ .0912262$$
$$= .1729917$$
$$\approx 17\%.$$ ∎

5.3 The Negative Binomial Distribution

Definition and Interpretation

A random variable X is said to have a negative binomial distribution with parameters r and p if its probability mass function is given by

$$p_X(x) = \frac{\Gamma(r+x)}{\Gamma(r)\Gamma(x+1)} p^r (1-p)^x, \qquad x = 0, 1, 2, \dots,$$

where $\Gamma(\cdot)$ denotes the gamma function.[8] Here, r is a positive real number and p is a real number in the interval $[0, 1]$. When r is a positive integer, the probability mass function reduces to the more familiar form

$$p_X(x) = \binom{r+x-1}{r-1} p^r (1-p)^x, \qquad x = 0, 1, 2, \dots,$$

using the fact that $\Gamma(x) = (x-1)!$ for all positive integers x. Possible graphs of p_X are given in Figure 5.4. The notation NegativeBinomial(r, p) signifies a negative binomial distribution with parameters r and p.

---

[8] The gamma function was introduced in Example 7 of §4.3.1 in connection with the gamma distribution. The gamma function is defined by $\Gamma(r) = \int_0^\infty z^{r-1} e^{-z} dz$ for $r \neq 0, -1, -2, \dots$ and has the property that $\Gamma(n) = (n-1)!$ for all positive integers n. For details, see Appendix A.

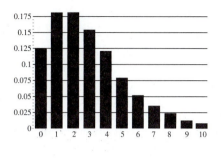 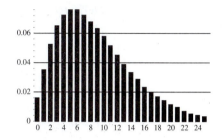

a. NegativeBinomial (3, .5) b. NegativeBinomial (3, .25)

FIGURE 5.4 Particular Negative Binomial Distributions

A negative binomial random variable X has the following interpretation when r is a positive integer: It is the number of failures that occur in a sequence of independent trials before the rth success is obtained, where the probability of success on an individual trial is p.[9]

However, in insurance applications, it arises in a completely different way. Indeed, it is the unconditional number of claim occurrences during a fixed time period when claim frequency is modeled using a Poisson distribution, but there is uncertainty in the true value of the Poisson parameter λ (i.e., the expected number of claims is uncertain), and this uncertainty is modeled using a gamma distribution.[10] In the language of mixtures, which we introduced in §4.1.12, the negative binomial distribution is simply a continuous mixture of Poissons with gamma mixing weights.

The reader may be curious why this distribution is referred to as "negative binomial" and whether there is some connection to the binomial distribution discussed in §5.1. The rationale for the name will become clearer after we have seen the form of the moment generating function for a negative binomial random variable.

The negative binomial distribution, as we have defined it, is sometimes referred to as the *Polya distribution* in the insurance literature. However, we will use the more familiar terminology in this book.

---

[9] Many authors define the negative binomial distribution to be the number of *trials* until the rth success is obtained rather than the number of *failures*. With this approach, the values of X with nonzero probability would be $x = r, r + 1, r + 2, \ldots$; that is, the distribution of probability would start at $x = r$ rather than at $x = 0$. Since we would like to use the negative binomial distribution to model uncertain frequencies, such as the number of claims on an insurance policy, we clearly should have nonzero probability masses assigned to all nonnegative integers, not just to the integers greater than or equal to r. For this reason, our definition of the negative binomial distribution is preferable. Our definition also allows us to interpret the negative binomial distribution as a particular mixture of Poisson distributions, as we will soon see. Moreover, our definition is consistent with the way that the negative binomial distribution is defined in the software package *Mathematica* (see Appendix F).

[10] The gamma distribution was introduced in Example 7 of §4.3.1 and will be discussed in detail in §6.1.2. For the purposes of this section, we need only know that a gamma random variable is one with density $f(\lambda) = \alpha^r \lambda^{r-1} e^{-\alpha\lambda} / \Gamma(r)$, $\lambda \geq 0$, where $\Gamma(\cdot)$ is the gamma function defined earlier.

Derivation of the Probability Mass Function

Most elementary probability books derive the probability mass function for a negative binomial random variable using combinatorial principles and its interpretation as the number of failures in a sequence of independent trials before the rth success. In the exercises, the reader will have an opportunity to derive p_X in this way. However, here we derive p_X by considering the Poisson probability mass function with an uncertain parameter and modeling the uncertainty in this parameter using a gamma distribution.

Hence, suppose that $(X|\Lambda = \lambda) \sim \text{Poisson}(\lambda)$ and suppose that Λ has the density function

$$f_\Lambda(\lambda) = \frac{\alpha^r \lambda^{r-1} e^{-\alpha\lambda}}{\Gamma(r)}, \qquad \lambda \geq 0,$$

where r and α are positive real numbers. (It is straightforward to verify that the formula for f_Λ defines a probability density function. See Appendix A.) Then, by the distributional form of the law of total probability, the unconditional probability mass function for X is

$$
\begin{aligned}
p_X(x) &= \int_0^\infty p_{X|\Lambda=\lambda}(x) \cdot f_\Lambda(\lambda) d\lambda \\
&= \int_0^\infty \left(\frac{\lambda^x e^{-\lambda}}{x!} \right) \cdot \left(\frac{\alpha^r \lambda^{r-1} e^{-\alpha\lambda}}{\Gamma(r)} \right) d\lambda \\
&= \frac{\alpha^r}{x! \Gamma(r)} \int_0^\infty \lambda^{r+x-1} e^{-(\alpha+1)\lambda} d\lambda \\
&= \frac{\Gamma(r+x)}{x! \Gamma(r)} \cdot \left(\frac{\alpha}{\alpha+1} \right)^r \cdot \left(\frac{1}{\alpha+1} \right)^x \int_0^\infty \frac{(\alpha+1)^{r+x} \lambda^{r+x-1} e^{-(\alpha+1)\lambda}}{\Gamma(r+x)} d\lambda.
\end{aligned}
$$

However,

$$\int_0^\infty \frac{(\alpha+1)^{r+x} \lambda^{r+x-1} e^{-(\alpha+1)\lambda}}{\Gamma(r+x)} d\lambda = 1$$

since the integrand of this integral has the same form as the density function for Λ (with $r + x$ taking the role of r and $\alpha + 1$ taking the role of α) and, hence, must also be a density function. Consequently, putting $p = \alpha/(\alpha+1)$ and using the fact that $\Gamma(x+1) = x!$ for all nonnegative integers x, we see that the probability mass function for X is

$$p_X(x) = \frac{\Gamma(r+x)}{\Gamma(x+1)\Gamma(r)} p^r (1-p)^x.$$

When r is a positive integer, $\Gamma(r+x) = (r+x-1)!$, $\Gamma(r) = (r-1)!$, and the formula for p_X reduces to the familiar form

$$p_X(x) = \binom{r+x-1}{r-1} p^r (1-p)^x.$$

The preceding derivation illustrates a computational technique that is quite useful for calculating the distribution of a mixture. The technique is to rewrite the integral that results from an application of the law of total probability as a constant times the integral of a function which is easily recognized to be a probability density and, hence, integrates to 1. This technique works for many different types of mixtures and will be used throughout the book without further elaboration.

Mean, Variance, and Higher Moments

The mean and variance of a negative binomial random variable X are, respectively,

$$E[X] = \frac{r(1-p)}{p},$$

$$Var(X) = \frac{r(1-p)}{p^2}.$$

The moment generating function of X is given by

$$M_X(t) = \left(\frac{p}{1-(1-p)e^t} \right)^r, \qquad t \in \mathbf{R}, (1-p)e^t < 1.$$

The formulas for the mean and variance can be readily obtained from the formula for the moment generating function. The derivation of the moment generating function is more delicate and will be postponed until §9.4, where we consider the calculation of moment generating functions for general mixtures. Notice, however, that the formula for $M_X(t)$ can be obtained *formally* from the moment generating function of a binomial distribution by substituting $-r$ and $1-1/p$ for n and p, respectively. This helps explain the reason for the terminology "negative binomial."

Note that the formula for the mean of a negative binomial distribution agrees with what our intuition tells us it should be. Indeed, if X represents the number of tails in a succession of coin tosses until the rth head is obtained, then according to our intuition, we should "expect" to flip the coin $1/p$ times to get our first head and, more generally, r/p times to get our rth head. That is, we should "expect" to get $(r/p) - r = r(1-p)/p$ tails before getting the rth head.

Note also that the mean and variance of a negative binomial random variable X are related by the inequality $Var(X) > E[X]$. This relationship should be contrasted to the inequality $Var(X) < E[X]$ for binomial random variables and the equality $Var(X) = E[X]$ for Poisson random variables. The fact that $Var(X) > E[X]$ for negative binomial random variables X should not be surprising when one considers that we derived the negative binomial probability mass function from the Poisson probability mass function by allowing the Poisson parameter to be uncertain. This additional uncertainty has the effect of increasing the variance while keeping the expected value the same, with the result that $Var(X) > E[X]$ for the unconditional distribution of X. The rationale for this increase in variance will become clearer when we discuss the formulas for calculating unconditional expectation and variance in §9.3.

Techniques for Calculating Probabilities

The simplest way to compute probabilities is with the recursive formula

$$\frac{p_X(x)}{p_X(x-1)} = \frac{r+x-1}{x} \cdot (1-p),$$

which follows directly from the definition of the probability mass function. This recursive formula can also be used to determine the character of the graph of the probability mass function for various values of r and p. This application of the formula is considered in the exercises.

Effect of Arithmetic Operations

An independent sum of negative binomial random variables with common parameter p is again negative binomial. In particular, if $X_1 \sim$ NegativeBinomial(r_1, p), $X_2 \sim$ NegativeBinomial(r_2, p), and X_1, X_2 are independent, then

$$X_1 + X_2 \sim \text{NegativeBinomial}(r_1 + r_2, p).$$

This property follows directly from the formula for the moment generating function of a negative binomial random variable given earlier and the general formula for the moment generating function of an independent sum. It also makes intuitive sense when one interprets the negative binomial random variable as the number of failures until the r_jth success in a sequence of independent trials with common success probability p. Note that the property only holds when the components of the sum are independent and have common parameter p. Intuition again suggests that this condition is not unreasonable.

Relationship with Other Distributions

The negative binomial distribution is related to the Poisson distribution and the gamma distribution in the following way:

If $(X|\Lambda = \lambda) \sim$ Poisson(λ) and $\Lambda \sim$ Gamma(r, α), then $X \sim$ NegativeBinomial$(r, \alpha/(\alpha + 1))$.

This relationship follows directly from our derivation of the negative binomial probability mass function and, as noted previously, has important applications in insurance.

Two other observations are worth making. First, the negative binomial distribution can be considered the discrete analog of the gamma distribution. In exercise 22 at the end of this chapter, the reader will have the opportunity to show this by deriving the gamma density by taking a limit of negative binomial probabilities. Second, the negative binomial distribution with $r = 1$ has the very special property that

$$\Pr(X > s + t | X > t) = \Pr(X > s) \qquad \text{for all positive integers } s, t.$$

A negative binomial distribution with $r = 1$ is known as a geometric distribution. The geometric distribution is discussed in detail in the next section. However, before considering the geometric distribution, we present a few worked examples for the negative binomial distribution.

EXAMPLE 1:   A couple with two girls has decided to keep having children until they have exactly two boys. The probability of a male birth is approximately 51%. Determine the probability that the couple will have at least two more girls before completing their family. What is the expected size of such a family?

Let X be the number of additional girls that the couple has before getting two boys. Then $X \sim$ NegativeBinomial$(2, .51)$. Hence, the desired probability is

$$\Pr(X \geq 2) = 1 - \Pr(X = 0) - \Pr(X = 1)$$

$$= 1 - (.51)^2 - \binom{2}{1}(.51)^2(.49)$$

$$\approx .485.$$

The expected size of such a family, counting the two girls that the couple already has and the two boys they plan to have, is

$$2 + E[X] + 2 = 2 + \frac{2(.49)}{.51} + 2 \approx 5.92.$$ ■

EXAMPLE 2:   A telemarketer selling subscriptions to a local newspaper has a sales quota of five subscriptions per night. Normally, only one in ten people contacted by telephone agrees to purchase a subscription to the newspaper. However, due to exceptional sales skills, this telemarketer's success rate is one in three. Determine the probability that this telemarketer will contact more than ten people who decline a subscription before reaching the nightly quota.

Let X be the number of decliners before the sales quota is reached. Suppose that individuals contacted by telephone make their subscription decisions independently. Then $X \sim$ NegativeBinomial(5, 1/3). Hence, the desired probability is

$$\Pr(X > 10) = 1 - \Pr(X \le 10)$$
$$= 1 - \sum_{x=0}^{10} \binom{4+x}{4} \left(\frac{1}{3}\right)^5 \left(\frac{2}{3}\right)^x$$
$$= 1 - .5959352$$
$$\approx .4041.$$ ■

EXAMPLE 3:   A midsized employer has recently changed carriers for the group health insurance plan that it provides to its employees. On the basis of its experience with similar employers, the new insurance carrier expects four hospitalization claims from this employee group during the next year. However, the carrier is not completely certain if the new employee group's experience will turn out to be similar to these other employee groups, and it decides to model the uncertainty in the expected number of hospitalizations using a Gamma(2, 0.5) density. Determine the probability that the carrier receives more than four hospitalization claims from the new group during the coming year. Assume that the number of hospitalizations on groups for which there is extensive claim experience has a Poisson model.

Let N be the number of hospitalization claims filed by members of the new group during the coming year. Then, from the given information,

$$(N|\Lambda = \lambda) \sim \text{Poisson}(\lambda) \text{ and } \Lambda \sim \text{Gamma}(2, 0.5).$$

Consequently, the unconditional distribution of N is negative binomial with parameters $r = 2$ and $p = \alpha/(\alpha + 1) = (0.5)/(1.5) = 1/3$ (i.e., $N \sim$ NegativeBinomial(2, 1/3)). Hence, the desired probability is

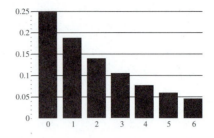

FIGURE 5.5   Geometric(.25)

$$\Pr(N > 4) = 1 - \Pr(N \leq 4)$$

$$= 1 - \sum_{n=0}^{4} \binom{n + 1}{1} \left(\frac{1}{3}\right)^2 \left(\frac{2}{3}\right)^n$$

$$= 1 - \frac{1}{9} - \frac{4}{27} - \frac{4}{27} - \frac{32}{243} - \frac{80}{729}$$

$$= \frac{256}{729}$$

$$= .351166$$

$$\approx 35\%.$$

5.4   The Geometric Distribution

Definition and Interpretation

A random variable X is said to have a geometric distribution with parameter p if its probability mass function is given by

$$p_X(x) = p(1 - p)^x, \qquad x = 0, 1, \ldots.$$

Here, p is a real number in the interval [0, 1]. A possible graph for p_X is given in Figure 5.5. The notation Geometric(p) signifies a geometric distribution with parameter p.

A geometric random variable X has the following interpretation: It is the number of failures that occur in a sequence of independent trials before the first success is obtained, where the probability of success on an individual trial is p.[11] In applications, the geometric distribution is often used to model uncertain frequencies, such as the frequency of claims on an auto insurance policy.

Note that the geometric distribution is a special case of the negative binomial distribution discussed in the previous section. The reason for the name "geometric" should be

---

[11] Many authors define the geometric distribution to be the number of *trials* until the first success is obtained rather than the number of *failures*. However, to maintain consistency with our previous definition of the negative binomial distribution, we will continue to define the geometric and the negative binomial distributions in terms of the number of failures. This definition is also consistent with the way in which the geometric distribution is defined in the software package *Mathematica* (see Appendix F).

clear from the form of p_X; indeed, the values $p_X(x)$, $x = 0, 1, 2, \ldots$ are simply terms in a geometric progression with common ratio $1 - p$.

Geometric distributions are characterized by the property that

$$\Pr(X > s + t | X > t) = \Pr(X > s) \qquad \text{for all positive integers } s, t.$$

Indeed, it is fairly straightforward to show that every *discrete*[12] distribution that satisfies this property is a geometric distribution with $p = \Pr(X = 0)$.

This property is usually described as the **lack of memory** or **memoryless property**.[13] The rationale for the terminology is clear when we consider the interpretation of the geometric distribution in the context of coin tosses, with success being the occurrence of a head. Indeed, since coin tosses are independent, the coin does not "remember" how many tails have occurred in the past. Thus, the distribution of the number of tails to be observed in the future until the first head occurs is exactly the same as if the coin had never been tossed.

From a graphical perspective, the memoryless property simply asserts that the graph of the probability mass function p_X is similar in the sense of geometry (i.e., proportional) to every graph that can be obtained from p_X by discarding the first n probability masses, where n can be any positive integer.

Derivation of the Probability Mass Function

The probability mass function for a geometric random variable X follows directly from its interpretation as the number of tails in a sequence of coin tosses until the first head is obtained. Indeed, the event $X = x$ is simply the event of obtaining $x - 1$ tails followed by a head, which from basic principles has probability $(1 - p)^{x-1} p$, where p is the probability of heads on a single toss. Hence,

$$p_X(x) = p(1 - p)^{x-1}, \qquad x = 0, 1, 2, \ldots$$

as claimed.

Mean, Variance, and Higher Moments

The mean and variance of a geometric random variable X are, respectively,

$$E[X] = \frac{(1 - p)}{p},$$

$$Var(X) = \frac{(1 - p)}{p^2}.$$

The moment generating function of X is given by

$$M_X(t) = \frac{p}{1 - (1 - p)e^t}, \qquad t \in \mathbf{R}, (1 - p)e^t < 1.$$

---

[12] Every *continuous* distribution that satisfies this property for all positive *real* numbers s, t is an exponential distribution with $\lambda = 1/E[X]$. The exponential distribution was introduced in Example 2 of §4.2.1 and will be discussed in detail in §6.1.1.

[13] The memoryless property is also referred to as the **ageless property** when the random variable X with this property represents a lifetime, since in this context, it asserts that the entity whose lifetime is given by the variable X does not age. See §6.1.1 for more details.

The formula for the moment generating function follows directly from the definition of moment generating function and the summation formula for a geometric series. Indeed,

$$M_X(t) = E[e^{tX}]$$

$$= \sum_{x=0}^{\infty} e^{tx} p(1-p)^x$$

$$= p \sum_{x=0}^{\infty} \{(1-p)e^t\}^x$$

$$= \frac{p}{1-(1-p)e^t}, \qquad (1-p)e^t < 1.$$

Formulas for the mean, variance, and also the skewness are then readily obtainable from $M_X(t)$. Details are left to the reader.

Techniques for Calculating Probabilities

The probability mass function is easy to evaluate. Hence, no special computational techniques are required.

Effect of Arithmetic Operations

An independent sum of geometric random variables with common parameter p is negative binomial. In particular, if $X_1, \ldots, X_r$ are independent identically distributed random variables such that $X_j \sim \text{Geometric}(p)$ for $j = 1, \ldots, r$, then $X_1 + \cdots + X_r \sim$ NegativeBinomial(r, p).

This property follows directly from the formula for the moment generating function of a geometric random variable and the general formula for the moment generating function of an independent sum. It also makes intuitive sense when one interprets the geometric random variable as the number of failures until the first success and the negative binomial random variable as the number of failures until the rth success in a sequence of independent trials with common success probability.

Relationship with Other Distributions

As we have already noted, the Geometric(p) distribution is the special case of the NegativeBinomial(r, p) distribution with $r = 1$. Hence, the relationships for the negative binomial distribution discussed in the previous section apply to the geometric distribution as well (taking r to be 1, of course). In the interest of brevity, we do not restate these relationships here.

One particular relationship for the geometric distribution is worth highlighting, however: The geometric distribution can be considered the discrete analog of the exponential distribution. In exercise 21 at the end of this chapter, the reader will have the opportunity to demonstrate this important fact by deriving the exponential density as a limit of probabilities of geometric type.

EXAMPLE 1:   A manufacturer is interested in the reliability of the light bulbs it produces. Light bulbs typically fail when they are switched on and not while they are functioning. Suppose that the probability of failure each time the light bulb is switched on is p and the performance on a given trial has no effect on future performance unless the bulb fails, in which case all future trials fail with certainty. The manufacturer is interested in the probability that the light bulb functions at least 100 times. For what values of p is this probability at least 90%?

Let X be the number of times that a given light bulb functions. Then, from the given information, $X \sim \text{Geometric}(p)$, and the probability of interest is $\Pr(X \geq 100)$. Note that success in the sense of the interpretation of a geometric random variable discussed earlier means that the light bulb fails. Now, from the formula for the probability mass function of a geometric random variable and the formula for the summation of a geometric series, we have

$$\Pr(X \geq 100) = \sum_{x=100}^{\infty} p(1 - p)^x$$

$$= p(1 - p)^{100} \sum_{n=0}^{\infty} (1 - p)^n$$

$$= p(1 - p)^{100} \frac{1}{1 - (1 - p)}$$

$$= (1 - p)^{100}.$$

Hence, the requirement that $\Pr(X \geq 100) \geq .90$ is equivalent to the requirement

$$(1 - p)^{100} \geq .90.$$

That is,

$$p \leq .0010531.$$

Hence, to ensure the desired reliability, the probability of failure on a given trial should be 0.1% or less. ∎

5.5   Exercises

1. In this question, we analyze the graphical properties of the binomial distribution.

   a. Qualitatively describe how the binomial distribution is affected by changes in the parameters n and p. In particular, describe what happens to the graph of the probability mass function as p approaches 0, as p approaches 1, and as n becomes large. What is the effect of replacing p by $1 - p$? Under what circumstances is the distribution symmetric? How do changes in n and p affect the mean, variance, and skewness of the distribution?

   b. The mass function bar charts in Figure 5.1 suggest that the mass function for a binomial random variable first increases and then decreases. By considering the ratio $p_X(x + 1)/p_X(x)$, show algebraically that the observed behavior of the mass function is correct if $1/(n + 1) < p < 1 - 1/(n + 1)$. What happens if $p \leq 1/(n + 1)$? What happens if $p \geq n/(n + 1)$?

   c. Show that the mode(s) m of the binomial distribution satisfy the condition

   $$(n + 1)p - 1 \leq m \leq (n + 1)p.$$

A **mode** for a discrete probability distribution is a value of x for which $p_X(x)$ is greatest. Under what circumstances is there more than one mode?

2. In this question, we analyze the graphical properties of the Poisson distribution.

   a. Qualitatively describe how the Poisson distribution is affected by changes in the parameter λ. In particular, describe what happens to the graph of the probability mass function as λ approaches 0, and as λ becomes large. How do changes in λ affect the mean, variance, and skewness of the distribution?

   b. The mass function bar charts in Figure 5.3 suggest that the mass function first increases and then decreases. By considering the ratio $p_X(x + 1)/p_X(x)$, show algebraically that the observed behavior of the mass functions is correct if $\lambda > 1$. Under what circumstances is the mass function strictly decreasing? Explain. Show that if the parameter λ of a Poisson distribution is a positive integer, then the distribution has exactly two modes, one each at $\lambda - 1$ and λ, but if the parameter λ is not a positive integer, then the mode is the unique integer m such that $\lambda - 1 < m < \lambda$.

3. In this question, we analyze the graphical properties of the negative binomial distribution.

   a. Qualitatively describe how the negative binomial distribution is affected by changes in the parameters r and p. In particular, describe what happens to the graph of the probability mass function as p approaches 0, as p approaches 1, and as r becomes large. How do changes in r and p affect the mean, variance, and skewness of the distribution?

   b. The mass function bar charts in Figures 5.4 and 5.5 suggest that the mass function increases and then decreases, or just decreases. By considering the ratio $p_X(x + 1)/p_X(x)$, show algebraically that the observed behavior of the mass functions is correct. Show that the mass function is strictly decreasing precisely when the condition $r(1 - p) \leq 1$ is satisfied.

   c. Show that the mode(s) m of the distribution satisfy the condition

   $$\frac{(r - 1)(1 - p)}{p} - 1 \leq m \leq \frac{(r - 1)(1 - p)}{p}.$$

4. In §5.1, we noted that the binomial distribution can be used as a model for the number of subjects in a random sample with a particular trait when the sample is selected from a fixed population with replacement. In this question, we develop an appropriate model when subjects are selected without replacement.

   a. Consider a population with two distinguishable types of subjects. Suppose that the population has α subjects of Type I and β subjects of Type II and every member of the population is one of these types. Suppose further that a random sample of size n is selected without replacement from this population, so no subject can be selected more than once. Let X be the number of subjects in the sample of Type I. Show that the distribution of X is given by

   $$p_X(x) = \frac{\binom{\alpha}{x}\binom{\beta}{n-x}}{\binom{\alpha+\beta}{n}}, \qquad x = 0, 1, \ldots, n,$$

   using basic combinatorial principles. A random variable X with a distribution of this form is said to have a **hypergeometric distribution.**

b. Show that

$$xp_X(x) = n \left(\frac{\alpha}{\alpha + \beta} \right) \cdot \frac{\binom{\alpha-1}{x-1}\binom{\beta}{n-x}}{\binom{\alpha-1+\beta}{n-1}}.$$

Deduce that

$$E[X] = n \cdot \left(\frac{\alpha}{\alpha + \beta} \right).$$

Is this consistent with your intuition? Explain.

c. Suppose that α and β are large relative to n. Show that

$$p_X(x) \approx \binom{n}{x} \alpha^x \beta^{n-x} (\alpha + \beta)^{-n}.$$

Deduce that the distribution of X can be approximated by Binomial$(n, \alpha/(\alpha + \beta))$. What does this result imply about the difference between sampling with replacement and sampling without replacement? Is this reasonable?

5. Consider a coin for which the probability of heads on a single toss is p. Let X be the number of tails until the rth head is obtained. In this question, you will derive formulas for p_X, $E[X]$, and $Var(X)$ from first principles.

a. Suppose that the rth head is obtained on the $(x + r)$th flip of the coin. Show that the number of different sequences of heads and tails leading to this outcome is given by

$$\binom{x + r - 1}{r - 1}.$$

Hint: Argue that the desired number is the number of ways of arranging x Ts and $(r - 1)$ Hs in a row.

b. Deduce that

$$p_X(x) = \binom{x + r - 1}{r - 1} p^r (1 - p)^x, \qquad x = 0, 1, \ldots.$$

c. Using the formula for p_X given in part b and the general formula $E[X^k] = \sum x^k p_X(x)$, show that

$$E[X] = \frac{r(1 - p)}{p},$$

$$Var(X) = \frac{r(1 - p)}{p^2}.$$

6. The number of claims N on a block of insurance policies is to be modeled using one of the following families of distributions: binomial, Poisson, or negative binomial. Estimates for $E[N]$ and $Var(N)$ are obtained from historical data. In each of the following cases, indicate which family of distributions is the best choice to model N.

a. $Var(N) = E[N]$
b. $Var(N) > E[N]$
c. $Var(N) < E[N]$.

7. A health insurance provider is trying to develop a probability model for the number of claims made against each of its policies in a given year. It plans to use this information when determining the level of premiums to charge its customers. One thousand policies of a similar type are examined, and the number of claims made against each policy in the previous year is observed. These observations are summarized in the following table:

Number of Claims Against a Given Policy	Number of Such Policies Among the Observed 1000
0	122
1	188
2	188
3	156
4	117
5	82
6	55
7	35
8	22
9	13
≥ 10	22
Total Observed	1000

Let N be the number of claims against a randomly chosen policy in the previous year. Our goal is to fit an approximate probability distribution to the data given in the table.

a. Construct a table for the relative frequency distribution of N using the data given. For example, the relative frequency of the event $N = 0$ is .122 because 122 of the 1000 policies examined had no claims made against them.

b. Sketch a graph of the relative frequency distribution of N using the table you constructed in part a. What common discrete distributions might be considered reasonable candidates to model these data?

c. Use the table you constructed in part a to calculate the mean and the variance of N implied by the data. Give your answers to one decimal place. What does the relationship between the mean and the variance appear to be? Is $E[N] > Var(N)$, $E[N] \approx Var(N)$, or $E[N] < Var(N)$? Does this narrow the list of "candidates" for modeling N that you gave in part b?

d. Fit a negative binomial distribution to the data using the numerical mean and variance computed in part c to determine the parameters r and p. That is, determine an algebraic formula for the probability mass function of the negative binomial distribution whose mean and variance are equal to the mean and variance implied by the data. Then use this formula to construct a table of predicted observed frequencies for the values of N given in the original table. Do you think that a negative binomial distribution is an appropriate model for this insurance company to use?

e. The given table for the observed frequencies of N has only one entry for all values greater than nine. However, the true distribution of N is defined for all nonnegative integers. In view of this observation, would you say that the mean computed in part c is an overstatement or an understatement of the truth? What about the computed variance? How might this knowledge affect the way we choose a particular negative binomial distribution to fit a given set of data?

f. A client with risk profile similar to the 1000 clients considered in the given study has just transferred her policy to this company. According to the model constructed earlier, what is the probability that this client will file more than two claims in the coming year?

8. The health insurance provider of the previous question has recently acquired a block of policies from a recently liquidated rival insurer. The insurer has discovered that the model developed in the previous question for predicting the number of claims on a given policy in a given year is inadequate for this new block of policies. You have been hired to develop a new model using the techniques of the previous question.

   You examine 1000 of these new policies and record the number of claims made against each policy in the previous year. These observations are summarized in the following table:

Number of Claims Against a Given Policy	Number of Such Policies Among the Observed 1000
0	50
1	149
2	224
3	224
4	168
5	101
6	50
7	21
8	8
9	3
≥ 10	2
Total Observed	1000

Let M be the number of claims against a randomly chosen policy from the acquired block of policies in the previous year.

a. Construct a table for the relative frequency distribution of M using the data given.

b. Sketch a graph of the relative frequency distribution of M using the table you constructed in part a. What common discrete distributions might be considered reasonable candidates to model these data?

c. Calculate the mean and variance of M implied by the data. What does the relationship between $E[M]$ and $Var(M)$ appear to be? What common discrete distribution(s) do you now think are reasonable candidates to model these data?

d. Fit a Poisson distribution to the data using the mean implied by the data to determine the parameter λ. That is, determine an algebraic formula for the probability mass function of the Poisson distribution whose mean equals the mean implied by the data. Then use this formula to construct a table of predicted observed frequencies for the values of M given in the original table. Do you think that a Poisson model is an appropriate model for the company to use for the acquired block of policies?

e. Qualitatively describe the differences between the distributions of M and N, where N is the distribution of question 7. What do these differences suggest about the overall health of the two groups of policyholders?

f. Suppose that both groups of policies are for dental insurance. The policies in one group cover the complete cost of preventive dental care (two visits per calendar year) and a portion of restorative dental procedures. The policies in the other group cover the complete cost of restorations, but only a portion of preventive dental care. Can you determine which group is which? If you were an insurer analyzing these data, which type of policy might be more prudent to offer?

9. For each of the following random variables X, determine the type of distribution (i.e., binomial, hypergeometric, etc.) that best models X. Where possible, give values for the parameters of the distribution chosen. Give reasons for your choice of distribution.

a. At a certain university, there are 2000 students enrolled in an introductory psychology course. The term papers in this course are graded by a team of teaching assistants; however, a sample of the papers is examined by the course professor for grading consistency. Experience suggests that 1% of all papers will be improperly graded. The professor selects 10 papers at random from the 2000 submitted and examines them for grading inconsistencies. X is the number of papers in the sample that are improperly graded.

b. The Sigma Chi fraternity at a certain college is holding a raffle to raise money to purchase computers for a neighboring grade school. Three prizes are to be awarded by drawing from a jar stubs of the tickets sold. Winning stubs are to be returned to the jar after each draw so that each ticket is eligible to win all three prizes. Of the 1000 tickets sold, 100 are purchased by faculty at the college. X is the number of prizes won by faculty members.

c. Calls to an Internet service provider are placed independently and at random. During evening hours, the service provider receives an average of 100 calls an hour. X is the number of calls received between 7:00 P.M. and 7:30 P.M. tonight.

d. An immunologist is studying blood disorders exhibited by people with rare blood types. It is estimated that 10% of the population has the type of blood being investigated. Volunteers whose blood type is unknown are tested until 100 people with the desired blood type are found. X is the number of people tested who do not have the desired rare blood type.

e. A representative of a certain telemarketing firm is given 500 numbers to call. Each number is to be called exactly once and the response is to be recorded. The response received on any given call is independent of the responses on the rest. Experience suggests that 15% of the calls will result in a favorable response. X is the number of favorable responses in the first 100 calls.

f. During peak periods, 20% of the calls made to a telephone information service cannot be completed. Each call to the telephone service has the same probability of being completed, independent of any other calls placed. A certain caller has decided to keep dialing the service's number until he is connected. X is the number of times he gets a busy signal.

g. A quality control supervisor on a certain production line chooses 100 items from a lot of 2000 manufactured items and examines them for defects. X is the number of defective items in the sample.

h. Of the customers at a certain grocery store, 20% use cash, 30% use checks, and the remaining 50% use credit or debit cards. The method of payment that any given customer uses is independent of what the other customers are using. A certain cashier has run out of coins for change and will have to close before the next customer wishing to use cash appears. X is the number of customers who can be served before this cashier must close.

i. Customers arrive at an automated teller machine independently and at random. During lunch hour, customers arrive at the machine at a rate of one per minute on average. X is the number of people who arrive between 12:15 P.M. and 12:30 P.M.

j. A personnel agency is interviewing candidates for two open positions. Management has decided to hire the first two candidates who possess the necessary skills. Experience suggests that only 30% of those interviewed will have the necessary qualifications. X is the number of candidates who must be interviewed before two suitable candidates are found.

10. Of the option contracts in a particular investment portfolio, 10% expire worthless. The contracts in this portfolio are sufficiently heterogeneous that we may assume that the values of distinct contracts at expiration are independent. The portfolio manager selects fifteen contracts from the portfolio at random and sells them for their current market price. What is the probability that at least two of these contracts expire worthless?

11. The moment generating functions $M_X(t)$ for particular discrete random variables X follow. In each case, identify the distribution (e.g., binomial, geometric, etc.) and specify its associated parameters.

a. $M_X(t) = \dfrac{1}{2 - e^t}$

b. $M_X(t) = \left(\dfrac{e^t + 1}{2}\right)^3$

c. $M_X(t) = \exp(e^t - 1)$

d. $M_X(t) = \left(\dfrac{1}{2 - e^t}\right)^3$

e. $M_X(t) = \left(\dfrac{3e^t + 1}{4}\right)^5$

12. The moment generating functions $M_X(t)$ for three discrete random variables X follow. For each of these distributions, determine the mean and variance of X and calculate $\Pr(X > 1)$ and $\Pr(X = 2)$.

a. $M_X(t) = \left(\dfrac{e^t + 3}{4}\right)^{10}$

b. $M_X(t) = \left(\dfrac{1}{4 - 3e^t}\right)^3$

c. $M_X(t) = \exp\left(2(e^t - 1)\right)$

13. On a particular type of insurance policy, at most two claims are possible in any given year. An insurer is interested in the claim occurrence behavior on a group of n newly acquired policies of this type. Let p_j be the probability that exactly j claims occur on a policy of the given type and suppose that claims on distinct policies are independent. Let X_j be the number of policies in the group of n on which exactly j claims occur. Show that the joint probability mass function of X_1 and X_2 is given by

$$p_{X_1,X_2}(x_1, x_2) = \frac{n!}{x_0!x_1!x_2!} p_0^{x_0} p_1^{x_1} p_2^{x_2},$$

where $x_0 = n - x_1 - x_2$ and $p_0 + p_1 + p_2 = 1$.

14. On a particular type of insurance policy, at most two claims are possible in a given year. There is a 10% chance that a policyholder with this type of policy will file two claims in a given year and a 25% chance that the policyholder will file one claim. A small insurance company has sold four policies of this type. Determine the probability that more than three claims are filed on these four policies in the coming year.

15. The price of a particular highly volatile stock either increases 25% or decreases 20% in any given week. The probability of an increase in any week is 55%, independent of the stock's performance on other weeks. The current price of the stock is $10. Determine the probability that the stock's price exceeds $15 four weeks from now. What is the stock's expected value two weeks hence? Four weeks hence?

16. A bank specializing in credit cards for university alumni has just purchased the rights to exclusively market its cards at a major northeastern university. On the basis of its experience at similar schools, it expects 100 delinquencies per month. However, since the demographic profile of the school is different from the other schools where the company markets its products, it is uncertain about the true number of delinquencies it should expect and decides to model this uncertainty using the density

$$f_\Lambda(\lambda) = (0.02)^2 \lambda e^{-0.02\lambda}.$$

Determine the probability that the bank has fewer than 50 delinquencies in the first month.

17. An insurer has three lines of business. Claims within each line and between lines are independent. The number of claims in each line has a negative binomial distribution. The respective parameters for the three lines are $r_1 = 2$, $p_1 = 1/4$; $r_2 = 10$, $p_2 = 1/4$; $r_3 = 5$, $p_3 = 1/4$. Determine the mean, variance, and skewness for the total number of claims on the three lines combined.

18. A temporary employment agency provides health insurance to its contract workers through a basic group health insurance policy. To qualify for coverage, one must have worked at least 100 hours in the past 4 weeks. Turnover at this agency is high because many of the workers often find permanent employment within the first 6 months of joining the agency's roster. However, the size and demographic composition of the agency's contract work force have remained fairly stable over time. Suppose that the agency's insurance carrier expects one hospitalization claim every 6 months. Determine the probability that there is at least one hospitalization claim in the next month. State any assumptions you make to determine this probability.

19. By considering the values of $p_X(0)$, $p_X(1)$, $p_X(2)$, $p_X(3)$, $p_X(4)$, $p_X(5)$, discuss the accuracy of the Poisson approximation to the binomial distribution in each of the following cases:

    a. $n = 10$, $p = .10$
    b. $n = 10$, $p = .01$
    c. $n = 50$, $p = .10$
    d. $n = 50$, $p = .01$

20. A group health insurance provider has just sold a group health plan to a new employer. On the basis of limited underwriting, there is a 30% chance the group will turn out to be a high utilizer. High-utilization groups average 50 claims per month, and low-utilization groups average 20 claims per month. The number of claims on a group for which the utilization is known follows a Poisson distribution. Determine the probability that the group submits fewer than 20 claims in the first month.

21. A certain bank would like to reduce the amount of time that its tellers are idle. You have been hired by this bank to develop an appropriate probability model for predicting the time a teller must wait for a customer to arrive.

    To simplify the analysis of the problem, you assume that there is only one teller and that the bank has just opened for business. Let T be the time until the first customer arrives and let f_T be a probability density function for T. Your objective is to determine an explicit algebraic formula for $f_T(t)$.

    As a first step toward this goal, you decide to develop a discrete probability model based on intervals (rather than exact values) of time. To be precise, you plan to develop a discrete model based on the intervals

    $$\left[0, \frac{t}{k}\right), \left[\frac{t}{k}, \frac{2}{k}t\right), \ldots, \left[\frac{k-1}{k}t, t\right)$$

    and then to deduce a formula for $f_T(t)$ using the approximation

    $$f_T(t) \approx \frac{\Pr\left(t - \frac{t}{k} \le T \le t\right)}{\frac{t}{k}}.$$

    Let t be fixed and let k be large enough so that the number of arrivals during any time interval of the type $\left[\frac{j-1}{k}t, \frac{j}{k}t\right)$, $j = 1, 2, \ldots, k$ is at most one. Let I_j be an indicator of an arrival in the interval $\left[\frac{j-1}{k}t, \frac{j}{k}t\right)$,

    $$I_j = \begin{cases} 1 & \text{if a customer arrives during } \left[\frac{j-1}{k}t, \frac{j}{k}t\right) \\ 0 & \text{otherwise.} \end{cases}$$

    Let λ be the expected number of arrivals during the first hour.
    You make the following assumptions:

    i. The probability of an arrival is the same in each of the intervals $\left[\frac{j-1}{k}t, \frac{j}{k}t\right)$.
    ii. The random variables I_j are mutually independent (i.e., an arrival in one subinterval has no effect on the probability of an arrival in another).
    iii. The expected number of arrivals in any interval of the type $(0, s)$ is λs.

Let p be the common value of $\Pr(I_j = 1)$.

a. Show that

$$\Pr\left(t - \frac{t}{k} \le T \le t\right) = (1-p)^{k-1}p.$$

Hint: Let X be the number of subintervals $\left[\frac{j-1}{k}t, \frac{j}{k}t\right)$ with no arrivals before the first customer arrives. Then $X \sim \text{Geometric}(p)$.

b. Show that $E[I_j] = p$ and $E[I_j] = \lambda\frac{t}{k}$. Deduce that

$$p = \lambda\frac{t}{k}.$$

c. Using your answers from parts a and b, show that

$$\frac{\Pr\left(t - \frac{t}{k} \le T \le t\right)}{\frac{t}{k}} = \frac{\left(1 - \frac{\lambda t}{k}\right)^k \cdot \lambda\frac{t}{k}}{\left(1 - \frac{\lambda t}{k}\right) \cdot \frac{t}{k}}.$$

Deduce that

$$f_T(t) = \lambda e^{-\lambda t}.$$

Hint: Show that $\lim\limits_{k\to\infty}\left(1 - \frac{\lambda t}{k}\right)^k = e^{-\lambda t}$ using the fact that $e = \lim\limits_{n\to\infty}\left(1 + \frac{1}{n}\right)^n$.

d. Using the formula for $f_T(t)$ derived in part c, show that T has the memoryless property:

$$\Pr(T > s + t \mid T > s) = \Pr(T > t)$$

for all $s, t \ge 0$.

e. By referring to the property derived in part d, explain why T is an appropriate model for the waiting time between customer arrivals, even though we defined T to be the waiting time until the first arrival.

22. The bank of the previous question is also interested in modeling the time until the second, third, and fourth customers arrive to determine the likelihood of having a large number of customers enter the bank within a short period of time. You have been asked to develop a model for the waiting time until the rth arrival.

Let T_r be the waiting time until the rth arrival. As in the previous question, you decide to develop a discrete model based on the intervals

$$\left[0, \frac{t}{k}\right), \left[\frac{t}{k}, \frac{2}{k}t\right), \ldots, \left[\frac{k-1}{k}t, t\right)$$

and then to deduce a formula for $f_{T_r}(t)$ using the approximation

$$f_{T_r}(t) \approx \frac{\Pr\left(t - \frac{t}{k} \le T_r \le t\right)}{\frac{t}{k}}.$$

Let t be fixed and let k be large enough so that at most one arrival can occur in any of the given intervals. As before, let I_j be an indicator of an arrival in the jth subinterval and let λ be the expected number of arrivals per hour. Further, suppose that the I_j and the subintervals $\left[\frac{j-1}{k}t, \frac{j}{k}t \right)$ satisfy the same three conditions stated in the previous question, and let p be the common value of $\Pr(I_j = 1)$.

a. Show that

$$\Pr\left(t - \frac{t}{k} \le T_r \le t \right) = \binom{k-1}{r-1} p^r (1-p)^{k-r}.$$

Hint: Show that the number of subintervals with no arrival until the rth customer arrival is NegativeBinomial(r, p).

b. Show that

$$p = \lambda \frac{t}{k}.$$

c. Using your answers from parts a and b, show that

$$\frac{\Pr\left(t - \frac{t}{k} \le T_r \le t \right)}{\frac{t}{k}}$$

$$= \frac{1}{(r-1)!} \cdot \frac{(k-1)\cdots(k-r+1)}{k^{r-1}} \cdot \frac{\lambda^r t^{r-1}}{1} \cdot \frac{\left(1 - \frac{\lambda t}{k}\right)^k}{\left(1 - \frac{\lambda t}{k}\right)^r}.$$

Deduce that

$$f_{T_r}(t) = \lambda e^{-\lambda t} \frac{(\lambda t)^{r-1}}{(r-1)!}.$$

d. Conclude that the gamma distribution is a continuous analog of the negative binomial distribution.

23. A certain automotive parts supplier would like to reduce the number of defective parts it ships to its customers without having to test every item it produces. You have been hired by this company to develop a probability model for predicting the fraction of defective parts in any given shipment.

  The company ships parts in lots of 10,000 and is prepared to test 100 parts in each lot for quality. The company wishes to ship lots with at most 1% defectives and plans to use the results of the test to determine whether or not a given lot should be shipped.

  Let P be the fraction of defective items in a given (fixed) lot. Note that P is a continuous random variable with values in [0, 1]. Without any test results, you reason that all subintervals of the same length are equally likely. Hence, you assume that the unconditional distribution of P is uniform on [0, 1]; that is, $f_P(p) = 1$ for all $p \in [0, 1]$.

  Let X be the number of defective items in a random sample of size 100 chosen from this lot of 10,000. From your knowledge of probability, you realize that if P is known to have the value p, then X has approximately a binomial distribution with parameters 100 and p; that is, $(X|P = p) \sim \text{Binomial}(100, p)$. Your objective is to determine the conditional distribution of P given $X = x$.

a. Explain why it is reasonable to assume that the distribution of $X|P = p$ is Binomial(100, p). What implicit assumption about the sampling is being made?

b. Show that the unconditional distribution of X is given by

$$\Pr(X = x) = \int_0^1 \binom{100}{x} p^x (1 - p)^{100-x} dp$$

for $x = 0, 1, \ldots, 100$.

c. Show that the conditional density of P given $X = x$ is

$$f_{P|X=x}(p) = \frac{p^x (1 - p)^{100-x}}{\int_0^1 p^x (1 - p)^{100-x} dp}.$$

d. Suppose that one defective is found. Determine the probability that the fraction of defectives in the lot is greater than 1%. *Hint:* Apply integration by parts to $\int_0^1 p(1 - p)^{99} dp$.

24. A casualty actuary is trying to predict the claim frequency for a new group of 100 policyholders. Experience with similar groups suggest that 25% of these policyholders will file at least one claim during their first year. However, since no data on this group are available, the actuary treats this 25% fraction with some skepticism and instead decides to model the fraction P using the density

$$f_P(p) = 3(1 - p)^2, \quad 0 < p < 1.$$

Suppose that no claims are filed during the first year. What is the probability that P exceeds 10%?

6

Special Continuous Distributions

In the previous chapter, we discussed four special discrete distributions that arise frequently in applications. In this chapter, we consider eight special continuous distributions.

Our discussion of these eight distributions is organized in the following way: In the first section, we consider three distributions—the exponential, the gamma, and the Pareto—which are frequently used to model uncertain sizes such as the size of an insurance claim or a monetary loss. In the second section, we consider two distributions—the Weibull and the DeMoivre—which are frequently used to model lifetimes. Finally, in the third section, we consider the three remaining distributions—the normal, the lognormal, and the beta—which arise in many contexts and do not fit easily into the other categories.

Our presentation in this chapter follows the style of the previous chapter. In particular, for each special distribution, we discuss: the definition and interpretation of the distribution; a derivation of the probability density function, where practical; formulas for calculating the mean, variance, and higher moments; techniques for calculating probabilities in practice; the effect of arithmetic operations on the distribution of probability; and the relationship of the distribution to other special distributions. We also give numerous examples throughout the chapter to illustrate how these distributions arise in practice.

6.1 Special Continuous Distributions for Modeling Uncertain Sizes

There are three distributions that are frequently used to model uncertain sizes: the exponential, the gamma, and the Pareto. As we will soon see, the gamma is an independent sum of exponentials, and the Pareto is a continuous mixture of exponentials with gamma mixing weights.

6.1.1 The Exponential Distribution

Definition and Interpretation

A random variable X is said to have an exponential distribution with parameter λ if its probability density function is given by

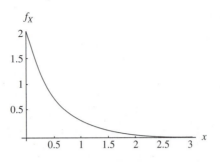

FIGURE 6.1 Exponential(2)

$$f_X(x) = \begin{cases} \lambda e^{-\lambda x} & x \geq 0, \\ 0 & x < 0. \end{cases}$$

Here, λ is a positive real number and is in fact the reciprocal of the mean of the distribution. A possible graph of f_X is given in Figure 6.1. The notation Exponential(λ) signifies an exponential distribution with parameter λ.

An exponential random variable X has two important interpretations:

1. It is the waiting time until the first arrival when arrivals are such that the number of arrivals in the time interval $[0, t]$ is Poisson(λt) for each t.
2. It is also the lifetime of an item that does not age.

These interpretations will become clearer after we have derived the probability density function for X.

Derivation of the Probability Density Function

The probability density function for the exponential distribution can be derived in several ways. We illustrate two methods that coincide with the two interpretations just stated.

Derivation 1 Consider a sequence of claim arrivals over time. Let T be the time until the first arrival and, for each t, let N_t be the number of arrivals in the time interval $[0, t]$. Suppose that $N_t \sim$ Poisson(λt) for each t.

Then for $t \geq 0$, ·

$$\begin{aligned} \Pr(T > t) &= \Pr(\text{first claim arrives after time } t) \\ &= \Pr(\text{no claim arrives in } [0, t]) \\ &= \Pr(N_t = 0) \\ &= e^{-\lambda t} \end{aligned}$$

since $N_t \sim$ Poisson(λt). On the other hand, for $t < 0$, it is trivially true that $\Pr(T > t) = 1$. Hence, the survival function of T is given by

$$S_T(t) = \begin{cases} e^{-\lambda t} & \text{for } t \geq 0, \\ 1 & \text{for } t < 0. \end{cases}$$

Consequently, the density function of T is

$$f_T(t) = \begin{cases} \lambda e^{-\lambda t} & \text{for } t \geq 0, \\ 0 & \text{for } t < 0, \end{cases}$$

which we recognize as the density of an exponential distribution with parameter λ.

Derivation 2 Consider an object that has a constant risk of failure. This means that the hazard function for this object is constant,

$$\lambda(t) = \lambda \qquad \text{for all } t \geq 0.$$

Hence, from the general formula for a survival function derived in §4.3.2,

$$S_T(t) = \exp\left(-\int_0^t \lambda dt\right) = e^{-\lambda t}, \qquad \text{for } t \geq 0.$$

This is precisely the formula for the survival function derived earlier. Hence, T has an exponential distribution with parameter λ.

Mean, Variance, and Higher Moments

The mean, variance, and skewness of an exponential random variable X are, respectively,

$$E[X] = \frac{1}{\lambda},$$

$$Var(X) = \frac{1}{\lambda^2},$$

$$\gamma_X = 2.$$

The moment generating function of X is given by

$$M_X(t) = \frac{\lambda}{\lambda - t}, \qquad t < \lambda,$$

and the cumulant generating function of X is given by

$$\psi_X(t) = \log \lambda - \log(\lambda - t), \qquad t < \lambda.$$

The formula for the moment generating function of an exponential random variable was derived in Example 6 of §4.3.1. Formulas for the mean, variance, and skewness are most easily obtained from ψ_X. Details are left to the reader.

Techniques for Calculating Probabilities

The survival function S_T is easy to evaluate. Hence, no special computational techniques are required.

Effect of Arithmetic Operations

There are two important arithmetic operations to consider: scalar multiples and sums.

A positive scalar multiple of an exponential random variable is again exponential. In particular, if $X \sim$ Exponential(λ) and a is a positive constant, then $aX \sim$ Exponential(λ/a). This property follows directly from the formula for the survival function derived earlier. It also makes intuitive sense when one interprets the exponential random variable as a waiting time or a lifetime.

On the other hand, an independent sum of identically distributed exponential random variables has a *gamma* distribution.[1] In particular, if $X_1 \sim$ Exponential(λ), $X_2 \sim$

---

[1] The gamma distribution was introduced in Examples 7 and 8 of §4.3.1, where this property was actually demonstrated using moment generating functions, and will be discussed in detail in §6.1.2.

Exponential(λ), and X_1, X_2 are independent, then $X_1 + X_2 \sim$ Gamma(2, λ). This property follows directly from the formula for the moment generating function of an exponential random variable given earlier and the general formula for the moment generating function of an independent sum. It also makes intuitive sense when one interprets the exponential and gamma random variables as waiting times (see §6.1.2 for details).

There is one other operation—the operation of taking the minimum of a collection of random variables—which is important for exponential random variables. This operation is not really an arithmetic operation per se; however, we discuss it here for lack of a better location.

The minimum of a collection of independent exponential random variables is again exponential. In particular, if $X_1 \sim$ Exponential(λ_1), $X_2 \sim$ Exponential(λ_2), and X_1, X_2 are independent, then $\min(X_1, X_2) \sim$ Exponential($\lambda_1 + \lambda_2$). This property has important consequences for the reliability of systems whose components fail independently and have exponentially distributed lifetimes (see §7.5 for details). It follows directly from the observations that (a) $\min(X_1, X_2) > x$ if and only if $X_1 > x$ and $X_2 > x$, and (b) $\Pr(X_1 > x, X_2 > x) = \Pr(X_1 > x) \Pr(X_2 > x)$ for any independent random variables X_1, X_2.

Relationship with Other Distributions

The exponential distribution is connected to the Poisson distribution in the following way:

$$\Pr(\text{Exponential}(\lambda) > t) = \Pr(\text{Poisson}(\lambda t) = 0)$$

and

$$\Pr(\text{Exponential}(\lambda) \leq t) = \Pr(\text{Poisson}(\lambda t) > 0).$$

The first of these identities follows directly from the definition of the respective probability distributions; the second is a consequence of the first.

These identities can also be understood in the context of a queuing system with a constant expected arrival rate. Indeed, from our earlier derivation of the probability mass function for an exponential random variable, we see that $\Pr(\text{Exponential}(\lambda) > t)$ represents the probability that the first arrival occurs after time t, and $\Pr(\text{Poisson}(\lambda t) = 0)$ represents the probability that there is no arrival in the time interval $[0, t]$; common sense dictates that these probabilities must be the same.

The exponential distribution can also be viewed as the continuous analog of the geometric distribution discussed in §5.4. There are several reasons for viewing these two distributions as analogs of one another. However, the most compelling reason is that they are both characterized by the important memoryless property:

$$\Pr(X > s + t \mid X > t) = \Pr(X > s) \qquad \text{for all positive real numbers } s, t.$$

Indeed, using the outline given in exercise 4 of §4.5, one can show that every *discrete* random variable that satisfies this property must have a Geometric(p) distribution for some p. Similarly, using techniques from mathematical analysis, one can show that every *continuous* random variable that satisfies this property must have an Exponential(λ) distribution for some λ.[2]

One can also derive the exponential density function as a limit of geometric probability masses, suggesting a close connection between the two distributions. For details on this derivation, see exercise 21 of §5.5.

We conclude our discussion of the exponential distribution with two numerical examples.

EXAMPLE 1:   Calls to an Internet service provider are made independently and randomly at a constant expected rate during the early morning hours. It is currently 2 A.M. and the service provider's lines are all open. What is the probability that all the lines will remain open for the next 5 minutes if the service provider expects ten calls over the next half hour?

Let T be the time in minutes until the next call is received. Since calls are made randomly and independently at a constant expected rate, an appropriate model for T is an exponential distribution. From the given information, the service provider expects to receive ten calls in the next 30 minutes (i.e., $1/3$ call per minute). Hence, $T \sim$ Exponential($1/3$) and the desired probability is

$$\Pr(T > 5) = e^{-5/3} \approx .1889. \qquad \blacksquare$$

EXAMPLE 2:   A professor takes early retirement on July 1, exactly 3 months after his 60th birthday, and purchases a term life insurance policy for $25,000. The maturity date of the policy coincides with his 65th birthday. According to a mortality table for similarly situated men, the probability that a male of exact age 60 dies before reaching his 61st birthday is .0137604. Determine the probability that the professor dies before reaching his 61st birthday if deaths are assumed to be exponentially distributed between ages.

Let T be the future lifetime of a 60-year-old male. Then from the given information, $\Pr(T \leq 1) = .0137604$. We are interested in determining $\Pr(T \leq 1 | T > \frac{1}{4})$.

The assumption of an exponential distribution of deaths means that

$$\Pr(T > t) = e^{-\lambda t}, \qquad t \in (0, 1).$$

Since $\Pr(T > 1) = 1 - .0137604 = .9862396$, $\lambda = -\log\{\Pr(T > 1)\} = 0.013856$. Consequently, the desired probability is

$$\Pr\left(T \leq 1 | T > \frac{1}{4}\right) = 1 - \Pr\left(T > 1 | T > \frac{1}{4}\right)$$

$$= 1 - \frac{\Pr(T > 1)}{\Pr(T > \frac{1}{4})}$$

$$= 1 - \frac{e^{-\lambda}}{e^{-\lambda/4}}$$

$$= 1 - e^{-(3/4)(0.013856)}$$

$$\approx .0103382.$$

---

[2] The demonstration of this fact requires tools in mathematical analysis that lie beyond the scope of this book. The interested reader should consult the book *Lectures on Functional Equations and Their Applications* by Janos Aczél, published by Academic Press, 1966.

We could also have obtained this using the memoryless property of the exponential. That is, $\Pr(T \le 1 | T > \frac{1}{4}) = \Pr(T \le \frac{3}{4}) = 1 - e^{-(3/4)(0.013856)} \approx .0103382$. ∎

6.1.2 The Gamma Distribution

Definition and Interpretation

A random variable X is said to have a gamma distribution with parameters r and λ if its probability density function is given by

$$f_X(x) = \begin{cases} \dfrac{\lambda^r x^{r-1} e^{-\lambda x}}{\Gamma(r)} & x \ge 0, \\ 0 & x < 0, \end{cases}$$

where $\Gamma(\cdot)$ denotes the gamma function.[3] Here, r and λ are both positive real numbers. Possible graphs of f_X are given in Figure 6.2. The notation Gamma(r, λ) signifies a gamma distribution with parameters r and λ.

From the definition of the gamma density, it is clear that the exponential distribution considered in the previous section is a special case of the gamma distribution. In particular, Exponential(λ) is simply Gamma$(1, \lambda)$. As we will soon see, the exponential and gamma random variables are related in many other important ways.

A gamma random variable X has the following interpretations when r is a positive integer:

1. It is the waiting time until the rth arrival when arrivals are such that the number of arrivals in the time interval $[0, t]$ is Poisson(λt) for each t.
2. It is also the sum obtained when r independent exponential random variables, each with parameter λ, are added together; that is, it is the variable Y given by $Y = X_1 + \cdots + X_r$, where the X_j are independent and identically distributed with $X_j \sim$ Exponential(λ).[4]

These interpretations will become clearer after we have derived the probability density function for X.

Derivation of the Probability Density Function

The probability density function for the gamma distribution can be derived in several ways. We give a derivation here based on the interpretation of the gamma random variable as a waiting time. In exercise 22 of §5.5, we outlined a distinctly different procedure for deriving the gamma density that relied on considering a limit of negative binomial probability masses. In Chapter 8, after we have discussed algebraic techniques for determining the distribution function of a sum of random variables, the reader will have the opportunity to derive the gamma density in yet another way by considering a sum of independent exponential random variables with common parameter λ.

---

[3] The gamma function is defined by $\Gamma(r) = \int_0^\infty z^{r-1} e^{-z} dz$ for $r \ne 0, -1, -2, \ldots$ and has the property that $\Gamma(n) = (n-1)!$ for all positive integers n. A discussion of the gamma function and its properties is contained in Appendix A.

[4] One could argue that this is a property rather than an interpretation. However, since sums of independent exponential random variables arise frequently in insurance in connection with aggregate losses, we state this fact here to highlight its importance.

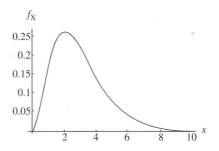

a. Gamma (3, 1)

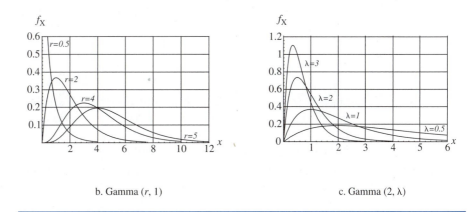

b. Gamma (r, 1)

c. Gamma (2, λ)

FIGURE 6.2 Particular Gamma Distributions

Hence, consider a sequence of claim arrivals over time. Let T be the time until the rth arrival and, for each t, let N_t be the number of arrivals in the time interval $[0, t]$. Suppose that $N_t \sim \text{Poisson}(\lambda t)$ for each t.

Then for $t \geq 0$,

$$\begin{aligned}
\Pr(T > t) &= \Pr(r\text{th claim arrives after time } t) \\
&= \Pr(\text{fewer than } r \text{ arrivals in } [0, t]) \\
&= \Pr(N_t = 0, 1, \ldots, r - 1) \\
&= e^{-\lambda t} + \sum_{n=1}^{r-1} \frac{(\lambda t)^n e^{-\lambda t}}{n!}
\end{aligned}$$

since $N_t \sim \text{Poisson}(\lambda t)$. On the other hand, for $t < 0$, it is trivially true that $\Pr(T > t) = 1$. Hence, the survival function of T is given by

$$S_T(t) = \begin{cases} \displaystyle\sum_{n=0}^{r-1} \frac{(\lambda t)^n e^{-\lambda t}}{n!} & \text{for } t \geq 0, \\ 1 & \text{for } t < 0. \end{cases}$$

Now, the density function for any continuous random variable is connected to its survival function by the relationship $f_T(t) = -S_T'(t)$. Hence, differentiating $-S_T$ and simplifying the resulting telescoping sum, we obtain (for $t \geq 0$)

$$f_T(t) = -S_T'(t)$$

$$= -\frac{d}{dt}\left\{ e^{-\lambda t} + \sum_{n=1}^{r-1} \frac{(\lambda t)^n e^{-\lambda t}}{n!} \right\}$$

$$= \lambda e^{-\lambda t} + \sum_{n=1}^{r-1} \frac{\lambda^n}{n!}(\lambda t^n e^{-\lambda t} - n t^{n-1} e^{-\lambda t})$$

$$= \frac{\lambda^r t^{r-1} e^{-\lambda t}}{(r-1)!}.$$

Consequently, the density function for T is

$$f_T(t) = \begin{cases} \dfrac{\lambda^r t^{r-1} e^{-\lambda t}}{(r-1)!} & \text{for } t \geq 0, \\ 0 & \text{for } t < 0. \end{cases}$$

Since $\Gamma(r) = (r-1)!$ when r is a positive integer, we see that this f_T has the form of the gamma density with parameters r and λ defined at the beginning of this section.

Mean, Variance, and Higher Moments

The mean, variance, and skewness of a gamma random variable X are, respectively,

$$E[X] = \frac{r}{\lambda},$$

$$Var(X) = \frac{r}{\lambda^2},$$

$$\gamma_X = \frac{2}{r^{1/2}}.$$

The moment generating function of X is given by

$$M_X(t) = \left(\frac{\lambda}{\lambda - t}\right)^r, \quad t < \lambda,$$

and the cumulant generating function of X is given by

$$\psi_X(t) = r \log \lambda - r \log(\lambda - t), \quad t < \lambda.$$

The formula for the moment generating function can be derived directly from the general definition of a moment generating function by considering the gamma density with parameters r and $\lambda - t$, as was done in Example 7 of §4.3.1. Formulas for the mean, variance, and skewness can then be obtained from the cumulant generating function ψ_X. Details are left to the reader.

Note that the skewness of a gamma distribution only depends on the parameter r and not on the parameter λ. In the exercises, the reader will have an opportunity to explore further why this is so.

Techniques for Calculating Probabilities

When r is a positive integer, probabilities can be computed fairly efficiently using the formula for the survival function derived earlier. Indeed, in this case,

$$S_X(x) = \sum_{n=0}^{r-1} \frac{(\lambda x)^n e^{-\lambda x}}{n!}, \qquad \text{for } x \geq 0.$$

This formula can also be derived directly from the density function f_X by using integration by parts successively. Details are left to the reader.

However, when r is not a positive integer, no such elementary formula for S_X exists. In this case, we must resort to some numerical technique. Most numerical techniques for computing gamma probabilities rely on the formula

$$F_X(x) = I_r(\lambda x),$$

where $I_r(\cdot)$ is the function given by

$$I_r(t) = \int_0^t \frac{z^{r-1} e^{-z}}{\Gamma(r)} dz.$$

This formula follows directly from a change of integration variable. Indeed, for $x \geq 0$,

$$\begin{aligned} F_X(x) &= \int_0^x \frac{\lambda^r t^{r-1} e^{-\lambda t}}{\Gamma(r)} dt \\ &= \int_0^{\lambda x} \frac{z^{r-1} e^{-z}}{\Gamma(r)} dz \\ &= I_r(\lambda x). \end{aligned}$$

Note that the formula $F_X(x) = I_r(\lambda x)$ reduces the computation of probability for the two-parameter family of gamma distributions to the computation of function values for the single-parameter family of functions I_r.

The function I_r, which is actually the distribution function of the gamma distribution with parameters r and 1, is known as the **incomplete gamma function.** This function has several important properties that are discussed in Appendix B.

Before the advent of computers, values of I_r were determined using tables and the recursion formula

$$I_r(t) - I_{r+1}(t) = \frac{t^r e^{-t}}{\Gamma(r+1)}.$$

However, values of I_r can now be computed using mathematical software packages such as *Mathematica*. Some of the more sophisticated hand-held calculators currently available also contain routines for evaluating incomplete gamma functions. Nevertheless, we have included tables of values for the incomplete gamma function in Appendix B, just in case you find yourself stranded on a desert island while reading this book!

Effect of Arithmetic Operations

There are two important arithmetic operations to consider: scalar multiples and sums.

A positive scalar multiple of a gamma random variable is again gamma. In particular, if $X \sim \text{Gamma}(r, \lambda)$ and a is a positive constant, then $aX \sim \text{Gamma}(r, \lambda/a)$. This property follows directly from the formula for the survival function derived earlier, and it explains why $F_X(x) = I_r(\lambda x)$ which is the property mentioned earlier for calculating probabilities.

An independent sum of gamma random variables has a gamma distribution. In particular, if $X_1 \sim \text{Gamma}(r_1, \lambda)$, $X_2 \sim \text{Gamma}(r_2, \lambda)$, and X_1, X_2 are independent, then $X_1 + X_2 \sim \text{Gamma}(r_1 + r_2, \lambda)$. This property follows directly from the formula for the moment generating function of a gamma random variable and the general formula for the moment generating function of an independent sum. It also makes intuitive sense when one interprets the gamma random variable to be a waiting time.

Relationship with Other Distributions

The gamma distribution is extremely important in applications, in part because it is connected with so many different distributions. We give a partial list here of some of the more important relationships that arise frequently in practice:[5]

1. If r is a positive integer, then

$$\text{Pr}(\text{Gamma}(r, \lambda) > t) = \text{Pr}(\text{Poisson}(\lambda t) \leq r - 1)$$

and

$$\text{Pr}(\text{Gamma}(r, \lambda) \leq t) = \text{Pr}(\text{Poisson}(\lambda t) > r - 1).$$

This property follows from the waiting time interpretation of the gamma distribution.

2. The gamma distribution can be considered the continuous analog of the negative binomial distribution (see exercise 22 in §5.5).

3. A continuous mixture of Poissons with gamma mixing weights is negative binomial. In particular, if $(N|\Lambda = \lambda) \sim \text{Poisson}(\lambda)$ and $\Lambda \sim \text{Gamma}(r, \alpha)$, then $N \sim \text{NegativeBinomial}(r, \alpha/(\alpha + 1))$. This property was demonstrated in our discussion of the negative binomial distribution in §5.3.

4. If the distribution of a Poisson parameter prior to knowledge of the value assumed by the Poisson variable is gamma, then the distribution of the Poisson parameter after knowledge of the value assumed by the Poisson variable is also gamma. In particular, if $(N|\Lambda = \lambda) \sim \text{Poisson}(\lambda)$ and $\Lambda \sim \text{Gamma}(r, \alpha)$, then $(\Lambda|N = n) \sim \text{Gamma}(r + n, \alpha + 1)$. This property will be derived in the exercises using Bayes' theorem, and it has important applications in insurance rating.

5. A continuous mixture of exponentials with gamma mixing weights is Pareto. In particular, if $(X|\Lambda = \lambda) \sim \text{Exponential}(\lambda)$ and $\Lambda \sim \text{Gamma}(s, \beta)$, then $X \sim \text{Pareto}(s, \beta)$. This property can be used to define the Pareto distribution, which is considered in §6.1.3.

6. A continuous mixture of gammas with gamma mixing weights is generalized Pareto. In particular, if $(X|\Lambda = \lambda) \sim \text{Gamma}(r, \lambda)$ and $\Lambda \sim \text{Gamma}(s, \beta)$, then $X \sim \text{GeneralizedPareto}(r, s, \beta)$. This property will be derived in the exercises, where a formal definition of the generalized Pareto random variable will be given.

7. If $(X|\Lambda = \lambda) \sim \text{Gamma}(r, \lambda)$ and $\Lambda \sim \text{Gamma}(s, \beta)$, then $(\Lambda|X = x) \sim \text{Gamma}(r + s, x + \beta)$. This property will be derived in the exercises and has important applications in insurance rating.

---

[5] Some of these relationships make reference to distributions that are considered later in this chapter or in the exercises. We list them here for the sake of completeness and for future reference.

8. The gamma distribution is related to the beta distribution (to be discussed in §6.3.3) in the following way: If $X \sim$ Gamma(r, λ), $Y \sim$ Gamma(s, λ), and X and Y are independent, then $X/(X + Y) \sim$ Beta(r, s). This property is important in insurance when considering the size of payouts on a particular block of business in relation to total payout. It is also important in industrial quality control.

9. The gamma distribution is related to the normal distribution (to be discussed in §6.3.1) in the following way: If $Z \sim$ Normal$(0, 1)$, then $Z^2 \sim$ Gamma$(1/2, 1/2)$. Hence, if $Z_1, \ldots, Z_n$ are independent, standard normal random variables, then $Z_1^2 + \cdots + Z_n^2 \sim$ Gamma$(n/2, 1/2)$. This property is important in the statistical estimation of the variance parameter for a Normal distribution. It is also important when determining the distribution of kinetic energy or the distribution of power in an electrical circuit (see §7.1).

10. Two important special cases of the gamma distribution are the exponential distribution (considered in §6.3.1) and the chi-square distribution:

   a. Gamma$(1, \lambda)$ is Exponential(λ)
   b. Gamma$(n/2, 1/2)$ is ChiSquare(n) when n is a positive integer.

   The chi-square distribution is important in the statistical estimation of a distribution's variance and also arises when determining the distribution of kinetic energy or the distribution of power in an electrical circuit (see §7.1).

We conclude our discussion of the gamma distribution with some worked examples.

EXAMPLE 1:   An insurance company has received notification of five pending claims. Claim settlement will not be complete for at least 1 year. An actuary working for the company has been asked to determine the size of the reserve fund that should be set up to cover these claims. Claims are independent and exponentially distributed with mean $2000. The actuary recommends setting up a claim reserve of $12,000. What is the probability that total claims will exceed the reserve fund?

Let X_j be the amount of the jth claim and let S be the total claim both measured in thousands of dollars. Then $S = X_1 + \cdots + X_5$. From the given information, $X_j \sim$ Exponential(0.5) and the X_j are independent. Hence, $S \sim$ Gamma$(5, .5)$.

Consequently, the required probability is

$$\Pr(S > 12) = \Pr\left(\frac{1}{2}S > 6\right)$$
$$= \Pr(\text{Gamma}(5, 1) > 6)$$
$$= 1 - I_5(6)$$
$$\approx 1 - .71494$$
$$= .28506.$$

Alternatively, using the formula for the survival function

$$\Pr(S > s) = \sum_{n=0}^{r-1} \frac{(\lambda s)^n e^{-\lambda s}}{n!}$$

we have

$$\Pr(S > 12) = \sum_{n=0}^{4} \frac{6^n e^{-6}}{n!} \approx .28505656. \qquad \blacksquare$$

EXAMPLE 2: Instantaneous surges of electrical current occur randomly and independently on a particular line at an expected rate of 0.1 per hour. The electrical system will fail after four surges. Determine the probability that the system is still functioning in 10 hours.

Let T_j be the time in hours between the $(j-1)$th and the jth surge, and put $T = T_1 + T_2 + T_3 + T_4$. Then T is the future lifetime of the system measured in hours, and the desired probability is $\Pr(T > 10)$. From the given information, each T_j can be modeled as Exponential(0.1) and the T_j are mutually independent. Hence, $T \sim$ Gamma(4, 0.1) and so the desired probability is

$$\begin{aligned} \Pr(T > 10) &= \Pr(\text{Gamma}(4, 0.1) > 10) \\ &= \Pr(\text{Gamma}(4, 1) > 1) \\ &= 1 - I_4(1) \\ &\approx 1 - .01899 \\ &= .98101. \end{aligned}$$

Alternatively, using the formula for the survival function, we have

$$\begin{aligned} \Pr(T > 10) &= \sum_{n=0}^{3} \frac{1^n e^{-1}}{n!} \\ &= \frac{8}{3} e^{-1} \\ &\approx .9810118. \end{aligned}$$

■

EXAMPLE 3: Determine a formula for the kth moment of a gamma random variable using the probability density function. Deduce formulas for the mean and variance.

Let X be a gamma random variable with parameters r and λ. Then the density function of X is given by

$$f_X(x) = \frac{\lambda^r x^{r-1} e^{-\lambda x}}{\Gamma(r)}, \qquad x \geq 0.$$

Hence, the kth moment of X is given by

$$\begin{aligned} E[X^k] &= \int_0^\infty x^k f_X(x) dx \\ &= \int_0^\infty \frac{\lambda^r x^{k+r-1} e^{-\lambda x}}{\Gamma(r)} dx. \end{aligned}$$

Writing the integrand of this integral in terms of the gamma density with parameters $r + k$ and λ and using the fact that all gamma densities integrate to 1 over the interval $(0, \infty)$, we have

$$\begin{aligned} E[X^k] &= \frac{1}{\lambda^k} \cdot \frac{\Gamma(r+k)}{\Gamma(r)} \int_0^\infty \frac{\lambda^{r+k} x^{r+k-1} e^{-\lambda x}}{\Gamma(r+k)} dx \\ &= \frac{1}{\lambda^k} \cdot \frac{\Gamma(r+k)}{\Gamma(r)}. \end{aligned}$$

Using the functional relationship $\Gamma(z) = (z-1)\Gamma(z-1)$, which holds for all z (see Appendix A), we have

$$E[X^k] = \frac{(r+k-1)(r+k-2)\cdots r}{\lambda^k}.$$

In particular,

$$E[X] = \frac{r}{\lambda}$$

and

$$Var(X) = E[X^2] - E[X]^2 = \frac{(r+1)r}{\lambda^2} - \left(\frac{r}{\lambda}\right)^2 = \frac{r}{\lambda^2}. \qquad \blacksquare$$

EXAMPLE 4:   You are waiting in line at the express checkout of the grocery store. Three customers are waiting ahead of you and one customer is currently being served. The checkout time for an individual customer has an exponential distribution with mean 20 seconds. Determine the probability that you will still be waiting for service to begin 2 minutes from now. What is the probability that you will be completely checked out and heading for the parking lot 2 minutes from now?

Let T_1 be the remaining service time in minutes for the customer currently being served; let T_2, T_3, T_4 be the service times in minutes for the three customers ahead of you; and let T_5 be your service time in minutes. Then your waiting time until service begins is $T_1 + T_2 + T_3 + T_4$, and the time until you are completely checked out is $T_1 + T_2 + T_3 + T_4 + T_5$. Let X, Y denote these two times, respectively. From the given information, $T_j \sim$ Exponential(3) for $j = 2, 3, 4, 5$. Since the exponential distribution has the memoryless property, it is also true that $T_1 \sim$ Exponential(3). (Note that T_1 represents the remaining service time, not the total service time, for the customer currently being served. However, since the total service time for this customer is exponentially distributed, the remaining service time must also be exponentially distributed by the memoryless property.) Consequently, if the T_j are independent, as we will assume, then $X \sim$ Gamma(4,3), $Y \sim$ Gamma(5,3), and the desired probabilities are

$$\Pr(X > 2) = \sum_{n=0}^{3} \frac{6^n e^{-6}}{n!} = 61e^{-6} \approx .1512$$

and

$$\Pr(Y \le 2) = 1 - \sum_{n=0}^{4} \frac{6^n e^{-6}}{n!} = 1 - 115e^{-6} \approx .7149$$

respectively. $\blacksquare$

6.1.3   The Pareto Distribution

Definition and Interpretation

A random variable X is said to have a Pareto distribution with parameters s and β if its survival function is given by

$$S_X(x) = \left(\frac{\beta}{\beta + x}\right)^s, \qquad x \ge 0,$$

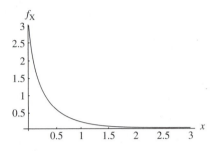

FIGURE 6.3 Pareto(3,1)

or equivalently, if its density function is given by

$$f_X(x) = \frac{s}{\beta} \left(1 + \frac{x}{\beta} \right)^{-s-1}, \qquad x \ge 0.$$

Here, s and β are positive real numbers. A possible graph of f_X is given in Figure 6.3. The notation Pareto(s, β) signifies a Pareto distribution with parameters s and β.

From Figure 6.3, it would appear that the Pareto distribution is very similar in character to the exponential distribution. This shouldn't be too surprising because the Pareto is actually a continuous mixture of exponentials with gamma mixing weights, as you will show in the exercises. However, an important difference between the two distributions is the size of the tail. Indeed, for any exponential-Pareto pair of distributions with the same means, the tail of the Pareto distribution will always be heavier than the tail of the corresponding exponential distribution.

The additional heaviness of the Pareto tail is actually a very desirable property of the distribution. Indeed, it is just what is needed to model many types of accident claims in casualty insurance.

Derivation of the Probability Density Function

The probability density function can be derived by considering a continuous mixture of exponential random variables with gamma mixing weights. The argument is similar to the one used to derive the probability mass function for the negative binomial distribution, and it will be developed in the exercises.

Mean, Variance, and Higher Moments

The only moments of the Pareto distribution that are finite are the ones of order less than s.

For $k < s$, the kth moment is given by

$$E[X^k] = \frac{\beta^k k!}{(s-1)(s-2) \cdots (s-k)}.$$

This formula is derived in the exercises.

Since the higher moments eventually become infinite, the moment generating function does not exist.

Techniques for Calculating Probabilities

The survival function is easy to evaluate. Hence, no special computational techniques are required.

Effect of Arithmetic Operations

A positive scalar multiple of a Pareto random variable is again Pareto. In particular, if $X \sim \text{Pareto}(s, \beta)$ and a is a positive constant, then $aX \sim \text{Pareto}(s, a\beta)$. There is no simple relationship for an independent sum of Pareto random variables.

Relationship with Other Distributions

The Pareto distribution is a continuous mixture of exponentials with gamma mixing weights. In particular, if $(X|\Lambda = \lambda) \sim \text{Exponential}(\lambda)$ and $\Lambda \sim \text{Gamma}(s, \beta)$, then $X \sim \text{Pareto}(s, \beta)$. This property is derived in the exercises.

We conclude our discussion of the Pareto distribution with a numerical example.

EXAMPLE 1: Historical data suggest that claim sizes for a particular type of insurance policy are exponentially distributed. Due to uncertainty in future inflation, one cannot be certain of the expected size of claims in the coming period. Suppose that the uncertainty in expected claim size is modeled by assuming that the exponential parameter λ has a Gamma(3,2) distribution. Determine the probability that a future claim exceeds 4. Claim sizes are assumed to be measured in thousands of dollars.

Let X be the claim size. By assumption, $(X|\Lambda = \lambda) \sim \text{Exponential}(\lambda)$ and $\Lambda \sim$ Gamma(3,2). Hence, the unconditional distribution of X is Pareto(3,2). Consequently, the desired probability is

$$\Pr(X > 4) = \left(\frac{2}{2 + 4} \right)^3 = \frac{1}{27}. \qquad \blacksquare$$

6.2 Special Continuous Distributions for Modeling Lifetimes

Two distributions arise frequently when modeling lifetimes: the Weibull and the DeMoivre, also known as the continuous uniform distribution. As we will soon see, the Weibull arises as a positive power of the exponential and, hence, generalizes the exponential distribution in a different way from the generalization given by the gamma distribution, whereas the DeMoivre is the continuous distribution that corresponds to the classical probability assumption of equally likely outcomes.

6.2.1 The Weibull Distribution

Definition and Interpretation

A random variable X is said to have a Weibull distribution with parameters α and β if its survival function is given by

$$S_X(x) = \exp\left\{ -\left(\frac{x}{\alpha} \right)^\beta \right\}, \qquad x \geq 0,$$

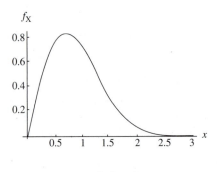

a. Weibull (1, 2)

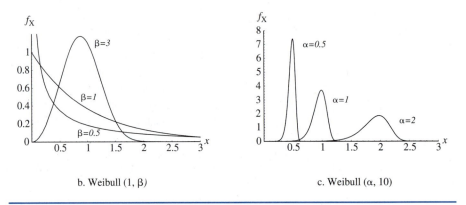

b. Weibull (1, β)   c. Weibull (α, 10)

FIGURE 6.4 Particular Weibull Distributions

or equivalently, if its density function is given by

$$f_X(x) = \frac{\beta}{\alpha} \left(\frac{x}{\alpha}\right)^{\beta-1} \exp\left\{-\left(\frac{x}{\alpha}\right)^{\beta}\right\}, \qquad x \geq 0.$$

Here, α and β are both positive real numbers. Possible graphs of f_X are given in Figure 6.4. The notation Weibull(α, β) signifies a Weibull distribution with parameters α and β.

A Weibull random variable X has two important interpretations:

1. It is the lifetime of an item whose instantaneous risk of failure (i.e., failure rate) is given by a power function; that is, $\lambda_X(t) = kt^{\beta-1}$ for some positive constants k and β.
2. It is a positive power of an exponential random variable. In particular, it is the random variable $\alpha Y^{1/\beta}$, where $Y \sim$ Exponential(1).

Either of these interpretations could be taken as the definition of a Weibull random variable.

When X has the interpretation of a lifetime, the parameter α is sometimes referred to as the **characteristic life.** The rationale for this terminology becomes clear when one realizes that α represents the life expectancy of X when $\beta = 1$, since for $\beta = 1$, the distribution reduces to an Exponential$(1/\alpha)$ distribution, and that regardless of the value of β, $\Pr(X > \alpha) = e^{-1}$; that is, $\Pr(X > \alpha)$ is independent of β.[6]

Derivation of the Probability Density Function

The density function for a Weibull random variable can be derived using either of the interpretations just given. The derivation of f_X from the assumed hazard function $\lambda_X(x) = kx^{\beta-1}$ is straightforward and follows directly from the relationships between survival and hazard functions developed in §4.3.2. Details are left to the reader. We do demonstrate here that any positive power of an exponential random variable is of Weibull type. It will then follow immediately that every Weibull distribution can be constructed by considering an appropriate power of some exponential random variable.

Hence, let W be an exponential random variable with parameter λ and consider $X = W^r$, where r is a positive real number. Notice that for all nonnegative x, $W^r > x \iff W > x^{1/r}$ since $W \geq 0$. Consequently, for $x \geq 0$,

$$
\begin{aligned}
S_X(x) &= \Pr(X > x) \\
&= \Pr(W^r > x) \\
&= \Pr(W > x^{1/r}) \\
&= e^{-\lambda x^{1/r}}.
\end{aligned}
$$

However, from the definition of the Weibull distribution, we see that this survival function is of Weibull type, with $\alpha = 1/\lambda^r$ and $\beta = 1/r$. Consequently, every positive power of an exponential random variable has a Weibull distribution, and moreover, every Weibull distribution can be generated in this way.

Mean, Variance, and Higher Moments

The moments of a Weibull random variable X are given by

$$
E[X^k] = \alpha^k \Gamma\left(1 + \frac{k}{\beta}\right),
$$

where $\Gamma(\cdot)$ denotes the gamma function (see Appendix A for details on the gamma function). In particular, the mean and variance are, respectively,

$$
E[X] = \alpha \Gamma\left(1 + \frac{1}{\beta}\right),
$$

$$
Var(X) = \alpha^2 \left\{ \Gamma\left(1 + \frac{2}{\beta}\right) - \left(\Gamma\left(1 + \frac{1}{\beta}\right) \right)^2 \right\}.
$$

The formulas for the moments follow immediately from the observation that $X^k = \alpha^k Y^{k/\beta}$ where $Y \sim$ Exponential(1) and the fact that $E[Y^s] = \Gamma(s+1)$ for $Y \sim$ Exponential(1) and $s > 0$ (see Example 3 of §6.1.2). Details are left to the reader.

Techniques for Calculating Probabilities

The survival function S_X is easy to evaluate. Hence, no special computational techniques are required.

---

[6] Note, however, that α only represents the life expectancy when $\beta = 1$. For values of β different from 1, $E[X] \neq \alpha$.

Effect of Arithmetic Operations

There are two important arithmetic operations to consider: scalar multiples and positive powers.

A positive scalar multiple of a Weibull random variable is again of Weibull type. In particular, if $X \sim$ Weibull(α, β) and a is a positive constant, then $aX \sim$ Weibull$(a\alpha, \beta)$. This property follows directly from the formula for the survival function of a Weibull random variable.

A positive power of a Weibull random variable is also of Weibull type. In particular, if $X \sim$ Weibull(α, β) and r is a positive real number, then $X^r \sim$ Weibull$(\alpha^r, \beta/r)$. The derivation of this property is anaolgous to the derivation of the Weibull density function given earlier. Details are left to the reader.

There is no simple relationship for an independent sum of Weibull random variables.

Relationship with Other Distributions

The Weibull distribution is intimately related with the exponential distribution, which is simply the special Weibull distribution with $\alpha = 1/\lambda$ and $\beta = 1$.

Indeed, if $Y \sim$ Exponential(1), then $\alpha Y^{1/\beta} \sim$ Weibull(α, β). On the other hand, if $X \sim$ Weibull(α, β), then $(X/\alpha)^\beta \sim$ Exponential(1). More generally, if $Y \sim$ Exponential(λ), then $\alpha(\lambda Y)^{1/\beta} \sim$ Weibull(α, β), while if $X \sim$ Weibull(α, β), then $(X/\alpha)^\beta/\lambda \sim$ Exponential(λ).

We conclude our discussion of the Weibull distribution with some worked examples.

EXAMPLE 1:   The lifetimes of automobile tires can be measured in terms of the number of miles traveled until the tread wears out. For a particular brand of tire, the lifetime measured in thousands of miles of road wear has a Weibull distribution with parameters $\alpha = 45$ and $\beta = 3$. Determine the probability that a randomly selected tire of this type wears out in the first 15,000 miles of use.

Let X be the lifetime of the tire measured in thousands of miles of road wear. Then the survival function of X is given by

$$S_X(x) = \exp\left\{-\left(\frac{x}{45}\right)^3\right\}, \qquad x > 0.$$

Hence, the desired probability is

$$\Pr(X \le 15) = 1 - \exp\left\{-\left(\frac{15}{45}\right)^3\right\}$$
$$= 1 - e^{-1/27}$$
$$\approx .0364. \qquad\blacksquare$$

EXAMPLE 2:   The time to failure of a particular brand of battery has a Weibull distribution. Tests indicate that 5% of batteries of this type fail within the first 2 hours of continuous use and 5% are still functional after 10 hours of continuous use. Determine the probability that a new battery of this type is still functional after 5 hours of continuous use.

Let X be the time to failure in hours of a randomly selected battery of the given type. Since X is assumed to have a Weibull distribution, the survival function of X is of the form

$$S_X(x) = \exp\left\{-\left(\frac{x}{\alpha}\right)^\beta\right\}, \quad x \geq 0.$$

From the given information, we have $S_X(2) = .95$ and $S_X(10) = .05$. Consequently, α and β can be determined from the equations

$$\exp\left\{-\left(\frac{2}{\alpha}\right)^\beta\right\} = .95,$$

$$\exp\left\{-\left(\frac{10}{\alpha}\right)^\beta\right\} = .05.$$

This system of equations can be solved by taking natural logarithms twice and using standard properties of logarithms. When we do this, we obtain the system

$$\beta(\log 2 - \log \alpha) = \log \log(.95)^{-1},$$
$$\beta(\log 10 - \log \alpha) = \log \log(.05)^{-1}.$$

From this system of equations, we find that

$$\beta = \frac{\log \log(.05)^{-1} - \log \log(.95)^{-1}}{\log 10 - \log 2}$$
$$\approx 2.5272077$$

and

$$\log \alpha = \frac{(\log 10)(\log \log(.95)^{-1}) - (\log 2)(\log \log(.05)^{-1})}{\beta(\log 2 - \log 10)}$$
$$\approx 1.8684345.$$

Hence, $\alpha \approx 6.4781470$ and $\beta = 2.5272077$. Consequently, the desired probability is

$$\Pr(X > 5) = \exp\left\{-\left(\frac{5}{6.4781470}\right)^{2.5272077}\right\}$$
$$= .5947. \qquad \blacksquare$$

EXAMPLE 3: Determine conditions under which a Weibull random variable has (a) an increasing failure rate function, (b) a decreasing failure rate function, (c) a constant failure rate function.

From §4.3.2, the failure rate (i.e., hazard rate) function for any random variable is connected to its survival function by the relationship

$$\lambda_X(x) = \frac{-S_X'(x)}{S_X(x)} = \frac{f_X(x)}{S_X(x)}.$$

Hence, for $X \sim$ Weibull(α, β), the failure rate function has the form

$$\lambda_X(x) = \frac{\beta}{\alpha}\left(\frac{x}{\alpha}\right)^{\beta-1}, \qquad x \geq 0.$$

From this formula, it follows that λ_X is increasing if $\beta > 1$, λ_X is decreasing if $\beta < 1$, and λ_X is constant if $\beta = 1$.

The flexibility of the Weibull distribution to include increasing, decreasing, or constant hazard rates is one of the features that makes it attractive as a model for the time until failure. ∎

EXAMPLE 4: The lifetime in years of a particular device that improves with age has a Weibull distribution with mean 8 and variance 320. Determine the probability that the device fails within the first 5 years of use.

Let X be the lifetime of the device measured in years and let α, β be the parameters of the Weibull distribution that X is assumed to have (i.e., $X \sim$ Weibull(α, β)). Since we are given the mean and variance of X, we will have to determine the parameters α, β from the general formulas

$$\mu_X = \alpha\Gamma\left(1 + \frac{1}{\beta}\right),$$

$$\sigma_X^2 = \alpha^2\left\{\Gamma\left(1 + \frac{2}{\beta}\right) - \left(\Gamma\left(1 + \frac{1}{\beta}\right)\right)^2\right\}.$$

Eliminating α from these equations, we have

$$\frac{\mu_X^2}{\mu_X^2 + \sigma_X^2} = \frac{\left(\Gamma\left(1 + \frac{1}{\beta}\right)\right)^2}{\Gamma\left(1 + \frac{2}{\beta}\right)}.$$

Hence, substituting the given information $\mu_X = 8$, $\sigma_X^2 = 320$, we see that β is given by

$$\frac{\left(\Gamma\left(1 + \frac{1}{\beta}\right)\right)^2}{\Gamma\left(1 + \frac{2}{\beta}\right)} = \frac{1}{6}.$$

In general, an equation such as this would have to be solved numerically. However, in this case, there is an exact solution, which we find by trial and error to be $\beta = 1/2$. (We know that β must be less than 1 since the device is assumed to improve with age and, hence, must have a decreasing failure rate.) Substituting this value for β into the formula for the mean—that is, $\mu_X = \alpha\Gamma(1 + 1/\beta)$—we find that $\alpha = 4$. Consequently, the desired probability is

$$\Pr(X < 5) = 1 - \exp\left\{-\left(\frac{5}{4}\right)^{1/2}\right\} \approx .6731. \qquad ∎$$

EXAMPLE 5: An actuary is trying to determine a model for claim size from historical data on a particular group of policies. The actuary observes that by considering the cube of the claim sizes, one obtains an exponential distribution with mean 2. Determine the probability that a future claim on a policy similar to the ones being studied will exceed 1.5.

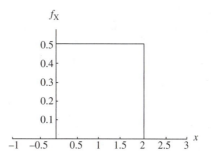

FIGURE 6.5 DeMoivre(2)

Let X be the claim size. We are given that $X^3 \sim$ Exponential(0.5). Put $Y = X^3$. Then $X = Y^{1/3}$, where $Y \sim$ Exponential(0.5). From the results of this section, we know that if $Y \sim$ Exponential(λ), then $Y^r \sim$ Weibull($1/\lambda^r$, $1/r$). Hence, $X \sim$ Weibull($2^{1/3}$, 3). Consequently, the desired probability is

$$\Pr(X > 1.5) = \exp\left\{ -\left(\frac{1.5}{2^{1/3}} \right)^3 \right\}$$

$$= e^{-1.6875}$$

$$\approx .1850. \qquad \blacksquare$$

EXAMPLE 6: A special type of Weibull distribution that arises frequently in engineering applications is the Weibull distribution with $\alpha = \sqrt{2}s$ and $\beta = 2$. The density function for such a Weibull distribution is given by

$$f_X(x) = \frac{x}{s^2} e^{-x^2/2s^2}, \qquad x \geq 0.$$

In engineering, this density is frequently referred to as the **Rayleigh density** and the corresponding probability distribution is known as the **Rayleigh distribution**. Rayleigh distributions can be used to analyze rocket landing errors and the distribution of misses around the target of a dartboard. They can also be used to model the amplitude of certain waveforms in electricity (see exercises 13 and 14 in §8.7). $\qquad \blacksquare$

6.2.2   The DeMoivre Distribution

Definition and Interpretation

A random variable X is said to have a DeMoivre (or alternatively, a continuous uniform) distribution with parameter ω if its probability density function is given by

$$f_X(x) = \begin{cases} \dfrac{1}{\omega} & \text{for } 0 < x < \omega, \\ 0 & \text{otherwise.} \end{cases}$$

Here, ω is a positive real number. A possible graph of f_X is given in Figure 6.5. The notation DeMoivre(ω) signifies a DeMoivre distribution with parameter ω.

The DeMoivre distribution is useful for modeling continuous quantities whose values we consider to be "equally likely" in the sense that all intervals of the same length have the same probability. The DeMoivre distribution is sometimes used to model uncertain parameters in probability models when we have no information that would lead us to believe that some values of the parameter are more likely than others (see exercise 23 from §5.5 for an illustration of this).

The DeMoivre distribution is also sometimes used to model lifetimes. In this situation, deaths or failures are said to be uniformly distributed. It is important to realize that a uniform distribution of deaths does not imply that the risk of death at each age is the same. Indeed, from the definition of f_X, one can readily determine that

$$S_X(x) = 1 - \frac{x}{\omega}, \qquad x \in (0, \omega)$$

and

$$\lambda_X(x) = \frac{1}{\omega - x}, \qquad x \in (0, \omega).$$

Hence, the risk of death increases at an increasing rate until age ω, at which time death is certain to occur.

The DeMoivre distribution is particularly convenient for modeling lifetimes because it has the property that the future lifetime at any age is also DeMoivre. Indeed, if total lifespans are modeled as DeMoivre(ω), then the future lifetime of a subject currently of age x is DeMoivre($\omega - x$). Details are left to the reader.

Derivation of the Probability Density Function

The form of the density function for a DeMoivre random variable is an immediate consequence of the interpretations just discussed.

Mean, Variance, and Higher Moments

The moments of a DeMoivre random variable X are given by

$$E[X^k] = \frac{\omega^k}{k+1}.$$

In particular, the mean and variance are, respectively,

$$E[X] = \frac{\omega}{2},$$

$$Var(X) = \frac{\omega^2}{12}.$$

The moment generating function of X is given by

$$M_X(t) = \frac{e^{t\omega} - 1}{t\omega}.$$

These formulas are straightforward to derive. Details are left to the reader.

Techniques for Calculating Probabilities

The survival function S_X is easy to evaluate. Hence, no special computational techniques are required.

Effect of Arithmetic Operations

A positive scalar multiple of a uniform distribution is again uniform. In particular, if $X \sim \text{DeMoivre}(\omega)$ and a is a positive constant, then $aX \sim \text{DeMoivre}(a\omega)$. However, an independent sum of uniform distributions can never be uniform. Indeed, for any independent uniform random variables X_1, X_2, we see from the formula for the moment generating function of a uniform distribution given earlier that $M_{X_1+X_2}$ does not have the form of the moment generating function of a uniform distribution, and hence, $X_1 + X_2$ cannot be uniform.

Relationship with Other Distributions

The special DeMoivre distribution with $\omega = 1$ is connected to every distribution in the following way:

For any random variable X, the random variable Y given by $Y = F_X(X)$ has a DeMoivre(1) distribution; that is, Y is uniformly distributed on $(0, 1)$.

This property allows one to generate observations for any distribution by first generating a random observation u from a DeMoivre(1) distribution and then applying the inverse transformation[7] $F_X^{-1}(u)$. Indeed, for a large number of independently chosen random observations $u_1, \ldots, u_n$ from DeMoivre(1), theory suggests that the values $F_X^{-1}(u_1), \ldots, F_X^{-1}(u_n)$ should have approximately the same distribution as X. This procedure of distribution sampling is often referred to as **Monte Carlo simulation**. It is a useful technique for determining approximate values for desired probabilities when a distribution function is not (or cannot be) explicitly given. The significance of this statement will become clearer after we consider the distributions of general sums and products in Chapter 8.

We conclude our discussion of the DeMoivre distribution with some worked examples.

EXAMPLE 1:   A professor takes early retirement on July 1, exactly 3 months after his 60th birthday, and purchases a term life insurance policy for $25,000. The maturity date of the policy coincides with his 65th birthday. According to a mortality table for similarly situated men, the probability that a man of exact age 60 dies before reaching his 61st birthday is .0137604. Determine the probability that the professor dies before reaching his 61st birthday if deaths are assumed to be uniformly distributed between ages.

Let T be the future lifetime of a 60-year-old male. Then, from the given information, $\Pr(T \leq 1) = .0137604$. We are interested in determining $\Pr(T \leq 1 \mid T > \frac{1}{4})$.

The uniform distribution of deaths assumption means that

$$\Pr(T \leq t) = t \cdot \Pr(T \leq 1)$$
$$= (.0137604)t, \quad t \in (0, 1).$$

Consequently, the desired probability is

---

[7] Some modification of this procedure is required for discrete or mixed distributions, but the ideas are basically the same.

$$\Pr\left(T \le 1 | T > \frac{1}{4}\right) = 1 - \Pr\left(T > 1 | T > \frac{1}{4}\right)$$

$$= 1 - \frac{\Pr(T > 1)}{\Pr(T > \frac{1}{4})}$$

$$= 1 - \frac{1 - .0137604}{1 - (\frac{1}{4})(.0137604)}$$

$$= .0103559.$$

Note that this value is slightly higher than the value obtained in Example 2 of §6.1.1, where the distribution of deaths between ages was assumed to be exponentially distributed. ∎

EXAMPLE 2: Let X be a random variable whose distribution function is continuous and strictly increasing. Show that the random variable Y given by $Y = F_X(X)$ has a DeMoivre(1) distribution.

Since F_X is continuous and strictly increasing, the inverse $F_X^{-1}(y)$ exists for all $y \in (0, 1)$. Moreover, $F_X(X) \le y \Longleftrightarrow X \le F_X^{-1}(y)$. Hence, for $y \in (0, 1)$, we have

$$\Pr(Y \le y) = \Pr(F_X(X) \le y)$$

$$= \Pr(X \le F_X^{-1}(y))$$

$$= F_X(F_X^{-1}(y))$$

$$= y.$$

Consequently, $Y \sim$ DeMoivre(1) as claimed.

Note that $F_X(X) \sim$ DeMoivre(1) for all random variables X. However, the proof is more complicated when F_X is not monotonic or continuous. ∎

EXAMPLE 3: The following random observations are made from a DeMoivre(1) distribution:

$$.21, .10, .85, .54, .31, .91, .05, .43, .76, .63.$$

Use these observations to simulate observations from a Weibull distribution with parameters $\alpha = 2, \beta = 3$.

Let X be a Weibull random variable with $\alpha = 2$ and $\beta = 3$. Then the random variable Y given by $Y = F_X(X)$ is uniformly distributed on $(0, 1)$, and so values of X can be simulated using the inverse transformation $F_X^{-1}(U)$, where $U \sim$ DeMoivre(1). Since

$$F_X(x) = 1 - \exp\left\{-\left(\frac{x}{2}\right)^3\right\},$$

the simulated values of X are given by

$$1 - \exp\left\{-\left(\frac{x}{2}\right)^3\right\} = u.$$

That is,

$$x = 2\{\log(1 - u)^{-1}\}^{1/3},$$

where u represents an observation from DeMoivre(1). Hence, the simulated values of X corresponding to the given observations from DeMoivre(1) are, respectively,

$$1.23546, \ 0.94462, \ 2.47587, \ 1.83830, \ 1.43719,$$
$$2.68068, \ 0.74311, \ 1.65059, \ 2.25173, \ 1.99616.$$

For example, the first simulated observation is

$$x = 2\{\log(1 - .21)^{-1}\}^{1/3} \approx 1.23546.$$

As a check on our calculations, note that if $x = 1.23546$, then $F_X(x) = 1 - \exp\{-(1.23546/2)^3\} = .21$. ∎

6.3   Other Special Distributions

Three special distributions that do not fit easily into any of the preceding categories are the normal, the lognormal, and the beta distributions.

The normal distribution arises in connection with *sums* of random variables, whereas the lognormal distribution arises in connection with *products*. The beta distribution arises in connection with fractional quantities, such as the fraction of total payouts that are due to a particular claim or the fraction of defectives that are present in a lot of manufactured items, and it can be used to model the probability parameter p in the binomial distribution when p is not known with certainty.

We now consider each of these special distributions in turn.

6.3.1   The Normal Distribution

Definition and Interpretation

A random variable X is said to have a normal distribution with parameters μ and σ if its probability density function is given by

$$f_X(x) = \frac{1}{\sqrt{2\pi}\,\sigma} e^{-(x-\mu)^2/2\sigma^2}, \qquad x \in \mathbf{R}.$$

Here, μ is the *mean* of the distribution and σ is the *standard deviation* of the distribution. Hence, μ and σ are real numbers with $\sigma > 0$. Possible graphs of f_X are given in Figure 6.6. The notation Normal(μ, σ) signifies a normal distribution with mean μ and standard deviation σ.

A normal distribution of particular importance is the normal distribution with $\mu = 0$ and $\sigma = 1$. This distribution is generally referred to as the **standard normal distribution,** and random variables with this distribution are generally referred to as **standard normal random variables.** The uppercase letter Z, with or without a subscript, is usually reserved in probability for standard normal random variables, and we will adhere to this convention throughout the book.[8]

---

[8] However, in using this convention, one must remember that there are many different random variables Z for which $Z \sim$ Normal(0,1), and one must be careful not to refer to a random variable with a Normal(0,1) distribution as *the* standard normal random variable. See §4.1.7 for a discussion of the differences between equality of random variables and equality of distributions.

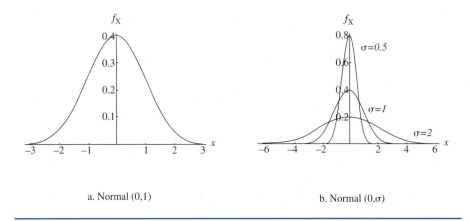

a. Normal (0,1) b. Normal (0,σ)

FIGURE 6.6 Particular Normal Distributions

Every normal distribution can be obtained from a standard normal random variable using an appropriate linear transformation. In particular, if $Z \sim \text{Normal}(0,1)$, then $\sigma Z + \mu \sim \text{Normal}(\mu, \sigma)$. On the other hand, if $X \sim \text{Normal}(\mu, \sigma)$, then

$$\frac{X - \mu}{\sigma} \sim \text{Normal}(0,1).$$

This property, which we will demonstrate later, is extremely important and allows one to calculate probabilities for any normal distribution using only the *standard* normal distribution function.

In view of the importance of the random variable $(X - \mu)/\sigma$ in calculating probabilities for X, this quantity is given a special name: It is called the **standard form of X.** This terminology is also used when X is not normally distributed. In that case, μ and σ represent the mean and standard deviation of X.[9]

The normal distribution can be interpreted in many ways. One interpretation is that it is the continuous analog of the binomial distribution (with $p = 1/2$). Another interpretation, which is closely related to the preceding one, is that it represents the distribution of possible measurements for a particular continuous quantity in a scientific experiment in which measurements are subject to error.

In investments and insurance, one generally interprets the normal distribution in the following way: It is the limiting distribution for the sum of any collection of independent and identically distributed random variables. This important property of the normal distribution is a major theorem in probability theory, known as the central limit theorem. It will be discussed briefly in this section and in more detail in Chapter 8.

Derivation of the Probability Density Function

The probability density function for the normal distribution can be derived by considering a limit of relative frequency densities from a binomial distribution, in much the same way that the exponential and gamma distributions were derived in the exercises of the previous chapter from relative frequency densities associated with the geometric and negative

---

[9] Note that the standard form of X is a standard normal random variable when $X \sim \text{Normal}(\mu, \sigma)$.

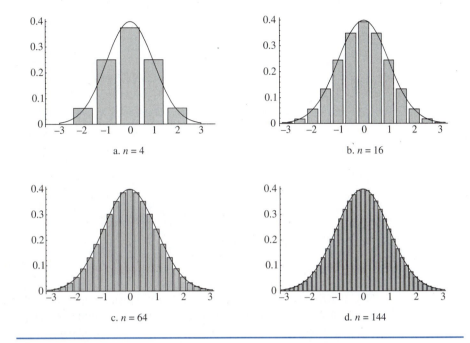

a. $n = 4$

b. $n = 16$

c. $n = 64$

d. $n = 144$

FIGURE 6.7 Relative Frequency Density Functions for Standardized Binomial$(n, .5)$ Distributions with the Standard Normal Distribution Superimposed

binomial distributions.[10] The mathematics involved in this derivation are considerably more sophisticated than in those other derivations. Thus, in the interest of simplicity, we only outline the main ideas here.

Let X_n be a Binomial$(n, .5)$ random variable and let Y_n be the standard form of X_n. Then

$$Y_n = \frac{X_n - \frac{n}{2}}{\sqrt{\frac{n}{4}}} = \frac{2X_n - n}{\sqrt{n}}.$$

One could easily specify a probability mass function for Y_n. However, for our present purposes, it is more convenient to consider a relative frequency *density* function (i.e., a function which measures probability per unit length). In that way, probabilities continue to be measured as areas in bar chart diagrams.

Figure 6.7 illustrates relative frequency density bar charts for the Y_n with $n = 4$, $n = 16$, $n = 64$, and $n = 144$. The graph of the standard normal distribution is superimposed on each of these charts to facilitate comparisons. From these graphs, it appears that

$$\text{Distribution of } Y_n \to \text{Normal}(0,1)$$

as $n \to \infty$.

From the relative frequency density functions for the Y_n, one can show analytically that the limiting density function as $n \to \infty$ is

---

[10] See exercises 21 and 22 from §5.5.

$$f(x) = \frac{1}{\sqrt{2\pi}} e^{-x^2/2}, \qquad x \in \mathbf{R}.$$

The general form of the normal density can then be obtained by applying a linear transformation to the standard normal distribution.

Mean, Variance, and Higher Moments

The mean, variance, and skewness of a normal random variable X are μ, σ^2, and 0, respectively. The skewness is clearly zero since the distribution is symmetric about its mean.

The moment and cumulant generating functions for X are given by

$$M_X(t) = \exp(\mu t + \frac{1}{2}\sigma^2 t^2),$$

$$\psi_X(t) = \mu t + \frac{1}{2}\sigma^2 t^2.$$

The formula for the moment generating function follows directly from the general definition of a moment generating function by considering the normal density with parameters $\mu + \sigma^2 t$ and σ. (This formula was derived in Example 1 of §4.3.1.) The formulas for the mean and variance are then readily deduced from the cumulant generating function.

One can also show that the central moments are given by

$$E[(X - \mu)^k] = \begin{cases} \dfrac{\sigma^k 2^{k/2}\Gamma(\frac{1}{2}(k+1))}{\sqrt{\pi}} & \text{for } k = 2, 4, 6, \ldots, \\ 0 & \text{for } k = 1, 3, 5, \ldots, \end{cases}$$

where $\Gamma(\cdot)$ denotes the gamma function. This derivation is more delicate and uses properties of the gamma function. Details are left to the reader.

Techniques for Calculating Probabilities

There is no simple algebraic formula for the distribution function of a normal distribution. Hence, probabilities must be calculated using numerical techniques.

As we noted earlier, the distribution function for an arbitrary normal random variable can be expressed in terms of the standard normal distribution function. Indeed, using an appropriate substitution, we have

$$F_X(x) = \frac{1}{\sqrt{2\pi}\sigma} \int_{-\infty}^{x} \exp(-(t - \mu)^2/2\sigma^2) dt$$

$$= \frac{1}{\sqrt{2\pi}} \int_{-\infty}^{(x-\mu)/\sigma} e^{-t^2/2} dt$$

$$= \Phi\left(\frac{x - \mu}{\sigma}\right),$$

where $\Phi(\cdot)$ is the distribution function for a standard normal random variable; that is, $\Phi(z) = \frac{1}{\sqrt{2\pi}} \int_{-\infty}^{z} e^{-t^2/2} dt$. Hence, calculation of normal distribution probabilities in general can be reduced to determining numerical values for the function Φ.

Tables of numerical values for $\Phi(z)$ are given in Appendix E. An illustration of the use of these tables is given in the examples at the end of this section.[11]

Effect of Arithmetic Operations

A linear transformation of a normal random variable is again normal. More precisely, if $X \sim \text{Normal}(\mu, \sigma)$ and a, b are constants with $a \neq 0$, then $aX + b \sim \text{Normal}(a\mu + b, |a|\sigma)$. In particular, $\sigma Z + \mu \sim \text{Normal}(\mu, \sigma)$ for any standard normal random variable Z, and $(X - \mu)/\sigma \sim \text{Normal}(0,1)$ for any normal random variable X with parameters μ and σ. This property follows directly from the definition of a normal distribution function and explains why $F_X(x) = \Phi\left((x - \mu)/\sigma\right)$.

A sum of independent normal random variables is again normal. In particular, if $X_1 \sim \text{Normal}(\mu_1, \sigma_1)$, if $X_2 \sim \text{Normal}(\mu_2, \sigma_2)$, and if X_1, X_2 are independent, then $X_1 + X_2 \sim \text{Normal}(\mu_1 + \mu_2, \sqrt{\sigma_1^2 + \sigma_2^2})$. This property follows directly from the formula for the moment generating function of a normal random variable and the general formula for the moment generating function of an independent sum.

A sum of *dependent* normal random variables can also be normal if their joint density function has a particular form. In particular, if X_1 and X_2 have the joint density

$$ f_{X_1, X_2}(x_1, x_2) = \frac{1}{2\pi\sigma_1\sigma_2\sqrt{1 - \rho^2}} \exp\left(-\frac{1}{2} \cdot \frac{1}{1 - \rho^2} Q(x_1, x_2)\right), $$

where

$$ Q(x_1, x_2) = \left(\begin{array}{cc} \frac{x_1 - \mu_1}{\sigma_1} & \frac{x_2 - \mu_2}{\sigma_2} \end{array}\right) \left(\begin{array}{cc} 1 & -\rho \\ -\rho & 1 \end{array}\right) \left(\begin{array}{c} \frac{x_1 - \mu_1}{\sigma_1} \\ \frac{x_2 - \mu_2}{\sigma_2} \end{array}\right), $$

then $X_1 \sim \text{Normal}(\mu_1, \sigma_1)$, $X_2 \sim \text{Normal}(\mu_2, \sigma_2)$, $Cov(X_1, X_2) = \rho\sigma_1\sigma_2$, and for any constants a, b, c,

$$ aX_1 + bX_2 + c \sim \text{Normal}\left(a\mu_1 + b\mu_2 + c, \sqrt{a^2\sigma_1^2 + 2ab\rho\sigma_1\sigma_2 + b^2\sigma_2^2}\right). $$

The density f_{X_1, X_2} is a generalization of the univariate normal density and is referred to as the *bivariate normal density*.

Relationship to Other Distributions

The most important property of the normal distribution is that it is the limiting distribution for any sum of independent and identically distributed random variables. This surprising property is illustrated in Figure 6.8 for sums of independent Poisson(1) random variables.

Recall that the Poisson distribution is positively skewed and the sum of n independent Poisson(1) random variables has a Poisson(n) distribution. Hence, it may seem strange to assert that the distribution of a sum of n independent Poisson(1) random variables approaches a normal distribution as $n \to \infty$. However, the graphs in Figure 6.8 suggest that this is indeed the case.

Since a binomial random variable with parameters n and p can be interpreted as a sum of n independent Bernoulli(p) random variables, this limiting property of the normal distribution provides us with a way to approximate probabilities for binomial distributions.

---

[11] Once again, we include tables of distribution values just in case you happen to be reading this book on a desert island without a computer.

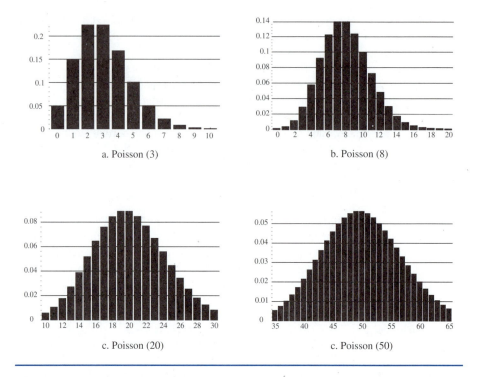

a. Poisson (3)

b. Poisson (8)

c. Poisson (20)

c. Poisson (50)

FIGURE 6.8 Distributions for Sums of Independent Poisson(1) Random Variables

Indeed, if $X \sim$ Binomial(n, p), then the distribution of $(X - np)/\sqrt{np(1 - p)}$ is approximately Normal$(0,1)$ when n is large. Thus, probability statements about X can be evaluated using the standard normal distribution function, as we noted in our discussion of the binomial distribution in §5.1.

When we use a continuous distribution, such as the normal to approximate a discrete distribution, such as the binomial, error can arise if the interval whose probability is to be approximated has probability masses located at its endpoints. Consider, for example, the random variable X with distribution Binomial$(6, .5)$ and let Y be a normal random variable with $\mu_Y = \mu_X = 3$ and $\sigma_Y = \sigma_X = \sqrt{3/2}$ (i.e., $Y \sim$ Normal$(3, \sqrt{3/2})$). The probability that $2 \leq X \leq 5$ is equal to the area of the shaded rectangles in Figure 6.9. However, the probability that $2 \leq Y \leq 5$ is only equal to the area under the normal density curve between the points $y = 2$ and $y = 5$, and it does not include all of the area in the shaded rectangles at positions 2 and 5 (Figure 6.9). The reason is that continuous distributions assign a probability of zero to all individual points, whereas discrete distributions do not. To "correct" for this error in the approximation, we should estimate $\Pr(2 \leq X \leq 5)$ using $\Pr(1.5 \leq Y \leq 5.5)$ instead of $\Pr(2 \leq Y \leq 5)$. This type of correction is generally referred to as a **correction for continuity.**

In general, if X is a discrete random variable with integer values and if Y is a continuous random variable whose distribution is approximately the same as the distribution of X, then for any integers a, b, the probability $\Pr(a \leq X \leq b)$ should be estimated in the following way:

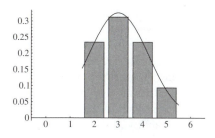

FIGURE 6.9   The Continuity Correction for Binomial(6, .5)

$$\Pr(a \le X \le b) \approx \Pr\left(a - \frac{1}{2} \le Y \le b + \frac{1}{2}\right).$$

To estimate probabilities of the type $\Pr(a < X < b)$, $\Pr(a < X \le b)$, $\Pr(a \le X < b)$, we suggest rewriting the desired probability in the form $\Pr(m \le X \le n)$ and then applying the continuity correction. For example,

$$\Pr(a < X \le b) = \Pr(a + 1 \le X \le b) \approx \Pr\left(a + \frac{1}{2} \le Y \le b + \frac{1}{2}\right).$$

A numerical example is given in a moment.

The normal distribution has several other interesting properties. We highlight only one of these: If Z_1 and Z_2 are independent standard normal random variables, then Z_1/Z_2 has a *Cauchy distribution,* that is, a distribution with density function

$$f_X(x) = \frac{1}{\pi} \cdot \frac{1}{1 + x^2}, \qquad x \in \mathbf{R}.$$

This distribution is symmetric about $x = 0$ and appears, from a graph of its density, to be very similar in character to the standard normal distribution. However, contrary to what a graph of f_X might suggest, the Cauchy distribution does not have a well-defined mean.[12] Consequently, if $X_1, \ldots, X_n$ are independent observations from a Cauchy distribution, then the distribution of the average $(X_1 + \cdots + X_n)/n$ does not become concentrated around zero, or any other value for that matter, as n increases. In fact, one can show that $(X_1 + \cdots + X_n)/n$ has a Cauchy distribution for all n (see exercise 15 of §8.7). On the surface, this might appear to be a contradiction of the insurance principle, discussed in Chapter 1. However, as we will see in §8.4 when we discuss the law of large numbers in detail, this simply illustrates that for the insurance principle to make sense, the mean of the distribution under consideration must actually exist. Hence, the insurance principle does not apply to the ratio Z_1/Z_2 when Z_1 and Z_2 are independent standard normal random variables.

We conclude our discussion of the normal distribution with several numerical examples.

EXAMPLE 1:   An auto insurance company is analyzing the claim frequency on a block of 250 policies. Historical data suggest that 10% of policyholders in this block will file

---

[12] This property of the Cauchy distribution, as well as others, will be discussed in §7.6. However, it is a good exercise for the reader to try to demonstate this fact now without consulting §7.6 first.

at least one claim in the coming coverage period. What is the probability that more than 12% of policyholders file at least one claim in the coming coverage period? Assume that claim occurrences are independent for distinct policyholders.

The reader may recognize this as Example 2 in §5.1. Here, we determine the probability using a normal approximation.

Let X be the number of policies in the block for which there is at least one claim in the coming period. An exact model for X is Binomial(250, .10), and so the exact value of the probability is

$$\Pr(X > 30) = \sum_{x=31}^{250} \binom{250}{x} (.10)^x (.90)^{250-x}.$$

This summation is difficult to evaluate, suggesting that an approximation for $\Pr(X > 30)$ may be a more practical way to proceed.

Note that $E[X] = (250)(.10) = 25$ and $Var(X) = (250)(.10)(.90) = 22.5$. Since the number of policies is large, we can use the normal approximation $X \approx$ Normal $(25, \sqrt{22.5})$. Hence, applying the correction for continuity and then standardizing, we have

$$
\begin{aligned}
\Pr(X > 30) &= \Pr(X \geq 30.5) \\
&= \Pr\left(\frac{X - 25}{\sqrt{22.5}} \geq \frac{30.5 - 25}{\sqrt{22.5}}\right) \\
&\approx \Pr(Z \geq 1.1595) \\
&= 1 - \Phi(1.1595),
\end{aligned}
$$

where $Z \sim$ Normal(0,1) and Φ is the standard normal distribution function. Using a statistical calculator or the tables in Appendix E, we find that[13] $\Phi(1.1595) \approx .876895$. Hence, $\Pr(X > 30) \approx .12315$; that is, there is about a 12% chance that more than 12% of policyholders will file at least one claim in the coming period.

Notice that if we had not applied the continuity correction, we would have obtained[14]

$$\Pr(X > 30) \approx \Pr(Z > 1.0541) \approx 1 - .854043 = .145957.$$

That is, we would have concluded that there is a 14.6% chance that more than 12% of policyholders will file at least one claim. Clearly, the correction for continuity makes a difference! ∎

EXAMPLE 2: The claim sizes in thousands of dollars for a particular type of policy have a Weibull distribution with mean 5 and variance 6. Fifty claims have been filed. However, the amounts of the claims will not be known until claim settlement, which is estimated to be a year from now. (This is not unusual in liability insurance.) Claim amounts arise from different sources and may be considered independent. Determine the probability that the total amount of these 50 claims exceeds $300,000.

Let X_j be the amount of the jth claim in thousands of dollars and let S be the total amount of all claims. By hypothesis, $E[X_j] = 5$ and $Var(X_j) = 6$. Hence, $E[S] =$

---

[13] If the reader is using the tables in Appendix E, this value can be obtained by applying linear interpolation to the values $\Phi(1.15) = .8749$ and $\Phi(1.16) = .8770$.

[14] From Appendix E, $\Phi(1.05) = .8531$ and $\Phi(1.06) = .8554$.

$E[X_1 + \cdots + X_{50}] = (50)(5) = 250$ and $Var(S) = Var(X_1 + \cdots + X_{50}) = (50)(6) = 300$.

The exact distribution for S is difficult to determine. We know that it isn't a Weibull distribution! However, since the number of terms in the sum is reasonably large and since the X_j are independent and identically distributed, we can determine the desired probability (approximately) using a normal distribution.

Hence, the desired probability is

$$\Pr(S > 300) = \Pr \left(\frac{S - 250}{\sqrt{300}} > \frac{300 - 250}{\sqrt{300}} \right)$$
$$\approx \Pr(Z > 2.8868)$$
$$= 1 - \Phi(2.8868)$$
$$\approx .0019.$$

That is, there is a 1.9% chance of aggregate claims exceeding $300,000.

Notice that we did not correct for continuity in this example. The reason is that S already has a continuous distribution. Consequently, it would be wrong to correct for an approximation error that doesn't actually arise. ∎

EXAMPLE 3:   According to electrical circuit theory, the voltage drop across a resistor is related to the current flowing through the resistor by the equation $V = IR$, where R is the resistance level measured in ohms, I is the current in amperes, and V is the voltage in volts. In alternating current systems, the direction and magnitude of the current change in a cyclical pattern. Hence, if the resistance level is held constant, the voltage will also vary in a cyclical pattern. Consequently, measurements of the voltage will have a distribution that is symmetric about the mean.

Suppose that the measured voltage in a certain electrical circuit has a normal distribution with mean 120 and standard deviation 2 and five measurements of the voltage are taken. Determine the probability that two of the measurements lie outside the range 118–122.

Let X_1, X_2, X_3, X_4, X_5 be the voltage measurements. By assumption, $X_j \sim$ Normal $(120, 2)$ for each j. Hence, the probability that the jth measurement lies in the range 118–122 is

$$\Pr(118 < X_j < 122) = \Pr \left(\frac{118 - 120}{2} < Z < \frac{122 - 120}{2} \right)$$
$$= \Pr(-1 < Z < 1)$$
$$= \Pr(Z \leq 1) - \Pr(Z \leq -1)$$
$$= \Pr(Z \leq 1) - \Pr(Z > 1)$$
$$= \Pr(Z \leq 1) - (1 - \Pr(Z \leq 1))$$
$$= 2 \Pr(Z \leq 1) - 1$$
$$\approx 2(.8413) - 1$$
$$= .6826.$$

Consequently, the probability that the jth measurement lies outside the range 118–122 is .3174.

Now let N be the number of voltage measurements that lie outside the range (118–122). Then $N \sim$ Binomial(5, .3174). Hence, the desired probability is

$$\Pr(N = 2) = \binom{5}{2}(.3174)^2(.6826)^3 \approx .3204.$$ ■

EXAMPLE 4:   Measurements in science and engineering are always subject to error. Consider a scientist who is trying to determine the value of a particular physical constant whose true but unknown value is k. Suppose that the scientist's measurement errors are normally distributed with mean 0 and standard deviation 1; that is, the scientist's measurements are of the form $k + E$, where $E \sim \text{Normal}(0,1)$. Intuition suggests that the scientist can improve her estimate of the physical constant by taking a large number of independent measurements and averaging the results.

Suppose that n independent measurements are made and let $E_1, \ldots, E_n$ be the respective measurement errors. Then the estimate of the physical constant that the scientist obtains by averaging the measurements is

$$\frac{1}{n}\{(k + E_1) + (k + E_2) + \cdots + (k + E_n)\}$$
$$= \frac{1}{n} \cdot nk + \frac{1}{n}(E_1 + \cdots + E_n)$$
$$= k + \overline{E}_n,$$

where $\overline{E}_n = (E_1 + \cdots + E_n)/n$. Hence, $\overline{E}_n$ is the error in the scientist's estimate of the physical constant.

Now since the E_j are independent standard normal random variables, $\overline{E}_n$ has a normal distribution with mean

$$E[\overline{E}_n] = \frac{1}{n}E[E_1 + \cdots + E_n] = 0$$

and with variance

$$Var(\overline{E}_n) = \frac{1}{n^2}\{Var(E_1) + \cdots + Var(E_n)\} = \frac{1}{n}.$$

Hence, $\overline{E}_n \sim \text{Normal}(0, 1/\sqrt{n})$. Consequently, the distribution of $\overline{E}_n$ becomes more concentrated around zero as n increases (see Figure 6.6b for a comparison of Normal(0, σ) for various σ). This means that the scientist's estimate of the physical constant improves with the number of observations taken, which intuition suggests should be the case. ■

EXAMPLE 5:   Suppose that the scientist of the previous example would like her estimate to be within 0.01 of the true value with probability 95%. How many measurements must be taken?

Continuing with the notation of the previous example, the scientist's requirement is

$$\Pr(-0.01 < \overline{E}_n < 0.01) \geq .95.$$

Since $\overline{E}_n = \frac{1}{\sqrt{n}}Z$, where $Z \sim \text{Normal}(0,1)$, this requirement is equivalent to

$$\Pr(-0.01\sqrt{n} < Z < 0.01\sqrt{n}) \geq .95,$$

which by the symmetry of Normal(0,1) is equivalent to the requirement that

$$\Pr(Z < 0.01\sqrt{n}) \geq .975.$$

From tables of the standard normal distribution, $\Phi(1.96) = .975$. Hence, the latter requirement is equivalent to

$$0.01\sqrt{n} \geq 1.96.$$

That is,

$$n \geq 38{,}416.$$

Consequently, if the scientist wishes the estimate to be within 0.01 of the true value with probability 95%, she must take at least 38,416 measurements! ∎

EXAMPLE 6: An actuary has received notification that 100 claims on a particular account have been filed but are still in the course of settlement. The actuary has been asked to determine the size of an appropriate claim reserve for these 100 claims. Historical data suggest that claims on this account are exponentially distributed and average about $300 per claim. The actuary recommends setting up a claim reserve of $31,000. Determine the probability that claim payments exceed this reserve. Assume that the claim sizes are independent.

Let X_j be the size of the jth claim in thousands of dollars and let S be the total amount of claims in thousands of dollars. Then, from the given information, $X_j \sim$ Exponential(10/3). Since a sum of independent exponentials with common parameter has a gamma distribution, an exact model for S is Gamma(100, 10/3). Consequently, the desired probability is

$$\Pr(S > 31) = \Pr\left(\frac{10}{3}S > \frac{310}{3}\right)$$

$$= 1 - \Pr\left(\frac{10}{3}S \leq \frac{310}{3}\right)$$

$$= 1 - I_{100}\left(\frac{310}{3}\right),$$

where $I_{100}(\cdot)$ is the incomplete gamma function with $r = 100$; that is, $I_{100}(\cdot)$ is the distribution function for Gamma(100, 1). The tables in Appendix B don't include values of $I_r(t)$ for $r \geq 10$ or $t > 15$. We could try to use the recursive formula for $I_r(t)$ stated in §6.1.2. However, this would be tedious. Using a statistical calculator, we find that

$$I_{100}\left(\frac{310}{3}\right) \approx .64167880$$

and thus,

$$\Pr(S > 31) \approx .35832120.$$

As an alternative, we could use a normal approximation. Since $X_j \sim$ Exponential(10/3), $E[X_j] = 0.3$ and $Var(X_j) = (0.3)^2 = 0.09$. Hence, $E[S] = (100)(.30) = 30$ and $Var(S) = (100)(0.09) = 9$. Consequently,

$$\Pr(S > 31) = \Pr\left(\frac{S - 30}{3} > \frac{31 - 30}{3}\right)$$

$$\approx \Pr(Z > 1/3)$$

$$\approx .3694.$$

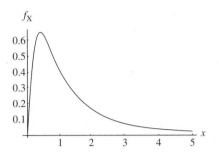

a. Lognormal (0,1)

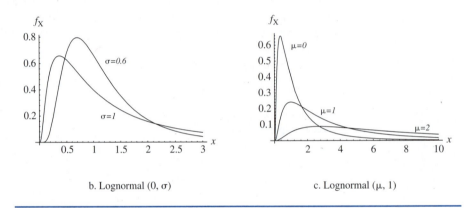

b. Lognormal (0, σ)   c. Lognormal (μ, 1)

FIGURE 6.10 Particular Lognormal Distributions

Notice that the normal approximation is quite close. It is also a whole lot easier to determine when we don't have access to a statistical calculator! ■

6.3.2 The Lognormal Distribution

Definition and Interpretation

A random variable X is said to have a lognormal distribution with parameters μ and σ if its *logarithm* has a normal distribution with mean μ and standard deviation σ (i.e., if $X = e^Y$, where $Y \sim \text{Normal}(\mu, \sigma)$). Equivalently, it is a random variable with probability density function

$$f_X(x) = \begin{cases} \dfrac{1}{x\sigma\sqrt{2\pi}} e^{-(\log x - \mu)^2/2\sigma^2} & x > 0, \\ 0 & x \le 0. \end{cases}$$

Here, μ and σ are real numbers that represent the mean and standard deviation of $\log X$ (not X). Possible graphs of f_X are given in Figure 6.10. The notation Lognormal(μ, σ) signifies a lognormal distribution with parameters μ and σ.

The lognormal distribution arises naturally as the limiting distribution of the product of any independent and identically distributed positive random variables. More precisely, for any sequence of independent and identically distributed positive random variables X_j, the partial products P_n defined by $P_n = X_1 \cdots X_n$ are such that

$$\text{Distribution of } \left(\frac{P_n}{e^{n\mu_Y}}\right)^{1/\sigma_Y \sqrt{n}} \to \text{Lognormal}(0, 1),$$

where $Y = \log X$, and convergence is in the sense of distributions.

When returns on an investment in small disjoint intervals of time are independent and identically distributed, the growth of the investment over time can be modeled using a lognormal distribution. Many models of stock prices are based on lognormal distributions.

Derivation of the Probability Density Function

The probability density function for a lognormal random variable can be derived from the assumption that $X = e^Y$, where $Y \sim \text{Normal}(\mu, \sigma)$, using the formula for the normal density stated in the previous section. Details are left to the reader.

Mean, Variance, and Higher Moments

The mean, variance, and skewness of a lognormal random variable X are, respectively,

$$E[X] = e^{\mu + \sigma^2/2},$$
$$Var(X) = E[X]^2 (e^{\sigma^2} - 1),$$
$$\gamma_X = (e^{\sigma^2} + 2)\sqrt{e^{\sigma^2} - 1}.$$

More generally, the kth moment is given by

$$E[X^k] = \exp(\mu k + \frac{1}{2}\sigma^2 k^2).$$

The formulas for the moments are most easily derived by considering the moment generating function of the associated normal distribution (i.e., the moment generating function of $\log X$). Details of this approach were discussed in §4.3.1. (See the discussion entitled "The moments of e^X.") Adaptation of these results to the lognormal case is left as an exercise for the reader.

Note that the arithmetic mean of X is not e^μ as one might naively think. However, from the formula $\log E_g[X] = E_a[\log X]$ developed in §4.2.1, we see that e^μ is an important average. Indeed, e^μ is the geometric mean of X.

Techniques for Calculating Probabilities

The simplest way to calculate probabilities for a lognormal random variable is to restate them as probabilities about the associated normal random variable and then use the techniques for calculating normal probabilities discussed in §6.3.1. For example, if $X \sim \text{Lognormal}(\mu, \sigma)$ and $a > 0$, then

$$\begin{aligned}
\Pr(X \le a) &= \Pr(e^Y \le a) \\
&= \Pr(Y \le \log a) \\
&= \Pr\left(\frac{Y - \mu}{\sigma} \le \frac{(\log a) - \mu}{\sigma}\right) \\
&= \Phi\left(\frac{(\log a) - \mu}{\sigma}\right),
\end{aligned}$$

where $Y = \log X \sim \text{Normal}(\mu, \sigma)$ and Φ is the standard normal distribution function.

Effect of Arithmetic Operations

A scalar multiple of a power of a lognormal random variable is again lognormal. To be precise, if $X \sim \text{Lognormal}(\mu, \sigma)$, then $aX^b \sim \text{Lognormal}((\log a) + b\mu, |b|\sigma)$ for any constants a, b with $a > 0$ and $b \ne 0$.

A product of independent lognormal random variables is again lognormal. To be precise, if $X_1 \sim \text{Lognormal}(\mu_1, \sigma_1)$, $X_2 \sim \text{Lognormal}(\mu_2, \sigma_2)$, and X_1, X_2 are independent, then $X_1 X_2 \sim \text{Lognormal}(\mu_1 + \mu_2, \sqrt{\sigma_1^2 + \sigma_2^2})$. In fact, a product of dependent lognormal random variables X_1, X_2 is again lognormal, provided that the joint distribution of $\log X_1$ and $\log X_2$ has a bivariate normal density (see §6.3.1 for details).

These properties are immediate consequences of the corresponding properties of normal distributions.

Relationship to Other Distributions

The most important property of the lognormal distribution is that it is the limiting distribution for any *product* of independent and identically distributed *positive* random variables. This property follows directly from the corresponding property with respect to sums of independent, identically distributed random variables for the normal distribution and can be considered a multiplicative form of the central limit theorem.

We conclude our discussion of the lognormal distribution with two numerical examples.

EXAMPLE 1: A particular high-technology stock whose current price is $100 either increases 5% or decreases 3% in any given day. There is a 50% chance that the stock price will increase on any given day independent of price movements on previous days. Determine the probability that the stock's price is more than five times its current price 100 trading days from now.

One could try to construct a binomial tree and determine the desired probability by considering the terminal probabilities. However, this approach could get quite tedious. As an alternative, we can approximate the terminal stock price using a lognormal distribution and calculate the probability accordingly.

Let X_j be the accumulation factor on the jth day and let S be the stock's price 100 days from now. Put $Y_j = \log X_j$. Then from the given information,

$$X_j = \begin{cases} 1.05 & \text{with probability } .50, \\ 0.97 & \text{with probability } .50; \end{cases}$$

and so

$$Y_j = \begin{cases} 0.0487902 & \text{with probability } .50, \\ -0.0304592 & \text{with probability } .50. \end{cases}$$

Consequently,

$$E[Y_j] = 0.0091655$$

and

$$Var(Y_j) = E[Y_j^2] - E[Y_j]^2 = 0.0015701 = (0.0396247)^2.$$

Recall that the multiplicative form of the central limit theorem asserts that

$$\text{Distribution of } \left(\frac{P_n}{e^{n\mu_Y}}\right)^{1/\sigma_Y\sqrt{n}} \to \text{Lognormal}(0, 1),$$

where $P_n = X_1 \cdots X_n$, $X_j \sim X$, and $Y = \log X$.
Hence, the desired probability is

$$\Pr(S > 500) = \Pr(X_1 \cdots X_{100} > 5)$$

$$= \Pr\left(\left(\frac{P_{100}}{e^{100\mu_Y}}\right)^{1/10\sigma_Y} > \left(\frac{5}{e^{100\mu_Y}}\right)^{1/10\sigma_Y}\right)$$

$$\approx \Pr(e^Z > 5.7467024)$$
$$= \Pr(Z > 1.7486262)$$
$$\approx .040212$$
$$\approx 4\%. \qquad \blacksquare$$

EXAMPLE 2: Losses from large fires can often be modeled using a lognormal distribution. Suppose that the average loss due to fire for buildings of a particular type is $25 million and the standard deviation of the loss is $10 million. Determine the probability that a large fire results in losses exceeding $40 million.

Let X be the size of the loss in millions of dollars and suppose that $X \sim$ Lognormal (μ, σ), where μ, σ are the mean and standard deviation of $\log X$. Let μ_X, σ_X denote the mean and standard deviation of X. We are given that $\mu_X = 25$ and $\sigma_X = 10$. From this information, we can determine the values of the parameters μ, σ for $\log X$ using the formulas

$$E[X] = e^{\mu+\sigma^2/2},$$
$$Var(X) = E[X]^2(e^{\sigma^2} - 1).$$

Indeed, from these formulas, we have

$$\frac{\sigma_X^2}{\mu_X^2} = \frac{E[X]^2(e^{\sigma^2} - 1)}{E[X]^2} = e^{\sigma^2} - 1$$

and from the given information, we have

$$\frac{\sigma_X^2}{\mu_X^2} = \frac{100}{625} = 0.16.$$

Hence, equating these two expressions, we have

$$\sigma^2 = \log(1.16) = 0.14842.$$

Substituting this value for σ^2 into the formula for $E[X]$, we find that

$$\mu = \log\mu_X - \frac{1}{2}\sigma^2 = \log 25 - \frac{1}{2}(0.14842) \approx 3.1446658.$$

Consequently, the desired probability is

$$\Pr(X > 40) = \Pr(\log X > \log 40)$$
$$= \Pr(\text{Normal}(3.1447, \sqrt{0.14842}) > \log 40)$$
$$= \Pr\left(Z > \frac{(\log 40) - 3.1447}{\sqrt{0.14842}}\right)$$
$$\approx \Pr(Z > 1.4125)$$
$$\approx 1 - .9211$$
$$= .0789.$$ ∎

6.3.3 The Beta Distribution

Definition and Interpretation

A random variable X is said to have a beta distribution with parameters r and s if its probability density function is given by

$$f_X(x) = \begin{cases} \dfrac{x^{r-1}(1-x)^{s-1}}{B(r, s)} & 0 < x < 1, \\ 0 & \text{otherwise}, \end{cases}$$

where $B(\cdot, \cdot)$ denotes the beta function[15] that is defined by $B(r, s) = \int_0^1 z^{r-1}(1-z)^{s-1}dz$. Here, r and s are both positive real numbers. Possible graphs of f_X are given in Figure 6.11. The notation Beta(r, s) signifies a beta distribution with parameters r and s.

The shape of the graph for the beta density is determined by the signs of $(r - 1)$ and $(s - 1)$: If $r < 1$ and $s < 1$, then the graph is U-shaped; if $(r - 1)(s - 1) < 0$, then the graph is J-shaped—either backward or forward depending on which parameter is less than 1. In all other cases, the graph is unimodal. If $r = s$, the graph is symmetric about $x = 1/2$.

The beta distribution is frequently used to model uncertain fractions. The following two relationships give some justification for this:

1. If $X \sim \text{Gamma}(r, \lambda)$, $Y \sim \text{Gamma}(s, \lambda)$, and X and Y are independent, then $X/(X + Y) \sim \text{Beta}(r, s)$.
2. If $(X|P = p) \sim \text{Binomial}(n, p)$ and $P \sim \text{Beta}(r, s)$, then $(P|X = x) \sim \text{Beta}(r + x, s + n - x)$.

As a special case of the second relationship, we obtain the following interpretation for Beta(r, s): It is the fraction of defective items in a large lot of manufactured items, given that a sample of size $r + s - 2$ chosen with replacement from this lot has exactly $r - 1$ defective items in it and we have no prior opinion about the fraction of defects in the entire lot (i.e., the prior distribution of P is Beta $(1,1)$, which is simply the uniform distribution on $(0,1)$). Hence, if a sample of size n is drawn from a population with replacement and the sample contains x defectives, then the fraction of defectives in the entire population has the distribution Beta$(x + 1, n - x + 1)$.

---

[15] The beta function and its properties are discussed in Appendix C.

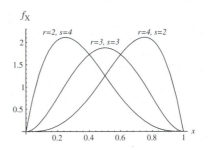

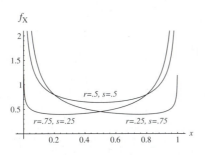
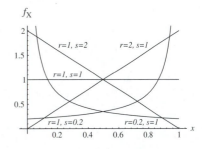

FIGURE 6.11   Beta(r, s) for Selected r, s

Derivation of the Probability Density Function

The beta density can be derived from the binomial distribution using Bayes' theorem. More precisely, one begins with a binomial random variable X with parameters n and p and assumes that the *prior* distribution of P is uniform on $(0, 1)$. The beta distribution then naturally arises as the *posterior* distribution of P (i.e., the distribution $P|X = x$). A derivation along these lines was developed in exercise 23 of §5.5. Details are left to the reader.

Mean, Variance, and Higher Moments

The moments of a beta random variable can be determined directly by integrating x^k with respect to the beta density and expressing the result in terms of beta functions. When we do this, we find that

$$E[X^k] = \frac{B(r + k, s)}{B(r, s)}.$$

In particular, the mean and variance are given by

$$E[X] = \frac{r}{r + s},$$

$$Var(X) = \frac{rs}{(r + s)^2(r + s + 1)}.$$

Details are left to the reader.

Techniques for Calculating Probabilities

The distribution function F_X is given by

$$F_X(x) = \begin{cases} 0 & x < 0, \\ I_x(r, s) & 0 \le x < 1, \\ 1 & x \ge 1, \end{cases}$$

where

$$I_x(r, s) = \int_0^x \frac{t^{r-1}(1-t)^{s-1}}{B(r, s)} dt, \qquad x \in (0, 1).$$

The function $I_x(r, s)$ is known as the **incomplete beta function.**[16]

In general, the incomplete beta function must be evaluated numerically or using tables. However, there are some special cases, where $I_x(r, s)$ has an elementary form:

1. $I_x(r, 1) = x^r$.
2. $I_x(1, s) = 1 - (1 - x)^s$.
3. When s is a positive integer,

$$I_x(r, s) = \sum_{i=0}^{s-1} \left[\frac{\Gamma(r + s)}{\Gamma(r + i + 1)\Gamma(s - i)} \right] x^{r+i}(1 - x)^{s-i-1}.$$

4. When r and s are both positive integers,

$$I_x(r, s) = \sum_{j=r}^{r+s-1} \binom{r + s - 1}{j} x^j (1 - x)^{r+s-1-j}.$$

The latter two formulas follow from the recurrence relation

$$I_x(r, s) = \frac{\Gamma(r + s)x^r(1 - x)^{s-1}}{\Gamma(r + 1)\Gamma(s)} + I_x(r + 1, s - 1),$$

which can be derived using integration by parts, and the relation

$$\Gamma(n) = (n - 1)!$$

for positive integers n.

Formula 4 actually asserts that

$$\Pr(\text{Beta}(r, s) \le x) = \Pr(\text{Binomial}(r + s - 1, x) \ge r)$$

when r and s are positive integers. This is a handy formula to remember and will be used without further explanation.

Effect of Arithmetic Operations

If $X \sim \text{Beta}(r, s)$, then $1 - X \sim \text{Beta}(s, r)$. This follows directly from the definition of the beta density. It also makes intuitive sense, given the interpretation of the beta random variable as the fraction of defectives in a large lot of manufactured items.

Relationship with Other Distributions

The beta distribution is related to the gamma distribution and the binomial distribution in the following way:

---

[16] The incomplete beta function and its properties are discussed in Appendix D.

1. If $X \sim$ Gamma(r, λ), $Y \sim$ Gamma(s, λ), and X and Y are independent, then $X/(X + Y) \sim$ Beta(r, s).
2. If $(X|P = p) \sim$ Binomial(n, p) and $P \sim$ Beta(r, s), then $(P|X = x) \sim$ Beta$(r + x, s + n - x)$.

The latter property is a direct consequence of Bayes' theorem. The former is derived most easily by considering the bivariate transformation $(X, Y) \mapsto (X + Y, X - Y)$; for simplicity, we omit its proof.

The beta distribution can also be considered a generalization of the DeMoivre distribution because Beta(1,1) is uniformly distributed on $(0, 1)$.

We conclude our discussion of the Beta distribution with some numerical examples.

EXAMPLE 1:   A quality control engineer is interested in the fraction of defectives present in a large lot of manufactured items. Due to cost constraints and the fact that the testing procedure can damage the item being tested, it is not possible to test every item in the lot. The engineer selects 100 items at random with replacement and discovers no defects. Determine the probability that more than 2% of the items in the entire lot are defective.

Let X be the fraction of items in the lot that are defective. Then X can be modeled using a beta distribution. Since no defectives were found in the sample of size 100, the parameters of this beta distribution are $r = 1$ and $s = 101$. Hence, the desired probability is

$$\Pr(X > .02) = \int_{.02}^{1} \frac{x^0(1 - x)^{100}}{B(1, 101)} dx.$$

Since

$$\int_{.02}^{1} (1 - x)^{100} dx = \left. \frac{-(1 - x)^{101}}{101} \right|_{.02}^{1} = \frac{1}{101}(.98)^{101}$$

and since

$$B(1, 101) = \frac{\Gamma(1)\Gamma(101)}{\Gamma(102)} = \frac{0!100!}{101!} = \frac{1}{101},$$

the desired probability is

$$\Pr(X > .02) = (.98)^{101} \approx .13. \qquad \blacksquare$$

EXAMPLE 2:   The electrical work on a particular construction site will take at least 10 days to complete and, in the worst case scenario, will not take more than 30 days. The average time to completion on similar projects is 18 days and the standard deviation in the time to completion is 4 days. Determine the probability that this project will take at least 20 days to complete using an appropriate beta distribution.

Let N be the number of days to complete the project and let $X = (N - 10)/20$. Suppose that we model X using a beta distribution. The parameters r, s of this beta distribution can be determined using the given information and the general formulas for the mean and variance of a beta distribution. Indeed, from the information $\mu_N = 18$, $\sigma_N = 4$, we have $\mu_X = 2/5$, $\sigma_X = 1/5$ and so r, s are given by the equations

$$\frac{2}{5} = \frac{r}{r+s},$$

$$\frac{1}{25} = \frac{rs}{(r+s)^2(r+s+1)}.$$

From the first of the these equations, we have $s = \frac{3}{2}r$. Substituting this into the second equation and solving the resulting cubic, we find that

$$r = 2, \quad s = 3.$$

Consequently, $X \sim \text{Beta}(2,3)$ and the desired probability is

$$\Pr(N \geq 20) = \Pr\left(X \geq \frac{1}{2}\right)$$

$$= \int_{1/2}^{1} \frac{x(1-x)^2}{B(2,3)} dx.$$

Using integration by parts, we have

$$\int_{1/2}^{1} x(1-x)^2 dx = -x \frac{(1-x)^3}{3}\bigg|_{1/2}^{1} - \int_{1/2}^{1} \frac{-(1-x)^3}{3} dx$$

$$= \frac{1}{3}\left(\frac{1}{2}\right)^4 - \frac{(1-x)^4}{12}\bigg|_{1/2}^{1}$$

$$= \frac{5}{192}.$$

Further,

$$B(2,3) = \frac{\Gamma(2)\Gamma(3)}{\Gamma(5)} = \frac{1!2!}{4!} = \frac{1}{12}.$$

Hence, the desired probability is

$$\Pr(N \geq 20) = \frac{5}{16}.$$ ∎

EXAMPLE 3:   The saturation of dissolved oxygen in a particular river can be modeled using a beta distribution with parameters $r = 3$ and $s = 2$. An environmental engineer takes a sample of water from this river and finds that the saturation fraction is less than 40%. What is the probability of this happening?

Let X be the saturation fraction of dissolved oxygen in the sample. Then from the given information, $X \sim \text{Beta}(3,2)$ and so the desired probability is

$$\Pr(X < .40) = \int_{0}^{.40} \frac{x^2(1-x)}{B(3,2)} dx.$$

Using integration by parts, we have

$$\int_0^{.40} x^2(1-x)dx = \frac{x^3}{3}(1-x)\bigg|_0^{.40} + \int_0^{.40} \frac{x^3}{3}dx$$

$$= \frac{7(.40)^4}{12}$$

and thus,

$$\Pr(X < .40) = 7(.40)^4 = .1792.$$

Alternatively, using the relationship between the beta and the binomial distributions, we have

$$\Pr(\text{Beta}(3,2) \le .40) = \Pr(\text{Binomial}(4, .40) \ge 3)$$

$$= \binom{4}{3}(.40)^3(.60)^1 + \binom{4}{4}(.40)^4$$

$$= 7(.40)^4$$

$$= .1792$$

as before.   ■

EXAMPLE 4:   An actuary is interested in the fraction P of policies that file at least one claim in a year. On the basis of historical data, the actuary models P as Beta(2,8) at the beginning of the year. At the end of the year, the actuary observes that in a particular group of 20 policies, only one policyholder filed a claim. Policies in this group are assumed to be independent. Determine the probability that P is at most 15%.

Let X be the number of policyholders in the group of 20 that file a claim and let P be the fraction of policies (overall) that file at least one claim. Then, at the beginning of the year, $P \sim \text{Beta}(2,8)$ and $(X|P = p) \sim \text{Binomial}(20, p)$. Hence, at the end of the year, $(P|X = 1) \sim \text{Beta}(3,27)$.

Consequently, the desired probability is

$$\Pr(P \le .15|X = 1) = \Pr(\text{Binomial}(29, .15) \ge 3)$$

$$= 1 - \sum_{n=0}^{2} \binom{29}{n}(.15)^n(.85)^{29-n}$$

$$= .8315735.$$ ■

6.4   Exercises

1. In this question, we analyze the graphical properties of the gamma distribution.

   a. Qualitatively describe how the gamma distribution is affected by changes in the parameters r and λ. In particular, describe what happens to the graph of the probability density function as r increases with λ held fixed and as λ increases with r held fixed. What happens when r approaches zero or λ approaches zero? How do changes in r and λ affect the mean, variance, and skewness of the distribution?

   b. The gamma distribution has two parameters: a "shape" parameter r and a "scale" parameter λ. By considering the effect that changes in r and λ have on the graph of the density function, explain why it is appropriate to refer to r as a shape parameter and λ as a scale parameter.

c. The density function plots in Figures 6.1 and 6.2 suggest that the density function for a gamma random variable increases and decreases or just decreases. By considering the derivative of the density function, show algebraically that the observed behavior of the density function is correct. What condition on the parameters r, λ must be satisfied for the density function to be strictly decreasing?

2. In this question, we analyze the graphical properties of the Weibull distribution.

a. Qualitatively describe how the Weibull distribution is affected by changes in the parameters α and β. In particular, describe what happens to the graph of the probability density function as α increases with β fixed and as β increases with α fixed. What happens when α approaches zero or β approaches zero? How do changes in α and β affect the mean, variance, and skewness of the distribution?

b. The Weibull distribution has two parameters: a "scale" parameter α and a "shape" parameter β. By considering the effect that changes in α and β have on the graph of the density function and on its algebraic formula, explain why it is appropriate to refer to α as a scale parameter and β as a shape parameter.

c. The density function plots in Figures 6.1 and 6.4 suggest that the density function for a Weibull random variable increases and then decreases or just decreases. By considering the derivative of the density function, show algebraically that the observed behavior of the density function is correct. *Hint:* Show that for $x > 0$,

$$f_X'(x) = \left\{ \alpha^\beta \left(1 - \frac{1}{\beta}\right) - x^\beta \right\} \left(\frac{\beta}{\alpha^\beta}\right)^2 x^{\beta-2} \exp\left\{-\left(\frac{x}{\alpha}\right)^\beta\right\}$$

and then proceed to analyze the quantity $x^\beta - \alpha^\beta(1 - \frac{1}{\beta})$. What condition on the parameters α, β must be satisfied for the density function to be strictly decreasing?

3. In this question, we analyze the graphical properties of the beta distribution.

a. The density function plots in Figure 6.11 suggest that the shape of the density function for a beta random variable is determined by the signs of $(r - 1)$ and $(s - 1)$ in the following way:

   i. If $r < 1$ and $s < 1$, then the graph is in the shape of a U. The graph is symmetric precisely when $r = s$.
   ii. If $r \geq 1$ and $s < 1$, then the graph is in the shape of a forward J.
   iii. If $r < 1$ and $s \geq 1$, then the graph is in the shape of a backward J.
   iv. If $r = 1$ and $s = 1$, then the graph is a straight line.
   v. In all other cases, the graph is unimodal.

   By considering the formulas for the density function and its derivative, show that the observed behavior is correct.

b. Show that if $r \geq 1$ and $s \geq 1$, then the mode of the distribution is $(r - 1)/(r + s - 2)$ unless $r = s = 1$, in which case the density function is constant and every point is a mode.

c. Qualitatively describe how the beta distribution is affected by changes in the parameters r and s. In particular, describe what happens to the graph of the probability density function in each of the following cases:

      i. $r < 1$ is fixed; s tends to 0 from above or 1 from below.

     ii. $s < 1$ is fixed; r tends to 0 from above or 1 from below.

    iii. $r \geq 1$ is fixed; s tends to 0 from above or 1 from below.

    iv. $s \geq 1$ is fixed; r tends to 0 from above or 1 from below.

     v. $r \geq 1$ is fixed; s tends to 1 from above or becomes arbitrarily large.

    vi. $s \geq 1$ is fixed; r tends to 1 from above or becomes arbitrarily large.

   vii. $r = s$, and the common value tends to 0 or becomes arbitrarily large.

How do changes in r and s affect the mean, variance, and skewness of the distribution?

4. The beta distributions with parameters r and s such that $r = s > 1$ have many of the same properites as the normal distributions with mean $1/2$:

  a. Their density functions are symmetric about $x = 1/2$ and have only one critical point in the interval $(0, 1)$.

  b. Their means, medians, and modes are all located at $x = 1/2$.

However, the beta distributions are only defined on the interval $[0, 1]$, whereas the normal distributions are defined for all values of x. We are interested in how well a beta distribution of the foregoing type can be approximated by a normal distribution with the same mean and variance.

For each of the beta distributions X that follow, calculate the exact values of $\Pr(|X - \frac{1}{2}| > \frac{1}{2})$, $\Pr(|X - \frac{1}{2}| > \frac{1}{3})$, $\Pr(|X - \frac{1}{2}| > \frac{1}{4})$, and $\Pr(|X - \frac{1}{2}| > \frac{1}{6})$ and compare them to the approximate values obtained when X is assumed to be a normal distribution with the same mean and variance. Based on these rudimentary calculations, do you think that this approximation is a reasonable one? For what values of r and s does the approximation appear to be best? Worst?

  a. Beta(2, 2)

  b. Beta(4, 4)

  c. Beta(6, 6)

5. For each of the following random variables X, determine the type of distribution (i.e., exponential, gamma, normal, etc.) that best models X. Where possible, give values for the parameters of the distribution chosen. Give reasons for your choice of distribution.

  a. Calls to an Internet service provider are placed independently and at random. During evening hours, the service provider receives an average of 100 calls an hour. X is the waiting time between consecutive calls tonight.

  b. A quality control supervisor on a certain production line chooses 100 items at random from a large lot of manufactured items and examines them for defects. Of the 100 items examined, 5 are determined to be defective. X is the fraction of defectives in the entire lot.

  c. Shares in a new mutual fund are offered for sale at an initial price of \$1 per share. After the initial offering, shares can be purchased and sold through the company at a price that is determined at the close of business on the day the order is received. Suppose that the day-to-day price fluctuations are independent and identically distributed random variables and that daily returns are just as likely to be positive as negative. X is the value of one share of this fund exactly 1 year after the initial public offering.

d. Switching circuits at a certain telephone exchange fail at a constant rate of three per month. X is the lifetime of a randomly chosen switching circuit at this exchange.

e. An intercity bus company offers express service between two cities that are 50 miles apart. A breakdown can occur anywhere along the travel route. There is no point on the route where a breakdown is more likely to occur than the other points. X is the location of the next breakdown.

f. As part of a grand opening promotion, a department store has advertised that every one-thousandth purchase made on opening day will be given to the customer for free. The store expects five purchases to be made every minute. X is the time from opening until the first purchase is given away.

g. The average height of professors at a certain college is 67 inches, and the mean square deviation from this average is 2 squared inches. X is the height of a randomly chosen professor.

h. The time limit for completing a certain test in probability is 1 hour. An average student takes 50 minutes to complete the test. The standard deviation from this average completion time is 6 minutes. X is the time it takes a randomly chosen student to complete the test.

i. A postal service outlet had two customer service representatives on duty—one experienced and one in training. On average, the experienced representative takes 1 minute to serve a customer, while the trainee takes 2 minutes. Both representatives are currently occupied. X is the time until the first representative becomes available.

j. An automotive engineer is attempting to design an exhaust system that will outlast the systems currently being used. Preliminary studies indicate that the failure rate of the new system after t years of use is proportional to $t^{3/2}$. X is the lifetime of a randomly chosen exhaust system of the new type.

k. A summer theater company has 30 cast members who meet daily for rehearsals during production season. All 30 members must be present for a full dress rehearsal, but only 20 members need be present for a partial dress rehearsal. The director of the company, who has recently become concerned with rehearsal attendance, has scheduled a full dress rehearsal for this evening and has advised the cast that the rehearsal will not begin until all 30 members are present. The director arrives at 6:30 P.M. Cast members subsequently arrive randomly and independently at a rate of 2 per minute. X is the fraction of the total time (i.e., the time for all 30 cast members to arrive) spent waiting for the last 10 cast members to arrive.

6. A health insurance provider is trying to develop a probability model for the dollar amount of the claims made against each of its policies in a given year. It plans to use this information when determining the level of premiums to charge its customers. One thousand policies of a similar type are examined and the dollar amount of the claims made against each policy in the previous year is observed. These observations are summarized in the following table:

Claim Amount Range	Number of Such Policies Among the Observed 1000
0–25	28
25–75	152
75–125	184
125–175	167
175–225	135
225–275	103
275–325	75
325–375	59
375–425	37
425–475	25
475–525	17
525–575	11
575–	7
Total Observed	1000

Let X be the amount in hundreds of dollars of the claims against a randomly chosen policy in the previous year. The insurance company's actuarial department believes that X has (approximately) a gamma distribution. Our goal is to determine the parameters of this distribution and then use the resulting distribution function to calculate probabilities.

a. Explain why it is reasonable to assume that the distribution of X is continuous.
b. Construct a table for the *relative frequency densities* of X at $X = 0, 0.5, 1, \ldots, 6$ using the data given. For example, the relative frequency density at $X = 1$ is approximately 0.368 since 184 of the 1000 policies examined had claim amounts between \$75 and \$125 and since the length of this interval in hundreds of dollars is 1/2. Can you explain why these numbers do not add up to 1?
c. Use the table you constructed in part b to calculate the mean and the variance of X implied by the data. Begin by calculating the first and second moments implied by the data using the formula $E[X^k] \approx \sum x^k f(x)\Delta x$, where $f(x)$ is the relative frequency density at x. Then deduce the implied mean and variance.
d. By equating means and variances, determine the parameters r, λ of the gamma distribution which fits the data. Round your answers for r and λ to the nearest whole number. Then construct a table of gamma densities for $X = 0, 0.5, 1, \ldots, 6$ and compare it to the table of relative frequency densities you constructed in part b. Does the gamma distribution constructed appear to fit the data?
e. In judging the fit of a continuous distribution to data, it is sometimes better to consider *cumulative relative frequencies* rather than relative frequency densities. Construct a table of cumulative relative frequencies at $X = 0.25, 0.75, 1.25, \ldots,$ 5.75 using the given data. Then construct a table of values at these points for the distribution function of the gamma distribution determined in part d. Be sure to use the whole number estimates for r and λ determined in part d. Does the gamma distribution constructed appear to fit the data under this criterion?

f. Using the gamma distribution determined in part d, calculate $\Pr(4 \le X \le 4.25)$, $\Pr(X \ge 6)$, and $\Pr(X \le 0.25)$.

7. The moment generating functions $M_X(t)$ for particular continuous random variables X follow. In each case, identify the distribution (e.g., normal, exponential, etc.) and specify its associated parameters.

a. $M_X(t) = \dfrac{1}{1-t}$

b. $M_X(t) = \left(\dfrac{1}{1-3t}\right)^{1/2}$

c. $M_X(t) = \exp(t^2/2)$

d. $M_X(t) = \exp(t + t^2)$

8. The moment generating functions $M_X(t)$ for three continuous random variables X follow. For each of these distributions, determine the mean and variance of X and calculate $\Pr(X > 1)$ and $\Pr(-1 < X < 1)$.

a. $M_X(t) = \exp\left(t + t^2/2\right)$

b. $M_X(t) = \left(\dfrac{1}{1-t}\right)^3$

c. $M_X(t) = \dfrac{3}{3-t}$

9. A service center on a major highway has two automated teller machines located next to each other. Customers wishing to obtain cash from one of these machines form a single queue and use the first machine to become available. On average, the service time at the machine on the left is 30 seconds, and the service time at the machine on the right (a newer model) is 20 seconds. The machines operate on separate power supplies and function independently. Suppose that both machines are currently in use.

a. How long should the person at the front of the line expect to wait before a machine becomes available?

b. What is the probability that the person at the front of the line will have to wait more than 15 seconds for a machine to become available?

c. How long should the third person in line expect to wait before a machine becomes available?

d. What is the probability that the third person in line will have to wait more than 30 seconds for a machine to become available?

e. The machine on the left has just become available. What is the probability that the person beginning a transaction at this machine will still be there 1 minute from now?

10. An insurance company receives 100 claims per day on average. Claims arrive independently and at random at the company office. Of the claims, 95% are for amounts less than $100 and are processed immediately; the remaining 5% are examined more closely to verify their accuracy and eligibility.

a. What is the probability of getting no claims over $100 in a given day?

b. What is the probability of getting at most two claims over $100 in a given day?

c. How many claims for amounts less than $100 should this company expect to receive in 5 business days?

11. An automated teller machine allows customers to make deposits, make withdrawals, and pay bills. Of the customers who use this machine, 70% make withdrawals only, 20% pay bills, and the remaining 10% make deposits. On average, it takes 15 seconds to make a withdrawal, 25 seconds to pay a bill, and 50 seconds to make a deposit. Suppose that the service times for the three customer types are exponentially distributed. What is the probability that the next customer will take more than 20 seconds?

12. A laser printer on a local area computer network is capable of printing eight pages per minute. Print requests arrive randomly and independently at an average rate of 20 per hour and are placed in a queue according to their arrival times. The average request results in ten printed pages. Suppose that the time it takes to process each request is exponentially distributed and that the print times for different jobs are independent.

   a. What is the probability that this printer will be idle for the next 5 minutes if the queue is currently empty?
   b. What is the probability that the next request will take more than 2 minutes to process?
   c. There is currently one job in the queue and it has been active for the last 5 minutes. What is the probability that this job will still be active 1 minute from now?
   d. There are currently five jobs in the queue—one active and four waiting to be processed. You submit a job to this printer and it becomes the sixth job in the queue. What is the probability that you will have to wait more than 5 minutes for the printer to begin processing your job? More than 10 minutes? Assume that the time between consecutive jobs in the queue is negligible.

13. A particular stock, whose current price is $100 either increases 2% or decreases 1% in any given day. There is a 50% chance that the stock price will increase on any given day, independent of previous price movements. Determine the probability that the stock's price is more than double its current price 50 trading days from now.

14. A particular stock, whose current price is $100, either increases $2 or decreases $1 in any given day. There is a 50% chance that the stock price will increase on any given day, independent of previous price movements. Determine the probability that the stock's price is greater than $145 fifty trading days from now.

15. For each of the following random variables X, determine the type of distribution (i.e., binomial, Poisson, etc.) that best models X. Where possible, give values for the parameters of the distribution chosen. Give reasons for your choice of distribution.

   a. A particular loan portfolio contains 500 securities. The risk of default on any given loan in the portfolio in the coming year is considered to be 5%, independent of the other securities in the portfolio. X is the number of defaults in the coming year.
   b. Claims arrive at an insurer randomly and independently at a rate of 25 per day. X is the number of claims in the next 5 days.
   c. The risk of prepayment on a particular mortgage is considered to be constant throughout the life of the mortgage. X is the lifetime of the mortgage.

d. The risk of prepayment on a particular mortgage is considered to be constant throughout the life of the mortgage. Ten years have passed since the mortgage contract was originally written. X is the remaining lifetime of the mortgage.

e. Calls to an investment broker are placed independently and at random. During the first hour of trading, the broker receives an average of 500 calls an hour. X is the number of calls received between 9:30 A.M. and 10:00 A.M.

f. A particular loan for $10,000 gives the borrower the option of repaying the loan in full at the end of each month. No partial repayments of the principal amount are permitted. In months where the borrower chooses not to repay the principal amount, an interest payment of $100 is required. In any given month, the probability of the principal being repaid is 10%, independent of the other months. X is the number of payments made before the loan is repaid.

g. A mortgage portfolio currently contains three types of mortgages: variable rate, 15-year fixed rate, and 30-year fixed rate. Company data suggest that the risk of prepayment on each of these mortgage types is constant; in particular, the data suggest that the company should expect one hundred variable rate mortgages, thirty-five 15-year mortgages, and ten 30-year mortgages to be prepaid in any given year. Borrowers are assumed to make the prepayment decision independently of one another. X is the time until the next mortgage is prepaid.

h. An actuary is trying to obtain an estimate for the fraction of policies in a given block of business that are in arrears. The actuary selects 300 policies at random and determines that 15 of the policyholders are behind in their premium payments. X is the fraction of policyholders in arrears in the entire block of business.

i. Claims arrive randomly and independently at an insurance company at a rate of 100 per day. Historical data suggest that 55% of the claims will be related to health insurance policies, 35% will be related to auto insurance policies, and the remaining 10% will be related to life insurance policies. X is the number of health-related claims in the next 2 days.

j. A particular loan for $10,000 requires the borrower to repay the principal in $5000 installments; however, the borrower may choose the months in which the payments are made. Interest at a rate of 12% per annum compounded monthly is charged on the outstanding loan and must be paid each month that the loan is alive. All payments (both principal and interest) are assumed to be made at the end of the month. In any given month, the probability that a $5000 payment on the principal is made is 5%, independent of the other months. X is the number of months in which payments of $50 or $100 are received.

k. Customers arrive randomly and independently at a bank at a rate of 25 per hour. X is the time until the fifth customer arrives.

l. A reinsurer agrees to pay 10% of every claim submitted to an insurer, up to a maximum of $10,000 (i.e., the reinsurer will pay no more than $10,000 to the insurer for any given claim). The probability that a given claim to the insurer exceeds $100,000 is 1%, independent of the other claims. X is the number of claims before a $10,000 payout is required.

16. Consider the random variable X which has a gamma distribution with parameters $r = 2$ and $\lambda = 3$.

a. Calculate $\Pr(X \geq 1)$ directly using the gamma density and successive applications of integration by parts.

b. Calculate $\Pr(X \geq 1)$ by reformulating the statement $X \geq 1$ in terms of a related Poisson random variable.

c. Calculate $\Pr(X \geq 1)$ by standardizing the distribution (i.e., considering the distribution of λX) and using tables for the incomplete gamma function.

d. Which of these methods is feasible when r is not a positive integer?

17. The continuously compounded rate of return on a particular stock has a normal distribution with mean 10% and standard deviation 20%. The stock's price is currently $50. Calculate the probability that the stock's price will be between $55 and $60 one year from now.

18. An actuary in a recently merged insurance company is trying to determine the fraction of policies in the new company on which there is a claim in a given year. Because the computer systems in the various departments are not yet compatible this information is not readily available. Hence, the actuary selects a sample of 100 policies at random and examines their claim history. Suppose that exactly 5 of these policies filed claims in the previous year. What is the probability that more than 10% of the policies in the entire company filed claims in the previous year? *Hint:* Use the fact that $\Pr(\text{Beta}(r, s) \leq x) = \Pr(\text{Binomial}(r + s - 1, x) \geq r)$.

19. The annualized continuously compounded rate of return on a particular investment has a normal distribution with mean 10% and standard deviation 20%.

a. What is the expected accumulation of such a $1 investment 1 year from now?

b. What is the expected accumulation of such a $10 investment 2 years from now?

Hint: Consider moment generating functions.

20. Suppose that $X \sim \text{Normal}(\mu, \sigma)$ and $Y = e^X$. In this question, you will determine a formula for the skewness of the lognormal random variable Y.

a. Determine a formula for $E[Y^k]$ using the moment generating function of X.

b. Using the formula derived in part a, show that

$$E[Y^k] = \mu_Y^k q^{\binom{k}{2}},$$

where $q = e^{\sigma^2}$ and $\mu_Y = e^{\mu + \frac{1}{2}\sigma^2}$. Deduce that

$$E[(Y - \mu_Y)^2] = \mu_Y^2 (q - 1)$$

and

$$E[(Y - \mu_Y)^3] = \mu_Y^3 (q - 1)^2 (q + 2).$$

c. Show that the skewness of $Y = e^X$ is given by

$$\gamma_Y = (q + 2)(q - 1)^{1/2}$$
$$= (e^{\sigma^2} + 2)\sqrt{e^{\sigma^2} - 1}.$$

d. You are evaluating two different investment prospects. The annual continuously compounded rate of return for each prospect has a normal distribution; however, the parameters of the two distributions are different. What can you say about the skewness of the distribution of the dollar accumulations for the two prospects?

21. In a random birth, the probability of a boy is about 51%. You are interested in the probability that at most half the newborns in a nursery are boys. In particular, you are interested in how this probability changes as the number of newborns increases. Let n be the number of newborns in the nursery.

    a. Using a normal approximation with continuity correction, determine the probability that at most $n/2$ of the newborns in the nursery are boys when:
       i. $n = 100$
       ii. $n = 500$
       iii. $n = 1000$
       What conclusion do you reach?
    b. Repeat the calculations of part a without correcting for continuity. Is it more important to correct for continuity when n is large or small? Why do you think this is so?

22. A gambler has two coins in his pocket: a fair coin and a biased one that comes up heads 55% of the time. The gambler chooses one of these coins at random and tosses it 1000 times. If at least 525 of the 1000 tosses are heads, the gambler concludes that he has the biased coin; otherwise, he concludes that he has the fair coin. What is the probability that the gambler reaches a false conclusion?

23. Each of the following random variables X can be modeled using a beta distribution with certain parameters r and s. For each X, indicate what you think should be true of the signs of $(r - 1)$, $(s - 1)$, and $(r - s)$. Give reasons for your answers.

    a. A 1-hour test is given to a group of 1000 students. Seventy percent of the students turn in their papers in the first 30 minutes. X is the time that a randomly chosen student spends writing the test.
    b. When a suspension bridge is subjected to an earthquake, fractures in the bridge are most likely to occur at locations where a support gives way. A particular section of a suspension bridge is supported at each end but nowhere else in the middle. X is the distance from one end of the section to the location where a fracture occurs.
    c. A car traveling between two cities can have a breakdown at any place on the road. The chance of a breakdown occuring in any given stretch of road is the same as any other stretch of similar length. X is the location, measured from the starting point, of a breakdown.
    d. A 1-hour test is given to a group of 1000 students. Only 10% of the students turn in their papers before time is called. X is the time that a randomly chosen student spends writing the test.
    e. A metal stamping machine produces circular tokens with a diameter of 25 mm. Depending on how the machine is calibrated, the diameter of any given coin could be off by as much as 1 mm in either direction; the diameter is just as likely to be over the average of 25 mm as it is to be under. X is the diameter of a randomly chosen coin.

24. A bus travels between two cities that are 100 miles apart. A breakdown can occur anywhere on the travel route; the chance of a breakdown occurring on any given stretch of road is the same as any other stretch of similar length. There are service garages in both cities and midway between the cities on the travel route. If a breakdown occurs, a tow truck is sent from the garage closest to the breakdown site.

a. What is the probability that a tow truck has to travel more than 10 miles to reach the bus?

b. Would it more efficient to locate the service garages at the 25-, 50-, and 75-mile points along the road rather than at the 0-, 50-, and 100-mile points?

25. A loggamma random variable is a random variable of the form $Y = e^X$, where $X \sim \text{Gamma}(r, \lambda)$. Determine a formula for the moments $E[Y^k]$ of a loggamma distribution in terms of the parameters r, λ. *Hint:* Consider $M_X(t)$, the moment generating function of X.

26. Suppose that $X \sim \text{Gamma}(r, \lambda)$.

a. Using the known formula for the moment generating function $M_X(t)$, determine formulas for $M_X^{(k)}(t)$ and $\psi_X^{(k)}(t)$. Deduce that

$$(\lambda - t)^r M_X^{(k)}(t) = \lambda^r \binom{r + k - 1}{r} \psi_X^{(k)}(t)$$

when $X \sim \text{Gamma}(r, \lambda)$.

b. Using the formulas derived in part a, determine formulas for the moments and the cumulants of X; that is, determine formulas for $E[X^k]$ and $\psi_X^{(k)}(0)$.

c. Determine a formula for the skewness of a gamma distribution.

d. Suppose that $X_1, X_2, \ldots, X_n$ are independent gamma distributions with $X_j \sim \text{Gamma}(r_j, \lambda)$. What is the skewness of $X_1 + \cdots + X_n$?

27. A portfolio consists of two investment securities. The continuously compounded rates of return in percentage terms on the securities are Normal(5, 10) and Normal(10, 20), respectively. One dollar is invested in each security. Determine the variance and the skewness of the dollar accumulation of the portfolio over the year, assuming the returns on the two securities are independent.

28. The losses L_1 and L_2 on two investments known to be money-losers have gamma distributions with respective parameters $r_1 = 2$, $\lambda_1 = 4$, $r_2 = 3$, $\lambda_2 = 5$. Determine the mean, variance and skewness of the combined loss $L_1 + L_2$, assuming the losses are independent.

29. In this question, you demontrate that the Pareto distribution is a continuous mixture of exponentials, and you derive the formulas for the moments of a Pareto random variable stated in §6.1.3. Hence, suppose that $(Y|\Lambda = \lambda) \sim \text{Exponential}(\lambda)$ and $\Lambda \sim \text{Gamma}(s, \beta)$.

a. Show that the survival function S_Y for the unconditional distribution of Y is given by

$$S_Y(y) = \left(\frac{\beta}{\beta + y}\right)^s, \qquad y \geq 0.$$

Hint: Use the law of total probability to show that

$$S_Y(y) = \int_0^\infty \Pr(Y > y | \Lambda = \lambda) f_\Lambda(\lambda) d\lambda$$

and then rewrite the integral as $(\beta/(\beta + y))^s$ times the integral of the gamma density with parameters s and $\beta + y$.

Deduce that the Pareto distribution is a continuous mixture of exponentials.

b. Show that

$$f_Y(y) = \frac{s}{\beta} \left(1 + \frac{y}{\beta}\right)^{-s-1}, \quad y \geq 0.$$

Deduce that

$$\int_0^\infty \frac{dz}{(1+z)^{s+1}} = \frac{1}{s}.$$

Hint: Put $z = y/\beta$.

c. Using integration by parts, show that

$$\int_0^\infty z^k (1+z)^{-s-1} dz = \frac{k}{s} \int_0^\infty z^{k-1}(1+z)^{-s} dz,$$

provided that $s > k$. Deduce that if $s > k$, then

$$\int_0^\infty z^k(1+z)^{-s-1} dz = \left(\frac{k}{s}\right)\left(\frac{k-1}{s-1}\right) \cdots \left(\frac{1}{s-k+1}\right) \cdot \left(\frac{1}{s-k}\right).$$

Hint: Use the foregoing recursion and the formula derived in part b.

d. Using the formula derived in part c, show that for $k < s$,

$$E[Y^k] = \frac{\beta^k k!}{(s-1)(s-2) \cdots (s-k)}.$$

Hint: Apply the substitution $y = \beta z$ to the integral for $E[Y^k]$.

30. In a particular block of business, there is a 25% chance of a claim being submitted on a given policy. On policies for which a claim is submitted, the claim size has a Pareto distribution with parameters $s = 3$, $\beta = 100$.

a. Determine the probability that a randomly chosen policyholder submits a claim for more than $50.

b. Determine the probability that the claim on a randomly chosen policy is at most $10.

c. Determine the distribution function and survival function for the claim size on a randomly chosen policy.

d. Determine the expected value and the variance of the claim size on a randomly chosen policy.

31. Customers arrive randomly and independently at a bank at a constant, but unknown, rate of λ per minute. There is a 50% chance that $\lambda = 2$, a 30% chance that $\lambda = 3$, and a 20% chance that $\lambda = 4$. No other values of λ are possible. Determine the probability that at most three customers arrive in the next minute.

32. Suppose that $(N|P = p) \sim$ Binomial(m, p) and $P \sim$ Beta(r, s). Show that the unconditional distribution of N has probability mass function

$$p_N(n) = \binom{m}{n} \frac{\Gamma(r+s)\Gamma(n+r)\Gamma(m-n+s)}{\Gamma(r)\Gamma(s)\Gamma(m+r+s)}.$$

Deduce that if r and s are positive integers, then

$$p_N(n) = \binom{m}{n} \frac{(r+s-1)!(n+r-1)!(m-n+s-1)!}{(r-1)!(s-1)!(m+r+s-1)!}.$$

Hint: Use the law of total probability to obtain an expression for $\Pr(N = n)$ and rewrite the integrand of the resulting integral as a multiple of the beta density with parameters $n + r$ and $m - n + s$. Then use the fact that $B(r, s) = \Gamma(r)\Gamma(s)/\Gamma(r + s)$.

33. Suppose that $(X|\Lambda = \lambda) \sim \text{Gamma}(r, \lambda)$ and $\Lambda \sim \text{Gamma}(s, \beta)$.

    a. By using the law of total probability, show that the unconditional random variable X has distribution function

    $$F_X(x) = \frac{1}{B(r, s)} \int_0^x \left(\frac{z}{\beta + z}\right)^{r-1} \left(\frac{\beta}{\beta + z}\right)^{s-1} \frac{\beta}{(\beta + z)^2} \, dz.$$

    Deduce that

    $$F_X(x) = I_{r,s}\left(\frac{x}{\beta + x}\right),$$

    where $I_{r,s}(x)$ is the incomplete beta function with parameters r and s; that is, $I_{r,s}(x) = \Pr(\text{Beta}(r, s) \le x)$.

    b. Using the result in part a, deduce that the unconditional density of X is

    $$f_X(x) = \frac{1}{B(r, s)} \cdot \frac{\beta^s x^{r-1}}{(\beta + x)^{r+s}}.$$

    A random variable X with a density function of this type is known as a **generalized Pareto distribution with parameters** r, s, β, and it is denoted by GeneralizedPareto(r, s, β). Explain why this terminology is appropriate.

34. Claims arrive randomly and independently at an insurance company at a constant, but unknown, rate of Λ per day. The insurer models the unknown arrival rate Λ as Gamma(5, 1). Calculate the probability that the insurer receives more than four claims in the next day.

35. A credit consultant is responsible for the lines of credit for ten large companies. The probability that any given company of the ten will use its line of credit in the coming month is P, independent of the other companies in the portfolio. However, the value of P is unknown. Suppose that the credit consultant models P as Beta(2, 3). Determine the probability that at most two of the ten companies use their line of credit in the coming month.

36. Claims on a newly acquired block of insurance policies are believed to follow an exponential distribution. However, the average claim size is not known with certainty. Calculate the probability that a given claim exceeds \$100 if the parameter of the exponential distribution is assumed to have the distribution Gamma(2, 100).

37. Suppose that $(N|\Lambda = \lambda) \sim \text{Possion}(\lambda)$ and $\Lambda \sim \text{Gamma}(r, \alpha)$. Show that $(\Lambda|N = n) \sim \text{Gamma}(r + n, \alpha + 1)$. *Hint:* Use Bayes' theorem to express the density of $\Lambda|N = n$ in terms of the densities of $N|\Lambda = \lambda$, Λ, and N. Then, by considering only the factors in λ, show that this expression is proportional to a gamma density with parameters $r + n$ and $\alpha + 1$.

38. Claims on a particular block of an insurer's business arrive randomly and independently at a constant, but unknown, rate of Λ per year. The insurer decides to model the parameter Λ using a gamma distribution. At the beginning of the year, the insurer chooses the parameters $r = 2$ and $\alpha = 0.1$. The insurer actually receives ten claims during the year.

    a. Based on the first year's experience, what model should the insurer assume for Λ in the second year?

    b. Based on the first year's experience, what is the probability that the insurer receives fewer than five claims in the second year?

39. Suppose that $(N|P = p) \sim$ Binomial(m, p) and $P \sim$ Beta(r, s). Show that $(P|N = n) \sim$ Beta$(n + r, m + s - n)$. *Hint:* Use Bayes' theorem to show that the density of $(P|N = n)$ is proportional to a beta density with parameters $n + r$ and $m + s - n$.

40. A credit consultant is responsible for the lines of credit for ten large companies. The probability that any given company of the ten will use its line of credit in the coming month is P, independent of the other companies in the portfolio. However, the value of P is unknown. The credit consultant decides to model P as Beta$(2, 3)$. In the first month, two of the ten companies use their lines of credit.

    a. Based on the first month's experience, what model should the credit consultant use for P in the second month?

    b. Based on the first month's experience, what is the probability that more than four companies will use their lines of credit in the second month?

41. Suppose that $(X|\Lambda = \lambda) \sim$ Gamma(r, λ) and $\Lambda \sim$ Gamma(s, β). Show that $(\Lambda|X = x) \sim$ Gamma$(r + s, x + \beta)$.

42. Claims on a newly acquired block of insurance policies are believed to follow an exponential distribution. However, the average claim size is not known with certainty. Prior to receiving any claims, the insurer assumes that the parameter of the exponential distribution has the distribution Gamma$(2, 100)$. The first claim submitted is for \$200. How does this observation affect the insurer's belief about the parameter of the claim size distribution?

43. The return on the stock of an oil company depends on the price of oil. During a normal year, the return has a normal distribution with parameters $\mu = 10\%$ and $\sigma = 5\%$; however, during a year when oil prices are high, the return has a normal distribution with parameters $\mu = 40\%$ and $\sigma = 10\%$. There is a 10% chance that oil prices will be high in the coming year. Determine the probability that the return on this stock is greater than 15%.

44. A portfolio consists of two stocks: 90% in Reduce-the-Risk Casualty and 10% in Urban Cowboys Inc. The returns on the two stocks are assumed to follow the distributions Normal$(15\%, 5\%)$ and Normal$(30\%, 25\%)$, respectively. Historical data suggest that the correlation between the stocks is $\rho = 0.2$. Calculate the probability that the return on the portfolio is greater than 20%.

45. The moment generating function of a particular random variable X is given by

$$M_X(t) = \frac{1}{4} + \frac{1}{2} \cdot \frac{1}{1 - t} + \frac{1}{4} \cdot \frac{2}{2 - t}.$$

    a. Determine a formula for the distribution function F_X of X.

    b. Determine the expected value and variance of X.

    c. Determine the probability that $X > 1/2$.

46. In a certain population, 10% of people have the eye disease glaucoma. For people with glaucoma, measurements of eye pressure will be normally distributed with a mean of 25 and a variance of 1; for people without glaucoma, the pressure will be

normally distributed with a mean of 20 and a variance of 1. Suppose that a person is selected at random from this population and the person's eye pressure is measured.

a. Determine a formula for the probability that the person has glaucoma when the measured eye pressure is x.

b. For what values of x is the probability in part a greater than $1/2$?

47. You are given two boxes: One contains nuts; and the other contains bolts. The diameters of the bolts are approximately normally distributed with mean 2 centimeters and standard deviation 0.03 centimeter; the diameters of the holes in the nuts are approximately normally distributed with mean 2.02 centimeters and standard deviation 0.04 centimeter. A bolt and a nut will fit together if the diameter of the hole in the nut is greater than the diameter of the bolt, and the difference between these diameters is not greater than 0.05 centimeter. You choose a bolt from one box and a nut from the other at random. What is the probability that they will fit together?

7

Transformations of Random Variables

It often happens in engineering or financial problems that the random variable of interest is a function of some other random quantity whose distribution is already known. For example, the kinetic energy of a moving object with mass m and uncertain velocity V is the random variable $K = \frac{1}{2}mV^2$, which is a function of the uncertain quantity V. Similarly, the accumulated value of \$1 at time t from now subject to an uncertain per annum continuously compounded rate of return X over the time interval $(0, t)$ is the random variable e^{Xt}, which is a function of the uncertain quantity X. In this chapter, we consider techniques for determining the distribution of a random variable that is a function of some other random quantity with known probability distribution.

We have already encountered several examples of transformations of random variables in this book. Indeed, we encountered the *linear transformation* $Y = aX + b$, where a, b are constants with $a \neq 0$, in connection with the calculation of probabilities for a normal distribution; we encountered the *power transformation* $Y = X^n$, where n is a positive integer, in connection with the calculation of the moments of a random variable (i.e., the numbers $E[X^n]$); and we encountered the *exponential transformation* $Y = e^X$ in connection with the dollar accumulation of an investment with uncertain rate of return. In this chapter, we encounter many more examples of transformations, particularly ones that arise in connection with engineering or financial problems.

Our presentation is organized in the following way: In §7.1, we discuss a general method for determining the distribution for a transformed variable of the type $Y = h(X)$ or $Y = h(X_1, \ldots, X_n)$, and we illustrate this method by considering six basic transformations, including power transformations, exponential transformations, and inverse transformations. In §7.2, we consider two methods for calculating the expected value of a transformed random variable, and we derive the formula for the expected value of a function of a random variable stated in §4.2.1 in a special case. We also discuss Jensen's inequality, which is a generalization of the arithmetic-geometric means inequality stated in §4.2.1, and we introduce the concept of **limited expected value,** which arises in connection with insurance contracts with caps and deductibles and in connection with limiters in electrical circuits. In §7.3, we consider the payout on insurance contracts with caps, deductibles, and coinsurance in detail. As particular examples, we consider the payout on claims under traditional health insurance plans and the value at maturity of an option on a stock or other investment security. In §7.4, we consider pricing and reserving techniques for some traditional life insurance and annuity products such as whole life insurance, term life insurance, endowment insurance, life annuities and pensions. In

§7.5, we consider the reliability of systems with multiple components or processes such as electrical circuits and microprocessors. The analysis of such systems requires an understanding of the distributions of the minimum and maximum of a collection of random variables. We conclude this chapter in §7.6 with a brief discussion of some trigonometric transformations that arise in the study of electricity and have important applications in electrical engineering.

Sections 7.1 and 7.2 cover core material and should be read fairly thoroughly, except where otherwise noted. On the other hand, Sections 7.3 through 7.6 cover optional topics and can be read selectively depending on the reader's interests.

7.1 Determining the Distribution of a Transformed Random Variable

In this section, we develop a general method for determining the distribution of a transformed random variable and illustrate its use by considering six special transformations that arise frequently in applications.

Types of Transformations

There are two types of transformations that arise frequently in engineering and financial applications: scalar-to-scalar transformations and vector-to-scalar transformations. A **scalar-to-scalar transformation** is a transformation of the form $Y = h(X)$, and a **vector-to-scalar transformation** is a transformation of the form $Y = h(X_1, X_2, \ldots, X_n)$. Vector-to-vector transformations, that is, transformations of the form

$$Y_1 = h_1(X_1, \ldots, X_n)$$
$$\vdots$$
$$Y_m = h_m(X_1, \ldots, X_n)$$

are also possible; however, determining the distribution of the transformed random vector $(Y_1, \ldots, Y_m)$ requires an understanding of and facility with the change of variables formula for multiple integrals, which we will not assume the reader has. Hence, in the interest of simplicity, we do not consider vector-to-vector transformations in detail in this book.

Scalar-to-Scalar Transformations

Consider the scalar-to-scalar transformation $Y = h(X)$. Determining the distribution function F_Y for such a Y amounts to calculating $\Pr(h(X) \le y)$ for each y. Since the meaning of the statement $h(X) \le y$ is that the random variable X assumes a value x such that $h(x) \le y$, the probability $\Pr(h(X) \le y)$ can be determined by integrating the density for X over the region $\{x \in \mathbf{R} : h(x) \le y\}$,

$$\Pr(h(X) \le y) = \int_{h(x) \le y} f_X(x)dx.$$

Hence, the distribution function for Y when $Y = h(X)$ is given by

$$F_Y(y) = \int_{h(x) \le y} f_X(x)dx.$$

When the transformation h is one-to-one and order preserving, the distribution function for Y can be written more explicitly as

$$F_Y(y) = \int_{-\infty}^{h^{-1}(y)} f_X(x)dx$$

because, in that case, the region of integration $\{x \in \mathbf{R} : h(x) \le y\}$ is simply $\{x \in \mathbf{R} : x \le h^{-1}(y)\}$. However, if the transformation h is not one-to-one or if it is not order preserving, the region of integration cannot be described so simply, and extra care must be taken when determining a formula for F_Y. We will illustrate in the examples that follow some of the difficulties that are typically encountered when h is not one-to-one or order preserving.

Vector-to-Scalar Transformations

Before we consider some examples of scalar-to-scalar transformations, let's determine a formula for the distribution function F_Y when Y is defined by the vector-to-scalar transformation $Y = h(X_1, \ldots, X_n)$. Note that the meaning of the statement $h(X_1, \ldots, X_n) \le y$ is that the random vector $(X_1, \ldots, X_n)$ assumes a vector value $(x_1, \ldots, x_n)$ such that $h(x_1, \ldots, x_n) \le y$. Hence, the distribution function for Y when $Y = h(X_1, \ldots, X_n)$ is given by

$$F_Y(y) = \int \cdots \int_{h(x_1,\ldots,x_n)\le y} f_{X_1,\ldots,X_n}(x_1, \ldots, x_n)dx_1 \cdots dx_n.$$

Determining the region of integration $\{(x_1, \ldots, x_n) \in \mathbf{R}^n : h(x_1, \ldots, x_n) \le y\}$ explicitly and performing the multivariate integration can be extremely complicated. Fortunately, most vector-to-scalar transformations that arise in engineering and financial applications are fairly elementary. In §7.5, we consider the special vector-to-scalar transformations $h(X_1, X_2) = \min(X_1, X_2)$ and $h(X_1, X_2) = \max(X_1, X_2)$, which arise when studying the reliability of systems with multiple components. Chapter 8 is devoted to an in-depth analysis of the special transformations $h(X_1, X_2) = X_1 + X_2$ and $h(X_1, X_2) = X_1X_2$.

Some Special Scalar-to-Scalar Transformations

In the remainder of this section, we consider six special scalar-to-scalar transformations that arise frequently in applications. For simplicity, we assume throughout that X has a continuous distribution with differentiable distribution function F_X. This assumption has no effect on the formulas for F_Y. However, it does enable us to simplify the formulas we give for the density f_Y and to focus on the main ideas in these examples.

Linear Transformations Suppose that $Y = aX + b$, where a and b are constants with $a \ne 0$. If $a > 0$, then the transformation is one-to-one and order preserving. Hence, in this case,

$$F_Y(y) = \Pr(Y \le y) = \Pr(aX + b \le y)$$

$$= \Pr\left(X \le \frac{y-b}{a}\right)$$

$$= F_X\left(\frac{y-b}{a}\right)$$

and so by the Chain rule,

$$f_Y(y) = \frac{d}{dy} F_X\left(\frac{y-b}{a}\right)$$

$$= \frac{1}{a} f_X\left(\frac{y-b}{a}\right).$$

On the other hand, if $a < 0$, then the transformation is no longer order preserving, although it still is one-to-one. Hence, in this latter case,

$$F_Y(y) = \Pr(Y \le y) = \Pr(aX + b \le y)$$

$$= \Pr\left(X \ge \frac{y-b}{a}\right)$$

$$= S_X\left(\frac{y-b}{a}\right) + \Pr\left(X = \frac{y-b}{a}\right),$$

and thus,

$$f_Y(y) = \frac{d}{dy}\left\{S_X\left(\frac{y-b}{a}\right) + \Pr\left(X = \frac{y-b}{a}\right)\right\}$$

$$= -\frac{1}{a} f_X\left(\frac{y-b}{a}\right).$$

Consequently, in either case,

$$f_Y(y) = \frac{1}{|a|} f_X\left(\frac{y-b}{a}\right).$$

Notice how the height of the density function is adjusted by the scaling factor $1/|a|$. The reason for this adjustment is that f_X and f_Y measure probabilities *per unit length*. Since the transformation $Y = aX + b$ changes the scale of the length measurements by a factor of $|a|$, the density function for Y must be adjusted accordingly. Note that such an adjustment is not necessary for the probability mass function when X and Y are discrete. Figure 7.1 illustrates the effect of the transformation $Y = 2X + 1$ on the density f_X given by

$$f_X(x) = \begin{cases} 1 + x & \text{for } -1 \le x \le 0, \\ 1 - x & \text{for } 0 \le x \le 1. \end{cases}$$

Power Transformations Suppose that $Y = X^n$, where n is a positive integer. If n is odd, then the transformation is one-to-one and order preserving. Hence, if n is odd,

$$F_Y(y) = \Pr(Y \le y) = \Pr(X^n \le y)$$

$$= \Pr(X \le y^{1/n})$$

$$= F_X(y^{1/n})$$

and by the Chain rule,

$$f_Y(y) = \frac{d}{dy} F_X(y^{1/n}) = \frac{1}{n} y^{1/n-1} f_X(y^{1/n}).$$

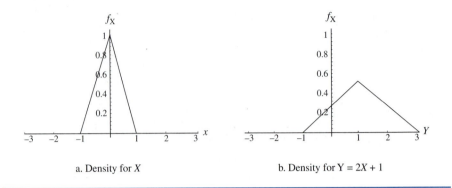

a. Density for X   b. Density for Y = 2X + 1

FIGURE 7.1 The Effect of a Linear Transformation on the Graph of the Density Function

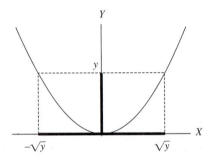

FIGURE 7.2 The Transformation $Y = X^2$

However, if n is even, the transformation is no longer one-to-one, and greater care must be exercised when determining F_Y. Figure 7.2 illustrates the situation for the transformation $Y = X^2$.

From Figure 7.2, it is clear that for n even, the statement $X^n \leq y$ (with $y \geq 0$) is equivalent to the statement $-y^{1/n} \leq X \leq y^{1/n}$. Hence, for n even and $y \geq 0$,

$$F_Y(y) = \Pr(Y \leq y) = \Pr(-y^{1/n} \leq X \leq y^{1/n})$$
$$= F_X(y^{1/n}) - F_X(-y^{1/n}) + \Pr(X = -y^{1/n})$$

and

$$f_Y(y) = \frac{1}{n} y^{1/n-1} \{ f_X(y^{1/n}) + f_X(-y^{1/n}) \}.$$

Consequently, for the transformation $Y = X^2$,

$$f_Y(y) = \frac{1}{2} y^{-1/2} \{ f_X(\sqrt{y}) + f_X(-\sqrt{y}) \}, \qquad y \geq 0.$$

Power transformations arise frequently in engineering applications. For example, the kinetic energy of a moving object with mass m and uncertain velocity V is $K = \frac{1}{2} m V^2$. More generally, power transformations arise in connection with nonlinear amplifiers where an electrical signal (subject to an uncertain noise component) is amplified according to the formula $Y = X^n$ for some odd integer n.

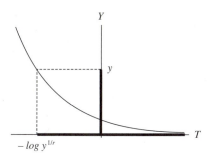

FIGURE 7.3 The Transformation $Y = e^{-rT}$ for $r > 0$

Exponential Transformations   Suppose that $Y = e^{Xt}$, where $t > 0$. This transformation is one-to-one and order preserving for all $t > 0$. Consequently,

$$F_Y(y) = \Pr(Y \le y) = \Pr(e^{Xt} \le y) = \Pr(X \le \log y^{1/t}) = F_X(\log y^{1/t}),$$

and thus,

$$
\begin{aligned}
f_Y(y) &= \frac{d}{dy} F_X(\log y^{1/t}) \\
&= \frac{1}{ty} \cdot f_X(\log y^{1/t}).
\end{aligned}
$$

This transformation arises when considering investment accumulation for an uncertain rate of return.

Negative Exponential Transformations   Suppose that $Y = e^{-rT}$, where $r > 0$. This transformation is one-to-one but not order preserving. Figure 7.3 illustrates the situation. From this figure, it is clear that the statement $e^{-rT} \le y$ is equivalent to the statement $T \ge - \log y^{1/r}$. Hence, for $y \ge 0$,

$$
\begin{aligned}
F_Y(y) &= \Pr(Y \le y) = \Pr(T \ge - \log y^{1/r}) \\
&= S_T(- \log y^{1/r}) + \Pr(T = - \log y^{1/r})
\end{aligned}
$$

and thus,

$$
\begin{aligned}
f_Y(y) &= \frac{d}{dy} \left\{ S_T(- \log y^{1/r}) + \Pr(T = - \log y^{1/r}) \right\} \\
&= \frac{1}{ry} \cdot f_T(- \log y^{1/r}).
\end{aligned}
$$

This transformation arises when considering the present value of \$1 payable on the death of a particular policyholder. Indeed, e^{-rT} is the present value of \$1 payable at the uncertain future time T from now discounted to the present at the constant continuously compounded per annum interest rate r.

Inverse Transformations   Suppose that $Y = 1/X$. This transformation is order reversing and not well-defined at the point $x = 0$. Since we are assuming that X has a continuous distribution of probability, the fact that the transformation is not defined at the origin

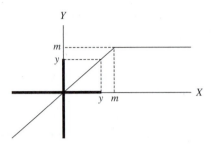

FIGURE 7.4 The Transformation $Y = \min(X, m)$

is not serious.[1] For simplicity, let's suppose that X is a *positive* random variable (i.e., probability is only distributed over the positive values of X). Then

$$S_Y(y) = \Pr(Y > y) = \Pr\left(\frac{1}{X} > y\right) = \Pr\left(X < \frac{1}{y}\right) = F_X\left(\frac{1}{y}\right)$$

and thus,

$$f_Y(y) = \frac{1}{y^2} \cdot f_X\left(\frac{1}{y}\right), \qquad y > 0.$$

The Special Transformation $Y = \min(X, m)$ This transformation arises in connection with caps and deductibles on insurance policies. More precisely, if X is the size of the claim filed, then for a policy with a maximum per claim payment of m, the reimbursement is

$$Y = \min(X, m),$$

and for a policy with a per claim deductible of d but no per claim cap, the reimbursement is

$$Y = (X - d)^+ = \begin{cases} X - d & \text{if } X > d, \\ 0 & \text{if } X \le d, \end{cases}$$
$$= X - \min(X, d).$$

The transformation $Y = \min(X, m)$ also arises in connection with limiters in electrical circuits. A *limiter* is a device that "limits" the flow of electrical current to a particular item such as a television tube so that surges in the electrical supply do not damage the item.

The transformation $Y = \min(X, m)$ is not one-to-one (Figure 7.4) and it is not given by an elementary algebraic expression, as were the previously considered special transformations. Hence, determining the set $\{x \in \mathbf{R} : \min(x, m) \le y\}$ explicitly is more difficult than in the previous examples. The form of this transformation suggests that we should consider the cases $y \ge m$ and $y < m$ separately.

---

[1] However, if X had a nonzero probability mass at $x = 0$, the distribution of $1/X$ would not be defined in the traditional sense.

Suppose first that $y \geq m$. Then since it is always the case that $\min(X, m) \leq m$, we must have $\Pr(\min(X, m) \leq y) = 1$; that is, $F_Y(y) = 1$ when $y \geq m$. On the other hand, suppose that $y < m$. Then the statement $\min(X, m) \leq y$ is equivalent to the statement $X \leq y$, and so $F_Y(y) = \Pr(\min(X, m) \leq y) = \Pr(X \leq y) = F_X(y)$.

Combining these observations, we see that the distribution function of Y is given by

$$F_Y(y) = \begin{cases} F_X(y) & \text{for } y < m, \\ 1 & \text{for } y \geq m. \end{cases}$$

Hence, in general, Y will have a mixed distribution which agrees with the distribution of X on $(-\infty, m)$ and which has a point mass of $S_X(m)$ at $Y = m$. If X already has a point mass at m, then the size of the point mass at $Y = m$ will be $S_X(m) + \Pr(X = m)$.

We now consider some specific examples of these special transformations.

EXAMPLE 1: Determine the distribution of Z^2, where Z is a standard normal random variable (i.e., $Z \sim \text{Normal}(0,1)$).

For the power transformation $Y = X^2$, we have

$$f_Y(y) = \frac{1}{2} y^{-1/2} \{ f_X(\sqrt{y}) + f_X(-\sqrt{y}) \}, \qquad y \geq 0.$$

From §6.3.1, the density function for a standard normal random variable X is

$$f_X(x) = \frac{1}{\sqrt{2\pi}} e^{-x^2/2}, \qquad x \in \mathbf{R}.$$

Hence, if $Y = Z^2$ with $Z \sim \text{Normal}(0,1)$, we have

$$f_Y(y) = \frac{1}{2} y^{-1/2} \left\{ \frac{1}{\sqrt{2\pi}} e^{-(\sqrt{y})^2/2} + \frac{1}{\sqrt{2\pi}} e^{-(-\sqrt{y})^2/2} \right\}$$

$$= \frac{1}{\sqrt{2\pi}} y^{-1/2} e^{-y/2}$$

$$= \frac{\left(\frac{1}{2}\right)^{1/2} y^{-1/2} e^{-y/2}}{\Gamma\left(\frac{1}{2}\right)}, \qquad y \geq 0,$$

where the latter equality follows from the identity $\Gamma(\frac{1}{2}) = \sqrt{\pi}$ (see Appendix A). Now the density function for Y has the form of a gamma density with parameters $r = 1/2$ and $\lambda = 1/2$ (see §6.1.2). Hence, $Y \sim \text{Gamma}(\frac{1}{2}, \frac{1}{2})$.

Consequently, $Z^2 \sim \text{Gamma}(\frac{1}{2}, \frac{1}{2})$ for any standard normal random variable Z. ∎

EXAMPLE 2: The power dissipated when a current passes through a resistor in an electrical circuit is given by $W = I^2 R$, where I is the current in amperes, R is the resistance in ohms, and W is the power in watts. Suppose that I is normally distributed with mean 0 and standard deviation 2 and that $R = 5$ ohms. Determine the probability that the power dissipated is greater than 24 watts.

From the given information, $I/2$ has a standard normal distribution. Hence, $W = 20Z^2$, where $Z \sim \text{Normal}(0,1)$. From the previous example, we know that $Z^2 \sim \text{Gamma}(\frac{1}{2}, \frac{1}{2})$. Hence, $W \sim \text{Gamma}(\frac{1}{2}, \frac{1}{40})$ and so the desired probability is

$$\Pr(W > 24) = \Pr\left(\frac{1}{40}W > 0.6\right)$$

$$= \Pr(\text{Gamma}(\tfrac{1}{2}, 1) > 0.6)$$

$$= 1 - I_{1/2}(0.6)$$

$$\approx 1 - .72668$$

$$= .27332.$$ ∎

EXAMPLE 3:   The velocity of a particle that moves at random along a straight line can be modeled using a normal distribution with mean 0 and standard deviation 3. The kinetic energy of a particle with mass m and velocity V is given by $K = \frac{1}{2}mV^2$. What is the probability that a particle of mass 10 has kinetic energy less than 18?

From the given information, $V/3$ has a standard normal distribution and $m = 10$. Hence, $K = 45Z^2$, where $Z \sim \text{Normal}(0,1)$. Since $Z^2 \sim \text{Gamma}(\frac{1}{2}, \frac{1}{2})$ whenever $Z \sim \text{Normal}(0,1)$, it follows that $K \sim \text{Gamma}(\frac{1}{2}, \frac{1}{90})$. Consequently, the desired probability is

$$\Pr(K \le 18) = \Pr\left(\frac{1}{90}K \le 0.2\right)$$

$$= \Pr(\text{Gamma}(\tfrac{1}{2}, 1) \le 0.2)$$

$$= I_{1/2}(0.2)$$

$$\approx .47291.$$ ∎

EXAMPLE 4:   Determine the distribution of $\min(X, m)$ when $X \sim \text{Exponential}(\lambda)$.

Put $Y = \min(X, m)$. Then the distribution function of Y is in general given by

$$F_Y(y) = \begin{cases} F_X(y) & \text{for } y < m, \\ 1 & \text{for } y \ge m. \end{cases}$$

Consequently, if $X \sim \text{Exponential}(\lambda)$, the distribution of Y is

$$F_Y(y) = \begin{cases} 1 - e^{-\lambda y} & \text{for } y < m, \\ 1 & \text{for } y \ge m. \end{cases}$$

This is simply the distribution function for the truncated exponential distribution with parameters λ and m introduced in Example 3 of §4.2.1. Hence, a truncated exponential distribution is the distribution that results from applying the transformation $Y = \min(X, m)$ to an exponential random variable. ∎

EXAMPLE 5:   Determine the distribution of X^n when $X \sim \text{Exponential}(\lambda)$ assuming $n > 0$.

Put $Y = X^n$. Then the distribution function of Y when $X \ge 0$ and $n > 0$ is in general given by

$$F_Y(y) = F_X(y^{1/n}).$$

Consequently, if $X \sim \text{Exponential}(\lambda)$,

$$F_Y(y) = \begin{cases} 1 - e^{-\lambda y^{1/n}} & \text{for } y > 0, \\ 0 & \text{for } y \le 0. \end{cases}$$

However, this is simply the distribution function for a Weibull distribution with parameters $\alpha = 1/\lambda^n$ and $\beta = 1/n$ (see §6.2.1). Hence, if $X \sim$ Exponential(λ), then $X^n \sim$ Weibull($1/\lambda^n$, $1/n$). ∎

EXAMPLE 6:   The negative exponential transformation is a composite of an exponential transformation and an inverse transformation. Indeed, if g_1 and g_2 are the transformations

$$g_1(X) = e^{rX}, \ r > 0,$$

$$g_2(X) = \frac{1}{X},$$

then the composite $g_2 \circ g_1$ is a negative exponential transformation:

$$(g_2 \circ g_1)(X) = e^{-rX}.$$

Let's check that the density of $Y = (g_2 \circ g_1)(X)$ is the same as determined previously.

Put $W = e^{rX}$ and $Y = 1/W$. Then from our earlier calculations, the density of W is given by

$$f_W(w) = \frac{1}{rw} \cdot f_X(\log w^{1/r})$$

and the density of Y is given by

$$f_Y(y) = \frac{1}{y^2} \cdot f_W\left(\frac{1}{y}\right).$$

Hence,

$$f_W\left(\frac{1}{y}\right) = \frac{y}{r} \cdot f_X\left(\log \left(\frac{1}{y}\right)^{1/r}\right)$$

and so the density of Y is

$$f_Y(y) = \frac{1}{ry} \cdot f_X(\log y^{-1/r})$$

$$= \frac{1}{ry} \cdot f_X(-\log y^{1/r}).$$

This is precisely the form of the density for Y obtained earlier when $Y = e^{-rT}$. ∎

7.2 Expectation of a Transformed Random Variable

In this section, we derive the formula for the expected value of a function of a random variable stated in §4.2.1. We also discuss a generalization of the arithmetic-geometric means inequality known as Jensen's inequality, and we introduce the concept of limited expected value for a random variable.

Formula for Calculating the Expected Value of a Function of a Random Variable

Suppose that h is a scalar-to-scalar transformation and let Y be a random variable given by $Y = h(X)$ for some X. According to the definition of arithmetic expectation, the expected value of Y is

$$E[Y] = \int_{-\infty}^{\infty} y \cdot f_Y(y) dy.$$

Hence, to determine $E[Y]$ using this formula, we must first determine the density of Y, which entails the use of the techniques discussed in §7.1. Fortunately, there is a way to calculate $E[Y]$ when given the density of X which does not require us to determine the density of Y at all. Indeed, as we noted in §4.2.1, the expected value of Y when $Y = h(X)$ can be calculated using the formula

$$E[h(X)] = \int_{-\infty}^{\infty} h(x) f_X(x) dx.$$

We now proceed to demonstrate the validity of this formula in two special cases: (a) h one-to-one and order preserving and (b) X discrete.

DEMONSTRATION WHEN H IS ONE-TO-ONE ORDER PRESERVING: Suppose that $Y = h(X)$ and h is a one-to-one order preserving transformation. Suppose further that X and Y are continuous random variables. Then for any y, the statement $Y \leq y$ is equivalent to the statement $X \leq h^{-1}(y)$, and so the density function of Y is given by

$$f_Y(y) = \frac{d}{dy} F_Y(y)$$

$$= \frac{d}{dy} \Pr(Y \leq y)$$

$$= \frac{d}{dy} \Pr(X \leq h^{-1}(y))$$

$$= \frac{d}{dy} F_X(h^{-1}(y))$$

$$= f_X(h^{-1}(y)) \cdot \frac{d}{dy} h^{-1}(y).$$

Hence,

$$E[Y] = \int_{-\infty}^{\infty} y f_Y(y) dy$$

$$= \int_{-\infty}^{\infty} y f_X(h^{-1}(y)) \left\{ \frac{d}{dy} h^{-1}(y) \right\} dy.$$

Consequently, applying the substitution $y = h(x)$ to the latter integral, we obtain

$$E[Y] = \int_{-\infty}^{\infty} h(x) f_X(x) \left\{ \frac{1}{h'(x)} \right\} h'(x) dx$$

$$= \int_{-\infty}^{\infty} h(x) f_X(x) dx.$$

A similar argument can be given when X and Y are not continuous by considering the generalized derivatives and integrals discussed in the appendix of Chapter 4. See §4.6 for details. ∎

DEMONSTRATION WHEN X IS DISCRETE: Suppose that $Y = h(X)$ and X is discrete. Note that we are not assuming anything about the form of h in this case. Then from the definition of expectation,

$$E[Y] = \sum_y y \cdot \Pr(Y = y).$$

Further, for each y, we have by the law of total probability that

$$\Pr(Y = y) = \sum_x \Pr(Y = y | X = x) \Pr(X = x)$$

$$= \sum_x \Pr(h(X) = y | X = x) \Pr(X = x)$$

$$= \sum_{x \in h^{-1}(y)} \Pr(X = x),$$

where $h^{-1}(y)$ is the set of x such that $h(x) = y$. Note that $\Pr(h(X) = y | X = x)$ is either 1 or 0 depending on whether $x \in h^{-1}(y)$ or $x \notin h^{-1}(y)$. Consequently,

$$E[Y] = \sum_y y \cdot \left\{ \sum_{x \in h^{-1}(y)} \Pr(X = x) \right\}$$

$$= \sum_y \sum_{x \in h^{-1}(y)} y \cdot \Pr(X = x).$$

Now since $\sum_{x \in h^{-1}(y)} \Pr(X = x) = 0$ if y is not in the image of h, we see that the only nonzero terms in the double sum are the ones for which $y = h(x)$ for some x. Moreover, we see that this double sum partitions the values of x according to image values. Consequently,

$$E[Y] = \sum_y \sum_{x \in h^{-1}(y)} h(x) \cdot \Pr(X = x)$$

$$= \sum_x h(x) \cdot \Pr(X = x),$$

which is the formula for $E[h(X)]$ claimed. ∎

The derivation of the formula for $E[h(X)]$ when X is continuous or mixed uses similar ideas. However, the details of the argument are considerably more complicated and require techniques in mathematical analysis that lie beyond the scope of this book.

There is a similar formula for calculating $E[Y]$ when Y is defined by a vector-to-scalar transformation, that is, when $Y = h(X_1, \ldots, X_n)$. Indeed, in that case,

$$E[h(X_1, \ldots, X_n)] = \int_{-\infty}^{\infty} \cdots \int_{-\infty}^{\infty} h(x_1, \ldots, x_n) f_{X_1, \ldots, X_n}(x_1, \ldots, x_n) dx_1 \cdots dx_n.$$

Details are left to the reader.

EXAMPLE 1:   Suppose that $Y = h(X)$, where $h(X) = X^2 + 1$ and $X \sim$ DeMoivre(3). Determine the expected value of Y directly from the definition of $E[Y]$ using the density of Y and then using the formula for $E[h(X)]$ and the density of X.

Recall from §6.2.2 that the DeMoivre(3) distribution has density

$$f_X(x) = \begin{cases} \frac{1}{3} & \text{for } 0 < x < 3, \\ 0 & \text{otherwise.} \end{cases}$$

Further, from §7.1, the density for $W = X^2$ is in general

$$f_W(w) = \frac{1}{2}w^{-1/2}\{f_X(\sqrt{w}) + f_X(-\sqrt{w})\}, \quad w \geq 0.$$

Hence, if $X \sim$ DeMoivre(3), $W = X^2$ has density

$$f_W(w) = \begin{cases} \frac{1}{6}w^{-1/2} & \text{for } 0 < w < 9, \\ 0 & \text{otherwise;} \end{cases}$$

and so $Y = X^2 + 1$ has density

$$f_Y(y) = \begin{cases} \frac{1}{6}(y-1)^{-1/2} & \text{for } 1 < y < 10, \\ 0 & \text{otherwise.} \end{cases}$$

Consequently, using integration by parts,

$$E[Y] = \int_{-\infty}^{\infty} y f_Y(y)\,dy$$

$$= \int_1^{10} y \cdot \frac{1}{6}(y-1)^{-1/2}\,dy$$

$$= \frac{1}{3}y(y-1)^{1/2}\Big|_1^{10} - \frac{1}{3}\int_1^{10}(y-1)^{1/2}\,dy$$

$$= (10 - 0) - \frac{2}{9}(y-1)^{3/2}\Big|_1^{10}$$

$$= 10 - 6$$

$$= 4.$$

Alternatively, using the density of X directly, we have

$$E[X^2 + 1] = \int_{-\infty}^{\infty}(x^2 + 1)f_X(x)\,dx$$

$$= \int_0^3 (x^2 + 1) \cdot \frac{1}{3}\,dx$$

$$= \left(\frac{1}{9}x^3 + \frac{1}{3}x\right)\Big|_0^3$$

$$= 4.$$

Notice how much simpler it was to calculate $E[X^2 + 1]$ using the density of X!   ■

Jensen's Inequality (Optional)

The formula for calculating $E[h(X)]$ just discussed allows us to determine $E[h(X)]$ using only the distribution of X. However, this does not mean that we will always be able to perform the necessary integration. Jensen's inequality is a result that provides us with an *estimate* for $E[h(X)]$ when h is a convex or concave function. In particular, it describes a relationship between $E[h(X)]$ and $h(E[X])$ for such transformations h which generalizes the arithmetic-geometric means inequality discussed in §4.2.1.

We begin with a statement of Jensen's inequality:

Jensen's Inequality

If $h'' \geq 0$ on its domain of definition, then

$$E[h(X)] \geq h(E[X]).$$

On the other hand, if $h'' \leq 0$ on its domain of definition, then

$$E[h(X)] \leq h(E[X]).$$

To derive these inequalities, consider the Taylor polynomial about μ_X with second degree remainder term,

$$h(X) = h(\mu_X) + (X - \mu_X)h'(\mu_X) + \frac{h''(\xi)}{2!}(X - \mu_X)^2,$$

where ξ is some constant. Taking expectations, we obtain

$$E[h(X)] = h(\mu_X) + \frac{h''(\xi)}{2}Var(X).$$

Hence, if $h'' \geq 0$ on its entire domain, then

$$E[h(X)] \geq h(E[X]),$$

while if $h'' \leq 0$ on its entire domain, then

$$E[h(X)] \leq h(E[X])$$

as asserted.

As a special case, consider the transformation $h(X) = e^X$. Then Jensen's inequality asserts that

$$E[e^X] \geq e^{E[X]},$$

or equivalently,

$$E[Y] \geq e^{E[\log Y]}.$$

Since $e^{E[\log Y]}$ is the geometric expectation of Y (i.e., $E_g[Y] = e^{E[\log Y]}$), we see that this special case of Jensen's inequality is simply the arithmetic-geometric means inequality.

Jensen's inequality can also be derived by considering the tangent line to the curve $Y = h(X)$ at the point $X = \mu_X$ and observing that the curve $Y = h(X)$ lies on only one side of this tangent (see Figure 7.5). Jensen's inequality arises frequently in insurance and investment problems, particularly in connection with the risk-preference and risk-aversion levels of individuals.

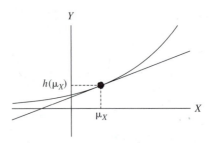

FIGURE 7.5 Graph of Jensen's Inequality when $h \geq 0$.

Limited Expected Value (Optional)

In §7.1, we considered the special transformation $Y = \min(X, m)$, and we discussed how this transformation arises in connection with insurance contracts with caps and deductibles. Recall from that discussion that if X represents a random loss, then $\min(X, m)$ is the amount of the loss covered by an insurance policy on which the maximum allowable payout is m, while

$$(X - d)^+ = \begin{cases} X - d & \text{if } X > d, \\ 0 & \text{if } X \leq d, \end{cases}$$
$$= X - \min(X, d)$$

is the amount of the loss covered when the policy has a deductible of d but no cap m.

Definition of Limited Expected Value

For any random variable X and any constant m, the **limited expected value** of X with respect to the limit m is the number $E[X; m]$ defined by

$$E[X; m] = E[\min(X, m)].$$

For a fixed random variable X, the quantity $E[(X; m)]$, when considered as a function of m, is called the **limited expected value function** for X.

From our discussion in §7.1, we know that if $Y = \min(X, m)$, then the distribution of Y is given by

$$F_Y(y) = \begin{cases} F_X(y) & \text{for } y < m, \\ 1 & \text{for } y \geq m. \end{cases}$$

Hence, in general, Y is a mixed random variable with the same distribution as X on the interval $(-\infty, m)$, but with a point mass of $S_X(m)$ at m.[2] Consequently, from

---

[2] If X already has a point mass at m, the size of the point mass at $Y = m$ will be $S_X(m) + \Pr(X = m)$.

the definition of expectation,[3]

$$E[Y] = \int_{-\infty}^{m} x \cdot f_X(x)\, dx + m \cdot S_X(m),$$

that is,

$$E[X; m] = \int_{-\infty}^{m} x \cdot f_X(x)\, dx + m \cdot S_X(m).$$

This formula for $E[X; m]$ can be rearranged to give another formula which emphasizes the relationship between $E[X; m]$ and $E[X]$. Indeed, by adding and subtracting $\int_{m}^{\infty} x \cdot f_X(x)\, dx$, we have

$$E[X; m] = E[X] - \int_{m}^{\infty} (x - m) f_X(x)\, dx.$$

This latter formula shows that $E[X; m]$ can never be greater than $E[X]$, as intuition would suggest.

When X is a nonnegative random variable, as is generally the case if X represents a loss or insurance claim, the formulas for $E[X; m]$ have a particularly simple form. Indeed, if X is nonnegative (i.e., $X \geq 0$ with certainty), then

$$E[X; m] = \int_{0}^{m} S_X(x)\, dx$$

and

$$E[X; m] = E[X] - e(m) S_X(m),$$

where

$$e(m) = \int_{m}^{\infty} (x - m) \frac{f_X(x)}{S_X(m)}\, dx.$$

The second of these formulas for $E[X; m]$ follows immediately from our preceding discussion. To see that the formula $E[X; m] = \int_{0}^{m} S_X(x)\, dx$ holds, note that by partial integration,

$$\int_{0}^{m} S_X(x)\, dx = x \cdot S_X(x)\big|_{0}^{m} - \int_{0}^{m} x \cdot (-f_X(x))\, dx$$

$$= m \cdot S_X(m) + \int_{0}^{m} x \cdot f_X(x)\, dx,$$

and that $\int_{0}^{m} x \cdot f_X(x)\, dx = \int_{-\infty}^{m} x \cdot f_X(x)\, dx$ since $f_X(x) = 0$ for all $x < 0$; consequently, if $X \geq 0$ with certainty,

$$\int_{0}^{m} S_X(x)\, dx = m \cdot S_X(m) + \int_{-\infty}^{m} x \cdot f_X(x)\, dx$$

$$= E[X; m]$$

---

[3] This formula and the ones that follow it in this section make sense for all types of random variables X, not just continuous ones, provided that the integrals are interpreted in the generalized sense discussed in the appendix to Chapter 4. However, for simplicity, the reader may wish to assume for the remainder of this section that X has a continuous distribution.

as claimed.

The quantity $e(m)$ introduced earlier has an interesting interpretation when X represents a lifetime. Indeed, $e(m)$ is the expected future lifetime for a person of age m whose total lifetime random variable is X. Hence, $e(m)$ is often referred to as the **mean residual lifetime** at age m with respect to the random variable X.

Before discussing the properties of the limited expected value function, let's summarize the formulas for calculating $E[X; m]$ just derived:

Formulas for Calculating the Limited Expected Value

1. For any X,

$$E[X; m] = \int_{-\infty}^{m} x \cdot f_X(x)\, dx + m \cdot S_X(m)$$

and

$$E[X; m] = E[X] - \int_{m}^{\infty} (x - m) f_X(x)\, dx.$$

2. For any nonnegative X,

$$E[X; m] = \int_{0}^{m} S_X(x)\, dx$$

and

$$E[X; m] = E[X] - e(m) \cdot S_X(m),$$

where $e(m)$ is the mean residual lifetime at age m with respect to X.

EXAMPLE 2:   Suppose that $X \sim \text{Exponential}(\lambda)$ and $Y = \min(X, m)$. Determine a formula for $E[Y]$ in terms of λ and m.

Since X is a nonnegative random variable, $E[Y]$ can be determined by integrating the survival function of X from 0 to m. Indeed,

$$E[Y] = E[X; m]$$
$$= \int_{0}^{m} S_X(x)\, dx$$
$$= \int_{0}^{m} e^{-\lambda x}\, dx$$
$$= \frac{1}{\lambda}(1 - e^{-\lambda m}).$$

Note that this is precisely the formula obtained in Example 3 of §4.2.1 and Example 1 of §4.3.2 for the expected value of a truncated exponential distribution with parameters λ and m. Clearly, the calculation of the expected value for a truncated exponential distribution is much easier using the formula $E[X; m] = \int_{0}^{m} S_X(x)\, dx$. ■

The limited expected value function has three important properties that follow directly from the formulas for $E[X; m]$ derived earlier:

> **Properties of the Limited Expected Value Function**
>
> For fixed X, the quantity $E[X; m]$ has the following properties when considered as a function of m:
>
> 1. $E[X; m]$ is increasing, continuous, and concave as a function of m.
> 2. $E[X; m] \rightarrow E[X]$ as $m \rightarrow \infty$.
> 3. $\frac{d}{dm} E[X; m] = S_X(m)$.

The proofs of these properties are left as an exercise for the reader.

7.3 Insurance Contracts with Caps, Deductibles, and Coinsurance (Optional)

In the previous section, we introduced the limited expected value function, and we mentioned briefly its connection to insurance contracts with caps and deductibles. In this section, we discuss insurance contracts with caps and deductibles in greater detail.

Preliminary Comments

Throughout this section, we assume that X represents the size of a random claim. We also use the notation $h_o(X)$ to represent the amount of the claim X covered by the owner of the contract (i.e., the policyholder) and $h_w(X)$ to represent the amount of the claim covered by the writer of the contract (i.e., the insurance company or some other party providing the insurance). For example, for a contract in which the insurance company pays 80% of each claim, $h_w(X) = (.80)X$ and $h_o(X) = (.20)X$. Note that h_o and h_w are functions that define how a given claim is to be shared between the policyholder and the insurance company. It should be clear that h_o and h_w are determined by the provisions of the contract and that

$$h_o(X) + h_w(X) = X.$$

In the discussion that follows, we consider five different types of contracts: excess-of-loss contracts, limited excess-of-loss contracts, proportional insurance contracts, contracts which are some combination of these three, and option contracts. For each contract, we specify the functions h_o, h_w and determine formulas for the policyholder's and the insurer's expected loss, using the limited expected value function. We also discuss how these formulas can be used to design a contract with a target per claim payout.

Excess-of-Loss Contracts

An **excess-of-loss contract** is an insurance contract in which the writer of the contract agrees to pay the purchaser of the contract the amount of the random claim X in excess of a fixed amount d. The amount d is known as a **deductible.**

From this definition, it is clear that the writer's portion of the claim is given by

$$h_w(X) = (X - d)^+,$$

where for any random variable W

$$W^+ = \begin{cases} 0 & \text{if } W \leq 0, \\ W & \text{if } W > 0, \end{cases}$$

and the policyholder's portion of the claim is given by

$$h_o(X) = \min(X, d).$$

Since $h_o(X) + h_w(X) = X$, it follows that

$$h_w(X) = X - \min(X, d).$$

Consequently, the amount of a random claim X that the insurer should expect to pay is

$$E[h_w(X)] = E[X] - E[X; d],$$

and the amount that the policyholder should expect to pay is

$$E[h_o(X)] = E[X; d].$$

EXAMPLE 1: An insurance company offers insurance against a particular type of loss. The particular contract has a deductible but no cap. Losses are assumed to have an exponential distribution with mean 2 measured in thousands of dollars. At what level should the deductible be set if the insurer would like the expected payment for any loss of this type to be $1500?

Let X be the size of a random loss in thousands of dollars. Then, by assumption, $X \sim \text{Exponential}(1/2)$. The insurer would like to choose the deductible d such that

$$E[h_w(X)] = 1.5.$$

From Example 2 of §7.2, we know that if $X \sim \text{Exponential}(\lambda)$, then

$$E[X; d] = \frac{1}{\lambda}(1 - e^{-\lambda d}).$$

Consequently,

$$\begin{aligned} E[h_w(X)] &= E[X] - E[X; d] \\ &= \frac{1}{\lambda}e^{-\lambda d} \\ &= 2e^{-d/2} \end{aligned}$$

and so the required deductible d is given by

$$2e^{-d/2} = \frac{3}{2},$$

that is,

$$d = 2\log(0.75)^{-1} \approx 0.5753641.$$

Thus, the required deductible is about $575. ∎

Limited Excess-of-Loss Contracts

A **limited excess-of-loss contract** is an insurance contract in which the writer of the contract agrees to pay the purchaser of the contract the amount of the random claim X

in excess of an amount d, but limits the size of the payout to m. In insurance jargon, this is often expressed by saying that the purchaser is covered for the **layer** m **xs** d.

From this definition, it is clear that the writer's portion of the claim is given by

$$h_w(X) = \begin{cases} 0 & \text{for } 0 \leq X < d, \\ X - d & \text{for } d \leq X < d + m, \\ m & \text{for } X \geq d + m. \end{cases}$$

$$= \min(X, d + m) - \min(X, d).$$

(The latter equality follows by considering cases.) Hence, the policyholder's portion of the claim is given by

$$h_o(X) = X + \min(X, d) - \min(X, d + m).$$

Consequently, the expected value of the insurer's payment is

$$E[h_w(X)] = E[X; d + m] - E[X; d],$$

and the expected amount that the policyholder must pay is

$$E[h_o(X)] = E[X] + E[X; d] - E[X; d + m].$$

EXAMPLE 2:   The insurance company of the previous example wants to cap its per claim payment at $5000. Continue to assume that losses are exponentially distributed with mean $2000 and that the insurer desires an expected per claim payment of $1500. What should be the size of the deductible in this case?

Let X be the size of a random loss in thousands of dollars. Then $X \sim \text{Exponential}(1/2)$ and the insurer's requirement is that

$$E[h_w(X)] = 1.5.$$

Here,

$$\begin{aligned} E[h_w(X)] &= E[X; d + m] - E[X; d] \\ &= \frac{1}{\lambda}\{1 - e^{-\lambda(d+m)}\} - \frac{1}{\lambda}\{1 - e^{-\lambda d}\} \\ &= \frac{1}{\lambda}e^{-\lambda d}(1 - e^{-\lambda m}) \\ &= 2e^{-d/2}(1 - e^{-m/2}). \end{aligned}$$

Since the cap is to be set at $5000, $m = 5$. Consequently, the required deductible d is given by

$$2e^{-d/2}(1 - e^{-5/2}) = \frac{3}{2},$$

that is,

$$d = 2\log\left(\frac{4(1 - e^{-5/2})}{3}\right) \approx 0.4040632.$$

Thus, the required deductible is about $404.

Notice that this deductible is lower than the one calculated in the previous example. Convince yourself why this should be so.   ∎

Proportional Insurance Contracts

A **proportional insurance contract** (also known as a **quota-share contract**) is one in which the writer of the contract agrees to pay a fixed fraction α of each claim.

From this definition, it is clear that the writer's portion of the claim is given by

$$h_w(X) = \alpha X,$$

and the policyholder's portion of the claim is given by

$$h_o(X) = (1 - \alpha)X.$$

Consequently, the expected value of the writer's payment is

$$E[h_w(X)] = \alpha E[X],$$

and the expected value of the policyholder's payment is

$$E[h_o(X)] = (1 - \alpha)E[X].$$

Combination Contracts

Most insurance contracts tend to combine the loss limiting features of excess-of-loss insurance with the proportionality feature of quota-share insurance. The loss limiting feature protects the policyholder against the adverse financial consequences of a catastrophic loss (e.g., total destruction of one's home by fire), while the proportionality feature helps the insurer limit the effect of moral hazard by aligning the policyholder's incentives with those of the insurer.

The most basic type of combination contract is one in which the writer agrees to pay a fraction α of the claim X in excess of a deductible d but limits the size of the payment to a maximum of m,

$$h_w(X) = \begin{cases} 0 & \text{for } X \le d, \\ \alpha(X - d) & \text{for } d < X \le d + m/\alpha, \\ m & \text{for } X > d + m/\alpha. \end{cases}$$

In terms of the function $\min(X, \cdot)$, the writer's payment for this contract is

$$h_w(X) = \alpha \left\{ \min\left(X, d + \frac{m}{\alpha} \right) - \min(X, d) \right\},$$

and hence, the expected size of the writer's payment is

$$E[h_w(X)] = \alpha \left\{ E\left[X; d + \frac{m}{\alpha} \right] - E[X; d] \right\}.$$

Combination contracts of this type in which there is only one proportionality factor α are known as **single layer contracts.**

Combination contracts with multiple layers are also possible. In a typical two layer contract, the writer will agree to pay the fraction α of the first a dollars of claim in excess of the deductible d and the fraction β of the remaining claim amount but will limit the total payment to a maximum of m. As a concrete illustration, the terms of the contract may require the writer to pay 50% of the first \$1000 in excess of a \$100 deductible and 25% of the remaining amount up to a maximum payment of \$1500. Then for a claim of size \$500, the payment would be $(.50)(500 - 100) = \$200$; for a claim of size \$1500, the payment would be $(.50)(1100 - 100) + (.25)(1500 - 1100) = \$600$; and for a claim of size $6000, the payment would be $1500, which is the maximum possible reimbursement.

A multiple layer contract can actually be considered a collection of single layer contracts. It is easy to convince oneself that this must be the case by considering the graph of h_w. Indeed, after a little reflection, it is clear that the graph of h_w for a contract with n layers can be constructed from the graphs of n particular single layer contracts by performing an appropriate summation.

To see this concretely, consider the two layer contract just discussed in which the writer pays 50% of the first $1000 in excess of a $100 deductible and 25% of the remaining amount up to a maximum of $1500. Clearly, this contract is equivalent to a portfolio consisting of the following single layer contracts:

1. A single layer contract in which the writer pays 50% of the claim amount in excess of a $100 deductible but limits the size of the payment to a maximum of $500;
2. A single layer contract in which the writer pays 25% of the claim amount in excess of a $1100 deductible but limits the size of the payment to a maximum of $1000.

The preceding discussion suggests that the expected payout on a combination contract with multiple layers is most easily determined by decomposing the multiple layer contract into a portfolio of single layer contracts and calculating the expected payout on each single layer contract using the formula

$$E[h_w(X)] = \alpha \left\{ E\left[X; d + \frac{m}{\alpha}\right] - E[X; d] \right\}$$

for single layer contracts derived previously. We illustrate this procedure in the following example.

EXAMPLE 3:   According to the terms of a particular insurance contract, the writer of the contract agrees to pay the purchaser 50% of the first $1000 of the claim in excess of a $100 deductible and 25% of the remaining claim amount up to a maximum of $1500. Determine a formula for the insurer's expected payment using the limited expected value function.

Note that this is the specific two layer contract discussed in the preceding paragraphs. Let $\mathcal{C}_1$ be the single layer contract that pays 50% of the claim in excess of $100 to a maximum of $500 and let $\mathcal{C}_2$ be the single layer contract that pays 25% of the claim in excess of $1100 to a maximum of $1000. Further, let $h_w^{\mathcal{C}_1}$ and $h_w^{\mathcal{C}_2}$ denote the writer's portion of the claim under contracts $\mathcal{C}_1$ and $\mathcal{C}_2$, respectively, and let h_w denote the writer's portion of the claim under the two layer contract given in the statement of the example. Then, from our previous comments,

$$h_w(X) = h_w^{\mathcal{C}_1}(X) + h_w^{\mathcal{C}_2}(X)$$
$$= (.50)\{\min(X, 1100) - \min(X, 100)\}$$
$$+ (.25)\{\min(X, 5100) - \min(X, 1100)\}$$

and so the insurer's expected payout is

$$E[h_w(X)] = (.50)\{E[X; 1100] - E[X; 100]\}$$
$$+ (.25)\{E[X; 5100] - E[X; 1100]\}$$
$$= (.25)E[X; 5100] + (.25)E[X; 1100] - (.50)E[X; 100].$$ ∎

EXAMPLE 4:   Under the terms of a particular comprehensive major medical plan, a plan member is reimbursed for 80% of allowable medical expenses incurred during the year after paying a $250 deductible. The maximum out-of-pocket expense in any given year that a plan member must pay is $2000. Suppose that aggregate annual allowable medical expenses for individual plan members have a Pareto distribution with parameters $s = 6$ and $\beta = 2$, where expenses are measured in thousands of dollars. Determine the amount of reimbursement that each plan member should expect to receive from this plan.

Let X be the total allowable expenses in thousands incurred during the year for a given policyholder and let Y be the amount reimbursed. Then

$$Y = \begin{cases} 0 & \text{if } 0 \leq X \leq \dfrac{1}{4}, \\[2mm] (.80)\left(X - \dfrac{1}{4}\right) & \text{if } \dfrac{1}{4} < X \leq 9, \\[2mm] X - 2 & \text{if } X \geq 9. \end{cases}$$

By assumption, $X \sim \text{Pareto}(6,2)$ (i.e., $f_X(x) = 3(1 + \frac{x}{2})^{-7}, x \geq 0$). Consequently, using integration by parts,

$$E[Y] = \int_0^{1/4} 0 \cdot f_X(x)\, dx + \int_{1/4}^9 (.80)\left(x - \frac{1}{4}\right) f_X(x)\, dx + \int_9^\infty (x - 2) f_X(x)\, dx$$

$$= (2.4) \int_{1/4}^9 \left(x - \frac{1}{4}\right)\left(1 + \frac{x}{2}\right)^{-7} dx + 3 \int_9^\infty (x - 2)\left(1 + \frac{x}{2}\right)^{-7} dx$$

$$= \left\{ (2.4)\left(x - \frac{1}{4}\right)\left(1 + \frac{x}{2}\right)^{-6}\left(-\frac{1}{3}\right)\Bigg|_{1/4}^9 - (2.4)\int_{1/4}^9 \left(-\frac{1}{3}\right)\left(1 + \frac{x}{2}\right)^{-6} dx \right\}$$

$$+ \left\{ 3(x - 2)\left(-\frac{1}{3}\right)\left(1 + \frac{x}{2}\right)^{-6}\Bigg|_9^\infty - 3\int_9^\infty \left(-\frac{1}{3}\right)\left(1 + \frac{x}{2}\right)^{-6} dx \right\}$$

$$\approx .1775932.$$

Thus, each plan member should expect to receive about $177.59. Details of the calculation are left to the reader. ∎

Option Contracts

We conclude this section by discussing a type of financial contract—an option—which has many of the characteristics of the insurance contracts that we have considered thus far.

An **option contract** is a contract in which the owner has the right (but not the obligation) to buy or sell a particular security for a particular price on a particular date or dates. A **call option** on a security S gives the owner the right to buy the security, and a **put option** gives the owner the right to sell the security. The specified price at which the security can be bought or sold is called the **exercise price**. The option is considered a **European option** if it can only be exercised on the maturity date; it is considered an **American option** if it can be exercised at any time up to and including the maturity date. The owner of the option is often said to have a **long position** in the option, whereas the writer of the option is said to have a **short position**.

Let $h_o^{(c)}(S; k)$ denote the value at maturity of a long position in a European call option with exercise price k on a stock whose price on the maturity date is S and let $h_w^{(c)}(S; k)$ be the value at maturity of a short position in this option. Similarly, let $h_o^{(p)}(S; k)$, $h_w^{(p)}(S; k)$, respectively, be the values at maturity of long and short positions in a European put option with exercise price k.

Then,

$$h_o^{(c)}(S; k) = \begin{cases} 0 & \text{if } S \leq k, \\ S - k & \text{if } S > k; \end{cases}$$
$$= (S - k)^+$$
$$= S - \min(S, k)$$
$$= \max(S - k, 0);$$

$$h_w^{(c)}(S; k) = \begin{cases} 0 & \text{if } S \leq k, \\ k - S & \text{if } S > k; \end{cases}$$
$$= -h_o^{(c)}(S; k);$$

$$h_o^{(p)}(S; k) = \begin{cases} k - S & \text{if } S \leq k, \\ 0 & \text{if } S > k; \end{cases}$$
$$= (k - S)^+$$
$$= k - \min(S, k)$$
$$= \max(k - S, 0);$$

$$h_w^{(p)}(S; k) = \begin{cases} S - k & \text{if } S \leq k, \\ 0 & \text{if } S > k; \end{cases}$$
$$= -h_o^{(p)}(S; k).$$

Note that for any option contract,

$$h_o(S) + h_w(S) = 0.$$

where $h_o(S)$, $h_w(S)$ denote the values of the long and short positions, respectively. This relationship differs from the relationship $h_o(X) + h_w(X) = X$ for insurance contracts in which X represents a random loss. Note further that the values of puts and calls on the same security with the same exercise price are connected by the relationship

$$h_o^{(c)}(S; k) - h_o^{(p)}(S; k) = S - k.$$

It should be clear from the formulas for $h_o^{(c)}$, $h_w^{(c)}$, $h_o^{(p)}$, $h_w^{(p)}$ that option contracts are similar in character to excess-of-loss and limited excess-of-loss insurance contracts. We leave it as an exercise for the reader to consider the reason for the similarity between call options and insurance contracts with deductibles and between put options and insurance contracts with caps.

7.4   Life Insurance and Annuity Contracts (Optional)

In the previous section, we considered insurance contracts in which the claim size is uncertain. In this section, we consider insurance contracts in which the claim size is certain, but the time of the claim is uncertain. As we will soon see, the negative exponential transformation $Y = e^{-rT}$, where T represents the time of the claim and r represents a continuously compounded interest rate, arises naturally when analyzing any contract of this type.

Whole Life Insurance

A **whole life insurance policy** is an insurance contract in which the insurer agrees to pay a specified amount on the death of a particular person. The specified amount is known as the **face value** of the policy, and the person on whose life the policy is written is called the **insured.** The person who receives the proceeds of the policy on the death of the insured is known as the **beneficiary.** For simplicity, we will assume a face value of $1 unless otherwise specified.

The question facing the insurer is this: How much should the insured person be charged today for the insurer's promise to pay $1 on the death of the insured? The amount is clearly less than $1: $1 today is worth more than $1 at any future date since $1 today can be deposited in the bank and earn interest. This effect is known as the **time value of money.** Indeed, at a continuously compounded rate of r per annum, $1 today will grow to e^{rt} at time t from now; hence, the value today of $1 to be received at future time t is e^{-rt}.

Unfortunately, the insurer does not know when the insured person will die. Hence, the insurer faces the risk that the premium charged today may be inadequate to honor the commitment to pay $1 on the insured's death.

+.5 Let T be the future lifetime of the insured and let P be the premium charged to-day.[4] Suppose that the per annum continuously compounded rate of interest is r.[5] Then the premium will be adequate to honor the insurer's future commitment if $P e^{rT} \geq 1$ (i.e., if $e^{-rT} \leq P$). Hence, the insurer would like to choose P such that $e^{-rT} \leq P$ with high probability.

Now $\Pr(e^{-rT} \leq P) = F_Y(P)$, where $Y = e^{-rT}$. Hence, the premium P can be determined by considering the distribution of the present value random variable $Y = e^{-rT}$. From §7.1, we know that for the negative exponential transformation $Y = e^{-rT}$,

$$F_Y(y) = S_T(- \log y^{1/r}) + \Pr(T = - \log y^{1/r}).$$

Consequently, if T is modeled using a continuous distribution, as we will assume, the premium P can be determined by considering the distribution function

$$F_Y(y) = S_T(- \log y^{1/r}), \qquad y \in (0, 1).$$

EXAMPLE 1: Suppose that $T \sim \text{Exponential}(1/10)$ and $r = 0.05$.[6] Determine the premium that should be charged today on a policy with face value $10,000 so that the

---

[4] Note that P in this context is a *deterministic* quantity, not a random variable. The reason for this departure from the convention that uppercase letters be used to represent random quantities is that lowercase p is commonly used to represent a probability. The notation P also agrees with what is used in the insurance literature.

[5] For simplicity, we assume that r is constant in this discussion and throughout the rest of this section.

[6] We have selected an exponential distribution in this example and others in this section to simplify the calculations and focus on the main ideas. However, the reader should keep in mind that the exponential distribution does not provide a realistic description of human mortality because under an exponential model, the risk of death (i.e., force of mortality) is the same for all ages. Hence, the exponential distribution should not be used to price life insurance in practice unless the period of time for which the life insurance is in effect is fairly short (e.g., one or two years), in which case the assumption of a constant force of mortality is not unrealistic.

probability that there are sufficient funds on hand at the time of the insured's death is 90%.

Put $Y = e^{-rT}$ and let P be the premium per dollar of face value. We require that P be chosen so that

$$\Pr(P \geq e^{-rT}) = .90.$$

Since $T \sim$ Exponential$(1/10)$ and $r = 0.05$, we have

$$
\begin{aligned}
F_Y(y) &= S_T(-\log y^{1/r}) \\
&= \exp\left\{ -\frac{1}{10}(-\log y^{1/(0.05)}) \right\} \\
&= \exp\{\log y^2\} \\
&= y^2.
\end{aligned}
$$

Hence, $\Pr(P \geq e^{-rT}) = F_Y(P) = P^2$, and so $\Pr(P \geq e^{-rT}) = .90 \Longleftrightarrow P^2 = .90 \Longleftrightarrow P \approx .9486833$. Consequently, the required premium is \$9486.83. ∎

The expected value of a present value random variable such as $Y = e^{-rT}$ is referred to in the insurance literature as the **actuarial present value.** Hence, the actuarial present value of \$1 payable at uncertain future time T is simply $E[e^{-rT}]$, where r is the constant per annum continuously compounded discount rate.

EXAMPLE 2:   Suppose that the premium charged on the whole life policy of the preceding example is the actuarial present value of the policy's face amount. Determine the premium in this case.

Recall that $T \sim$ Exponential$(1/10)$ and $r = 0.05$. The actuarial present value of \$1 payable on the insured's death is

$$
\begin{aligned}
E[e^{-rT}] &= \int_0^\infty e^{-rt} \lambda e^{-\lambda t} dt \\
&= \frac{\lambda}{r + \lambda} \\
&= \frac{0.10}{0.05 + 0.10} \\
&= \frac{2}{3}.
\end{aligned}
$$

Hence, the premium on a \$10,000 face value policy is \$6666.67.

Note that the premium based on the actuarial present value is considerably smaller than the premium based on the 90th percentile of the present value random variable. ∎

Premiums can also be determined by considering policies in groups rather than considering individual policies in isolation. This approach is illustrated in the next example.

EXAMPLE 3:   Consider a group of 100 individuals all exactly age 60. Suppose that each individual in the group owns a whole life policy with face value \$10,000. Suppose further that each individual's future lifetime is Exponential$(1/10)$ and that future lifetimes for members of the group are mutually independent. Benefits for these policies are to be paid out of an investment fund earning 5% per annum compounded continuously. Determine the present size of this fund so that the probability that sufficient funds are on hand to pay benefits at the time of each insured's death is 90%.

Let T_j be the future lifetime of the jth insured. By assumption, the T_j are mutually independent and $T_j \sim$ Exponential(1/10). Let L be the present value of the insurer's promises and let m be the amount of funds currently set aside by the insurer to pay for these promises. Then

$$L = 10,000\{e^{-rT_1} + \cdots + e^{-rT_{100}}\}$$

and the requirement on m is that

$$\Pr(L \leq m) = .90.$$

Since L is a sum of 100 independent, identically distributed random variables, it can be approximated by a normal distribution (see §6.3.1). The number m can then be taken to be the 90th percentile of this normal distribution. To determine the appropriate normal distribution to use, we must first determine the mean and variance of e^{-rT}.

Hence, put $Y = e^{-rT}$. Since $T \sim$ Exponential(1/10), the first and second moments of Y are, respectively,

$$E[Y] = \int_0^\infty e^{-t/20} \cdot \frac{1}{10} e^{-t/10} dt = \frac{0.10}{0.10 + 0.05} = \frac{2}{3},$$

$$E[Y^2] = \int_0^\infty \left(e^{-t/20}\right)^2 \frac{1}{10} e^{-t/10} dt = \frac{0.10}{0.10 + 0.10} = \frac{1}{2},$$

and thus, the mean and variance of e^{-rT} are, respectively, $E[Y] = 2/3$ and $Var(Y) = E[Y^2] - E[Y]^2 = 1/2 - (2/3)^2 = 1/18$.

Returning to the aggregate liability L, we have

$$E[L] = (10,000)(100E[Y]) = \frac{2}{3}(1,000,000)$$

and

$$Var(L) = (10,000)^2(100Var(Y)) = (10,000)^2 \left(\frac{50}{9}\right).$$

Hence,

$$\Pr(L \leq m) = \Pr\left(\frac{L - E[L]}{SD(L)} \leq \frac{m - E[L]}{SD(L)}\right)$$

$$\approx \Phi\left(\frac{m - E[L]}{SD(L)}\right),$$

where $\Phi(\cdot)$ is the distribution function for Normal(0,1). Now, from tables, $\Phi(1.2818) \approx .90$. Consequently, the requirement $\Pr(L \leq m) = .90$ is approximately equivalent to the requirement

$$\frac{m - E[L]}{SD(L)} = 1.2818.$$

Solving for m, we find that $m = 696,878.98$. Therefore, the insurer requires that the fund be of size \$696,878.98 (i.e., \$6968.79 per policy).

Notice that this per policy amount is considerably less than the \$9486.83 premium computed using a 90% probability requirement on a single policy but somewhat larger than the \$6666.67 premium computed using actuarial present values. The reason for these differences will become clearer after we have discussed the law of large numbers and the central limit theorem in the next chapter. ∎

Term Life Insurance

A **term life insurance policy** is an insurance contract in which the insurer agrees to pay a specified amount on the death of the insured person if the insured dies during a specified period of time (e.g., within the next n years). If the insured survives until the end of the term, no benefits are paid. Premiums paid for protection during the term are not refunded either.

We can analyze this contract in the same way we analyzed whole life policies by considering a present value random variable. Hence, let T be the future lifetime of the insured and let Y be the present value of the insurer's promise to pay \$1 on the death of the insured if death occurs within n time units from now. Then

$$Y = \begin{cases} e^{-rT} & \text{if } 0 \le T \le n, \\ 0 & \text{if } T > n. \end{cases}$$

Consequently, if T has a continuous distribution, then the present value random variable Y has a mixed distribution with a point mass equal to $\Pr(T > n)$ at $y = 0$ and a continuous distribution of probability on the interval $e^{-rn} \le y \le 1$. In particular,

$$F_Y(y) = \begin{cases} 0 & \text{for } y < 0, \\ S_T(n) & \text{for } 0 \le y < e^{-rn}, \\ S_T(-\log y^{1/r}) & \text{for } e^{-rn} \le y < 1, \\ 1 & \text{for } y \ge 1. \end{cases}$$

The expected value of Y (i.e., the actuarial present value of \$1 payable on the death of the insured if death occurs within n time units from now) is

$$E[Y] = \int_0^n e^{-rt} f_T(t) dt.$$

EXAMPLE 4: Suppose that $T \sim \text{Exponential}(1/10)$ and $r = 0.05$. Determine the premium that should be charged today on a 5-year term insurance policy with face value \$10,000 if the insurer would like the probability to be 90% that sufficient funds will be on hand to pay benefits. What is the premium if the insurer charges the policyholder the actuarial present value?

Let P be the premium per dollar of face value and let Y be the present value of a \$1 death benefit payable at the insured's death if death occurs within 5 years. We require that P be chosen so that

$$\Pr(P \ge Y) = .90.$$

Since $T \sim \text{Exponential}(1/10)$ and $r = 0.05$, we have

$$F_Y(y) = \begin{cases} 0 & \text{for } y < 0, \\ S_T(n) & \text{for } 0 \le y < e^{-rn}, \\ S_T(-\log y^{1/r}) & \text{for } e^{-rn} \le y < 1, \\ 1 & \text{for } y \ge 1; \end{cases}$$

$$= \begin{cases} 0 & \text{for } y < 0, \\ e^{-1/2} & \text{for } 0 \le y < e^{-1/4}, \\ y^2 & \text{for } e^{-1/4} \le y < 1, \\ 1 & \text{for } y \ge 1. \end{cases}$$

From this formula for F_Y, we see that

$$\Pr(P \ge Y) = .90 \Longleftrightarrow P^2 = .90 \Longleftrightarrow P \approx .9486833.$$

Consequently, the required premium is $9486.83. This is exactly the same premium determined for the whole life policy considered earlier in Example 1 of this section.

The actuarial present value of the insurer's liability is

$$10,000E[Y] = 10,000 \int_0^5 e^{-t/20} \frac{1}{10} e^{-t/10} dt$$

$$= 1000 \left(-\frac{20}{3} e^{-3t/20} \right) \Big|_0^5$$

$$= \frac{20,000}{3}(1 - e^{-3/4})$$

$$\approx 3517.56.$$

Hence, the premium is considerably less if the insurer charges the actuarial present value. Notice that the actuarial present value for a term insurance is less than the actuarial present value for a whole life policy (see Example 2). A little reflection reveals that this will always be the case. ∎

Endowment Insurance

An **endowment insurance policy** is an insurance contract in which the insurer agrees to pay a specified amount on the death of the insured person or upon the survival of the insured to the end of a given term, whichever comes first. Endowment insurance contracts used to be popular savings vehicles in the United States, particularly with regard to college savings plans, until changes in tax law made them unattractive. However, insurance contracts of this type are still quite popular in parts of Europe and Asia.

As before, let T be the future lifetime of the insured and let Y be the present value of the insurer's promise to pay $1 on the death of the insured or upon the insured's survival to time n from now, whichever comes first. Then

$$Y = \begin{cases} e^{-rT} & \text{if } 0 \le T \le n, \\ e^{-rn} & \text{if } T > n. \end{cases}$$

$$= \max(e^{-rT}, e^{-rn}).$$

Consequently, if T has a continuous distribution, then Y has a mixed distribution with a point mass equal to $\Pr(T > n)$ at $y = e^{-rn}$ and a continuous distribution of probability on the interval $e^{-rn} \le y \le 1$. In particular,

$$F_Y(y) = \begin{cases} 0 & \text{for } y < e^{-rn}, \\ S_T(-\log y^{1/r}) & \text{for } e^{-rn} \leq y < 1, \\ 1 & \text{for } y \geq 1. \end{cases}$$

Note the similarities and differences between this distribution and the distribution of the present value random variable for a term insurance policy.

The actuarial present value of $1 payable on the death of the insured or upon the insured's survival to time n, whichever comes first, is

$$E[Y] = \int_0^n e^{-rt} f_T(t)dt + e^{-rn} S_T(n).$$

EXAMPLE 5: Determine the actuarial present value for a 5-year endowment insurance with face value $10,000 if $T \sim$ Exponential$(1/10)$ and $r = 0.05$.

From the preceding formula, the actuarial present value is

$$10,000 \left\{ \int_0^5 e^{-t/20} \frac{1}{10} e^{-t/10} dt + e^{-5/20} \cdot e^{-5/10} \right\}$$

$$= 10,000 \left\{ \frac{2}{3}(1 - e^{-3/4}) + e^{-3/4} \right\}$$

$$\approx 8241.22.$$

Notice that the actuarial present value of a 5-year endowment insurance policy is greater than the actuarial present value of a 5-year term insurance policy and the actuarial present value of a whole life policy (see Example 4 and Example 2). ■

Life Annuities

A **life annuity** is a contract in which the insurer agrees to make a regular stream of payments of a specified size to the policyowner until the policyowner's death. Annuities provide insurance against outliving one's savings.

For the purposes of this discussion, it is convenient to assume that the insurance company makes payments *continuously* at a rate of $1 per year until the insured's death. This simplifying assumption is often made in actuarial calculations.

Let T be the future lifetime of the insured and let r be the (constant) per annum continuously compounded rate of interest. Let Y be the present value of $1 per year payable continuously until the insured's death. Then

$$Y = \frac{1 - e^{-rT}}{r}.$$

One can derive this formula using the summation formula for a geometric series. To illustrate this, suppose that $T = n$, where n is a whole number of years, and payments of $1/m$ are made m times per year beginning at time $1/m$ from now. Suppose further that i is the per annum interest rate compounded m times per year. Then the present value of this payment stream is

$$\frac{1}{m}\left\{\left(1+\frac{i}{m}\right)^{-1}+\left(1+\frac{i}{m}\right)^{-2}+\cdots+\left(1+\frac{i}{m}\right)^{-mn}\right\}$$

$$=\frac{1}{m}\left(1+\frac{i}{m}\right)^{-1}\left\{\frac{1-\left((1+\frac{i}{m})^{-1}\right)^{mn}}{1-\left(1+\frac{i}{m}\right)^{-1}}\right\}$$

$$=\frac{1}{i}\cdot\left\{1-\left(\left(1+\frac{i}{m}\right)^{-m}\right)^{n}\right\}$$

$$\to\frac{1}{r}\cdot(1-e^{-rn})\quad\text{as } m\to\infty.$$

The derivation for other times T is similar.

From the formula $Y=(1-e^{-rT})/r$, we see that if T has a continuous distribution on $(0,\infty)$, then Y has a continuous distribution on $(0,1/r)$. Moreover, for $y\in(0,1/r)$,

$$
\begin{aligned}
F_Y(y) &= \Pr((1-e^{-rT})/r\le y)\\
&= \Pr(e^{-rT}\ge 1-ry)\\
&= \Pr(T\le -\frac{1}{r}\log(1-ry))\\
&= F_T(-\log(1-ry)^{1/r})
\end{aligned}
$$

and the expected value of Y is

$$
\begin{aligned}
E[Y] &= \int_0^\infty \frac{1-e^{-rt}}{r}f_T(t)dt\\
&= \frac{1}{r}-\frac{1}{r}\int_0^\infty e^{-rt}f_T(t)dt.
\end{aligned}
$$

The latter formula gives an important relationship between the actuarial present value of a (continuous) life annuity and the actuarial present value of a whole life policy on the same life. Indeed, if we let $\overline{a}_x$ denote the actuarial present value of a continuous life annuity of \$1 per year to a person currently age x and we let $\overline{A}_x$ denote the actuarial present value of a whole life policy that pays \$1 on the death of a person currently age x, and if the future lifetime distribution for the annuity and the whole life policy is the same, then $\overline{a}_x$ and $\overline{A}_x$ are connected by the formula

$$1=r\overline{a}_x+\overline{A}_x.$$

This formula has an interesting economic interpretation. Indeed, it asserts that \$1 today is equivalent to a contract that pays interest continuously at a rate of r per annum until the contract holder's death, at which time a payment of \$1 (representing the principal amount of the investment) is made to the policyholder's estate.

EXAMPLE 6: Suppose that $T\sim\text{Exponential}(1/10)$ and $r=0.05$. Determine the actuarial present value of a continuous life annuity paying \$10,000 per year.

One could calculate the actuarial present value directly from the definition. However, it's easier to calculate $\overline{A}_x$ and use the relationship between $\overline{a}_x$ and $\overline{A}_x$. Indeed,

$$\overline{A}_x=\int_0^\infty e^{-t/20}\frac{1}{10}e^{-t/10}dt=\frac{2}{3}.$$

Hence,

$$\bar{a}_x = \frac{1 - \overline{A}_x}{r} = \frac{1 - \frac{2}{3}}{0.05} = \frac{20}{3} = 6\frac{2}{3}.$$

Consequently, the actuarial present value for an annuity paying \$10,000 per annum is \$66,666.67. ∎

7.5 Reliability of Systems with Multiple Components or Processes (Optional)

In this section, we discuss the reliability of electrical and mechanical systems with multiple components. The analysis of such systems requires an understanding of the distributions of the minimum and maximum of a collection of random variables. We begin by determining general formulas for the distribution of a minimum and a maximum when the random variables are all independent. We then consider the reliability of systems that are configured in series, parallel, or a combination of the two. Finally, we consider applications of the minimum and maximum transformations to scheduling and contract tendering. Additional applications are developed in the exercises in §7.7.

Minimum and Maximum of a Collection of Independent Random Variables

Let $X_1, \ldots, X_n$ be a collection of independent random variables and let $X_{\min}$ and $X_{\max}$ be the random variables defined by

$$X_{\min} = \min(X_1, \ldots, X_n),$$
$$X_{\max} = \max(X_1, \ldots, X_n).$$

Then the distributions of $X_{\min}$ and $X_{\max}$ are given by

$$F_{X_{\min}}(x) = 1 - (1 - F_{X_1}(x)) \cdots (1 - F_{X_n}(x))$$

and

$$F_{X_{\max}}(x) = F_{X_1}(x) \cdots F_{X_n}(x),$$

respectively.

To see this, note that for any x,

$$X_{\min} > x \text{ if and only if } X_j > x \text{ for all } j$$

and

$$X_{\max} \le x \text{ if and only if } X_j \le x \text{ for all } j.$$

Hence, since the X_j are mutually independent,

$$\Pr(X_{\min} > x) = \Pr(X_1 > x) \cdots \Pr(X_n > x),$$

that is,

$$S_{X_{\min}}(x) = S_{X_1}(x) \cdots S_{X_n}(x),$$

and

$$\Pr(X_{\max} \leq x) = \Pr(X_1 \leq x) \cdots \Pr(X_n \leq x),$$

that is,

$$F_{X_{\max}}(x) = F_{X_1}(x) \cdots F_{X_n}(x).$$

Consequently, the distribution functions of $X_{\min}$ and $X_{\max}$ have the form claimed.

Minimum and Maximum of Independent Random Variables

The **survival function** for a **minimum** of independent random variables is the product of the individual survival functions:

$$S_{X_{\min}}(x) = S_{X_1}(x) \cdots S_{X_n}(x).$$

The **distribution function** for a **maximum** of independent random variables is the product of the individual distribution functions:

$$F_{X_{\max}}(x) = F_{X_1}(x) \cdots F_{X_n}(x).$$

When there are only two random variables under consideration, then the minimum and maximum are connected by the relationship

$$\min(X_1, X_2) + \max(X_1, X_2) = X_1 + X_2.$$

This relationship is useful for calculating moments when one of $\min(X_1, X_2)$, $\max(X_1, X_2)$ has a much simpler distribution than the other and the distribution of the sum $X_1 + X_2$ is easily determined.

EXAMPLE 1:   Determine the distributions of $X_{\min}$ and $X_{\max}$ when the X_j are exponentially distributed.

Suppose that $X_j \sim \text{Exponential}(\lambda_j)$ and the X_j are independent. Then the survival function of $X_{\min}$ is given by

$$\begin{aligned} S_{X_{\min}}(x) &= S_{X_1}(x) \cdots S_{X_n}(x) \\ &= e^{-\lambda_1 x} \cdots e^{-\lambda_n x} \\ &= e^{-(\lambda_1 + \cdots + \lambda_n)x}. \end{aligned}$$

Hence, $X_{\min} \sim \text{Exponential}(\lambda_1 + \cdots + \lambda_n)$.

On the other hand, the distribution function for $X_{\max}$ is given by

$$\begin{aligned} F_{X_{\max}}(x) &= F_{X_1}(x) \cdots F_{X_n}(x) \\ &= (1 - e^{-\lambda_1 x}) \cdots (1 - e^{-\lambda_n x}). \end{aligned}$$

Hence, the distribution of $X_{\max}$ is not of exponential type.   ■

Series and Parallel Systems

There are two basic configurations underlying all electrical and mechanical systems: series and parallel. A **series configuration** is one in which all components of the system

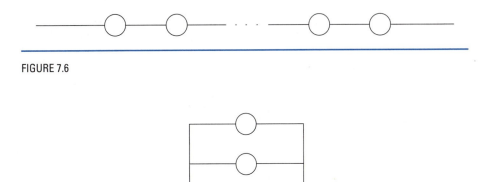

FIGURE 7.6

FIGURE 7.7

must be operational for the system to operate. On the other hand, a **parallel configuration** is a configuration in which only one of the components of the system needs to be operational for the entire system to operate. An example of a series configuration is a flashlight requiring multiple batteries in which one discharged battery makes the flashlight inoperable. An example of a parallel configuration is a multiple server computer system in which computation requests are automatically directed to a server that is free or are put into a priority queue.

Series configurations can be visually represented as shown in Figure 7.6. In this diagram, we can think of electrical current flowing from left to right through the system and being halted at a component such as a discharged battery that is inoperable. On the other hand, parallel configurations can be visually represented as shown in Figure 7.7. In this diagram, electrical current that is flowing from left to right will only be halted if all components are inoperable.

The reliability of systems configured in series and parallel can be analyzed using the minimum and maximum transformations discussed at the beginning of this section. Indeed, if $T_1, \ldots, T_n$ represent the lifetimes of the individual components of a system, then the lifetime of the system as a whole is T_{min} if the components are configured in series, and it is T_{max} if the components are configured in parallel. If the components of such systems fail independently, as we will assume throughout this section, then their reliabilities can be determined using the formulas for $F_{T_{min}}$ and $F_{T_{max}}$ derived earlier.

Most systems have a combination of series and parallel configurations. For example, the system illustrated in Figure 7.8 is a parallel configuration of two series configurations, whereas the system in Figure 7.9 is a series configuration of two parallel configurations. An important result in engineering is that every system of practical interest can be decomposed into series and parallel parts. Hence, the reliability of any system can be analyzed by considering appropriate maximums and minimums.

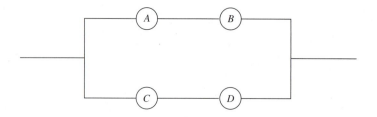

FIGURE 7.8 Parallel Configuration of Two Series Configurations

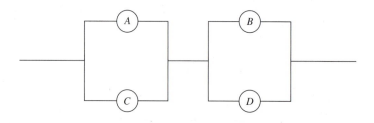

FIGURE 7.9 Series Configuration of Two Parallel Configurations

EXAMPLE 2: A particular flashlight requires four batteries to operate. The batteries in this flashlight are configured in series. Suppose that the lifetimes of individual batteries are independent and have Weibull distributions with parameters $\alpha = 2$ and $\beta = 3$, where lifetimes are measured in hours. Determine the probability that the flashlight is still functional after 2 hours of continuous use.

Let T_1, T_2, T_3, T_4 be the respective lifetimes in hours of the individual batteries in the flashlight. Then $T_{\min} = \min(T_1, T_2, T_3, T_4)$ represents the amount of time that the flashlight is functional. By hypothesis,

$$S_{T_j}(t) = \exp\left\{-\left(\frac{t}{2}\right)^3\right\}$$

for each j. Since the T_j are independent, it follows that

$$S_{T_{\min}}(t) = S_{T_1}(t)S_{T_2}(t)S_{T_3}(t)S_{T_4}(t)$$

$$= \exp\left\{-\left(\frac{t}{2}\right)^3\right\} \cdot \exp\left\{-\left(\frac{t}{2}\right)^3\right\} \cdot \exp\left\{-\left(\frac{t}{2}\right)^3\right\} \cdot \exp\left\{-\left(\frac{t}{2}\right)^3\right\}$$

$$= \exp\left\{-4\left(\frac{t}{2}\right)^3\right\}$$

$$= e^{-t^3/2}.$$

Hence, the desired probability is

$$\Pr(T_{\min} > 2) = e^{-4} \approx .0183.$$ ∎

EXAMPLE 3: Computers with parallel processors allow several sets of instructions to be executed at the same time. Suppose that a certain type of task can be split into three subtasks and each of these subtasks is submitted to a different processor. Suppose further that the processing times for the subtasks are exponentially distributed with mean 4 seconds. Determine the distribution of the processing time for the original task.

Let T_1, T_2, T_3 be the processing times in seconds for the individual subtasks and suppose that the T_j are independent. Then the processing time for the original task is $T_{max} = \{T_1, T_2, T_3\}$. Since $T_j \sim$ Exponential(0.25) for all j, the distribution of T_{max} is given by

$$F_{T_{max}}(t) = F_{T_1}(t) F_{T_2}(t) F_{T_3}(t)$$
$$= (1 - e^{-t/4})^3.$$

∎

EXAMPLE 4: By considering mean processing times, compare the parallel processor of the previous example to a sequential processor in which a task is split into three subtasks and the subtasks are executed sequentially.

Continue to assume that $T_j \sim$ Exponential(0.25). Let T_S be the processing time using a sequential processor. Then $T_S = T_1 + T_2 + T_3$. Hence, $T \sim$ Gamma(3, 0.25) and so the mean processing time with a sequential processor is $E[T_S] = 3/(0.25) = 12$ seconds. The mean processing time for the parallel processor is $E[T_{max}]$, where $T_{max} = \max(T_1, T_2, T_3)$. Using the formula for $F_{T_{max}}$ derived in the previous example and the general formula for the expectation of a continuous random variable, we find that $E[T_{max}] = 7\frac{1}{3}$. Hence, on average, the parallel processor is about 1.6 times faster.

One can also show that the standard deviation in the processing time is smaller with the parallel processor. Hence, computers with parallel processors are considerably more efficient than computers that only perform calculations sequentially. ∎

EXAMPLE 5: Customers waiting for service at a post office form a single line according to the order in which they arrive. There are three customer service representatives at the counter, and each is currently busy serving a customer. Suppose that the service times are independent and exponentially distributed. The mean service time for one of the customer service representatives is 20 seconds, and the mean service times for the others are both 30 seconds. Determine the distribution of the waiting time for the person waiting for service at the front of the line. How long should this person expect to wait?

Let T_1, T_2, T_3 be the service times measured in minutes. By hypothesis, $T_1 \sim$ Exponential(3), $T_2 \sim$ Exponential(2), and $T_3 \sim$ Exponential(2). The waiting time for the customer at the front of the line is $T_{min} = \min(T_1, T_2, T_3)$. Since the T_j are independent and exponentially distributed, it follows from Example 1 that $T_{min} \sim$ Exponential(3 + 2 + 2) (i.e., $T_{min} \sim$ Exponential(7)). Hence, the customer at the front of the line should expect to wait 1/7 minute (i.e., about 9 seconds). ∎

EXAMPLE 6: Consider the system in Figure 7.10. Suppose that the lifetimes of the components A, B, C are independent and exponentially distributed with means $\mu_A = 3$, $\mu_B = 4$, $\mu_C = 6$. Determine the distribution of the lifetime of the system as a whole.

Let T_A, T_B, T_C denote the lifetimes of the components A, B, C, respectively, and let T denote the lifetime of the system as a whole. Then

$$T = \min(\max(T_A, T_B), T_C).$$

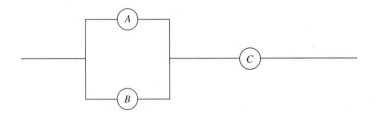

FIGURE 7.10

Put $Y = \max(T_A, T_B)$. Then since T_A, T_B, T_C are independent, the distribution of Y is given by

$$F_Y(t) = F_{T_A}(t)F_{T_B}(t)$$

and so the distribution of T is given by

$$S_T(t) = S_Y(t)S_{T_C}(t)$$
$$= \{1 - F_{T_A}(t)F_{T_B}(t)\}S_{T_C}(t),$$

that is,

$$F_T(t) = 1 - (1 - F_{T_A}(t)F_{T_B}(t))(1 - F_{T_C}(t)).$$

Now, by hypothesis, $T_A \sim \text{Exponential}(1/3)$, $T_B \sim \text{Exponential}(1/4)$, $T_C \sim \text{Exponential}(1/6)$. Hence, the distribution function of T is given by

$$F_T(t) = 1 - \{1 - (1 - e^{-t/3})(1 - e^{-t/4})\}e^{-t/6}$$
$$= 1 - (e^{-t/3} + e^{-t/4} - e^{-7t/12})e^{-t/6}$$
$$= 1 - e^{-t/2} - e^{-5t/12} + e^{-3t/4}. \qquad \blacksquare$$

Additional Applications of the Minimum and Maximum Transformations

We conclude this section with some additional illustrations of the use of minimum and maximum transformations in probability.

EXAMPLE 7:   A public administrator is seeking bids for the construction of a new public works building. Preliminary analysis suggests that bids for this project will fall between $20 million and $30 million. Suppose that bids are uniformly distributed over this range and are independent for distinct contractors. The administrator would like the cost of the building to be no more than $23 million. How many bids should the administrator plan on soliciting for there to be a 75% probability of getting a bid that is no more than $23 million?

Let $X_1, \ldots, X_n$ be the sizes of the individual bids measured in millions of dollars and let $X_{\min}$ be the lowest bid. By hypothesis, each X_j is uniformly distributed on $(20, 30)$,

$$f_{X_j}(x) = \begin{cases} \frac{1}{10} & \text{for } 20 < x < 30, \\ 0 & \text{otherwise.} \end{cases}$$

Hence, since the X_j are independent, the survival function of $X_{\min}$ is given by

$$S_{X_{\min}}(x) = S_{X_1}(x) \cdots S_{X_n}(x)$$

$$= \begin{cases} 1 & \text{for } x < 20, \\ \left(\frac{30-x}{10}\right)^n & \text{for } 20 < x < 30, \\ 0 & \text{for } x > 30. \end{cases}$$

The administrator's requirement is that

$$\Pr(X_{\min} \leq 23) \geq 0.75$$

that is,

$$S_{X_{\min}}(23) \leq 0.25.$$

Hence, the required number of bids is given by

$$\left(\frac{30-23}{10}\right)^n \leq 0.25$$

that is,

$$n \geq \frac{\log(0.25)}{\log(0.70)} \approx 3.887.$$

Thus, the administrator should solicit four independent bids. ∎

EXAMPLE 8: Two subcontractors have been hired to build different parts of an experimental aircraft. The expected completion times for the two subcontractors are 10 months and 12 months, respectively. Both parts are necessary before final assembly of the aircraft can take place. How long should the primary contractor expect to wait to complete final assembly if at least one of the subcontractors is expected to be finished in 6 months?

Let T_1, T_2 be the completion times in months for the two subcontractors and suppose that T_1, T_2 are independent. From the given information,

$$E[T_1] = 10,$$
$$E[T_2] = 12,$$
$$E[\min(T_1, T_2)] = 6.$$

Since $\min(T_1, T_2) + \max(T_1, T_2) = T_1 + T_2$, it follows that

$$E[\max(T_1, T_2)] = E[T_1] + E[T_2] - E[\min(T_1, T_2)]$$
$$= 10 + 12 - 6$$
$$= 16.$$

Hence, the main contractor should expect to wait 16 months before final assembly can occur. ∎

7.6 Trigonometric Transformations (Optional)

There are many phenomena in engineering and science that exhibit wavelike behavior. For example, alternating current in an electrical circuit oscillates in a sinusoidal pattern. The reader may recall from studies in trigonometry that the general form of the sine wave is

$$x(t) = A \sin(\omega t + \Theta),$$

where t represents time, A is the amplitude, $2\pi/\omega$ is the period, and Θ is the phase shift. In many applications involving sine waves, the amplitude and the phase shift will both be uncertain. The uncertainty in the phase shift Θ arises from the uncertainty of the time origin for the wave, which arises whenever one does not have complete control over the process generating the wave.

For simplicity, we assume in this section that the amplitude is constant and we focus on determining the distribution of $x(t)$ for each t when only Θ is uncertain. If we have no prior opinion about the value of Θ as we usually will not, then it is appropriate to assume that Θ is uniformly distributed on $(-\pi, \pi)$. Since the sine function is periodic with period 2π and since ωt is assumed to be constant, the distribution of $x(t)$ will be the same for all t. Hence, for simplicity, let's consider the distribution of $x(0)$ and let's assume that $A = 1$. Then, the problem reduces to determining the distribution of $\sin \Theta$ under the assumption that Θ is uniformly distributed on $(-\pi, \pi)$.

EXAMPLE 1:   Put $X = \sin \Theta$ and suppose that Θ has the density function

$$f_\Theta(\theta) = \begin{cases} \dfrac{1}{2\pi} & \text{for } -\pi < \theta < \pi, \\ 0 & \text{otherwise.} \end{cases}$$

Show that X has the density function

$$f_X(x) = \begin{cases} \dfrac{1}{\pi} \cdot \dfrac{1}{\sqrt{1-x^2}} & \text{for } -1 < x < 1, \\ 0 & \text{otherwise.} \end{cases}$$

Since $-1 \le \sin \theta \le 1$ for all θ, it is clear that there is no positive probability outside the interval $x \in [-1, 1]$. Hence, we need only consider x such that $-1 \le x \le 1$. Suppose first that $x \in [0, 1]$. Then $X \le x \iff \sin \Theta \le x \iff -\pi < \Theta \le \arcsin x$ or $\pi - \arcsin x \le \Theta < \pi$. Consider a graph of $\sin \theta$ for $\theta \in (-\pi, \pi)$. Hence,

$$\begin{aligned} \Pr(X \le x) &= \Pr(-\pi < \Theta \le \arcsin x) + \Pr(\pi - \arcsin x \le \Theta < \pi) \\ &= F_\Theta(\arcsin x) - F_\Theta(-\pi) + F_\Theta(\pi) - F_\Theta(\pi - \arcsin x) \end{aligned}$$

and so

$$\begin{aligned} f_X(x) &= \frac{d}{dx} \Pr(X \le x) \\ &= \frac{d}{dx} F_\Theta(\arcsin x) - \frac{d}{dx} F_\Theta(\pi - \arcsin x) \\ &= f_\Theta(\arcsin x) \cdot \frac{1}{\sqrt{1-x^2}} + f_\Theta(\pi - \arcsin x) \cdot \frac{1}{\sqrt{1-x^2}} \\ &= \frac{1}{\pi} \cdot \frac{1}{\sqrt{1-x^2}}. \end{aligned}$$

The latter equality follows from the fact that $f_\Theta(\theta) = 1/2\pi$ for all $\theta \in (-\pi, \pi)$. We obtain an identical formula for $f_X(x)$ when $-1 < x < 0$. Consequently, the density function for X is given by

$$f_X(x) = \begin{cases} \dfrac{1}{\pi} \cdot \dfrac{1}{\sqrt{1-x^2}} & \text{for } -1 < x < 1, \\ 0 & \text{otherwise} \end{cases}$$

as claimed. Details are left to the reader.

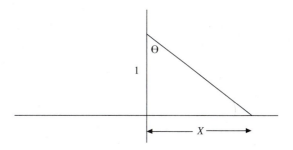

FIGURE 7.11 The Transformation $X = \tan \Theta$

EXAMPLE 2: A related trigonometric transformation of some interest is $X = \tan \Theta$, where Θ is uniformly distributed on $(-\frac{\pi}{2}, \frac{\pi}{2})$. The quantity $X = \tan \Theta$ can be interpreted as the horizontal position of impact for a laser beam originating one unit above the origin and at angle Θ to the vertical (Figure 7.11). The density function for X in this case is given by

$$f_X(x) = \frac{1}{\pi} \cdot \frac{1}{1 + x^2}, \qquad x \in \mathbf{R}$$

as the reader is invited to demonstrate.

A random variable X with a density function of this type is said to have a **Cauchy distribution.** An interesting property of the Cauchy distribution is that its mean does not exist, *even though the distribution is symmetric about the origin!* To see this, note that

$$\int_0^\infty x \cdot \frac{1}{1 + x^2} dx = \frac{1}{2} \log(1 + x^2) \Big|_0^\infty = \infty$$

and

$$\int_{-\infty}^0 x \cdot \frac{1}{1 + x^2} dx = \frac{1}{2} \log(1 + x^2) \Big|_{-\infty}^0 = -\infty.$$

Another interesting property of the Cauchy distribution is that the average of a set of independent observations taken from a Cauchy distribution also has a Cauchy distribution. That is, $\overline{X}_n = (X_1 + \cdots + X_n)/n$ has a Cauchy distribution whenever the X_j are independent Cauchy random variables. This means that the variability in $\overline{X}_n$ does not decrease with increasing n when the X_j are taken from a Cauchy distribution, contrary to what our intuition (i.e., the insurance principle) suggests should happen! (See Chapter 1.) In fact, this example does not show that the insurance principle (i.e., the law of large numbers) is invalid because, as we will see in the next chapter, the law of large numbers only holds for distributions whose mean exists. ■

7.7 Exercises

1. In medical imaging such as computer tomography, the relationship between detector readings Y and body absorptivity X is given by $Y = e^X$. Determine the density function of Y when X has a normal distribution.

2. In homomorphic image processing, images are enhanced by applying nonlinear transformations to the image functions. Suppose that an image is enhanced using the transformation $Y = \log X$. Determine the density function of Y when X has an exponential distribution.

3. Consider the transformation given by

$$Y = \begin{cases} X & 0 < X \le \frac{1}{2}, \\ 1 - X & \frac{1}{2} < X < 1, \\ 0 & \text{otherwise.} \end{cases}$$

Determine the density function of Y when X is uniformly distributed on $(0, 2)$.

4. Consider the transformation

$$Y = \begin{cases} \sqrt{X} & X \ge 0, \\ 0 & X < 0. \end{cases}$$

Determine the density function of Y when X is a standard normal random variable.

5. Determine a formula for the limited expected value function for a Pareto random variable. Sketch a graph of $E[X; m]$ as a function of m when $X \sim \text{Pareto}(s, \beta)$.

6. Let X be a random variable and consider the quantity $E[X; m]$ as a function of m. Show that

  a. $E[X; m]$ is increasing, continuous, and concave as a function of m;
  b. $E[X; m] \to E[X]$ as $m \to \infty$;
  c. $\dfrac{d}{dm} E[X; m] = S_X(m)$.

7. Let X be a nonnegative random variable and put $Y = \min(X, m)$. Determine formulas for $E[Y^k]$ which generalize the formula $E[X; m] = \int_0^m S_X(x)\, dx$.

8. Let X be a positive random variable, and put $Y = \log X$. Determine a formula for the distribution function of Y in terms of the distribution function of X. Specify the density function for Y when X is continuous.

9. A joint life insurance policy provides insurance on a group of lives in such a way that the benefit is paid when the first death in the group occurs. Hence, if $T_1, \ldots, T_n$ are the future lifetimes of the group members, then $\min(T_1, \ldots, T_n)$ is the time from now when the benefit is paid. Suppose that the future lifetimes of the group members are exponentially distributed with respective parameters $\lambda_1, \ldots, \lambda_n$ and suppose that future lifetimes are independent. Determine the distribution of $\min(T_1, \ldots, T_n)$.

10. Determine the actuarial present value for $1 of benefit on a joint life policy if future lifetimes are exponentially distributed and mutually independent.

11. A survivorship life insurance policy provides insurance on a group of lives in such a way that the benefit is paid when the final death in the group occurs. Hence, if $T_1, \ldots, T_n$ are the future lifetimes of the group members, then $\max(T_1, \ldots, T_n)$ is the time from now when the benefit is paid. Suppose that $T_j \sim \text{Exponential}(\lambda_j)$ and the T_j are mutually independent. Determine the distribution of $\max(T_1, \ldots, T_n)$.

12. Determine the actuarial present value for $1 of benefit on a survivorship life insurance policy if future lifetimes are exponentially distributed and mutually independent.

13. What is the relationship between $E[\sqrt{X}]$ and $\sqrt{E[X]}$? What is the relationship between $E[\log X]$ and $\log(E[X])$?

14. According to the terms of a particular insurance contract, the insurer agrees to pay 70% of eligible claim expenses in excess of a $500 deductible with a maximum payment of $5000. Claims in thousands of dollars are assumed to follow a Pareto distribution with parameters $s = 3$ and $\beta = 2$. Let Y be the policyholder's reimbursement on a random claim.

   a. Determine a formula for the distribution function of Y.
   b. What reimbursement should the policyholder expect?

15. Sketch graphs of the insurer's payment and the policyholder's residual loss as a function of the claim size for each of the following contract types:

   a. excess-of-loss
   b. limited excess-of-loss
   c. single layer combination contract
   d. double layer combination contract

16. Suppose that $Y = \alpha X$, where $\alpha > 0$. Determine formulas for the moment and cumulant generating functions of Y in terms of the moment and cumulant generating functions of X. Deduce formulas for the mean, variance, and skewness of Y.

17. According to the terms of a particular insurance contract, the insurer agrees to pay 80% of the first $2000 of eligible claim expenses in excess of a $200 deductible and 50% of any remaining claim expenses, with a maximum possible payment of $10,000.

   a. Decompose this double layer contract into a portfolio of single layer contracts.
   b. Suppose that claims are exponentially distributed with mean $4000. Determine the policyholder's expected reimbursement.

18. Under the terms of a comprehensive major medical insurance plan, the insurer agrees to pay 80% of eligible medical expenses during the year after a deductible of $500. The maximum out-of-pocket expense for the plan member is $1500 in any given year. Suppose that medical expenses for plan members are exponentially distributed with mean $1000. Determine the average reimbursement.

19. Suppose that the insurer of the previous question would like the average reimbursement per member to be no greater than $200.

   a. What adjustment to the maximum out-of-pocket plan member expense should be made to accomplish this if the deductible is to remain at $500?
   b. What adjustment to the deductible should be made to accomplish this if the maximum out-of-pocket plan member expense is to remain at $1500?

20. A whole life insurance policy pays $10,000 on the death of the insured. The future lifetime of the insured is uniformly distributed on $(0, 20)$. Determine the premium that should be charged today under each of the following criteria assuming a continuously compounded interest rate of 6% per annum:

   a. The probability that the insurer is able to meet its obligations is 95%.
   b. The premium is the actuarial present value.

21. An insurer has a portfolio of 200 whole life policies, each with a face value of $10,000. Each policy is written on a single life, and the future lifetime for each insured is assumed to be DeMoivre(20). The policies are also assumed to be independent. Determine the size of the fund that the insurer must have if the probability

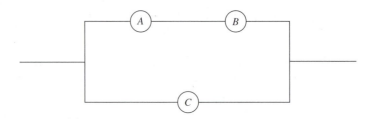

FIGURE 7.12

of being able to meet all obligations is to be 95%. How does this compare with the aggregate premium determined in the previous question? Assume a continuously compounded interest rate of 6% per annum.

22. A retiree who is currently 75 years old has $200,000 in an investment account and would like to purchase an annuity from an insurance company to reduce the risk of outliving this savings. Suppose that the retiree's future lifetime is modeled using a DeMoivre(20) distribution and the current long-term interest rate is 6% per annum compounded continuously. Determine the amount the retiree will receive each remaining year of life if annuity payments are made continuously.

23. The premium for a whole life policy can be very expensive if it is paid in a lump sum at policy inception. An alternative is to pay the premium continuously over the lifetime of the insured. Unfortunately, if the premium is paid in this way, the amount of premium received depends on the uncertain future lifetime of the insured and, hence, is itself uncertain. One way to determine such periodic premiums is to equate the actuarial present value of benefits with the actuarial present value of the premium payment stream.

    Consider the whole life policy of question 20 in which $10,000 is paid on the death of the insured, the insured's future lifetime is modeled as DeMoivre(20), and $r = 6\%$ per annum compounded continuously. Determine the required annual premium payable continuously by equating the actuarial present value of benefits and the actuarial present value of payments.

24. Consider the whole life policy of the previous question. Determine the annual premium payable continuously such that the probability that the present value of premium payments exceeds the present value of benefit payments is 95%.

25. Consider the system in Figure 7.12. Suppose that the lifetimes of the components A, B, C are independent and exponentially distributed with means $\mu_A = 2$, $\mu_B = 3$, $\mu_C = 4$ measured in hours. Determine the probability that the system fails in the first 3 hours of operation. What is the expected life of the system?

26. Consider the system in Figure 7.13. Suppose that the lifetimes of the components A, B, C, and D are independent and exponentially distributed with means $\mu_A = 4$, $\mu_B = 3$, $\mu_C = 2$, $\mu_D = 1$. Determine the distribution function for the lifetime of the system as a whole. What is the expected life of the system?

27. For each of the systems in Figures 7.14, 7.15, and 7.16, determine a formula for the distribution function for the life of the entire system assuming that individual components fail independently and have exponentially distributed lifetimes (with possibly different means).

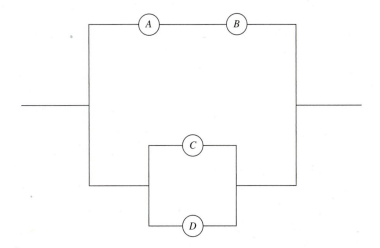

FIGURE 7.13

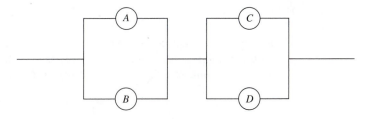

FIGURE 7.14

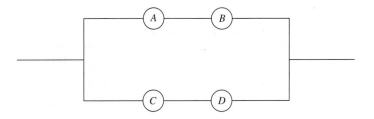

FIGURE 7.15

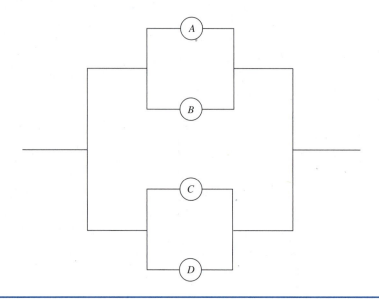

FIGURE 7.16

28. In analog-to-digital conversion, an analog waveform is sampled, quantized, and coded. A *quantizer* is a function that assigns to each sample value x a value y from a generally finite set of predetermined values. Consider the quantizer defined by

$$g(x) = [x] + 1,$$

where $[x]$ denotes the greatest integer less than or equal to x. Suppose that X has a standard normal distribution and put $Y = g(X)$. Specify the distribution of Y. Ignore values of Y for which the probability is essentially zero.

29. An amplifier on a particular electrical circuit transforms the signal in accordance with $Y = X^3$. Suppose that X has a standard normal distribution. Determine a formula for the probability density of Y.

30. Consider the trigonometric transformation $X = \tan \Theta$. Show that if Θ is uniformly distributed on $(-\pi, \pi)$, then the density of X is given by

$$f_X(x) = \frac{1}{\pi} \cdot \frac{1}{1 + x^2}, \qquad x \in \mathbf{R}.$$

8

Sums and Products of Random Variables

We have already encountered sums and products of random variables on several occasions throughout this book. However, until now, we have not discussed in any great detail how one determines the *distribution* of a given sum or product.

In this chapter, we develop some general methods for determining the distribution of a sum or product of random variables, and we illustrate the use of these methods with numerous examples. We also demonstrate the validity of the formulas for the expectation of a sum or product stated in Chapter 4, and we discuss the concepts of covariance and correlation in detail. Finally, we consider two important theorems on the limiting behavior of independent sums and products: the law of large numbers and the central limit theorem. The law of large numbers is a fundamental result that provides a mathematical justification for the principle of risk sharing in insurance (the principle upon which all of insurance is based). The central limit theorem is a fundamental result that justifies the widespread use of the normal distribution in probability modeling. We discuss both of these important results in detail.

8.1   Techniques for Calculating the Distribution of a Sum

In the discussion of moment generating functions in §4.3.1, we saw that it is sometimes possible to determine the distribution of an *independent* sum using the relationship

$$M_{X+Y}(t) = M_X(t) \cdot M_Y(t)$$

for independent X, Y and the fact that moment generating functions uniquely characterize distributions. While this method of determining the distribution of a sum can be extremely useful,[1] it is important to realize that it is only effective if we can recognize the distribution whose moment generating function is $M_X(t) \cdot M_Y(t)$, and it only works when X and Y are independent.

In this section, we discuss two general methods for determining the distribution of a sum. The first method relies on the observation that a sum is simply a vector-to-scalar transformation of the type discussed in the previous chapter. Indeed, the sum of the random variables $X_1, \ldots, X_n$ is simply the transformation

---

[1] Indeed, we used this method extensively in our discussion of special distributions in Chapters 5 and 6.

$$h(X_1, \ldots, X_n) = X_1 + \cdots + X_n.$$

Hence, the distribution of $S = X_1 + \cdots + X_n$ can be determined using the general formula

$$F_S(s) = \int \cdots \int_{h(x_1, \ldots, x_n) \leq s} f_{X_1, \ldots, X_n}(x_1, \ldots, x_n) dx_1 \cdots dx_n,$$

provided that we can perform the required integration.

The second method, on the other hand, relies on successive applications of the law of total probability. To be precise, if $S = X_1 + \cdots + X_n$, then

$$F_S(s) = \Pr(S \leq s)$$

$$= \int_{x_n} \Pr(S \leq s | X_n = x_n) \cdot f_{X_n}(x_n) dx_n$$

$$= \int_{x_n} \Pr(X_1 + \cdots + X_{n-1} \leq s - x_n | X_n = x_n) \cdot f_{X_n}(x_n) dx_n$$

$$= \int_{x_n} F_{X_1 + \cdots + X_{n-1} | X_n = x_n}(s - x_n) \cdot f_{X_n}(x_n) dx_n.$$

Hence, by recursively applying this procedure to the conditional distributions $X_1 + \cdots + X_{n-1} | X_n = x_n$; $X_1 + \cdots + X_{n-2} | X_{n-1} = x_{n-1}, X_n = x_n$; $\cdots$, we can, in principle, determine the unconditional distribution of $X_1 + \cdots + X_n$.

We also discuss the **method of convolutions**, which is a special case of the second method and only applies when $X_1, \ldots, X_n$ are mutually independent.

To simplify the discussion that follows, we restrict our attention to sums of two random variables from now on, unless otherwise indicated. Hence, let X and Y be (possibly dependent) random variables and let S be the random variable given by $S = X + Y$.

8.1.1   Using the Joint Density

As we just noted, one method for determining the distribution of $S = X + Y$ is to interpret the sum as the transformation $S = h(X, Y)$, where $h(X, Y) = X + Y$. Then, from the discussion in §7.1, the distribution function of S is given by the formula

$$F_S(s) = \iint_{x+y \leq s} f_{X,Y}(x, y) dx\, dy.$$

Hence, to determine F_S explicitly, we need only calculate the double integral $\iint_{x+y \leq s} f_{X,Y}(x, y) dx\, dy$ for each s.

Now if the joint density function $f_{X,Y}$ is given by a single algebraic expression for all x and y, then the calculation of the double integral is relatively straightforward and can be accomplished by performing iterated integration in one of the following ways:

$$\iint_{x+y \leq s} f_{X,Y}(x, y) dx\, dy = \int_{x=-\infty}^{\infty} \int_{y=-\infty}^{s-x} f_{X,Y}(x, y) dy\, dx$$

or

$$\iint_{x+y \leq s} f_{X,Y}(x, y) dx\, dy = \int_{y=-\infty}^{\infty} \int_{x=-\infty}^{s-y} f_{X,Y}(x, y) dx\, dy.$$

In the first formula, the variable y is integrated for each fixed x and then the variable x is integrated. In the second formula, the variable x is integrated for each fixed y and then the variable y is integrated. Note that the order of integration can affect the difficulty of the calculation, but it does not affect the ultimate value of the double integral.

However, if the joint density function $f_{X,Y}$ is defined in different ways on different portions of the x-y plane, then the calculation of the double integral $\iint_{x+y \le s} f_{X,Y}(x, y)\, dx\, dy$ becomes much more complicated. Indeed, in this case, one must break up the integral $\iint_{x+y \le s} f_{X,Y}(x, y)\, dx\, dy$ into a sum of double integrals in such a way that each of the integrals in this sum is calculated over a region for which there is a single algebraic expression for $f_{X,Y}$. To complicate things even further, the shapes of the regions of integration for the integrals in such a sum can change with s.

As a simple example, consider a density function $f_{X,Y}$, which is zero outside of the unit square $U = \{(x, y) \in \mathbf{R}^2 : 0 \le x \le 1, 0 \le y \le 1\}$ and nonzero inside this square. That is, $f_{X,Y}(x, y) = 0$ for $(x, y) \notin U$, but $f_{X,Y}(x, y) \ne 0$ for $(x, y) \in U$. Then, for any given s, the integration of $f_{X,Y}$ over the half-plane $x + y \le s$ can be accomplished by integrating separately over each of the regions.

$$R_1 = \{(x, y) \in \mathbf{R}^2 : x + y \le s \text{ and } (x, y) \in U\},$$
$$R_2 = \{(x, y) \in \mathbf{R}^2 : x + y \le s \text{ and } (x, y) \notin U\}.$$

In this example, the integral of $f_{X,Y}$ over R_2 is always zero since $f_{X,Y}(x, y) = 0$ for $(x, y) \notin U$, with the result that

$$F_S(s) = \iint_{R_1} f_{X,Y}(x, y)\, dx\, dy$$

and only one double integration is required. However, the shape of the region R_1 can still change with s. Indeed, if $s \in (0, 1)$, then R_1 is triangular, whereas if $s \in (1, 2)$ then R_1 is trapezoidal (Figure 8.1). Hence, even in this simple example, we see how complicated the calculation of F_S can be when the joint density $f_{X,Y}$ is not defined by a single algebraic expression for all x, y.

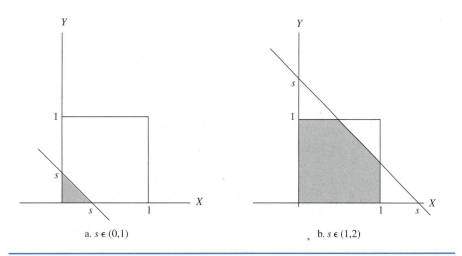

a. $s \in (0,1)$ b. $s \in (1,2)$

FIGURE 8.1 The Shape of the Region R_1 for various s.

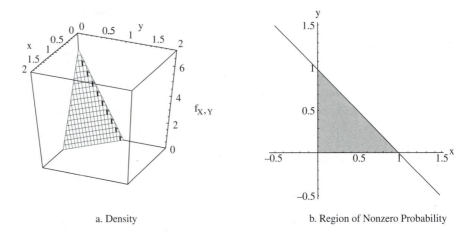

a. Density   b. Region of Nonzero Probability

FIGURE 8.2   Joint Density and Region of Nonzero Probability for Example 1.

In most of the situations that we will encounter in practice, the joint density will be similar in character to the simple example just discussed. That is, the density $f_{X,Y}$ will be defined by a single formula *on the region of nonzero probability*, but the region of nonzero probability need not be the entire plane. In such cases, the calculation of the distribution function for S can be accomplished using the following steps:

1. Identify the region of nonzero probability for the joint density of X and Y.
2. Identify the intervals of s values for which the shapes of the regions $\{(x, y) : x + y \le s\} \cap \{(x, y) : f_{X,Y}(x, y) \ne 0\}$ are similar and different.
3. For each of the intervals of s values identified in 2 with similarly shaped regions of integration, perform the appropriate integration.
4. Verify that the resulting function F_S has the properties of a distribution function, such as F_S is increasing, $\lim_{s \to \infty} F_S(s) = 1$, and so forth.

We now illustrate these steps by determining the distribution of $S = X + Y$ for two explicitly defined pairs X, Y.

EXAMPLE 1:   Suppose that X and Y are random variables with joint density

$$f_{X,Y}(x, y) = \begin{cases} 6(1 - x - y) & \text{for } x + y \le 1, x \ge 0, y \ge 0, \\ 0 & \text{otherwise.} \end{cases}$$

Determine the distribution of the sum $S = X + Y$.

Graphs of the joint density and the region of nonzero probability are given in Figure 8.2. Note that X and Y are *dependent*.

From Figure 8.2b, it is clear that the calculation of $F_S(s)$ for the sum $S = X + Y$ may be broken down into three cases: (a) $s < 0$; (b) $0 \le s < 1$; (c) $s \ge 1$. Since all of the probability is contained in the triangular region defined by $x + y \le 1, x \ge 0, y \ge 0$, it is immediately clear that $F_S(s) = 0$ for all $s < 0$ and $F_S(s) = 1$ for all $s \ge 1$. Hence, we only need consider the case $0 \le s < 1$ in detail.

Now if s is a number in the interval $[0, 1)$, then

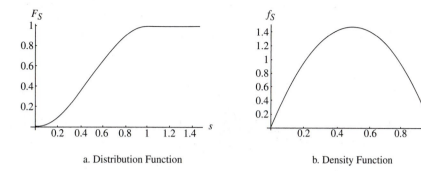

a. Distribution Function   b. Density Function

FIGURE 8.3 Distribution of the Sum Determined in Example 1.

$$\Pr(S \le s) = \iint_{x+y \le s} 6(1 - x - y)\, dx\, dy$$

$$= \int_{x=0}^{s} \int_{y=0}^{s-x} 6(1 - x - y)\, dy\, dx$$

$$= \int_{0}^{s} 6 \left\{ y - xy - \frac{y^2}{2} \right\} \Bigg|_{0}^{s-x} dx$$

$$= \int_{0}^{s} 3(s - x)(2 - s - x)\, dx$$

$$= \int_{0}^{s} 3(x^2 - 2x + s(2 - s))\, dx$$

$$= 3 \left\{ \frac{x^3}{3} - x^2 + s(2 - s)x \right\} \Bigg|_{0}^{s}$$

$$= s^2(3 - 2s).$$

Consequently,

$$F_S(s) = \begin{cases} 0 & \text{for } s < 0, \\ s^2(3 - 2s) & \text{for } 0 \le s < 1, \\ 1 & \text{for } s \ge 1; \end{cases}$$

and so

$$f_S(s) = \begin{cases} 6s(1 - s) & \text{for } 0 \le s < 1, \\ 0 & \text{otherwise.} \end{cases}$$

The graphs of F_S and f_S are shown in Figure 8.3. From Figure 8.3a, it is clear that F_S has the properties of a distribution function. Hence, there is no obvious mistake in the formula derived for F_S. ∎

EXAMPLE 2: Suppose that X and Y are random variables with joint density $f_{X,Y}$ given by

$$f_{X,Y}(x, y) = \begin{cases} 1 & \text{for } 0 \le x \le 1, 0 \le y \le 1, \\ 0 & \text{otherwise.} \end{cases}$$

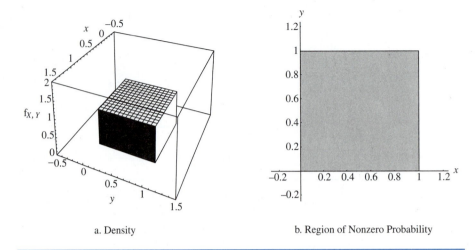

a. Density   b. Region of Nonzero Probability

FIGURE 8.4 Joint Density and Region of Nonzero Probability for Example 2.

Determine the distribution of the sum $S = X + Y$.

Graphs of the joint density and the region of nonzero probability are given in Figure 8.4. Note that X are Y are uniformly distributed[2] over the unit square.

From Figure 8.4b, it is clear that the calculation of $F_S(s)$ for the sum $S = X + Y$ in this example requires four cases: (a) $s < 0$; (b) $0 \le s < 1$; (c) $1 \le s < 2$; (d) $s \ge 2$. As in the previous example, it is clear that $F_S(s) = 0$ for all $s < 0$ and $F_S(s) = 1$ for all $s \ge 2$. Hence, we only need consider the cases $0 \le s < 1$ and $1 \le s < 2$ in detail.

Since the density $f_{X,Y}$ is constant and equal to 1, calculation of $F_S(s)$ in this example reduces to the calculation of areas in the x-y plane. Indeed, for $s \in [0, 1)$, $F_S(s)$ is simply the shaded area in Figure 8.5a, and for $s \in [1, 2)$, $F_S(s)$ is the shaded area in Figure 8.5b. Hence, for $s \in [0, 1)$, $F_S(s) = s^2/2$, and for $s \in [1, 2)$, $F_S(s) = s^2/2 - (s - 1)^2 = -(s - 2)^2/2 + 1$.

Consequently,

$$F_S(s) = \begin{cases} 0 & \text{for } s < 0, \\ \frac{s^2}{2} & \text{for } 0 \le s < 1, \\ -\frac{1}{2}(s - 2)^2 + 1 & \text{for } 1 \le s < 2, \\ 1 & \text{for } s \ge 2; \end{cases}$$

and thus,

$$f_S(s) = \begin{cases} s & \text{for } 0 \le s < 1, \\ 2 - s & \text{for } 1 \le s < 2, \\ 0 & \text{otherwise.} \end{cases}$$

---

[2] A **uniform** distribution is one whose density function is constant on the region of nonzero probability.

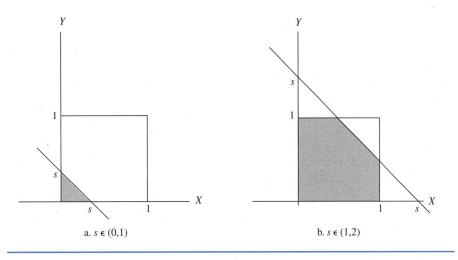

a. $s \in (0,1)$ b. $s \in (1,2)$

FIGURE 8.5   Regions of Integration for Example 2.

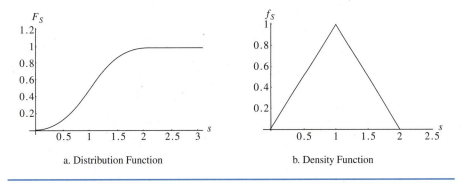

a. Distribution Function   b. Density Function

FIGURE 8.6   Distribution of the Sum Determined in Example 2.

The graphs of F_S and f_S are illustrated in Figure 8.6. From Figure 8.6a, it is clear that F_S has the properties of a distribution function. Hence, there is no obvious mistake in the formula derived for F_S. ∎

8.1.2   Using the Law of Total Probability

A second method for calculating the distribution of a sum relies on the distributional form of the law of total probability discussed in §4.1.10. To determine the distribution of S using this method, we condition on one of the variables X, Y and then perform a single variable integration using the density function for that variable.

To be precise, by conditioning on X, we have

$$F_S(s) = \Pr(S \le s)$$

$$= \int_x \Pr(S \le s | X = x) \cdot f_X(x)dx$$

$$= \int_x \Pr(X + Y \le s | X = x) \cdot f_X(x)dx$$

$$= \int_x \Pr(Y \le s - x | X = x) \cdot f_X(x)dx$$

$$= \int_x F_{Y|X=x}(s - x)f_X(x)dx.$$

On the other hand, by conditioning on Y and using similar steps, we have

$$F_S(s) = \int_y F_{X|Y=y}(s - y)f_Y(y)dy.$$

Note that the integrations in these two formulas need not be of the same difficulty. Indeed, it is quite likely that one of the formulas will be more easily evaluated than the other.

These two formulas for $F_S(s)$ also follow from the double integration formulas

$$\iint_{x+y \le s} f_{X,Y}(x, y)dx \, dy = \int_{x=-\infty}^{\infty} \int_{y=-\infty}^{s-x} f_{X,Y}(x, y)dy \, dx,$$

$$\iint_{x+y \le s} f_{X,Y}(x, y)dx \, dy = \int_{y=-\infty}^{\infty} \int_{x=-\infty}^{s-y} f_{X,Y}(x, y)dx \, dy$$

discussed in §8.1.1, using the relationship

$$f_{X,Y}(x, y) = f_{X|Y=y}(x) \cdot f_Y(y) = f_{Y|X=x}(y) \cdot f_X(x)$$

connecting joint, marginal, and conditional densities discussed in §4.1.9. Indeed,

$$\int_{x=-\infty}^{\infty} \int_{y=-\infty}^{s-x} f_{X,Y}(x, y)dy \, dx = \int_{-\infty}^{\infty} \int_{y=-\infty}^{s-x} f_{Y|X=x}(y)f_X(x)dy \, dx$$

$$= \int_{-\infty}^{\infty} F_{Y|X=x}(s - x)f_X(x)dx,$$

and similarly,

$$\int_{y=-\infty}^{\infty} \int_{x=-\infty}^{s-y} f_{X,Y}(x, y)dx \, dy = \int_{-\infty}^{\infty} F_{X|Y=y}(s - y)f_Y(y)dy.$$

Hence, one could view the formulas for $F_S(s)$ obtained using the law of total probability as simply special cases of the double integration formulas presented in §8.1.1 in which one of the variables has been integrated out.

Before proceeding further, let's illustrate the use of these formulas for F_S by considering the two examples discussed in §8.1.1 again.

EXAMPLE 1:   Suppose that X and Y are random variables with joint density

$$f_{X,Y}(x, y) = \begin{cases} 6(1 - x - y) & \text{for } x + y \le 1, \ge 0, y \ge 0, \\ 0 & \text{otherwise.} \end{cases}$$

Determine the distribution of the sum $S = X + Y$.

To calculate F_S using the law of total probability, we require formulas for the conditional distribution functions and the marginal densities. Hence, we begin by determining the marginal and conditional densities. By performing appropriate integrations, we find that the marginal and conditional densities are

$$f_X(x) = \begin{cases} 3(1-x)^2 & \text{for } 0 \le x \le 1, \\ 0 & \text{otherwise;} \end{cases}$$

$$f_Y(y) = \begin{cases} 3(1-y)^2 & \text{for } 0 \le y \le 1, \\ 0 & \text{otherwise;} \end{cases}$$

$$f_{X|Y=y}(x) = \begin{cases} \frac{2(1-x-y)}{(1-y)^2} & \text{for } 0 \le x \le 1-y, \\ 0 & \text{otherwise;} \end{cases} \quad (0 \le y \le 1)$$

$$f_{Y|X=x}(y) = \begin{cases} \frac{2(1-x-y)}{(1-x)^2} & \text{for } 0 \le y \le 1-x, \\ 0 & \text{otherwise;} \end{cases} \quad (0 \le x \le 1).$$

Hence,

$$F_{X|Y=y}(x) = \begin{cases} 0 & \text{for } x < 0, \\ 1 - \left(1 - \frac{x}{1-y}\right)^2 & \text{for } 0 \le x < 1-y, \\ 1 & \text{for } x \ge 1-y, \end{cases} \quad (0 \le y \le 1)$$

and

$$F_{Y|X=x}(y) = \begin{cases} 0 & \text{for } y < 0, \\ 1 - \left(1 - \frac{y}{1-x}\right)^2 & \text{for } 0 \le y < 1-x, \\ 1 & \text{for } y \ge 1-x. \end{cases} \quad (0 \le x \le 1)$$

Now we can calculate F_S by conditioning on X or by conditioning on Y. Let's try to calculate F_S by conditioning on X, that is, using the formula

$$F_S(s) = \int_x F_{Y|X=x}(s-x) \cdot f_X(x)dx.$$

Notice that the marginal density f_X is zero outside the interval $[0, 1]$. Hence, the integral in the formula for $F_S(s)$ need only be performed over the interval $[0, 1]$,

$$F_S(s) = \int_0^1 F_{Y|X=x}(s-x) \cdot 3(1-x)^2 dx.$$

Unfortunately, the conditional distribution function $F_{Y|X=x}$ is not defined by a single algebraic expression. As a result, the calculation of $F_S(s)$ is still fairly complicated.

From the formula for $F_{Y|X=x}$ given earlier, we have, for $x \in (0, 1)$,

$$F_{Y|X=x}(s-x) = \begin{cases} 0 & \text{for } s-x < 0, \\ 1 - \left(1 - \frac{s-x}{1-x}\right)^2 & \text{for } 0 \le s-x < 1-x, \\ 1 & \text{for } s-x \ge 1-x; \end{cases}$$

$$
= \begin{cases}
0 & \text{for } x > s, \\
1 - \left(1 - \frac{s-x}{1-x}\right)^2 & \text{for } x \leq s < 1, \\
1 & \text{for } s \geq 1.
\end{cases}
$$

Hence, it is immediately clear that for $s \geq 1$

$$
F_S(s) = \int_0^1 1 \cdot 3(1-x)^2 dx = 1.
$$

Moreover, it is clear (since the integral is only calculated over the interval $[0, 1]$) that for $s < 0$

$$
F_S(s) = \int_0^1 0 \cdot 3(1-x)^2 dx = 0.
$$

Consequently, it appears that the only values of s for which the calculation of $F_S(s)$ is nontrivial are $s \in [0, 1)$.

Now, if s is a number in the interval $[0, 1)$, we have

$$
\begin{aligned}
F_S(s) &= \int_0^1 F_{Y|X=x}(s-x) \cdot 3(1-x)^2 dx \\
&= \int_0^s \left\{ 1 - \left(1 - \frac{s-x}{1-x}\right)^2 \right\} \cdot 3(1-x)^2 dx + \int_s^1 0 \cdot 3(1-x)^2 dx \\
&= \int_0^s 3\{(1-x)^2 - (1-s)^2\} dx \\
&= \{-(1-x)^3 - 3(1-s)^2 x\}\big|_0^s \\
&= -(1-s)^3 - 3(1-s)^2 s + 1 \\
&= -(1-s)^2(1+2s) + 1 \\
&= 3s^2 - 2s^3 \\
&= s^2(3-2s).
\end{aligned}
$$

Consequently,

$$
F_S(s) = \begin{cases}
0 & \text{for } s < 0, \\
s^2(3-2s) & \text{for } 0 \leq s < 1, \\
1 & \text{for } s \geq 1.
\end{cases}
$$

This is precisely the formula for F_S that we derived in Example 1 of §8.1.1, as should be the case. Notice, however, that the complexity of the calculations here turned out to be considerably greater than before. This suggests that determining the distribution of a sum by conditioning on one of the variables and using the law of total probability may not be a very practical method. However, it would be premature to reach such a conclusion just yet. Indeed, we will see later that there are situations where use of the law of total probability is the only practical approach to take. ■

EXAMPLE 2:   Suppose that X and Y are random variables with joint density

$$f_{X,Y}(x, y) = \begin{cases} 1 & \text{for } 0 \le x \le 1, 0 \le y \le 1, \\ 0 & \text{otherwise.} \end{cases}$$

Determine the distribution of the sum $S = X + Y$.

To calculate F_S using the law of total probability, we require formulas for the conditional distribution functions and marginal densities. As it happens, the random variables X and Y are independent here. Indeed, since X and Y are uniformly distributed over the unit square, we have (by the graphical interpretation of marginal densities as projections or by brute force calculation) that

$$f_X(x) = \begin{cases} 1 & \text{for } 0 \le x \le 1, \\ 0 & \text{otherwise;} \end{cases}$$

$$f_Y(y) = \begin{cases} 1 & \text{for } 0 \le y \le 1, \\ 0 & \text{otherwise;} \end{cases}$$

and thus,

$$f_{X,Y}(x, y) = f_X(x) \cdot f_Y(y) \quad \text{for all } x, y.$$

Consequently, X and Y are independent, and so the conditional distributions are simply marginal distributions,

$$F_{X|Y=y} = F_X \quad \text{for all } y$$

and

$$F_{Y|X=x} = F_Y \quad \text{for all } x.$$

Since X and Y are independent, the calculation of $F_S(s)$ is simplified greatly. Indeed,

$$F_S(s) = \int_x F_Y(s - x) f_X(x) dx$$

and

$$F_S(s) = \int_y F_X(s - y) f_Y(y) dy.$$

Let's try calculating F_S using the first of these two expressions.

From the formula for f_Y, we have

$$F_Y(y) = \begin{cases} 0 & \text{for } y < 0, \\ y & \text{for } 0 \le y < 1, \\ 1 & \text{for } y \ge 1, \end{cases}$$

and so

$$F_Y(s - x) = \begin{cases} 0 & \text{for } x > s, \\ s - x & \text{for } s - 1 < x \le s, \\ 1 & \text{for } x \le s - 1. \end{cases}$$

Consequently,

$$F_S(s) = \int_{-\infty}^{\infty} F_Y(s - x) f_X(x) dx$$

$$= \int_{-\infty}^{s-1} 1 \cdot f_X(x) dx + \int_{s-1}^{s} (s - x) f_X(x) dx + \int_{s}^{\infty} 0 \cdot f_X(x) dx$$

$$= \int_{-\infty}^{s-1} f_X(x) dx + \int_{s-1}^{s} (s - x) f_X(x) dx.$$

Since f_X is zero outside the interval $[0, 1]$, the latter expression for $F_S(s)$ indicates that $F_S(s)$ can be determined explicitly by considering separately the cases $s < 0, 0 \leq s < 1$, $1 \leq s < 2$, and $s \geq 2$. Indeed,

$$F_S(s) = \begin{cases} 0 & \text{for } s < 0, \\ \int_0^s (s - x) dx & \text{for } 0 \leq s < 1, \\ \int_0^{s-1} dx + \int_{s-1}^1 (s - x) dx & \text{for } 1 \leq s < 2, \\ 1 & \text{for } s \geq 2. \end{cases}$$

Performing the integrations and simplifying, we obtain

$$F_S(s) = \begin{cases} 0 & \text{for } s < 0, \\ \frac{s^2}{2} & \text{for } 0 \leq s < 1, \\ 1 - \frac{1}{2}(s - 2)^2 & \text{for } 1 \leq s < 2, \\ 1 & \text{for } s \geq 2, \end{cases}$$

which is precisely the formula for $F_S(s)$ derived in Example 2 of §8.1.1.   ∎

8.1.3   Convolutions

When X and Y are independent, the formulas for the distribution function of $S = X + Y$ developed in §8.1.2 assume a particularly simple form. Indeed, in this case, $F_{Y|X=x} = F_Y$ for all x, $F_{X|Y=y} = F_X$ for all y, and the formulas for F_S given in §8.1.2 become

$$F_S(s) = \int_x F_Y(s - x) f_X(x) dx$$

and

$$F_S(s) = \int_y F_X(s - y) f_Y(y) dy.$$

The integrals in these formulas for $F_S(s)$ are examples of *convolution integrals*. A **convolution integral** is any integral of the form

$$\int f(x) g(s - x) dx$$

in which the integrand is a product of two functions whose arguments sum to a number that is independent of the variable of integration.

The notion of convolution integral allows us to define a convolution *operator* on probability distribution functions in the following way: For any two distribution functions F and G, let $F \star G$ be the function given by

$$(F \star G)(s) = \int_y F(s - y)g(y)dy,$$

where g is the density function corresponding to G. The function $F \star G$ is known as the **convolution of F and G.**

It is straightforward to show that $F \star G$ has the properties of a distribution function. Moreover, by interchanging the order of integration, it is immediate that $F \star G = G \star F$; that is, $\int_y F(s - y)g(y)dy = \int_x G(s - x)f(x)dx$. One can also show that $(F \star G) \star H = F \star (G \star H)$ for all distribution functions F, G, H. Hence, the convolution operator $\star$ is well-defined, commutative, and associative on the set of distribution functions.

The convolution operator allows us to simplify our description of the distribution of an independent sum of random variables. Indeed, if X and Y are independent and if $S = X + Y$, then the distribution function of S is simply the *convolution* of the distribution function of X and the distribution function of Y,

$$F_S = F_X \star F_Y.$$

More generally, if $X_1, \ldots, X_n$ are mutually independent and if $S = X_1 + \cdots + X_n$, then

$$
\begin{aligned}
F_S(s) &= \int F_{X_1 + \cdots + X_{n-1}}(s - x_n) f_{X_n}(x_n)dx_n \\
&= \iint F_{X_1 + \cdots + X_{n-2}}(s - x_n - x_{n-1}) f_{X_{n-1}}(x_{n-1}) f_{X_n}(x_n)dx_{n-1}dx_n \\
&\vdots \\
&= \int \cdots \int F_{X_1}(s - x_n - \cdots - x_2) f_{X_2}(x_2) \cdots f_{X_n}(x_n)dx_2 \cdots dx_n;
\end{aligned}
$$

that is,

$$
\begin{aligned}
F_S &= (F_{X_1 + \cdots + X_{n-1}}) \star F_{X_n} \\
&= (F_{X_1 + \cdots + X_{n-2}}) \star F_{X_{n-1}} \star F_{X_n} \\
&\vdots \\
&= F_{X_1} \star \cdots \star F_{X_n}.
\end{aligned}
$$

When $X_1, \ldots, X_n$ are independent and identically distributed with $X_j \sim X$, the distribution function of S is called the **n-fold convolution of X** and is denoted by $F_X^{\star n}$. Note that, with this notation, $F_X^{\star 1} = F_X$. Moreover, if we adopt the convention that the empty sum is zero, then $F_X^{\star 0}$ is the unit step function at the origin,

$$
\begin{aligned}
F_X^{\star 0}(x) &= \Pr(0 \le x) \\
&= \begin{cases} 1 & \text{if } x \ge 0, \\ 0 & \text{if } x < 0. \end{cases}
\end{aligned}
$$

Convolutions of this type arise frequently in insurance applications.

8.2 Distributions of Products and Quotients

In the previous section, we discussed three methods for determining the distribution of a sum: (a) use the joint density function; (b) use the law of total probability; and (c) when $X_1, \ldots, X_n$ are independent, use convolutions or moment generating functions.

The methods based on convolutions and moment generating functions are clearly not applicable when considering products and quotients. However, the other two methods can be adapted without too much difficulty. In practice, the distributions of products and quotients are generally determined by interpreting products as transformations of the type $h(X_1, \ldots, X_n) = X_1 \cdots X_n$ and quotients as transformations of the type $h(X, Y) = X/Y$ and by appealing to the formulas for the distributions of transformed variables developed in §7.1. We illustrate this procedure in the following example.

EXAMPLE 1: Suppose that X and Y are random variables with joint density

$$f_{X,Y}(x, y) = \begin{cases} 1 & \text{for } 0 \le x \le 1, 0 \le y \le 1, \\ 0 & \text{otherwise.} \end{cases}$$

Determine the distribution of XY and X/Y.

Put $W = XY$ and $V = X/Y$. Note that the range of possible values of W is $[0, 1]$ and the range of possible values of V is $(0, \infty)$. Since X and Y are uniformly distributed on the unit square, determining the distribution functions of W and V reduces to calculating areas in the x-y plane. Indeed,

$$F_W(w) = \Pr(W \le w) = \iint_{\substack{xy \le w \\ 0 \le x \le 1 \\ 0 \le y \le 1}} dx\, dy$$

and

$$F_V(v) = \Pr(V \le v) = \iint_{\substack{x/y \le v \\ 0 \le x \le 1 \\ 0 \le y \le 1}} dx\, dy.$$

Let's consider W first. From the preceding observations, $\Pr(W \le w)$ is simply the area of the region defined by $xy \le w, 0 \le x \le 1, 0 \le y \le 1$. This is the region in the x-y plane lying under the hyperbola $y = w/x$ and inside the square $(x, y) \in [0, 1] \times [0, 1]$. From a graph, it is clear that this region consists of the rectangle $(x, y) \in [0, w] \times [0, 1]$ and the region under the hyperbola $y = w/x$ from $x = w$ to $x = 1$. Consequently,

$$\Pr(W \le w) = w + \int_w^1 \frac{w}{x} dx$$

$$= w + w \log x \Big|_w^1$$

$$= w - w \log w.$$

Hence, the distribution function of $W = XY$ is given by

$$F_W(w) = \begin{cases} 0 & \text{for } w \le 0, \\ w(1 - \log w) & \text{for } 0 < w < 1, \\ 1 & \text{for } w \ge 1; \end{cases}$$

and thus, the density function of $W = XY$ is

$$f_W(w) = \begin{cases} \log w^{-1} & \text{for } 0 < w < 1, \\ 0 & \text{otherwise.} \end{cases}$$

Note that F_W has the properties of a distribution function. Hence, there are no obvious mistakes in our calculation.

Now consider $V = X/Y$. From our earlier comments, $\Pr(V \leq v)$ is simply the area inside the unit square lying above the line $y = x/v$. This region is triangular when $v \leq 1$ and can be decomposed into a right triangle and a rectangle when $v > 1$. Hence, the area can be easily determined using simple geometry. Performing these calculations, we find that

$$\Pr(V \leq v) = \begin{cases} \frac{v}{2} & \text{for } 0 < v < 1, \\ 1 - \frac{1}{2} \cdot \frac{1}{v} & \text{for } v \geq 1. \end{cases}$$

Hence, the distribution and density functions for $V = X/Y$ are given by

$$F_V(v) = \begin{cases} 0 & \text{for } v \leq 0, \\ \frac{v}{2} & \text{for } 0 < v \leq 1, \\ 1 - \frac{1}{2} \cdot \frac{1}{v} & \text{for } v \geq 1, \end{cases}$$

and

$$f_V(v) = \begin{cases} 0 & \text{for } v \leq 0, \\ \frac{1}{2} & \text{for } 0 < v < 1, \\ \frac{1}{2} v^{-2} & \text{for } v \geq 1. \end{cases}$$

Since F_V has the properties of a distribution function, there are no obvious mistakes in our calculation. ∎

In the exercises, the reader will have the opportunity to determine the distributions of some special products and quotients that arise frequently in engineering.

8.3 Expectations of Sums and Products

In this section, we derive the formulas for the expectation of a sum or product stated without proof in §4.2.1. We also give a proof of the Cauchy-Schwarz inequality and discuss the properties of covariance and correlation in detail.

8.3.1 Formulas for the Expectation of a Sum or Product

Recall that for any random variables X and Y, we claimed that

$$E[X + Y] = E[X] + E[Y]$$

and that for any *independent* X and Y, we claimed that

$$E[XY] = E[X] \cdot E[Y].$$

These formulas are easy to prove using the general formula for the expectation of a function of a random variable derived in §7.2.

Indeed, for any X and Y, we have

$$E[X+Y] = \int\int (x+y) f_{X,Y}(x,y) dx\, dy$$

$$= \int\int x f_{X,Y}(x,y) dy\, dx + \int\int y f_{X,Y}(x,y) dx\, dy$$

$$= \int x f_X(x) dx + \int y f_Y(y) dy$$

$$= E[X] + E[Y],$$

and for any X and Y that are independent, we have

$$E[XY] = \int\int xy f_{X,Y}(x,y) dx\, dy$$

$$= \int\int xy f_X(x) f_Y(y) dx\, dy$$

$$= \left(\int x f_X(x) dx\right)\left(\int y f_Y(y) dy\right)$$

$$= E[X] \cdot E[Y],$$

since $f_{X,Y}(x,y) = f_X(x) f_Y(y)$ for any independent random variables X and Y.

An important special case of the latter formula is the formula for the moment generating function of an independent sum. Indeed, if X and Y are independent, then e^{tX} and e^{tY} are independent for all t, and so

$$M_{X+Y}(t) = E[e^{t(X+Y)}]$$

$$= E[e^{tX} e^{tY}]$$

$$= E[e^{tX}] E[e^{tY}]$$

$$= M_X(t) M_Y(t).$$

Hence, the moment generating function of an independent sum is the product of the moment generating functions of the components.

Note that the formula $E[XY] = E[X]E[Y]$ need not hold if X and Y are dependent. For a concrete example of this, see the discussion in §4.2.1.

8.3.2   The Cauchy-Schwarz Inequality

The Cauchy-Schwarz inequality asserts that

$$E[XY]^2 \le E[X^2] E[Y^2]$$

for all X and Y, provided that the expressions $E[XY]$, $E[X^2]$, $E[Y^2]$ make sense. We will derive this important inequality by considering the quantities $E[(aX + bY)^2]$, $E[(aX - bY)^2]$ and by using the fact that expectation is a linear operator (i.e., $E[aX + bY] = aE[X] + bE[Y]$).

Note that

$$E[(aX + bY)^2] = a^2 E[X^2] + 2ab E[XY] + b^2 E[Y^2]$$

and

$$E[(aX - bY)^2] = a^2 E[X^2] - 2ab E[XY] + b^2 E[Y^2].$$

Note further that $E[(aX + bY)^2] \geq 0$ and $E[(aX - bY)^2] \geq 0$ for all a, b because $(aX + bY)^2$ and $(aX - bY)^2$ are always nonnegative. Hence,

$$-a^2 E[X^2] - b^2 E[Y^2] \leq 2ab E[XY] \leq a^2 E[X^2] + b^2 E[Y^2]$$

for all a, b.

Consequently, putting $a = \sqrt{E[Y^2]}$ and $b = \sqrt{E[X^2]}$ in the latter inequality and dividing by $2\sqrt{E[X^2]E[Y^2]}$, we obtain

$$-\sqrt{E[X^2]E[Y^2]} \leq E[XY] \leq \sqrt{E[X^2]E[Y^2]};$$

that is,

$$E[XY]^2 \leq E[X^2]E[Y^2]$$

as claimed.

The Cauchy-Schwarz inequality is a handy result for estimating $E[XY]$ when X and Y are dependent and the exact computation of $E[XY]$ is difficult. In §4.3.1, we used the Cauchy-Schwarz inequality to show that the graph of the cumulant generating function for a random variable is always concave.

8.3.3   Covariance and Correlation

In §4.2.2, we introduced the concept of covariance for a pair of random variables in connection with the variance of a sum. The covariance is actually an important concept in its own right. Indeed, it can be interpreted as a measure of association between random variables. In this subsection, we discuss the properties of covariance in greater detail and introduce the concept of *correlation,* which is simply a scaled covariance. We also give an inequality for the standard deviation of a sum which is important in the risk–reward analysis of investment portfolios.

Covariance as a Measure of Association

Recall that the **covariance** between two random variables X, Y is the number $Cov(X, Y)$ given by

$$Cov(X, Y) = E[(X - \mu_X)(Y - \mu_Y)].$$

This quantity generalizes the concept of variance and has the following algebraic properties, which follow directly from the definition of covariance:

Properties of Covariance

1. $Cov(X, Y) = Cov(Y, X)$.
2. $Cov(X, X) = Var(X)$.
3. $Cov(aX + b, cY + d) = ac\, Cov(X, Y)$ for all constants a, b, c, d.
4. $Cov(\sum X_i, \sum Y_j) = \sum_i \sum_j Cov(X_i, Y_j)$.
5. $Cov(X, Y) = E[XY] - E[X]E[Y]$.

The covariance also has an important interpretation as a measure of association between X and Y. To see this interpretation, consider the random variable $Q = (X - \mu_X)(Y - \mu_Y)$.

Notice that for any given pair of values (x, y) for (X, Y), the value assumed by Q will be positive if and only if x and y are either both above or both below the respective averages μ_X, μ_Y, and the value assumed by Q will be negative if and only if one of x, y is above its average and the other is below its average,

$$Q > 0 \iff (X > \mu_X \text{ and } Y > \mu_Y) \text{ or } (X < \mu_X \text{ and } Y < \mu_Y),$$
$$Q < 0 \iff (X > \mu_X \text{ and } Y < \mu_Y) \text{ or } (X < \mu_X \text{ and } Y > \mu_Y).$$

Hence, if the distribution of probability for X and Y is such that above average values of X and Y are likely to occur together and below average values of X and Y are likely to occur together, then the average value of Q should be positive (i.e., $E[Q] > 0$). On the other hand, if the distribution of probability for X and Y is such that above average values of X and below average values of Y or below average values of X and above average values of Y are likely to occur together, then the average value of Q should be negative (i.e., $E[Q] < 0$). Since $E[Q]$ is simply the covariance of X and Y defined earlier, we see that the sign of the covariance is a measure of the tendency for above or below average values of X to be associated with above or below average values of Y. Hence, covariance has the interpretation of being a measure of association as claimed.

Figure 8.7 illustrates the regions of nonzero probability for three different uniform distributions. Recall that a uniform distribution is one whose density function is constant on the region of nonzero probability. Note that covariance does not give any information about the relative locations of *particular* values x, y of X, Y; it only describes the association between X and Y *on average*. Indeed, it is quite possible (Figure 8.7a) for a pair of values (x, y) with $x > \mu_X$ and $y < \mu_Y$ to have nonzero probability, but for $Cov(X, Y) > 0$. Similarly, it is quite possible (Figure 8.7b) for a pair of values (x, y) with $x > \mu_X$ and $y > \mu_Y$ to have nonzero probability, but for $Cov(X, Y) < 0$.

The interpretation of the statement $Cov(X, Y) = 0$ is that there is no tendency on average for above or below average values of X to be associated with above or below average values of Y. This does not mean that X and Y are independent. Indeed, for the distribution in Figure 8.7c, it is clear that X and Y are dependent, even though $Cov(X, Y) = 0$. On the other hand, if X and Y are independent, then the covariance between X and Y must be zero. The reason is that $E[XY] = E[X]E[Y]$ for all independent random variables X, Y (see §8.3.1).

Correlation

As we have just seen, the *sign* of the covariance provides an indication of the tendency for above or below average values of one random variable to be associated with above or below average values of another random variable. However, it does not provide a measure of the *degree* of the association. Indeed, if the scales of X and Y are changed, then the magnitude of the covariance will generally also change, even though the degree of association between X and Y may not:

$$Cov(kX, kY) = k^2 Cov(X, Y)$$
$$\neq Cov(X, Y) \quad \text{unless } k^2 = 1.$$

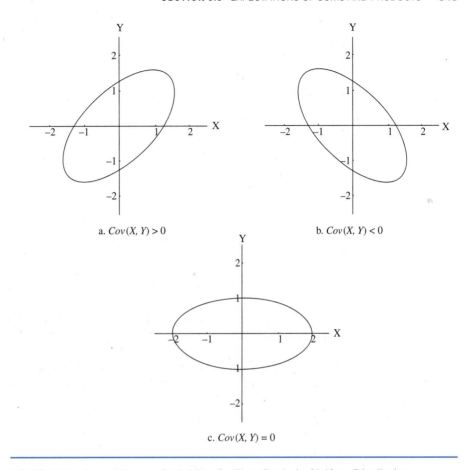

a. $Cov(X, Y) > 0$

b. $Cov(X, Y) < 0$

c. $Cov(X, Y) = 0$

FIGURE 8.7 Regions of Nonzero Probability for Three Particular Uniform Distributions

The *correlation* is a statistic that provides a measure of both the sign and the degree of association between two random variables.

Definition of Correlation

For any random variables X, Y, the **correlation** between X and Y is the number $\rho_{X,Y}$ defined by

$$\rho_{X,Y} = \frac{Cov(X, Y)}{\sigma_X \sigma_Y}.$$

The correlation is a measure of association with the following properties:

Properties of Correlation

1. $-1 \leq \rho_{X,Y} \leq 1$.
2. If $Y = aX + b$, then

$$\rho_{X,Y} = \begin{cases} 1 & \text{if } a > 0, \\ -1 & \text{if } a < 0. \end{cases}$$

3.

$$\rho_{aX+b,cY+d} = \frac{ac}{|ac|} \rho_{X,Y}.$$

Property 1, which is particularly important, follows from the Cauchy-Schwarz inequality. Indeed, applying the Cauchy-Schwarz inequality to $X - \mu_X$ and $Y - \mu_Y$, we have

$$E[(X - \mu_X)(Y - \mu_Y)]^2 \leq E[(X - \mu_X)^2] \cdot E[(Y - \mu_Y)^2]$$

from which $-1 \leq \rho_{X,Y} \leq 1$ follows immediately. Properties 2 and 3 are also straightforward to demonstrate.

An Inequality for the Standard Deviation of a Sum

Using the formula for the variance of a sum and the fact that $|\rho_{X,Y}| \leq 1$, we can derive an inequality for the standard deviation of a sum that turns out to have significant implications for risk–reward analysis.

Recall that

$$Var(aX + bY) = a^2 Var(X) + 2ab\, Cov(X, Y) + b^2 Var(Y).$$

Using the substitution

$$Cov(X, Y) = \rho_{X,Y}\sigma_X\sigma_Y$$

and completing squares, we can obtain the following two identities:

$$Var(aX + bY) = (a\sigma_X + b\sigma_Y)^2 + 2ab(\rho_{X,Y} - 1)\sigma_X\sigma_Y,$$
$$Var(aX + bY) = (a\sigma_X - b\sigma_Y)^2 + 2ab(\rho_{X,Y} + 1)\sigma_X\sigma_Y.$$

Suppose that $ab \geq 0$. Then, since $-1 \leq \rho_{X,Y} \leq 1$, we have from the two identities that

$$(a\sigma_X - b\sigma_Y)^2 \leq Var(aX + bY) \leq (a\sigma_X + b\sigma_Y)^2.$$

Consequently,

$$|a\sigma_X - b\sigma_Y| \leq \sigma_{aX+bY} \leq |a\sigma_X + b\sigma_Y|.$$

On the other hand, if $ab \leq 0$, then using a similar argument, we have

$$|a\sigma_X + b\sigma_Y| \leq \sigma_{aX+bY} \leq |a\sigma_X - b\sigma_Y|.$$

Hence, in any case, we have

$$\min\{|a\sigma_X + b\sigma_Y|, |a\sigma_X - b\sigma_Y|\} \leq \sigma_{aX+bY} \leq \max\{|a\sigma_X + b\sigma_Y|, |a\sigma_X - b\sigma_Y|\}.$$

This inequality has important consequences in risk–reward analysis, which we now proceed to describe. Hence, suppose that X and Y represent the returns on two different investments and a, b represent the weights of the respective investments in a portfolio consisting only of these two investments. Then $aX + bY$ is the return on the portfolio and $a + b = 1$. If a and b are both nonnegative (i.e., no short selling allowed), then the foregoing inequality for σ_{aX+bY} becomes

$$|a\sigma_X - b\sigma_Y| \leq \sigma_{aX+bY} \leq a\sigma_X + b\sigma_Y.$$

Further,

$$\mu_{aX+bY} = a\mu_X + b\mu_Y,$$

regardless of the values of a and b. Hence,

$$0 \leq \sigma_{\alpha X+(1-\alpha)Y} \leq \alpha\sigma_X + (1-\alpha)\sigma_Y$$

and

$$\mu_{\alpha X+(1-\alpha)Y} = \alpha\mu_X + (1-\alpha)\mu_Y$$

for all $\alpha \in (0, 1)$.

Suppose that the returns X and Y are such that

$$\mu_X \leq \mu_Y \quad \text{and} \quad \sigma_X \leq \sigma_Y.$$

That is, the investment with return Y has higher expected return but greater risk. Then as α decreases from 1 to 0 (i.e., as more weight is given to the riskier investment), the expected return on the portfolio increases from μ_X to μ_Y. At the same time, the riskiness of the portfolio as measured by its standard deviation changes from σ_X to σ_Y. However, since

$$\sigma_{\alpha X+(1-\alpha)Y} \leq \alpha\sigma_X + (1-\alpha)\sigma_Y,$$

the riskiness of the portfolio need not increase in tandem with expected return. Indeed, it may be possible (and generally is) for $\sigma_{\alpha X+(1-\alpha)Y}$ to decrease with decreasing α when α is in particular subintervals of $[0, 1]$. Consequently, by adding a high-risk investment with high expected return to a low-risk investment with low expected return, it may be possible to increase return and decrease risk simultaneously.

This observation is one of the principal justifications for diversification of investment portfolios. In Chapter 10, we will discuss the Nobel Prize winning portfolio selection model of Harry Markowitz, which provides a systematic procedure for selecting optimal investment portfolios.

8.4 The Law of Large Numbers

In this section, we discuss one of the most important theorems in probability: the law of large numbers. As the reader will soon appreciate, this theorem is really just a mathematical formulation of the insurance principle that was introduced in Chapter 1.

In keeping with the philosophy of the earlier parts of this book, we develop the law of large numbers through an extended concrete example.

8.4.1 Motivating Example: Premium Determination in Insurance

Consider a group of n insurance policies. Let X_j be the amount of claim expense paid by the insurer to the jth policyholder during a particular future time period and let $S = X_1 + \cdots + X_n$. Then S is the total payments made by the insurer on this group of policies during the period of insurance.

Suppose that the X_j are independent and identically distributed and the insurer charges each policyholder the same premium at the beginning of the insured period. Let P be the premium charged per policy. Clearly, the insurer would like the total premium collected to exceed (or at worst equal) the total payments made with high probability. That is, the insurer would like $\Pr(nP \geq S)$ to be close to 1.

Our goal in this extended example is to analyze the behavior of P with respect to n under varying assumptions about $\Pr(nP \geq S)$.

CONCRETE PROBLEM: Suppose that each X_j has an exponential distribution with mean[3] 5 and the premium P is set in such a way that $\Pr(nP \geq S) = .95$. Determine P as a function of n. What happens to the premium P as the size of the insured group becomes large?

From the given information, $X_j \sim \text{Exponential}(1/5)$ for all j. Moreover, the X_j are mutually independent, by assumption. Hence, the mean and variance of the total payment are, respectively,

$$E[S] = E[X_1] + \cdots + E[X_n] = 5n,$$
$$Var(S) = Var(X_1) + \cdots + Var(X_n) = 25n.$$

Suppose that n is sufficiently large that the distribution of S can be approximated by a normal distribution. Then

$$\Pr(S \leq nP) = \Pr\left(\frac{S - \mu_S}{\sigma_S} \leq \frac{nP - \mu_S}{\sigma_S}\right)$$

$$= \Pr\left(\frac{S - \mu_S}{\sigma_S} \leq \left(\frac{P}{5} - 1\right)\sqrt{n}\right)$$

$$\approx \Pr\left(Z \leq \left(\frac{P}{5} - 1\right)\sqrt{n}\right)$$

$$= \Phi\left(\left(\frac{P}{5} - 1\right)\sqrt{n}\right),$$

where $\Phi(\cdot)$ is the distribution function for a standard normal random variable. Consequently, the requirement $\Pr(nP \geq S) = .95$ is approximately equivalent to the requirement that $\Phi((\frac{P}{5} - 1)\sqrt{n}) = .95$.

Now $\Phi(1.645) = .95$. (See tables in Appendix E or use a statistical calculator.) Moreover, $\Phi(\cdot)$ is monotonically increasing. Hence, the requirement that $\Pr(nP \geq S) = .95$ is approximately equivalent to the requirement that

---

[3] For computational simplicity, we have chosen small numbers in this illustration. If one assumes that payments are measured in thousands of dollars, then the numbers become more realistic.

TABLE 8.1   Values of the Premium P for Selected n

n	P
100	5.8225000
250	5.5201947
500	5.3678332
1000	5.2600973
2000	5.1839166
5000	5.1163191

$$\left(\frac{P}{5} - 1\right)\sqrt{n} = 1.645.$$

Solving this latter equation for P, we obtain

$$P = 5 + \frac{8.225}{\sqrt{n}}.$$

Values of P for particular n are given in Table 8.1.

From Table 8.1 and the formula for P just derived, it is clear that $P \to 5$ as $n \to \infty$, although the rate of convergence is fairly slow. Notice that the value to which P converges is actually the mean of each X_j,

$$P \to E[X_j] \qquad \text{as } n \to \infty.$$

This is no accident. Indeed, we will see in §8.4.2 that this must occur whether each X_j is exponentially distributed or not.[4] ■

COMMENT: In the preceding discussion, we assumed that the distribution of S can be approximated by a normal distribution. However, the use of a normal approximation was not critical to our conclusion. Indeed, we would have reached precisely the same conclusion if we had determined P using the exact distribution of S. To see that this is so, let's consider the distribution of S and the requirement that $\Pr(nP \geq S) = .95$ once again.

Since $X_j \sim \text{Exponential}(1/5)$ for all j and since the X_j are mutually independent, we know from our discussion of the gamma distribution in §6.1.2 that $S \sim \text{Gamma}(n, 1/5)$ and so $S/n \sim \text{Gamma}(n, n/5)$. Consequently, the requirement that $\Pr(nP \geq S) = .95$ is simply the requirement that

$$\Pr(\text{Gamma}(n, n/5) \leq P) = .95.$$

Hence, P is the 95th percentile of a $\text{Gamma}(n, n/5)$ distribution.

From graphs of the densities of $\text{Gamma}(n, n/5)$ for various n, it is clear that the distribution of $\text{Gamma}(n, n/5)$ becomes more concentrated around the value 5 as n increases (Figure 8.8). Hence, $P \to 5$ as $n \to \infty$, which is the conclusion we reached before.

---

[4] The only requirement is that the X_j be independent and identically distributed and that $E[X_j]$ be finite.

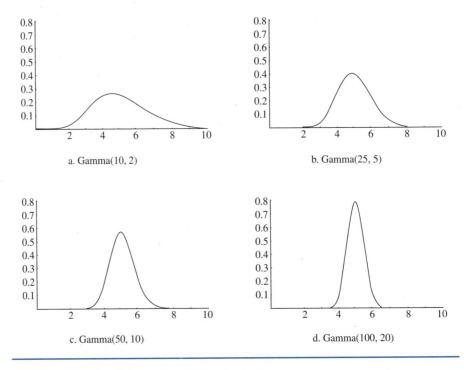

FIGURE 8.8 Behavior of Gamma$(n, n/5)$ density as n Increases

Note that the quantity S/n, which appears in the preceding discussion, represents the (arithmetic) average of the payments that will actually be made over the insured period. This average is uncertain at the beginning of the insured period; however, its expected value is (not surprisingly) the same as the expected value of each X_j. That is, $E[S/n] = E[X_j]$, assuming that the X_j are identically distributed. From the graphs in Figure 8.8, it appears that the distribution of S/n becomes more concentrated around the mean value $E[X_j]$ as $n \to \infty$. In fact, this will generally be the case for the (random) average $(X_1 + \cdots + X_n)/n$ when the X_j are independent and have the same distribution, as we will show in §8.4.2. ∎

It should be clear from the graphs in Figure 8.8 that the premium P will approach $E[X_j]$ under any requirement of the type $\Pr(nP \geq S) = \alpha$, not just the requirement with $\alpha = .95$ which we have been discussing. We conclude our discussion of this extended example by demonstrating this fact algebraically under the assumption that the distribution of S is approximately normal.

Hence, suppose that the X_j are independent and identically distributed with $X_j \sim X$ and the premium P is to be set such that $\Pr(nP \geq S) = \alpha$, where $\frac{1}{2} < \alpha < 1$. Suppose further that n is sufficiently large that the distribution of S can be approximated by a normal distribution. Then

$$E[S] = E[X_1] + \cdots + E[X_n] = n \cdot E[X],$$
$$Var(S) = Var(X_1) + \cdots + Var(X_n) = n \cdot Var(X)$$

and so

$$\Pr(nP \geq S) = \Pr(S \leq nP)$$

$$= \Pr\left(\frac{S - \mu_S}{\sigma_S} \leq \frac{P - E[X]}{SD(X)}\sqrt{n}\right)$$

$$\approx \Phi\left(\frac{P - E[X]}{SD(X)}\sqrt{n}\right),$$

where $\Phi(\cdot)$ is the distribution function for a standard normal random variable. Hence, the requirement $\Pr(nP \geq S) = \alpha$ is equivalent to the requirement that $\Phi((P - E[X])\sqrt{n}/SD(X)) = \alpha$.

Let z_α be the 100αth percentile of the standard normal distribution; that is, $\Phi(z_\alpha) = \alpha$. Then the requirement that $\Pr(nP \geq S) = \alpha$ is equivalent to the requirement that

$$\frac{P - E[X]}{SD(X)}\sqrt{n} = z_\alpha.$$

Solving for P, we obtain

$$P = E[X] + \frac{z_\alpha SD(X)}{\sqrt{n}}.$$

In particular, $P \rightarrow E[X]$ as $n \rightarrow \infty$.

8.4.2 Statement and Proof of the Law

We are now in a position to give a formal statement and proof of the law of large numbers.[5]

The Law of Large Numbers

Let $X_1, X_2, \ldots$ be a sequence of independent, identically distributed random variables with $X_j \sim X$ and suppose that $E[X]$ is finite. Put $\mu = E[X]$ and for each n let $\overline{X}_n$ be the random variable given by $\overline{X}_n = (X_1 + \cdots + X_n)/n$. Then for every $\varepsilon > 0$,

$$\Pr(|\overline{X}_n - \mu| \geq \varepsilon) \rightarrow 0 \quad \text{as } n \rightarrow \infty.$$

Note that $\overline{X}_n$ is the arithmetic average of the random variables $X_1, \ldots, X_n$. The mean and variance of $\overline{X}_n$ (when the X_j are independent and identically distributed) are, respectively,

---

[5] The statement of the law of large numbers that we give here is a weak form of the law. A stronger statement can actually be made. However, the weak form of the law is sufficient for most applications.

$$E[\overline{X}_n] = E[(X_1 + \cdots + X_n)/n] = \frac{1}{n}E[X_1 + \cdots + X_n]$$
$$= \frac{1}{n}\{E[X_1] + \cdots + E[X_n]\}$$
$$= \frac{1}{n} \cdot n\mu$$
$$= \mu$$

and

$$Var(\overline{X}_n) = Var((X_1 + \cdots + X_n)/n) = \frac{1}{n^2}Var(X_1 + \cdots + X_n)$$
$$= \frac{1}{n^2}\{Var(X_1) + \cdots + Var(X_n)\}$$
$$= \frac{\sigma^2}{n},$$

where $\mu = E[X_j]$ and $\sigma^2 = Var(X_j)$. Hence, $Var(\overline{X}_n) \to 0$ as $n \to \infty$, suggesting that the distribution of $\overline{X}_n$ becomes more like a point mass at $x = \mu$. This phenomenon was illustrated in Figure 8.8 for random variables X_j with $X_j \sim$ Exponential$(1/5)$.

The statement of the law of large numbers just given is actually a little stronger than the assertion that $Var(\overline{X}_n) \to 0$ as $n \to \infty$. Indeed, the law of large numbers holds even when the X_j do not have finite variance, whereas the assertion $Var(\overline{X}_n) \to 0$ has no meaning when $Var(X_j)$ is not defined. However, both statements encapsulate the intuitive observation that the distribution of $\overline{X}_n$ becomes concentrated around the mean value μ as $n \to \infty$.

The proof of the law of large numbers is fairly straightforward, particularly if one assumes that the X_j have finite variance, as we will do.

Proof of the Law of Large Numbers

Let $\mu = E[X_j]$, let $\sigma^2 = Var(X_j)$, and let ε be an arbitrary positive number. Recall that Chebyshev's inequality (§4.2.2) asserts that if X is a random variable with finite mean and finite variance, then for any $k > 0$,

$$\Pr(|X - \mu_X| \ge k) \le \frac{\sigma_X^2}{k^2}.$$

Applying this inequality with $X = \overline{X}_n$, $k = \varepsilon$, we have

$$\Pr(|\overline{X}_n - \mu| \ge \varepsilon) = \frac{1}{\varepsilon^2} \cdot \frac{\sigma^2}{n}.$$

Hence,

$$\Pr(|\overline{X}_n - \mu| \ge \varepsilon) = \frac{1}{\varepsilon^2} \cdot \frac{\sigma^2}{n} \to 0 \quad \text{as } n \to \infty.$$

Since ε was assumed to be arbitrary, the statement of the law of large numbers is proved.

■

8.4.3 Some Misconceptions Surrounding the Law of Large Numbers

We conclude this section by considering two common misconceptions associated with the law of large numbers that arise in insurance and investment contexts. The first misconception is that risk sharing eliminates the possibility of catastrophic loss. The second is that accumulated winnings from a sequence of independent and identical bets become more certain the longer we play. Let's consider each of these misconceptions in turn.

Risk Sharing in Insurance

As we noted in Chapter 1, insurance is based on the premise that individuals faced with large and unpredictable losses can reduce the financial effects of such losses by forming a group and sharing the losses incurred by the group as a whole. This principle can be justified mathematically using the law of large numbers.

Indeed, if $X_1, \ldots, X_n$ are the respective losses faced by n different individuals over the coming period and if each individual agrees to pay $(X_1 + \cdots + X_n)/n$ toward the aggregate loss $X_1 + \cdots + X_n$, then according to the law of large numbers, the amount that each person will wind up paying becomes more predictable as the size of the group increases, provided that the X_j are independent and have the same distribution. That is, the distribution of $(X_1 + \cdots + X_n)/n$ becomes more concentrated around μ, where μ is the loss that each individual expects to incur in the absence of insurance.

However, this does not mean that a person in a risk-sharing arrangement will never have to pay an unexpectedly large amount. Indeed, any loss that an individual faces before entering into a risk-sharing arrangement is still possible after entering into such an arrangement.[6] However, the probability that an individual will have to pay an unexpectedly large amount is greatly reduced when the individual participates in a risk-sharing arrangement. The reason is that the distribution of $\overline{X}_n$ becomes concentrated around μ as $n \to \infty$.

Consequently, we see that insurance does not eliminate the *possibility* of having to pay an unexpectedly large amount; rather, it reduces the *probability* of having to do so.

Accumulated Winnings

Another common misconception associated with the law of large numbers arises in gambling and investment problems. Many people mistakenly believe that accumulated winnings from a sequence of independent and identical bets become more certain the greater the number of bets. However, this is not so.

To see why, consider a sequence of independent, identically distributed random variables $X_1, X_2, \ldots$ with $X_j \sim X$ and suppose that $E[X]$ and $Var(X)$ are finite. Put $S_n = X_1 + \cdots + X_n$. Then

$$Var(S_n) = Var(X_1 + \cdots + X_n) = nVar(X) \to \infty \qquad \text{as } n \to \infty.$$

Hence, if X_j represents our winnings on the jth repetition, we see that the uncertainty in our accumulated winnings—that is, $Var(S_n)$—actually increases with the number of repetitions.

---

[6] Consider, for example, the situation where everyone in the group incurs a loss of exactly the same size.

For many people, this result seems counterintuitive, and it seems to contradict their understanding of the law of large numbers. The mistake that these people make is that they think the law of large numbers is about *sums* when in reality it is about *averages*. Indeed, the law of large numbers asserts that the average $(X_1 + \cdots + X_n)/n$ becomes more certain as $n \to \infty$, but it makes no such claim about the sum $X_1 + \cdots + X_n$.

In the exercises, you will have the opportunity to explore some other fallacies associated with the law of large numbers.

8.5   The Central Limit Theorem

In this section, we give a formal statement and proof of the central limit theorem. This theorem is important because it allows us to approximate the distribution of any sum of independent, identically distributed random variables using a normal distribution and, consequently, justifies the widespread use of the normal distribution in probability modeling. We have already given several examples of normal approximations in this book, and we have seen graphical illustrations of the theorem in §6.3.1. Hence, our current discussion of the central limit theorem focuses on a formal statement and proof.

Central Limit Theorem

Let $X_1, X_2, \ldots$ be a sequence of independent and identically distributed random variables with $X_j \sim X$ and let $\mu = E[X]$ and $\sigma^2 = Var(X)$. For each n, let $S_n = X_1 + \cdots + X_n$. Then the distribution of $(S_n - n\mu)/\sigma \sqrt{n}$ tends to a standard normal distribution as $n \to \infty$. That is, for each a,

$$\Pr\left(\frac{S_n - n\mu}{\sigma \sqrt{n}} \le a\right) \to \Phi(a) \quad \text{as } n \to \infty,$$

where $\Phi(\cdot)$ is the distribution function of a standard normal random variable.

The proof of the central limit theorem that we give here relies on the fact that moment and cumulant generating functions uniquely characterize probability distributions.[7]

Proof of the Central Limit Theorem

Let $X_1, X_2, \ldots$ be a sequence of independent and identically distributed random variables with $X_j \sim X$. By replacing X_j with $X_j - \mu$ if necessary, we can assume that each X_j has mean 0. Hence, suppose that $E[X] = 0$ and $Var(X) = \sigma^2$.

Consider the quantity

$$Q = \frac{X_1 + \cdots + X_n}{\sigma \sqrt{n}}.$$

---

[7] This proof requires the extra, and unnecessary, assumption that the moment generating function of X is finite in a neighborhood of the origin.

Since $M_{aY}(t) = M_Y(at)$ for any random variable Y and any real number a, and since $M_{Y_1+\cdots+Y_n}(t) = M_{Y_1}(t) \cdots M_{Y_n}(t)$ whenever $Y_1, \ldots, Y_n$ are independent, we see that

$$M_Q(t) = M_{X_1}\left(\frac{t}{\sigma\sqrt{n}}\right) \cdots M_{X_n}\left(\frac{t}{\sigma\sqrt{n}}\right) = M_X\left(\frac{t}{\sigma\sqrt{n}}\right)^n.$$

Consequently,

$$\psi_Q(t) = \frac{\psi_X\left(\dfrac{t}{\sigma\sqrt{n}}\right)}{\dfrac{1}{n}}.$$

Applying l'Hôpital's rule to the latter fraction, we find that

$$\lim_{n\to\infty} \psi_Q(t) = \lim_{n\to\infty} \frac{\psi_X\left(\dfrac{t}{\sigma\sqrt{n}}\right)}{\dfrac{1}{n}}$$

$$= \lim_{n\to\infty} \frac{\left(-\dfrac{1}{2}n^{-3/2}\dfrac{t}{\sigma}\right)\psi_X'\left(\dfrac{t}{\sigma\sqrt{n}}\right)}{-\dfrac{1}{n^2}}$$

$$= \lim_{n\to\infty} \frac{\dfrac{t}{2\sigma}\psi_X'\left(\dfrac{t}{\sigma\sqrt{n}}\right)}{n^{-1/2}}.$$

Applying l'Hôpital's rule again to the latter limit, we find that

$$\lim_{n\to\infty} \frac{\dfrac{t}{2\sigma}\psi_X'\left(\dfrac{t}{\sigma\sqrt{n}}\right)}{n^{-1/2}} = \lim_{n\to\infty} \frac{\left(\dfrac{t}{2\sigma}\right)\left(-\dfrac{1}{2}n^{-3/2}\dfrac{t}{\sigma}\right)\psi_X''\left(\dfrac{t}{\sigma\sqrt{n}}\right)}{\left(-\dfrac{1}{2}n^{-3/2}\right)}$$

$$= \frac{1}{2}\cdot\frac{t^2}{\sigma^2}\cdot\lim_{n\to\infty}\psi_X''\left(\frac{t}{\sigma}n^{-1/2}\right).$$

However,

$$\lim_{n\to\infty}\psi_X''\left(\frac{t}{\sigma}n^{-1/2}\right) = \psi_X''(0) = \sigma_X^2 = \sigma^2.$$

Hence,

$$\lim_{n\to\infty}\psi_Q(t) = \frac{1}{2}t^2.$$

Now the cumulant generating function for a standard normal random variable Z is given by $\psi_Z(t) = \frac{1}{2}t^2$ (see §6.3.1). Hence, we see that the distribution of Q approaches the distribution of Z as $n \to \infty$. However, the distribution of Q is simply the distribution of $(S_n - n\mu)/\sigma\sqrt{n}$. Consequently, the distribution of $(S_n - n\mu)/\sigma\sqrt{n}$ tends to a standard normal distribution as $n \to \infty$, which is the result we set out to prove. ∎

8.6 Normal Power Approximations (Optional)

The central limit theorem provides a theoretical justification for using the normal distribution to approximate the distribution of a sum of independent and identically distributed random variables. Unfortunately, it does not give us a bound on the error in such an approximation, nor does it tell us the number of terms in the sum needed for the approximation to be reasonably accurate.

In many insurance applications, particularly those related to casualty insurance, the distributions of the individual losses tend to be very skewed with the result that the distribution of the aggregate loss often departs significantly from a normal distribution. The reason for the large skewness in the individual distributions is that casualty insurance contracts generally insure against catastrophic events that have a low probability of occurrence.

In this section, we consider a generalization of the normal approximation, known as the *normal power approximation*, which is quite effective in approximating sums of highly skewed distributions.

A **normal power approximation** for the distribution of a random variable S is an approximation of the form

$$S \approx a_0 + a_1 Z + a_2 Z^2 + \cdots + a_r Z^r,$$

where $Z \sim \text{Normal}(0, 1)$. Here, the notation $X \approx Y$ should be interpreted to mean that the *distribution* of X is approximately equal to the *distribution* of Y, not that X and Y are approximately equal as random variables. From §6.1.2, we know that $Z^2 \sim \text{Gamma}(\frac{1}{2}, \frac{1}{2})$. Hence, any normal power approximation with $r \geq 2$ can incorporate a nonzero skewness into the approximating distribution if desired.

For simplicity, we restrict our attention to *quadratic* normal power approximations (i.e., approximations with $r = 2$). Such approximations provide sufficiently accurate results in most insurance applications and are completely determined by the mean, variance, and skewness. We begin with a precise statement of the desired approximation.

Normal Power Approximation Formula

Suppose that S is a random variable with $\gamma_S \in (0, 1)$. Then

$$\frac{S - \mu_S}{\sigma_S} \approx Z + \frac{\gamma_S}{6}(Z^2 - 1)$$

and for $s \geq \mu_S$,

$$F_S(s) \approx \Phi\left(-\frac{3}{\gamma_S} + \sqrt{\frac{9}{\gamma_S^2} + 1 + \frac{6}{\gamma_S} \cdot \frac{s - \mu_S}{\sigma_S}}\right),$$

where $Z \sim \text{Normal}(0, 1)$ and $\Phi(\cdot)$ is the standard normal distribution function.

Note that the approximate formula for $F_S(s)$ is only valid for $s \geq \mu_S$. One could give a formula for the values of s with $s < \mu_S$ as well. However, in insurance applications, only the values of s with $s \geq \mu_S$ are generally of interest.

In the remainder of this section, we outline the main ideas in the derivation of this approximation formula. Details of the derivation are left as an exercise for the reader.

Derivation of the Normal Power Approximation Formula

Consider a random variable S whose standard form $(S - \mu_S)/\sigma_S$ has a distribution that departs from the standard normal distribution. To allow for such a departure, we assume an approximation of the form

$$\frac{S - \mu_S}{\sigma_S} \approx Z + kZ^2 + l,$$

where k and l are constants to be determined. Note that the kZ^2 term introduces skewness into the approximation. The constant term l is included so that it is possible for $Z + kZ^2 + l$ to have mean 0, which is the mean of $(S - \mu_S)/\sigma_S$.

The constants k and l can be determined by equating the first three moments of $(S - \mu_S)/\sigma_S$ and $Z + kZ^2 + l$ to first order in k and l. Since $E[Z^j] = 0$ for all odd j and $E[Z^2] = 1$ and $E[Z^4] = 3$, the first three moments of $Z + kZ^2 + l$ are exactly

$$E[Z + kZ^2 + l] = k + l,$$
$$E[(Z + kZ^2 + l)^2] = 1 + E[(kZ^2 + l)^2],$$
$$E[(Z + kZ^2 + l)^3] = 9k + 3l + E[(kZ^2 + l)^3].$$

Hence, equating the first three moments of $(S - \mu_S)/\sigma_S$ and $Z + kZ^2 + l$ to first order in k and l, we have

$$k + l = 0,$$
$$1 = 1,$$
$$9k + 3l = \gamma_S.$$

Solving this system for k and l, we find that

$$k = -l = \frac{\gamma_S}{6}.$$

Consequently, we obtain the approximation

$$\frac{S - \mu_S}{\sigma_S} \approx Z + \frac{\gamma_S}{6}(Z^2 - 1)$$

as claimed.

To determine the formula for $F_S(s)$, note that the quadratic equation

$$z + \frac{\gamma_S}{6}(z^2 - 1) = \frac{s - \mu_S}{\sigma_S}$$

has solutions

$$z_0 = -\frac{3}{\gamma_S} - \sqrt{\frac{9}{\gamma_S^2} + 1 + \frac{6}{\gamma_S} \cdot \frac{s - \mu_S}{\sigma_S}},$$

$$z_1 = -\frac{3}{\gamma_S} + \sqrt{\frac{9}{\gamma_S^2} + 1 + \frac{6}{\gamma_S} \cdot \frac{s - \mu_S}{\sigma_S}}.$$

Hence, for $s \geq \mu_S$ and $\gamma_S > 0$,

$$Pr(S \leq s) = Pr(z_0 \leq Z \leq z_1).$$

(Consider the graph of the quadratic as a function of Z. Note that the conditions $s \geq \mu_S$ and $\gamma_S > 0$ ensure that the roots z_0, z_1 are real and distinct.)

Now, for $\gamma_S \in (0, 1)$, $z_0 < -3$ and so $\Phi(z_0) \approx 0$. Consequently, for $\gamma_S \in (0, 1)$ and $s \geq \mu_S$,

$$F_S(s) \approx Pr(Z \leq z_1)$$

$$= \Phi\left(-\frac{3}{\gamma_S} + \sqrt{\frac{9}{\gamma_S^2} + 1 + \frac{6}{\gamma_S} \cdot \frac{s - \mu_S}{\sigma_S}}\right),$$

which is the approximation for $F_S(s)$ that was claimed. ∎

8.7 Exercises

1. Suppose that X is uniformly distributed on $(0, 2)$, Y is uniformly distributed on $(0, 3)$, and X and Y are independent. Determine the distribution functions for the following random variables:

   a. $X + Y$
   b. $X - Y$
   c. XY
   d. X/Y

2. Determine the threefold convolution of X, where X is uniformly distributed on $(0, 1)$.

3. The joint density of X and Y is given by

$$f_{X,Y}(x, y) = \begin{cases} 1 & \text{for } x + 2y \leq 2, x \geq 0, y \leq 0; \\ 0 & \text{otherwise.} \end{cases}$$

   Determine the distribution of $X + Y$ and the distribution of $X - Y$.

4. The joint density of X and Y is given by

$$f_{X,Y}(x, y) = \begin{cases} e^{-(x+y)} & \text{for } x > 0, y > 0; \\ 0 & \text{otherwise.} \end{cases}$$

   Determine the distribution of $X + Y$.

5. Suppose that X and Y are independent continuous random variables with probability density functions

$$f_X(x) = 2e^{-2x}, \quad x \geq 0;$$
$$f_Y(y) = 3e^{-3y}, \quad y \geq 0.$$

   Determine the probability that $X > Y$.

6. Suppose that X and Y are independent random variables and that both X and Y are uniformly distributed on the interval $(0, 1)$. Determine the probability that $1/2 \leq X + Y \leq 3/2$.

7. In a particular basketball league, the standard deviation in the distribution of players' heights is 2 inches. Twenty-five players are selected at random and their heights are measured. Give an estimate for the probability that the average height of the players in this sample of 25 is within 1 inch of the league average height.

8. For each of the following pairs of random variables X, Y, indicate

   a. whether X and Y are dependent or independent;

   b. whether X and Y are positively correlated, negatively correlated, or uncorrelated.

      i. X and Y are uniformly distributed on the disk $\{(x, y) \in \mathbf{R}^2 : 0 \leq x^2 + y^2 \leq 1\}$.

      ii. X and Y are uniformly distributed on the square $\{(x, y) \in \mathbf{R}^2 : -1 \leq x \leq 1, -1 \leq y \leq 1\}$.

      iii. X and Y are uniformly distributed on the parallelogram $\{(x, y) \in \mathbf{R}^2 : x - 1 \leq y \leq x + 1, -1 \leq x \leq 1\}$.

      iv. X and Y are uniformly distributed on the diamond $\{(x, y) \in \mathbf{R}^2 : |x| + |y| \leq 1, |x| \leq 1\}$.

      v. X and Y are uniformly distributed on the parallelogram $\{(x, y) \in \mathbf{R}^2 : -x - 1 \leq y \leq -x + 1, -1 \leq x \leq 1\}$.

9. Suppose that X_1 and X_2 are independent, identically distributed exponential random variables. Determine the probability density function for $X_1 - X_2$.

10. Suppose that X and Y are independent random variables with $X \sim \text{Exponential}(1)$ and $Y \sim \text{Normal}(0, 1)$. Determine the density function for $X^{1/2}Y$.

11. Suppose that X and Y are independent random variables with respective densities

$$f_X(x) = \frac{1}{\sqrt{2\pi}}e^{-x^2/2}, \qquad x \in \mathbf{R};$$

$$f_Y(y) = ye^{-y^2/2}, \qquad y \geq 0;$$

and put $W = XY$. Show that W has the density function

$$f_W(w) = \frac{1}{2}e^{-|w|}, \qquad w \in \mathbf{R}.$$

A random variable with density of this type is said to have a **Laplacian distribution**. Laplacian distributions arise frequently in the computer analysis of speech and images. We introduced the Laplacian distribution in Example 4 of §4.2.2.

12. Suppose that X and Y are independent normal random variables with mean 0 and standard deviation 1. Put $W = X/Y$. Show that the density of W is given by

$$f_W(w) = \frac{1}{\pi(1 + w^2)}, \qquad w \in \mathbf{R}.$$

A random variable W with this density is said to have a **Cauchy distribution**.

13. In §7.6, we showed that if $X = \sin \Theta$ and Θ is uniformly distributed on $(-\pi, \pi)$, then the density of X is given by

$$f_X(x) = \begin{cases} \frac{1}{\pi\sqrt{1-x^2}} & \text{for } -1 < x < 1, \\ 0 & \text{otherwise.} \end{cases}$$

In this question, we consider the distribution of the more general waveform $A \sin(\omega t + \Theta)$ when both A and Θ are random variables.

Suppose that A has a Weibull distribution with parameters $\alpha = \sqrt{2}$ and $\beta = 2$. That is, the density function of A is given by

$$f(y) = ye^{-y^2/2}, \qquad y \geq 0.$$

This particular Weibull density is known as the *Rayleigh density* in electrical engineering. Suppose further that Θ is uniformly distributed on $(-\pi, \pi)$ and A and Θ are independent. Show that for each t, the quantity $A \sin(\omega t + \Theta)$ has a standard normal distribution. *Hint:* Convince yourself that $\sin(\omega t + \Theta)$ has the same distribution as $\sin \Theta$ when Θ is uniformly distributed on $(-\pi, \pi)$ and then consider the product XY, where $X = \sin \Theta$ and $Y = A$. Note that the conclusion of this question justifies the use of the normal distribution when sampling from waveforms of the type that frequently arise in electrical engineering.

14. There are many problems in engineering where it is necessary to determine the distribution of quantities such as $X^2 + Y^2$ or $\sqrt{X^2 + Y^2}$. For example, the kinetic energy of a particle constrained to move in two dimensions is simply $\frac{1}{2}m(V_1^2 + V_2^2)$, where V_1, V_2 are the velocity components. On the other hand, the distance of a particle from the origin is simply $\sqrt{X^2 + Y^2}$, where X, Y are the location coordinates for the particle.

    Suppose that X and Y are independent, identically distributed normal random variables with mean 0 and variance σ^2.

    a. Show that $X^2 + Y^2$ has an exponential distribution with mean $2\sigma^2$.
    b. Show that $\sqrt{X^2 + Y^2}$ has a Weibull distribution with parameters $\alpha = \sqrt{2}\sigma$ and $\beta = 2$. Weibull distributions of this type are known as *Rayleigh distributions* in engineering. The Rayleigh distribution was introduced in Example 6 of §6.2.1.

15. A random variable X with density function

$$f_X(x) = \frac{1}{\pi(1 + x^2)}, \qquad x \in \mathbf{R}$$

is said to have a **Cauchy distribution.** The Cauchy distribution arises in many different contexts. For example, if Z_1 and Z_2 are independent, standard normal random variables, the quotient Z_1/Z_2 has a Cauchy distribution (see question 12). Similarly, if Θ is uniformly distributed on $(-\pi, \pi)$, then $\tan \Theta$ has a Cauchy distribution (see §7.6).

    Suppose that $X_1, \ldots, X_n$ are independent Cauchy random variables. Show that the random variable $\overline{X}_n$ given by $\overline{X}_n = (X_1 + \cdots + X_n)/n$ also has a Cauchy distribution. Deduce that the distribution of $\overline{X}_n$ does not become more concentrated as n increases. How do you reconcile this observation with the law of large numbers?

16. An investment portfolio consists of two risky securities S_1 and S_2. Let d_1 and d_2 be the dollar amounts invested in S_1 and S_2, respectively, and let X_1 and X_2 be the uncertain simple rates of return on S_1 and S_2, respectively, over the coming time period. Then the fractions invested in the two securities at the beginning of the time period are, respectively,

$$\alpha_1 = \frac{d_1}{d_1 + d_2} \text{ and } \alpha_2 = \frac{d_2}{d_1 + d_2},$$

and thus, the return on the portfolio over the coming period is

$$R = \alpha X_1 + (1 - \alpha)X_2,$$

where α is the fraction invested in S_1.

a. Show that

$$\mu_R = \mu_{X_2} + \alpha(\mu_{X_1} - \mu_{X_2}),$$
$$\sigma_R^2 = \alpha^2 \sigma_{X_1}^2 + 2\alpha(1 - \alpha)\rho_{X_1 X_2}\sigma_{X_1}\sigma_{X_2} + (1 - \alpha)^2\sigma_{X_2}^2.$$

Deduce that $\sigma_R > 0$, except possibly when $\rho_{X_1 X_2} = \pm 1$. *Hint:* Complete the square.

b. By eliminating α from the equations in part a, show that if $\mu_{X_1} \neq \mu_{X_2}$, then

$$\sigma_R^2 = \frac{(\mu_R - \mu_{X_2})^2}{(\mu_{X_2} - \mu_{X_1})^2}\sigma_{X_1}^2 - 2\frac{(\mu_R - \mu_{X_1})(\mu_R - \mu_{X_2})}{(\mu_{X_2} - \mu_{X_1})^2}\rho_{X_1 X_2}\sigma_{X_1}\sigma_{X_2}$$
$$+ \frac{(\mu_R - \mu_{X_1})^2}{(\mu_{X_2} - \mu_{X_1})^2}\sigma_{X_2}^2.$$

Deduce that

$$\sigma_R^2 = A(\mu_R - \mu_0)^2 + \sigma_0^2,$$

where

$$A = \frac{(\sigma_{X_1} - \sigma_{X_2})^2 + 2(1 - \rho_{X_1 X_2})\sigma_{X_1}\sigma_{X_2}}{(\mu_{X_1} - \mu_{X_2})^2},$$

$$\mu_0 = \frac{\mu_{X_1}\sigma_{X_2}^2 - (\mu_{X_1} + \mu_{X_2})\rho_{X_1 X_2}\sigma_{X_1}\sigma_{X_2} + \mu_{X_2}\sigma_{X_1}^2}{(\sigma_{X_1} - \sigma_{X_2})^2 + 2(1 - \rho_{X_1 X_2})\sigma_{X_1}\sigma_{X_2}}$$

and σ_0^2 is some nonnegative number. (You needn't determine σ_0^2 explicity. Just use the fact that $\sigma_R^2 \geq 0$.)

c. Using the equation

$$\sigma_R^2 = A(\mu_R - \mu_0)^2 + \sigma_0^2,$$

show that σ_0^2 is the variance of the minimum variance portfolio and that μ_0 is the expected return on the minimum variance portfolio. Deduce that

$$\sigma_0^2 = \frac{\sigma_{X_1}^2\sigma_{X_2}^2(1 - \rho_{X_1 X_2}^2)}{(\sigma_{X_1} - \sigma_{X_2})^2 + 2(1 - \rho_{X_1 X_2})\sigma_{X_1}\sigma_{X_2}}.$$

Hint: Use your answer to part b or minimize the quadratic

$$h(\alpha) = \alpha^2 \sigma_{X_1}^2 + 2\alpha(1 - \alpha)\rho_{X_1 X_2}\sigma_{X_1}\sigma_{X_2} + (1 - \alpha)^2\sigma_{X_2}^2$$

using standard calculus techniques.

d. Sketch a possible graph in the σ-μ plane for the curve defined by

$$\sigma^2 = A(\mu - \mu_0)^2 + \sigma_0^2, \qquad \sigma \geq 0.$$

Specify the axis of symmetry and the asymptotes. What type of curve is this (ellipse, hyperbola, parabola)? Indicate the location of the minimum variance portfolio on your graph.

e. Sketch possible graphs for the curve

$$\sigma^2 = A(\mu - \mu_0)^2 + \sigma_0^2, \qquad \sigma \ge 0$$

in each of the following special cases: $\rho_{X_1 X_2} = 1$; $\rho_{X_1 X_2} = -1$; $\rho_{X_1 X_2} = 0$; $\mu_{X_1} = \mu_{X_2}$; $\sigma_{X_1} = \sigma_{X_2}$. Indicate the location of the minimum variance portfolio in each case.

17. An investment portfolio consists of two securites whose returns over the coming year are random variables X and Y, respectively. Analysis of these two securities suggests that $\mu_X = 10\%$, $\sigma_X = 5\%$, $\mu_Y = 20\%$, $\sigma_Y = 15\%$, $\rho = .30$.

a. What fraction of the portfolio should be invested in the security with return X if the portfolio variance is to be minimized?

b. What is the expected return on the portfolio with minimum variance?

18. A conservative investor is given a choice of two securities: a low-risk, low-return security and a higher-risk, higher-return security. The investor may invest in either security or in a combination of the two. Being conservative, this investor feels that the best choice is to invest all her assets in the low-risk security. Is this the most prudent thing to do? Explain how a combination portfolio could simultaneously lower risk and increase return. Under what circumstances would risk be minimized by investing everything in the low-risk security? *Hint:* Consider the condition $\rho_{X_1 X_2} \le \min(\sigma_{X_1}/\sigma_{X_2}, \sigma_{X_2}/\sigma_{X_1})$.

19. Paul offers John 2-to-1 odds on a $1000 bet on the toss of a coin (i.e., John gains $2000 if he wins the bet, but must pay $1000 if he loses). John refuses, arguing that he cannot afford the potential $1000 loss. However, finding the odds in his favor too good to pass up, John makes the following counteroffer: He proposes that they make 100 such bets.

a. Is John's reasoning sound? Explain using the language of probability.

b. Overhearing Paul's offer to John, Matt (who also can't afford to lose $1000) makes a different counteroffer to Paul: He proposes that they make 1000 bets each with Paul's $2 against his $1. Is Matt's reasoning sound? Explain using the language of probability.

c. Paul, being a high-roller, refuses Matt's offer, but accepts John's offer to make 100 bets of $1000 per bet at 2-to-1 odds. Explain how people of modest means like Matt and John could get around the riskiness of this tempting offer.

20. A risk-averse person owns a house worth $100,000. There is a 0.25% chance that the house will be struck by fire in the coming year. If fire does strike, the loss will be total. Insurance that provides full coverage in the event of a loss is available for a cost of $500. Let H be the value of the house and let I be the value of the insurance policy 1 year from now.

a. Show that

$$H = \begin{cases} 100{,}000 & \text{if no fire occurs,} \\ 0 & \text{if fire occurs;} \end{cases}$$

and

$$I = \begin{cases} -500 & \text{if no fire occurs,} \\ 99{,}500 & \text{if fire occurs.} \end{cases}$$

Calculate μ_H, σ_H, μ_I, and σ_I. Would you consider H and I to be individually risky investments?

b. Calculate the expected value and variance of the portfolio consisting of the house H and the insurance policy I. Would you consider the combined portfolio to be a risky investment?

c. What do your answers to parts a and b suggest about the riskiness of securities? Is it meaningful to speak of the riskiness of an asset in isolation?

21. An insurer faces aggregate claims S and has collected aggregate premiums $k\mu_S$, where $k > 1$. By using a normal approximation, show that the probability of insolvency (i.e., that claims exceed receipts) is approximately

$$1 - \Phi\left((k-1)\frac{\mu_S}{\sigma_S}\right).$$

Deduce that, for fixed $k > 1$, the ratio μ_S/σ_S is a rudimentary measure of insurer solvency. What aspect of this measure might be considered objectionable?

22. Two insurers face independent and identically distributed aggregate losses L_1, L_2 and have collected respective aggregate premiums $kE[L_1]$, $kE[L_2]$, where $k > 1$ is the same for both insurers. The insurers believe that they can improve their chances of remaining solvent by sharing premiums and losses equally. Let $L = (L_1 + L_2)/2$ be the loss incurred by each insurer with coinsurance. In this question, you determine whether the insurers are more or less secure with coinsurance.

a. Determine formulas for μ_L, σ_L in terms of μ_{L_1}, σ_{L_1} and μ_{L_2}, σ_{L_2}, respectively. Deduce formulas for the ratio μ_L/σ_L.

b. By considering your answer to the previous question, determine whether the insurers are more or less secure with coinsurance.

c. Can you think of another way that an insurer can reduce the risk inherent in a pool of risks? *Hint:* Why do many insurers sell shares?

23. An insurer has a portfolio of m independent property insurance policies. The insurer's loss on each policy has a Pareto distribution with parameters $s = 3$ and $\beta = 2$ measured in thousands. The insurer charges each policyholder the same premium P. The insurer would like to be 95% confident of its ability to meet its obligations. Determine the premium P as a function of m.

24. A life insurer has a group of m term life insurance policies, each with a face value of \$25,000. The policies in the group are written on single and distinct lives that are assumed to be independent. The probability of a death claim on any given policy in the group during the coming year is .01, and the 1-year continuously compounded interest rate is 5%. Let P be the premium charged to each policyholder for term insurance over the coming year. Suppose that the insurer would like to be 99% confident that total premiums exceed total claims. Determine P as a function of m.

25. The law of large numbers (and our discussion of it in this chapter) would seem to suggest that for large groups of independent risks with identically distributed losses, the premium that each member of the group should be charged is close to the expected loss per member, regardless of the desired level of confidence of being able to meet all obligations. In this question, you consider what happens if the premium collected is exactly equal to the expected loss.

Let $X_1, \ldots, X_n$ be independent, identically distributed losses and let $S = X_1 + \cdots + X_n$ be the aggregate loss. Suppose that the premium collected from each policyholder is $P = E[X_j]$. Determine the probability of insolvency; that is, $\Pr(S > nP)$. Are you surprised by your answer? How do you reconcile your answer with the observation made at the beginning of the question?

26. Suppose that $S \sim \text{Lognormal}(0, 0.25)$. Use the normal power approximation formula derived in §8.6 to obtain an estimate for $\Pr(S \le 1.05)$. How does this compare with the exact value of this probability? How does this compare with the value obtained using a normal approximation?

27. An insurer has a portfolio of 100 insurance contracts. The insurer's losses on these contracts are independent and identically distributed. The distribution of the loss on each contract is Exponential$(1/5)$ measured in thousands of dollars. A premium payment of \$5050 is collected from each policyholder. Determine the probability that aggregate losses exceed total premiums using a normal power approximation.

9

Mixtures and Compound Distributions

We have already encountered mixtures on several occasions throughout this book. Indeed, in §4.1.10, we introduced the notion of a mixture in connection with the distributional form of the law of total probability, and in §5.3 and §6.1.3, we saw that the negative binomial random variable can be considered a mixture of Poisson random variables with gamma mixing weights and the Pareto random variable can be considered a mixture of exponential random variables with gamma mixing weights. In this chapter, we discuss the properties of mixtures in detail and consider an important type of mixture known as a *compound distribution*, which arises in connection with claims aggregation in insurance.

Our presentation is organized in the following way: In §9.1, we give a precise definition of the concept of mixture and consider how mixtures are related to marginal distributions and mixed distributions. In §9.2, we discuss some important examples of mixtures that arise frequently in insurance applications. In §9.3 and §9.4, we state and derive formulas for calculating the mean, variance, and moment generating function of a mixture, and we illustrate the use of these formulas with numerical examples. Finally, in §9.5, we consider compound distributions in detail and discuss some particular compound models that are frequently used to model aggregate claims during a fixed time period.

9.1 Definitions and Basic Properties

A random variable X is said to be a **mixture** if the distribution function of X has one of the following forms:

1. $F_X(x) = \sum_i k_i F_{X_i}(x)$ for some sequence of random variables $X_1, X_2, \ldots$ and some sequence of positive numbers k_i such that $\sum k_i = 1$;
2. $F_X(x) = \int_{-\infty}^{\infty} F_{X|P=p}(x) f_P(p) dp$ for some family of random variables $X \mid P = p$ indexed by the real numbers and some nonnegative function f_P such that $\int_{-\infty}^{\infty} f_P(p) \, dp = 1$.

In the first case, X is said to be a **discrete mixture** of the random variables X_i with mixing weights k_i, and in the second case, X is said to be a **continuous mixture** of the $X \mid P = p$ with mixing weights given by the function f_P. It is also common to refer to a *distribution* as a mixture if its associated distribution function has one of the indicated forms.

363

Discrete mixtures often arise when the risk class of a policyholder is uncertain and the number of risk classes is *discrete*, whereas continuous mixtures often arise when a risk parameter for a distribution of losses is uncertain and the uncertain parameter is *continuous*. This is illustrated more clearly in the following two examples.

EXAMPLE 1: An insurer has two groups of policyholders: good risks and bad risks. The distributions for the size of a random loss are known for each risk class. A new customer, whose risk class is not known with certainty, has just purchased a policy from the insurer. What distribution should be used to model random losses for this new customer?

The appropriate distribution is a mixture of the distributions for each risk class. Let X_1 be the size of a random loss for a policyholder known to be a good risk, let X_2 be the size of a random loss for a policyholder known to be a bad risk, and let X be the size of a random loss for the new customer. Further, let C be an indicator of the risk class of the new customer as follows:

$$C = \begin{cases} 1 & \text{if the customer is a good risk,} \\ 2 & \text{if the customer is a bad risk.} \end{cases}$$

Then, by the law of total probability,

$$\Pr(X \leq x) = \Pr(X \leq x \mid C = 1) \Pr(C = 1) + \Pr(X \leq x \mid C = 2) \Pr(C = 2)$$
$$= \Pr(X_1 \leq x) \Pr(C = 1) + \Pr(X_2 \leq x) \Pr(C = 2).$$

Hence, the distribution function of X is given by

$$F_X(x) = p_1 F_{X_1}(x) + p_2 F_{X_2}(x),$$

where $p_j = \Pr(C = j)$. Consequently, the random loss for the new customer is a discrete mixture of X_1 and X_2 with mixing weights p_1 and p_2. ∎

EXAMPLE 2: Consider a group of policyholders whose individual random losses are believed to follow an exponential distribution. Recall that the mean of an exponential distribution with parameter λ is $1/\lambda$ (see §6.1.1). Hence, the parameter λ of the exponential distribution is a risk parameter that determines, and is determined by, the mean of the distribution.

Now, in practice, the value of the mean (and hence, the value of the parameter λ) can never be known with certainty because any estimate for the mean (or for λ) will be based on limited historical experience. Moreover, the mean loss for the future time period is affected by monetary inflation in the economy, which is inherently uncertain. Hence, the parameter λ itself may be considered a random variable, at least under the Bayesian interpretation of probability. Moreover, it may be considered to be a continuous random variable because there is a continuum of possible values that the historical mean can assume over any given period of observation.

Write Λ in place of λ to emphasize that the exponential parameter is now being considered a random variable and let X denote the random loss incurred by a policyholder in the given group. Then, by the law of total probability,

$$\Pr(X \leq x) = \int \Pr(X \leq x \mid \Lambda = \lambda) f_\Lambda(\lambda) d\lambda$$

$$= \int F_{X \mid \Lambda = \lambda}(x) f_\Lambda(\lambda) d\lambda,$$

where $X \mid \Lambda = \lambda \sim \text{Exponential}(\lambda)$, by assumption. Hence, the random loss X is a continuous mixture of exponentials with mixing density f_Λ. ∎

We will consider other important examples of mixtures in §9.2. In the remainder of this section, we discuss some basic properties of mixtures. We also review the difference between sums and mixtures that we discussed in §4.1.12.

Mixtures as Marginal Distributions

Mixtures are actually marginal distributions. Indeed, if X is a discrete mixture of $X_1, X_2, \ldots$ with respective mixing weights $k_1, k_2, \ldots$, then X is a scalar component in the vector (X, Y) with joint distribution given by

$$p_Y(j) = \begin{cases} k_j & \text{for } j = 1, 2, \ldots \\ 0 & \text{otherwise;} \end{cases}$$
$$f_{X|Y=j}(x) = f_{X_j}(x).$$

On the other hand, if X is a continuous mixture of $X \mid P = p$ with mixing density f_P, then it is clearly true from the definition of what it means to be a continuous mixture that X is a scalar component in the random vector (X, P).

This interpretation of mixtures as marginal distributions is very important and provides us with insight into how mixtures arise in practice. Recall that the marginal distribution of X with respect to the random vector (X, Y) is the distribution of probability for the scalar component X *in the absence of information about Y* (i.e., it is the *unconditional* distribution of X). Hence, a mixture is simply an unconditional distribution. This suggests that mixtures will arise whenever a random quantity X depends on another random quantity Y and it is desirable to know the distribution of X in the absence of information about Y.

You may be wondering why we need a new name for a concept we have already discussed. The reason has to do with perspective. Marginal distributions are usually considered in conjunction with (and subordinate to) a joint probability distribution; moreover, the marginal distributions in such a joint distribution are generally considered to be on an equal footing with one another. On the other hand, mixtures are usually considered *in isolation from* the joint distribution; moreover, a mixture represents a *particular* marginal distribution whose importance surpasses all the others. Consequently, even though mixtures and marginal distributions are mathematically the same thing, there is an important distinction in the usage of the two terms.

Mixed Distributions Are Mixtures

Note that "mixed" distributions, which we defined in §4.1.3 to be distributions with both continuous and discrete parts, are actually a special type of mixture. Indeed, if X is a mixed random variable, then as demonstrated in the appendix to Chapter 4,

$$F_X(x) = (1 - k)F_C(x) + kF_D(x)$$

for some continuous random variable C, some discrete random variable D, and some number $k \in (0, 1)$. Hence, a mixed random variable is a discrete mixture of a continuous and a discrete random variable.

Compound Distributions

Some authors use the term *compound distribution* in place of *mixture* to emphasize the fact that mixtures model an uncertainty that has been compounded. The rationale for this terminology becomes clear when one considers the continuous mixture discussed earlier in Example 2. However, we will reserve the term compound distribution for a special type of mixture—namely, a random sum of independent and identically distributed random variables. A random sum is one in which the number of terms in the sum is uncertain. We will discuss random sums and compound distributions more formally in §9.2.

The Difference Between Sums and Mixtures

Because of the way in which mixtures are defined, it is easy to confuse them with sums. However, it is important to keep in mind that mixtures are not the same as sums. Indeed, for any random variables X_1, X_2 and any number $\alpha \in (0, 1)$, the weighted sum $\alpha X_1 + (1 - \alpha)X_2$ is generally not the same as the mixture of X_1 and X_2 with mixing weights α, $1 - \alpha$; that is, it is generally not the case that $F_{\alpha X_1 + (1-\alpha)X_2}(x) = \alpha F_{X_1}(x) + (1 - \alpha)F_{X_2}(x)$ for all x. We better understand why this is so by considering a concrete example that emphasizes the differences in the meanings of these two concepts.

Hence, let X_1 and X_2 represent simple returns for the securities S_1 and S_2 over a given period and consider the following two portfolios:

- Portfolio $\mathcal{P}_1$ consists of precisely one of the securities S_1, S_2, but it is not known which one; the probability of the security being S_1 is α.
- Portfolio $\mathcal{P}_2$ consists of both securities, with the fraction invested in S_1 being α.

Further, let R_1, R_2 be the returns on the portfolios $\mathcal{P}_1$, $\mathcal{P}_2$, respectively. Then R_1 is the *mixture* of the returns X_1, X_2 with *mixing* weights α, $1 - \alpha$, while R_2 is the *sum* of the returns X_1, X_2 with *portfolio* weights α, $1 - \alpha$. Clearly, the returns on these two portfolios will generally be different.

Consequently, we see that the distribution of a weighted sum is not the same as a weighted sum of distributions.

9.2   Some Important Examples of Mixtures Arising in Insurance

We have already seen how mixtures can arise when the risk class of a policyholder or the risk parameter in a loss distribution is not known with certainty. Mixtures also arise naturally in connection with the insurer's payout on a given contract or group of contracts.

The insurer's payout on any given contract is determined by

1. the occurrence or nonoccurrence of a loss;
2. the size of a loss in relation to contract caps, deductibles, and other contract features;
3. the frequency of losses if more than one loss is possible.

Since each of these factors can be (and generally is) uncertain, the distribution of the insurer's payout in the absence of knowledge of the occurrence, size, or frequency of losses will be a mixture, as we now demonstrate through the following examples.

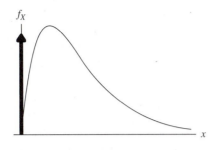
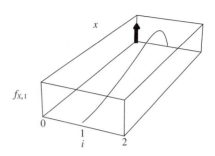

a. Payout Distribution as a Mixture

b. Joint Distribution of the Payout
and a Loss Indicator

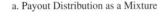

FIGURE 9.1 Payout Distribution When Loss Occurrence Is Uncertain

Insurer's Payout When Loss Occurrence Is Uncertain

Consider an insurance policy that indemnifies the policyholder for the full amount of a random loss the first time it occurs during a fixed period of time. Let L be the size of the uncertain loss and let X be the insurer's payout random variable. Then

$$X = \begin{cases} 0 & \text{if no loss occurs,} \\ L & \text{if a loss occurs.} \end{cases}$$

Hence, X is a mixture of a point mass at the origin (representing no loss occurrence) and the loss random variable L (Figure 9.1a).[1]

We can see this more clearly by introducing the loss indicator random variable I defined by

$$I = \begin{cases} 0 & \text{if no loss occurs,} \\ 1 & \text{if a loss occurs} \end{cases}$$

and observing that $X \mid I = 1$ is equal to L. Then X is the marginal distribution in the joint distribution of X and I as illustrated in Figure 9.1b. Note that the probability mass at the origin in this figure is equal to $\Pr(I = 0)$, which is the probability of no loss occurring, and that the area under the curve $f_{X,I}$ along the line $I = 1$ is equal to $\Pr(I = 1)$, which is the probability that a loss occurs. Note further that the distribution of X (shown in Figure 9.1a) can be obtained from the joint distribution (shown in Figure 9.1b) by simply projecting all of the probability onto the x axis.

We can obtain a formula for the distribution function for X by applying the law of total probability and by considering separately the cases $x < 0$ and $x \geq 0$. Indeed,

$$F_X(x) = \Pr(X \leq x \mid I = 0)\Pr(I = 0) + \Pr(X \leq x \mid I = 1)\Pr(I = 1)$$
$$= \begin{cases} (1-p) + p \cdot F_L(x) & \text{for } x \geq 0, \\ 0 & \text{for } x < 0, \end{cases}$$

where p is the probability of a claim occurrence; that is, $p = \Pr(I = 1)$. Details are left to the reader.

---

[1] Note that we are assuming here and in the subsequent examples that the loss distributions are continuously distributed.

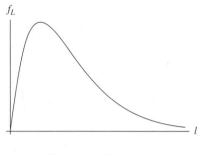
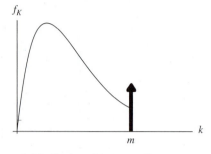

a. Distribution of the Loss

b. Distribution of the Indemnified Amount

FIGURE 9.2 Relationship Between the Incurred Loss and the Indemnified Amount for a Contract with a Cap

Indemnified Amount of a Loss in the Presence of a Cap or Deductible

Consider an insurance policy that indemnifies the policyholder for the amount of a random loss up to or in excess of a fixed amount, as the case may be. Let L be the size of the uncertain loss and let K be the indemnified amount.

Contract with a Cap m Suppose that the indemnified amount is capped at m. Then

$$K = \begin{cases} L & \text{if loss is at most } m, \\ m & \text{if loss is greater than } m. \end{cases}$$

Hence, K is a mixture of a point mass at m (representing losses greater than m) and a particular loss random variable that agrees with L on $(0, m)$ but is zero otherwise (see Figure 9.2b).

Contract with a Deductible d Suppose that the indemnified amount is subject to a deductible d. Then

$$K = \begin{cases} 0 & \text{if loss is less than } d, \\ L - d & \text{if loss is at least } d; \end{cases}$$
$$= (L - d)^+.$$

Hence, K is a mixture of a point mass at the origin (representing losses less than d) and a particular loss random variable whose distribution is obtained from the distribution of L by removing the portion below d and then scaling (Figure 9.3b).

Contract with a Cap m and a Deductible d Suppose that the indemnified amount is subject to both a deductible d and a cap m. Then

$$K = \begin{cases} 0 & \text{if loss is less than } d, \\ L - d & \text{if loss is at least } d \text{ but not greater than } d + m, \\ m & \text{if loss is greater than } d + m. \end{cases}$$

Hence, K is a mixture of a point mass at the origin (representing losses less than d), a point mass at m (representing losses greater than $d + m$), and a particular loss

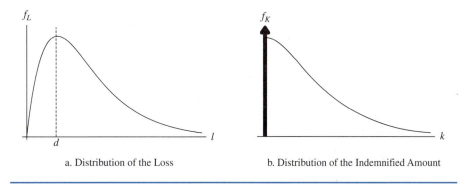

a. Distribution of the Loss   b. Distribution of the Indemnified Amount

FIGURE 9.3 Relationship Between the Incurred Loss and the Indemnified Amount for a Contract with a Deductible

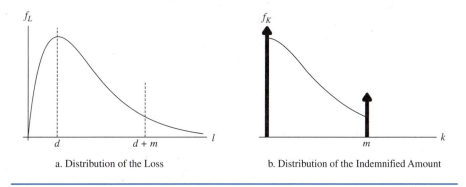

a. Distribution of the Loss   b. Distribution of the Indemnified Amount

FIGURE 9.4 Relationship Between the Incurred Loss and the Indemnified Amount for a Contract with a Cap and a Deductible

random variable whose distribution is obtained from the distribution of L by removing the portions below d and above $d + m$ and then scaling (Figure 9.4b).

Insurer's Payout When Loss Occurrence Is Uncertain and Indemnified Amount Is Subject to a Cap or Deductible

Consider an insurance policy that indemnifies the policyholder for the amount of a random loss up to or in excess of a fixed amount, as the case may be, the first time it occurs during a given period of time. Let L be the size of the uncertain loss (when it occurs), let K be the indemnified amount of the loss, and let X be the insurer's payout random variable. Then the payout X will be a mixture of mixtures because it depends on

1. whether or not a loss occurs;
2. whether or not the loss (when it occurs) is above or below the cap or deductible.

Contract with a Cap m   Suppose that the indemnified amount is capped at m. Then

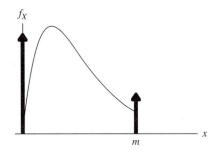

FIGURE 9.5   Payout Distribution for a Contract with a Cap When Loss Occurrence Is Uncertain

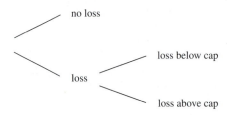

FIGURE 9.6   Tree Diagram for Insurer's Payout When Indemnified Loss Is Subject to a Cap

$$X = \begin{cases} 0 & \text{if no loss occurs,} \\ K & \text{if a loss occurs,} \end{cases}$$

and

$$K = \begin{cases} L & \text{if loss is at most } m, \\ m & \text{if loss is greater than } m. \end{cases}$$

Hence, X is a mixture of a point mass at the origin (representing no loss occurring) and the indemnified amount K, which itself is a mixture of a point mass at m (representing a loss greater than m) and a particular loss random variable that agrees with L on $(0, m)$ but is zero otherwise. Consequently,

$$X = \begin{cases} 0 & \text{if no loss occurs,} \\ L & \text{if a loss occurs and it is at most } m, \\ m & \text{if a loss occurs and it is greater than } m. \end{cases}$$

Thus, X is a mixture of a point mass at the origin, a point mass at m, and a particular loss random variable that agrees with L on $(0, m)$ but is zero otherwise (Figure 9.5).

In cases where a random variable is a mixture of mixtures, it is often helpful to construct a tree diagram to keep track of all the cases and subcases. A tree diagram for the payout on the contract just discussed appears in Figure 9.6.

Contract with a Deductible d   Suppose that the indemnified amount is subject to a deductible d. Then

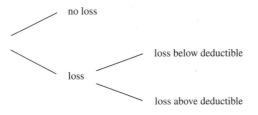

FIGURE 9.7 Tree Diagram for Insurer's Payout When Indemnified Loss Is Subject to a Deductible

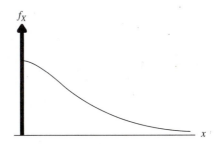

FIGURE 9.8 Payout Distribution for a Contract with a Deductible When Loss Occurrence Is Uncertain

$$X = \begin{cases} 0 & \text{if no loss occurs,} \\ K & \text{if a loss occurs,} \end{cases}$$

and

$$K = \begin{cases} 0 & \text{if loss is less than } d, \\ L - d & \text{if loss is at least } d. \end{cases}$$

Hence, X is a mixture of a point mass at the origin (representing no loss occurring) and the indemnified amount K which itself is a mixture of a point mass at the origin (representing a loss less than d) and a particular loss random variable whose distribution is obtained from the distribution of L by removing the portion below d and then scaling. (See Figure 9.7 for a tree diagram of the situation.)

Consequently,

$$X = \begin{cases} 0 & \text{if no loss occurs or if a loss occurs and it is less than } d, \\ L - d & \text{if a loss occurs and it is at least } d. \end{cases}$$

Hence, X is actually a mixture of a point mass at the origin and a particular loss random variable whose distribution is obtained from the distribution of L by removing the portion below d and then scaling (Figure 9.8).

In practice, the underwriter of such a contract will not know the distribution of the loss random variable L completely and will not know the exact probability of a loss occurrence, since losses below the deductible d will not be reported to the insurer. Hence, it is common to model X in such cases by conditioning on the occurrence of a claim rather than the occurrence of a loss. Consequently,

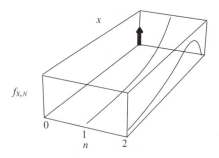

FIGURE 9.9 Joint Distribution of Payout and Claim Frequency When at Most Two Claims Are Possible

$$X = \begin{cases} 0 & \text{if no claim occurs,} \\ B & \text{if a claim occurs,} \end{cases}$$

where B is the size of a claim when it occurs. Note that B cannot have a positive probability mass at the origin since claims of size zero are not reported.

Insurer's Payout When Claim Frequency Is Uncertain

Consider an insurance policy on which the number of claims as well as the size of claims is uncertain. Let N be the uncertain number of claims, let W_j be the size of the jth claim, and let X be the insurer's *aggregate* payout random variable.

At Most Two Claims Possible   Suppose that the number of possible claims is zero, one, or two. Then

$$X = \begin{cases} 0 & \text{if no claim occurs,} \\ W_1 & \text{if one claim occurs,} \\ W_1 + W_2 & \text{if two claims occur.} \end{cases}$$

Hence, X is a mixture of a point mass at the origin (representing the payout when no claims occur) and the distributions of W_1 and $W_1 + W_2$. This becomes clearer by considering a graph of the joint density of X and N (Figure 9.9).

Arbitrary Number of Claims Possible   Suppose that the number of possible claims can be any nonnegative number. Then the payout X is a discrete (infinite) mixture of $0, W_1, W_1 + W_2, \ldots$, where 0 denotes the certain constant zero, with respective mixing weights $p_N(0), p_N(1), p_N(2), \ldots$. As a shorthand notation, one generally writes such an X as the "random sum"

$$X = W_1 + \cdots + W_N.$$

Keep in mind, however, that the distribution of such an X is unconditional even though the notation $X = W_1 + \cdots + W_N$ may suggest otherwise.

Notice how the uncertainty in N compounds the uncertainty in the aggregate payout X. For this reason, the distributions of such random sums are known as *compound distributions*.

More precisely, a random variable X is said to have a **compound distribution** if

$$X = W_1 + \cdots + W_N,$$

where the W_j are independent and identically distributed and each W_j is independent of N.

Notice that if each W_j is equal to the constant k, then $X = kN$ (i.e., X is a multiple of the claim number). On the other hand, if N is known with certainty to be the constant k, then X is the finite sum $W_1 + \cdots + W_k$.

Compound distributions are extremely important in insurance and will be considered in §9.5 in greater detail.

9.3 Mean and Variance of a Mixture

The mean and variance of a mixture can be determined directly from the generalized density function of the mixture by performing an appropriate integration (see Example 3 in §4.2.1 and Example 2 in §4.2.2). However, they can also be determined using the conditional means and variances associated with the underlying joint distribution, for which the mixture is a marginal distribution. In this section, we state and prove the formulas for calculating the mean and variance of a mixture in this way, and we give a concrete illustration of their use.

Before stating the formulas for mean and variance, we need to introduce the following new notation. For any random variables X and Y, let $E[X \mid Y]$ denote the expected value of $X \mid Y = y$,

$$E[X \mid Y] = \int_x x \cdot f_{X|Y=y}(x)dx$$

and let $Var(X \mid Y)$ denote the variance of $X \mid Y = y$,

$$Var(X \mid Y) = \int_x (x - E[X \mid Y])^2 \cdot f_{X|Y=y}(x)dx.$$

Note that for each fixed value of Y, the quantities $E[X \mid Y]$, $Var(X \mid Y)$ are numbers. However, in the absence of information about Y, the quantities $E[X \mid Y]$, $Var(X \mid Y)$ are random variables. In fact, the quantities $E[X \mid Y]$, $Var(X \mid Y)$ are transformations (i.e., functions) of the random variable Y. Consequently, it makes sense to inquire about the expectation and variance of both $E[X \mid Y]$ and $Var(X \mid Y)$ with respect to the distribution of Y.

In the formulas for the expectation and variance of a mixture that follow, three quantities arise: $E[E[X \mid Y]]$, $E[Var(X \mid Y)]$, $Var(E[X \mid Y])$. Note that each of these quantities is a number and that the outer expectation/variance is taken with respect to the distribution of Y. To emphasize this fact, we will sometimes write $E_Y[E[X \mid Y]]$, $E_Y[Var(X \mid Y)]$, $Var_Y(E[X \mid Y])$ in place of $E[E[X \mid Y]]$, $E[Var(X \mid Y)]$, $Var(E[X \mid Y])$, respectively.

Formulas for the Unconditional Mean and Variance

Suppose that X is a mixture with mixing variable Y. Then the unconditional mean and variance of X can be determined by the formulas

$$E[X] = E_Y[E[X \mid Y]],$$
$$Var(X) = E_Y[Var(X \mid Y)] + Var_Y(E[X \mid Y]).$$

We can understand these formulas better if we consider X to be a random loss for a policyholder whose risk class is not known with certainty and Y to be the risk class of the policyholder. Then the formula for the expectation asserts that the average loss per policy on a group of policies for which the risk classes of the individual policyholders is not known with certainty can be determined by finding the average loss for each risk class and then averaging these averages in a way which reflects the distribution of policyholders by risk class (i.e., the average loss is an average of averages). On the other hand, the formula for variance encapsulates the fact that the uncertainty in the loss arises from two sources: inherent uncertainty when the risk class is known and uncertainty due to the fact that the risk class is not known. Indeed, the formula for variance asserts that the "total variation" is simply the average of the variations for each risk class plus the variation in the averages.

Another concrete example that is helpful when trying to understand the meaning of these formulas is given in exercise 1 in §9.6.

In light of these formulas for $E[X]$ and $Var(X)$, the quantities $E[X]$ and $Var(X)$ are sometimes referred to as the *total expectation* and the *total variance* of X, respectively. When we wish to distinguish $E[X]$ and $Var(X)$ from $E[X \mid Y]$ and $Var(X \mid Y)$, we will follow this convention.

We now derive the formulas for $E[X]$ and $Var(X)$.

Derivation of the Total Expectation Formula

From the definition of expectation,

$$E[X] = \int x f_X(x) dx.$$

Hence, since

$$f_X(x) = \int_y f_{X\mid Y=y}(x) \cdot f_Y(y) dy,$$

we have

$$E[X] = \int_x x \left\{ \int_y f_{X|Y=y}(x) \cdot f_Y(y)dy \right\} dx$$

$$= \int_y \left\{ \int_x x f_{X|Y=y}(x)dx \right\} f_Y(y)dy$$

$$= \int_y E[X \mid Y]f_Y(y)dy$$

$$= E_Y[E[X \mid Y]]$$

as claimed.

Derivation of the Total Variance Formula

The formula for total variance follows by successively using the general formula $Var(U) = E[U^2] - E[U]^2$ with U replaced by X, $X \mid Y = y$, and $E[X \mid Y]$ and by using the formula for total expectation.

Indeed, from the general formula for variance, we have

$$Var(X) = E[X^2] - E[X]^2,$$
$$E[X^2 \mid Y] = Var(X \mid Y) + E[X \mid Y]^2,$$

and

$$Var_Y(E[X \mid Y]) = E_Y[E[X \mid Y]^2] - (E_Y[E[X \mid Y]])^2.$$

Moreover, from the formula for total expectation, we have

$$E[X^2] = E_Y[E[X^2 \mid Y]]$$

and

$$E[X] = E_Y[E[X \mid Y]].$$

Hence,

$$Var(X) = E[X^2] - E[X]^2$$
$$= E_Y[E[X^2 \mid Y]] - (E_Y[E[X \mid Y]])^2$$
$$= E_Y[Var(X \mid Y) + E[X \mid Y]^2] - (E_Y[E[X \mid Y]])^2$$
$$= E_Y[Var(X \mid Y)] + E_Y[E[X \mid Y]^2] - (E_Y[E[X \mid Y]])^2$$
$$= E_Y[Var(X \mid Y)] + Var_Y(E[X \mid Y])$$

as claimed.

Understanding the technical details of these derivations is not really important. However, knowing how to apply the formulas for total expectation and variance in practice is important. The following numerical example illustrates how these formulas can be used to calculate the mean and variance of the loss on an insurance policy for which the risk class of the policyholder is uncertain.

EXAMPLE 1: The policyholders of an insurance company fall into one of two classes. The loss distributions for each class are given in the following table:

Class 1		Class 2	
Size of Loss	Probability	Size of Loss	Probability
1,000	20%	1,000	70%
5,000	50%	5,000	20%
10,000	30%	10,000	10%

There are 30% of policyholders in class 1, while the remaining 70% are in class 2. Let L_1, L_2 denote the loss incurred by a randomly selected policyholder from class 1 and class 2, respectively, and let L denote the loss incurred by a randomly selected policyholder whose risk class is unknown. Our objective is to determine the total expectation $E[L]$ and the total variance $Var(L)$.

Let C denote the risk class of a randomly selected policyholder. Then

$$p_C(c) = \begin{cases} .30 & \text{if } c = 1, \\ .70 & \text{if } c = 2. \end{cases}$$

From the formulas for total expectation and variance, we see that we need to determine the distributions for each of the random variables $E[L \mid C]$ and $Var(L \mid C)$. Since C has only two possible values,

$$E[L \mid C] = \begin{cases} E[L_1] & \text{if } C = 1, \\ E[L_2] & \text{if } C = 2, \end{cases}$$

and

$$Var(L \mid C) = \begin{cases} Var(L_1) & \text{if } C = 1, \\ Var(L_2) & \text{if } C = 2. \end{cases}$$

Consequently, the means of $E[L \mid C]$ and $Var(L \mid C)$ are, respectively,

$$E[E[L \mid C]] = E[L_1] \cdot \Pr(C = 1) + E[L_2] \cdot \Pr(C = 2)$$
$$= (.30)E[L_1] + (.70)E[L_2]$$

and

$$E[Var(L \mid C)] = Var(L_1) \cdot \Pr(C = 1) + Var(L_2) \cdot \Pr(C = 2)$$
$$= (.30)Var(L_1) + (.70)Var(L_2).$$

Similarly, using the general formula $Var(U) = E[U^2] - E[U]^2$ with $U = E[L \mid C]$, we see that the variance of $E[L \mid C]$ is given by

$$Var(E[L \mid C]) = E[E[L \mid C]^2] - (E[E[L \mid C]])^2$$
$$= \{E[L_1]^2 \cdot \Pr(C = 1) + E[L_2]^2 \cdot \Pr(C = 2)\} - (E[E[L \mid C]])^2$$
$$= \{(.30)E[L_1]^2 + (.70)E[L_2]^2\} - \{(.30)E[L_1] + (.70)E[L_2]\}^2.$$

Consequently, to determine the total expectation $E[L]$ and the total variance $Var(L)$ from the formulas derived in this section, we need only calculate $E[L_1]$, $E[L_2]$, $Var(L_1)$, and $Var(L_2)$.

From the loss data in the given table, we have

$$E[L_1] = (1{,}000)(.20) + (5{,}000)(.50) + (10{,}000)(.30)$$
$$= 5{,}700,$$

$$E[L_2] = (1{,}000)(.70) + (5{,}000)(.20) + (10{,}000)(.10)$$
$$= 2{,}700,$$

$$Var(L_1) = E[L_1^2] - E[L_1]^2$$
$$= \{(1{,}000)^2(.20) + (5{,}000)^2(.50) + (10{,}000)^2(.30)\} - (5{,}700)^2$$
$$= 10{,}210{,}000$$

and

$$Var(L_2) = E[L_2^2] - E[L_2]^2$$
$$= \{(1{,}000)^2(.70) + (5{,}000)^2(.20) + (10{,}000)^2(.10)\} - (2{,}700)^2$$
$$= 8{,}410{,}000.$$

Consequently,

$$E[E[L \mid C]] = (.30)(5{,}700) + (.70)(2{,}700)$$
$$= 3600,$$

$$E[Var(L \mid C)] = (.30)(10{,}210{,}000) + (.70)(8{,}410{,}000)$$
$$= 8{,}950{,}000,$$

$$Var(E[L \mid C]) = \{(.30)(5{,}700)^2 + (.70)(2{,}700)^2\}$$
$$- \{(.30)(5{,}700) + (.70)(2{,}700)\}^2$$
$$= 1{,}890{,}000.$$

Thus, the total expectation and variance are, respectively,

$$E[L] = E[E[L \mid C]] = 3600$$

and

$$Var(L) = E[Var(L \mid C)] + Var(E[L \mid C])$$
$$= 8{,}950{,}000 + 1{,}890{,}000$$
$$= 10{,}840{,}000.$$

Note that the total variance in this example is bigger than both $Var(L_1)$ and $Var(L_2)$, which suggests that the uncertainty in the risk class is significant. Nevertheless, since $E[(Var(L \mid C)]$ is much greater than $Var(E[L \mid C])$, it is clear that the uncertainty in the loss size is still more significant than the uncertainty in the risk class, at least in this example.

By considering the standard deviation of the loss L in relation to the expected loss $E[L]$, we can get a sense of the risk in underwriting the contracts of this example. Since $\sigma_L = \sqrt{Var(L)} = 200\sqrt{271} \approx 3300$ (to two significant figures) while $E[L] = 3600$, it is clear that the risk is substantial! ∎

As the preceding example illustrates, the formulas for total expectation and variance, although conceptually quite elementary, can be very tricky to apply in practice. Since these formulas appear frequently in insurance applications, it is very important to be completely at ease with them. Hence, we recommend doing a large number of the exercises in §9.6.

9.4   Moment Generating Function of a Mixture

The moment generating function of a mixture is easy to determine using the formula for total expectation given in the previous section. Indeed, if X is a mixture with mixing random variable Y, then by conditioning on Y and using the formula for total expectation, the moment generating function for X is

$$M_X(t) = E[e^{tX}]$$
$$= E_Y[E[e^{tX} \mid Y]]$$
$$= E_Y[M_{X|Y}(t)],$$

where $M_{X|Y}$ denotes the moment generating function for the random variable $X \mid Y = y$.

Hence, if X is a discrete mixture of $X_1, X_2, \ldots$ with respective mixing weights $k_1, k_2, \ldots$, then the moment generating function of X is given by

$$M_X(t) = \sum k_i M_{X_i}(t),$$

whereas if X is a continuous mixture $X \mid P = p$, where p is a continuous parameter, with mixing weights given by the density function f_P, then the moment generating function of X is given by

$$M_X(t) = \int_{-\infty}^{\infty} M_{X|P}(t) f_P(p) dp.$$

We illustrate the use of these formulas in the following examples.

EXAMPLE 1: Suppose that X is a mixture of a point mass at a and an exponential distribution with parameter λ. Determine a formula for the moment generating function of X. What is the probability that $X = a$?

The moment generating function of X is given by

$$M_X(t) = k_1 e^{at} + k_2 \frac{\lambda}{\lambda - t},$$

where k_1, k_2 are the respective mixing weights. Note that k_1, k_2 are positive numbers such that $k_1 + k_2 = 1$. Since Exponential(λ) is continuous, the probability that $X = a$ is simply the size of the point mass at a, which is k_1. ∎

EXAMPLE 2: Determine a formula for the moment generating function of a negative binomial random variable using the fact that the negative binomial distribution is a mixture of Poisson distributions with gamma mixing weights.

Recall that if $(X \mid \Lambda = \lambda) \sim \text{Poisson}(\lambda)$ and $\Lambda \sim \text{Gamma}(r, \alpha)$, then $X \sim$ Negative-Binomial (r, p), where $p = \alpha/(\alpha + 1)$. Hence, using the formula for the moment generating function of a Poisson random variable and the formula for the gamma density (see §5.2 and §6.1.2), we see that

$$M_X(t) = E_\Lambda[M_{X|\Lambda}(t)]$$

$$= E_\Lambda[\exp(\lambda(e^t - 1))]$$

$$= \int_0^\infty e^{\lambda(e^t-1)} \cdot \frac{\alpha^r \lambda^{r-1} e^{-\alpha\lambda}}{\Gamma(r)} d\lambda$$

$$= \alpha^r (\alpha + 1 - e^t)^{-r} \int_0^\infty \frac{(\alpha + 1 - e^t)^r \lambda^{r-1} e^{-(\alpha+1-e^t)\lambda}}{\Gamma(r)} d\lambda$$

$$= \left(\frac{\alpha}{\alpha + 1 - e^t} \right)^r$$

$$= \left(\frac{p}{1 - (1 - p)e^t} \right)^r,$$

where $p = \alpha/(\alpha + 1)$. This is precisely the form of the moment generating function for a negative binomial random variable stated in §5.3. Details are left to the reader. ∎

9.5 Compound Distributions

Recall from §9.2 that a random variable X has a **compound distribution** if X is a random sum of the type

$$X = W_1 + \cdots + W_N,$$

where the W_j are independent and identically distributed and each W_j is independent of N. Compound models arise in insurance applications in the following contexts:

1. X represents the total claims during a fixed time period for a single policy on which more than one claim is possible. Casualty insurance policies are generally of this type.
2. X represents the total claims during a fixed time period for a group of policies that are independent and of a similar type. The individual policies in the group may or may not allow multiple claims.

In these contexts, N represents the random number of claims received during the time period and W_j represents the size of the jth claim.

In this section, we determine formulas for the distribution function, the moment generating function, and the moments of a compound distribution, and we illustrate the use of these formulas in some important special cases. In particular, we discuss the compound Poisson distribution, the compound mixed Poisson distribution, and the compound negative binomial distribution, which are frequently used to model aggregate claims in insurance. To emphasize the interpretation of a compound model as an aggregate claim, we use the letter C in place of X throughout this section to represent the random sum $W_1 + \cdots + W_N$.

9.5.1 General Formulas

We begin by stating some general formulas for the distribution function, the moment generating function, and the mean, variance, and skewness of a compound distribution. We assume throughout that $C = W_1 + \cdots + W_N$, where the W_j are independent and identically distributed with $W_j \sim W$ and where each W_j is independent of N.

Distribution Function

The distribution function of C is given by

$$F_C(c) = \sum_{n=0}^{\infty} F_{W^{*n}}(c) \cdot p_N(n),$$

where W^{*n} is the n-fold convolution of W (see §8.1.3 for a discussion of convolutions). Hence, C is a discrete mixture of W^{*0}, W^{*1}, W^{*2}, . . . with respective mixing weights $p_N(0)$, $p_N(1)$, $p_N(2)$, ...

The joint density for C and N is illustrated in Figure 9.10a when N has a Poisson distribution and W has an exponential distribution. The corresponding density for C is illustrated in Figure 9.10b. Note that in this illustration, C has a point mass at $c = 0$ and a continuous distribution of probability for $c > 0$; note further that the density function for the continuous part of C does not approach zero as $c \to 0$.

The derivation of the formula for F_C is straightforward using the law of total probability and the fact that the W_j are independent, identically distributed, and independent of N. Indeed,

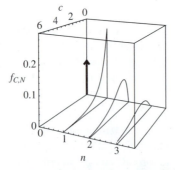

a. Joint Distribution of Aggregate Claim
and Claim Frequency (Truncated Graph)

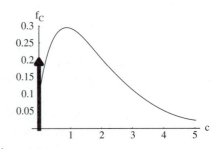

b. Distribution of Aggregate Claim

FIGURE 9.10 Distribution of Aggregate Claim When Claim Frequency Is Poisson and Claim Sizes Are Exponential

$$F_C(c) = \Pr(C \le c)$$

$$= \sum_{n=0}^{\infty} \Pr(C \le c \mid N = n) \cdot \Pr(N = n)$$

$$= \sum_{n=0}^{\infty} \Pr(W_1 + \cdots + W_N \le c \mid N = n) \cdot \Pr(N = n)$$

$$= \sum_{n=0}^{\infty} \Pr(W_1 + \cdots + W_n \le c) \cdot \Pr(N = n)$$

$$= \sum_{n=0}^{\infty} F_{W*n}(c) \cdot p_N(n)$$

as claimed. Note that $\Pr(W_1 + \cdots + W_N \le c \mid N = n) = \Pr(W_1 + \cdots + W_n \le c)$ because the W_j are independent of N. If this had not been the case, we would not have been able to drop the condition $N = n$, and the resulting distribution would not have been the n-fold convolution of W.

Moment and Cumulant Generating Functions

The moment and cumulant generating functions of C are given by

$$M_C(t) = M_N(\log M_W(t))$$

and

$$\psi_C(t) = \psi_N(\psi_W(t)),$$

respectively. Note that the formula for the cumulant generating function has a nicer form than the moment generating function and is more easily remembered.

The derivation of the moment generating function follows directly from the formula for the moment generating function of a mixture and the assumptions on N and the W_j. Indeed,

$$\begin{aligned}
M_C(t) &= E[e^{Ct}] \\
&= E_N[E[e^{Ct} \mid N]] \\
&= E_N[E[e^{(W_1 + \cdots + W_N)t} \mid N]] \\
&= E_N[E[e^{W_1 t}] \cdots E[e^{W_N t}]] \\
&= E_N[E[e^{Wt}]^N] \\
&= E_N[M_W(t)^N] \\
&= E_N[e^{N \log M_W(t)}] \\
&= M_N(\log M_W(t))
\end{aligned}$$

as claimed. Note the use of the fact that the expectation of a product of independent random variables is the product of the expectations. Taking logarithms, we obtain

$$\psi_C(t) = \psi_N(\psi_W(t)),$$

which is the desired formula for the cumulant generating function of C.

Mean, Variance, and Skewness

The mean, variance, and skewness of C are given by

$$E[C] = \mu_N \mu_W,$$

$$Var(C) = \mu_N \sigma_W^2 + \sigma_N^2 \mu_W^2,$$

$$\gamma_C = \frac{\gamma_N \sigma_N^3 \mu_W^3 + 3\sigma_N^2 \mu_W \sigma_W^2 + \mu_N \gamma_W \sigma_W^3}{(\mu_N \sigma_W^2 + \sigma_N^2 \mu_W^2)^{3/2}}.$$

These formulas can be derived from the cumulant generating function of C using the general formula for mean, variance, and skewness given in §4.3.1. Details are left to the reader.

9.5.2   Special Compound Distributions

We now consider some special compound distributions that are frequently used to model aggregate claims in insurance. All of the assumptions on N and the W_j stated in §9.5.1 continue to hold.

The Compound Poisson Distribution

The compound Poisson distribution arises when the claim number random variable N has a Poisson distribution with *constant* parameter λ. In this situation, the formulas for the distribution function of C, the moment and cumulant generating functions of C, and the mean, variance, and skewness of C simplify considerably:

$$F_C(c) = \sum_{n=0}^{\infty} F_{W*n}(c) \cdot \frac{\lambda^n e^{-\lambda}}{n!},$$

$$M_C(t) = \exp\{\lambda(M_W(t) - 1)\},$$

$$\psi_C(t) = \lambda(M_W(t) - 1),$$

$$E[C] = \lambda E[W],$$

$$Var(C) = \lambda E[W^2],$$

$$\gamma_C = \frac{1}{\lambda^{1/2}} \cdot \frac{E[W^3]}{E[W^2]^{3/2}}.$$

Details are left to the reader.

An important property of the compound Poisson distribution is that an independent sum of compound Poisson random variables is compound Poisson. More precisely, suppose that $Y_1, Y_2, \ldots, Y_k$ are independent compound Poisson distributions with respective arrival intensity parameters λ_j and respective claim amount distributions $W^{(j)}$,

$$Y_1 = W_1^{(1)} + \cdots + W_{N_1}^{(1)},$$

$$Y_2 = W_1^{(2)} + \cdots + W_{N_2}^{(2)},$$

$$\vdots$$

$$Y_k = W_1^{(k)} + \cdots + W_{N_k}^{(k)},$$

where $N_j \sim \text{Poisson}(\lambda_j)$ and $W_i^{(j)} \sim W^{(j)}$. Then the random variable C given by $C = Y_1 + \cdots + Y_k$ has a compound Poisson distribution with intensity parameter

$$\lambda = \sum_{j=1}^{k} \lambda_j$$

and with claim amount distribution W, where W is a mixture of $W^{(1)}, \ldots, W^{(k)}$ with respective mixing weights $\lambda_1/\lambda, \ldots, \lambda_k/\lambda$; that is, W has distribution function

$$F_W(w) = \sum_{j=1}^{k} \frac{\lambda_j}{\lambda} F_{W_j}(w).$$

This fact can be proved by showing that the cumulant generating function of C is of compound Poisson type and then appealing to the uniqueness of cumulant generating functions. Details are left to the reader.

We can gain a better understanding of why the sum of a finite number of independent compound Poisson random variables has the form that it does by interpreting the Y_j to be aggregate claims on k different independent groups that are submitting claims simultaneously to the same insurer. With this interpretation, $W_i^{(j)}$ represents the ith claim received from group j. Let C be the total amount of claims for all k groups and model C as $C = W_1 + W_2 + \cdots + W_N$, where N is the total number of claims received for all k groups and W_i is the size of the ith claim when claims from all k groups are considered together. Since $N_1, N_2, \ldots, N_k$ are independent Poisson random variables with respective parameters $\lambda_1, \lambda_2, \ldots, \lambda_k$, it is clear that N must have a Poisson distribution with parameter $\lambda = \lambda_1 + \cdots + \lambda_k$ (see §5.2). Hence, C must have a compound Poisson distribution.[2] Now W_i represents the ith claim received by the insurer. Since we do not know a priori from which group the ith claim is going to come, W_i must be a mixture of $W^{(1)}, W^{(2)}, \ldots, W^{(k)}$. Clearly, the higher the value of λ_j relative to $\lambda = \lambda_1 + \cdots + \lambda_k$, the more likely a given claim is to come from group j. This suggests that $\lambda_1/\lambda, \ldots, \lambda_k/\lambda$ are the appropriate mixing weights for W. Consequently, we see (at least intuitively) that the sum of k independent compound Poisson random variables is compound Poisson with the form described explicitly in the previous paragraph.

The Compound Mixed Poisson Distribution

The compound mixed Poisson distribution arises when the claim number random variable N has a mixed Poisson distribution. A mixed Poisson random variable is a random variable that is a mixture of Poisson random variables (i.e., a random variable of the type $N \sim \text{Poisson}(\Lambda)$, where Λ is a random variable that specifies the mixing weights).

The formulas for the cumulant generating function and the mean, variance, and third central moment of C in this case are as follows:

---

[2] We are implicitly appealing to the independence of the $W_i^{(j)}$, the N_j, and the Y_j here.

$$\psi_C(t) = \psi_\Lambda(M_W(t) - 1),$$
$$E[C] = \mu_\Lambda \mu_W,$$
$$Var(C) = \sigma_\Lambda^2 \mu_W^2 + \mu_\Lambda E[W^2],$$
$$E[(C - \mu_C)^3] = \gamma_\Lambda \sigma_\Lambda^3 \mu_W^3 + 3\sigma_\Lambda^2 \mu_W E[W^2] + \mu_\Lambda E[W^3].$$

Note that these formulas reduce to the corresponding formulas for the compound Poisson distribution when Λ is a constant. Details are left to the reader.

The Compound Negative Binomial Distribution

The compound negative binomial distribution is a special type of compound mixed Poisson distribution in which the mixing variable Λ has a gamma distribution. In this case, the claim number random variable N has a negative binomial distribution (see §5.3). This explains why the distribution is referred to as a compound negative binomial distribution.

The formulas for the moment generating function and the mean, variance, and third central moment of C in this case are as follows:

$$M_C(t) = \left(\frac{p}{1 - (1 - p)M_W(t)} \right)^r$$
$$E[C] = \mu_N \mu_W,$$
$$Var(C) = \frac{1}{r} \cdot \left(\mu_N \mu_W \right)^2 + \mu_N E[W^2],$$
$$E[(C - \mu_C)^3] = \frac{2}{r^2} \left(\mu_N \mu_W \right)^3 + \frac{3}{r} \cdot \mu_N^2 \mu_W E[W^2] + \mu_N E[W^3],$$

where $\mu_N = r(1 - p)/p$. Details are left to the reader.

9.6   Exercises

1. Consider six targets on a wall. Let X be the distance from some fixed origin where a dart lands and let Y be the number of the target at which the dart-thrower is aiming. Suppose that a large number of darts is thrown by the same dart-thrower and that, for each throw, the target is determined by the roll of a die.

   a. Give interpretations for each of the quantities $E[X \mid Y]$, $Var(X \mid Y)$, $E[E[X \mid Y]]$, $E[Var(X \mid Y)]$, $Var(E[X \mid Y])$.

   b. Give interpretations for the formulas $E[X] = E[E[X \mid Y]]$ and $Var(X) = E[Var(X \mid Y)] + Var(E[X \mid Y])$.

2. The policyholders of an insurance company fall into one of two classes. The loss distributions for each class are given in the following table:

Class 1		Class 2	
Size of Loss	Probability	Size of Loss	Probability
1,000	45%	1,000	10%
5,000	35%	5,000	50%
10,000	20%	10,000	40%

Of the policyholders, 20% are in class 1, while the remaining 80% are in class 2. Let L_1, L_2 denote the loss incurred by a randomly selected policyholder from class 1 and class 2, respectively, and let L denote the loss incurred by a randomly selected policyholder whose risk class is unknown.

a. Calculate $E[L_1]$ and $E[L_2]$.

b. Calculate $Var(L_1)$ and $Var(L_2)$.

c. Let X be defined by

$$X = \begin{cases} E[L_1] & \text{with probability .20,} \\ E[L_2] & \text{with probability .80.} \end{cases}$$

Calculate $Var(X)$.

d. Using your answers to parts a, b, and c, calculate $E[L]$ and $Var(L)$.

3. A finance company provides credit to people with a poor or nonexistent credit history. The finance company's clients are of three types: those with no credit history, those with a history of late payments, and those who have filed for personal bankruptcy at some point in the past. The annual loss to the company on a randomly selected account of a given type is exponentially distributed with parameters determined by the risk class of the client. Historical data indicate that the average loss for the three risk classes are $100, $500, and $1000, respectively. Suppose that 20% of the company's clients have no credit history, 50% have a history of late payments, and 30% have a bankruptcy in their past.

a. What is the expected loss on a randomly selected account?

b. What is the variance of the loss on a randomly selected account?

c. What is the probability that the company loses more than $500 on a randomly selected account?

d. To cover its expected losses, the company assesses each client an administration fee. If the company wishes to recoup its expected loss and also have a safety margin equal to 1 standard deviation for each customer, how much should it collect from each client?

4. Consider an insurance policy that indemnifies the policyholder for the full amount of a random loss the first time it occurs during a fixed period of time. Let L be the size of the uncertain loss and let X be the insurer's payout random variable. Suppose that the probability of a claim occurrence is p. Determine formulas for the mean and variance of X in terms of μ_L, σ_L, and p.

5. Consider an insurance policy that indemnifies the policyholder for the amount of a random loss up to or in excess of a fixed amount, as the case may be. Let L be the size of the uncertain loss and let K be the indemnified amount. Determine a formula for the distribution function of K in terms of the distribution function of L in each of the following situations:

a. The insurer caps the indemnified amount at m.

b. The indemnified amount is subject to a deductible d but no cap.

c. The indemnified amount is subject to both a deductible d and a cap m.

6. Consider an insurance policy that indemnifies the policyholder for the amount of a random loss exceeding a deductible d the first time the loss occurs during a given period of time. Let L be the size of the uncertain loss (when it occurs), let K be the indemnified amount of the loss, let p be the probability of a loss occurring, and let X

be the payment the insurer makes to a given policyholder. Determine a formula for the distribution function of X in terms of p, d, and the distribution function of L.

7. Suppose that $X = BI$, where $I \sim$ Bernoulli(p) and $B \geq 0$.

a. By conditioning on I and using the law of total probability, show that

$$F_X(x) = \begin{cases} (1 - p) + p \cdot F_B(x) & \text{for } x \geq 0, \\ 0 & \text{for } x < 0. \end{cases}$$

Deduce that

$$S_X(x) = \begin{cases} p \cdot S_B(x) & \text{for } x \geq 0, \\ 1 & \text{for } x < 0. \end{cases}$$

Sketch possible graphs for F_X and S_X, assuming that the distribution of B is continuous.

b. Sketch a possible graph for the generalized density function of X, assuming that the distribution of B is continuous. Label and interpret all probability masses.

c. By conditioning on I and using the formulas for unconditional expectation and variance given in §9.3, show that

$$E[X] = p\mu_B,$$
$$Var(X) = p(1 - p)\mu_B^2 + p\sigma_B^2.$$

d. By conditioning on I and using the formula for unconditional expectation, show that

$$M_X(t) = (1 - p) + p \cdot M_B(t).$$

Deduce that

$$E[X^k] = pE[B^k] \quad \text{for } k = 1, 2, \ldots.$$

e. Using the formula for the third central moment in terms of the moments about zero and the formula in part d, show that

$$E[(X - \mu_X)^3] = pE[B^3] - 3p^2 E[B^2]E[B] + 2p^3 E[B]^3.$$

Deduce that

$$E[(X - \mu_X)^3] = p\gamma_B\sigma_B^3 + 3p(1 - p)\mu_B\sigma_B^2 + p(1 - p)(1 - 2p)\mu_B^3.$$

8. Suppose that $X = BI$, where $B \sim$ Exponential(λ) and $I \sim$ Bernoulli(p).

a. Show that

$$F_X(x) = \begin{cases} (1 - p) + p(1 - e^{-\lambda x}) & \text{for } x \geq 0, \\ 0 & \text{for } x < 0; \end{cases}$$

$$S_X(x) = \begin{cases} pe^{-\lambda x} & \text{for } x \geq 0, \\ 1 & \text{for } x < 0; \end{cases}$$

$$M_X(t) = (1 - p) + p\frac{\lambda}{\lambda - t}.$$

b. Show that

$$\mu_X = \frac{p}{\lambda},$$

$$\sigma_X^2 = \frac{p(2-p)}{\lambda^2},$$

$$\gamma_X = \frac{2(p^2 - 3p + 3)}{p^{1/2}(2-p)^{3/2}}.$$

9. Suppose that $X = BI$, where $B \sim \text{Gamma}(r, \lambda)$ and $I \sim \text{Bernoulli}(p)$.

a. Show that

$$F_X(x) = \begin{cases} (1-p) + pI_r(\lambda x) & \text{for } x \geq 0, \\ 0 & \text{for } x < 0; \end{cases}$$

$$S_X(x) = \begin{cases} p(1 - I_r(\lambda x)) & \text{for } x \geq 0, \\ 1 & \text{for } x < 0; \end{cases}$$

$$M_X(t) = (1-p) + p\left(\frac{\lambda}{\lambda - t}\right)^r,$$

where $I_r(t)$ is the incomplete gamma function with parameter r; that is, $I_r(t) = \Pr(\text{Gamma}(r, 1) \leq t)$.

b. Show that

$$\mu_X = \frac{rp}{\lambda},$$

$$\sigma_X^2 = \frac{rp}{\lambda^2}(1 + r(1-p)),$$

$$\gamma_X = \frac{(r+1)(r+2) - 3pr(r+1) + 2p^2r^2}{\sqrt{pr}(r(1-p)+1)^{3/2}}.$$

10. In a particular block of business, there is a 15% chance of a claim being submitted on any given policy. On policies for which a claim is submitted, the claim size has an exponential distribution with mean \$100.

a. Determine the probability that a randomly chosen policyholder submits a claim for more than \$50.

b. Determine the probability that the claim on a randomly chosen policy is at most \$25.

c. Suppose that the insurer plans to charge each policyholder the amount $\mu + \sigma$, where μ, σ are the mean and standard deviation of the payout on a randomly chosen policy. How much must each policyholder pay?

11. Suppose that $X = BI$, where $B \sim \text{Exponential}(\Lambda)$, $\Lambda \sim \text{Gamma}(s, \beta)$ and $I \sim \text{Bernoulli}(p)$.

a. Show that

$$F_X(x) = \begin{cases} (1-p) + p\left\{1 - \left(\frac{\beta}{\beta+x}\right)^s\right\} & \text{for } x \geq 0, \\ 0 & \text{for } x < 0; \end{cases}$$

$$S_X(x) = \begin{cases} p\left(\frac{\beta}{\beta+x}\right)^s & \text{for } x \geq 0, \\ 1 & \text{for } x < 0. \end{cases}$$

b. Show that

$$\mu_X = p \cdot \frac{\beta}{s-1},$$

$$\sigma_X^2 = \frac{2p\beta^2}{(s-1)(s-2)}\left\{1 - \frac{p}{2}\left(\frac{s-2}{s-1}\right)\right\}.$$

Determine a formula for the skewness of X in terms of the parameters p, s, β.

12. The policyholders of an insurance company are of two types. Type 1 policyholders have a 25% chance of filing a claim, and type 2 policyholders have a 40% chance of filing a claim. On policies for which a claim is filed, the distribution of the claim size is given by the following table:

Class 1		Class 2	
Claim Size	Probability	Claim Size	Probability
1,000	20%	1,000	70%
5,000	50%	5,000	20%
10,000	30%	10,000	10%

Of the policyholders, 30% are of type 1, and the remaining 70% are of type 2.

a. Calculate the expected value and variance of the payout on a randomly selected policy of type 1.
b. Calculate the expected value and variance of the payout on a randomly selected policy of type 2.
c. Calculate the expected value and variance of the payout on a randomly selected policy whose type is unknown.
d. What is the probability that the payout on a randomly selected policy is at least $5000?

13. A risk-averse individual is subject to a random loss L, which has an exponential distribution with parameter λ. The individual purchases excess-of-loss coverage with deductible d. In this question, you will see that *because of the memoryless property of the exponential distribution*, the distribution of the claim size B payable by the insurer (when a loss occurs) is also exponentially distributed with parameter λ. In particular, the distribution of B is independent of the deductible d. However, you will also see that the claim submission frequency parameter p, which indicates the probability of a claim being submitted, is dependent on d, as intuition suggests it should be.

a. Put $W = (L - d)^+$. Show that

$$F_W(w) = \begin{cases} 1 - e^{-\lambda d} & \text{for } w = 0, \\ 1 - e^{-\lambda(d+w)} & \text{for } w > 0. \end{cases}$$

What is the meaning of the probability mass at $w = 0$?

b. Using the memoryless property of the exponential distribution, show that the distribution of the claim payout B (when there is a payout) is Exponential(λ).

c. Suppose that the probability that the policyholder incurs a loss is q. Show that the probability that the insurer makes a payment to the policyholder is $p = qe^{-\lambda d}$.

14. An auto insurance company provides collision limited excess-of-loss coverage for the layer \$10,000 xs \$250 (i.e., the owner is responsible for the first \$250 in damages and the insurer will pay at most \$10,000 above the \$250 deductible). The cost of repairing a damaged vehicle is assumed to have an exponential distribution with mean \$4000. The insurer's records suggest that 10% of policyholders will file claims during a given year. Assuming that each policyholder files at most one such claim during the year, determine the probability that the insurer's payout on a randomly chosen payout exceeds \$2000. *Suggestion:* Let X denote the payout in thousands of dollars.

15. An insurance portfolio contains n policies. The probability of a claim on a given policy is p, independent of the other policies in the portfolio, and the claim amount distribution on policies for which a claim is filed is B. Let S be the aggregate claims on the portfolio.

a. Show that

$$E[S] = np\mu_B,$$
$$Var(S) = np(1-p)\mu_B^2 + np\sigma_B^2,$$
$$\gamma_S = \frac{1}{n^{1/2}} \frac{p\gamma_B\sigma_B^3 + 3p(1-p)\mu_B\sigma_B^2 + p(1-p)(1-2p)\mu_B^3}{(p(1-p)\mu_B^2 + p\sigma_B^2)^{3/2}}.$$

b. Show that $M_S(t) = (1 - p + pM_B(t))^n$.

c. Suppose that $n = 10{,}000$, $p = .20$, $\mu_B = 1000$, $\sigma_B = 1000$, and $\gamma_B = 50$. Calculate the approximate probability that the aggregate claims on the portfolio exceed \$2.1 million using a normal approximation and using a quadratic normal power approximation.

16. A casualty insurer has a portfolio of 2000 homeowner policies. On each policy, there is a 5% chance of a claim. At most one claim is possible on any given policy during the year. The insurer decides to model claim sizes B measured in hundreds of thousands of dollars on policies for which a claim has occurred using a truncated exponential distribution:

$$F_B(b) = \begin{cases} 0 & \text{for } b < 0, \\ 1 - e^{-\lambda b} & \text{for } 0 \le b < m, \\ 1 & \text{for } b \ge m. \end{cases}$$

Here, m is the maximum allowable payout in \$100,000 lots, and λ is the parameter of the distribution. Historical data suggest that the values of λ and m can be taken to be 2 and 5, respectively.

The insurer would like to ensure that aggregate premiums exceed aggregate payouts at least 95% of the time. If the insurer charges each policyholder the same premium, how much must each policyholder pay?

17. A casualty insurer has a portfolio of 2500 homeowner policies. On 500 of these policies, there is a 10% chance of a claim, while on the remaining 2000 policies, the probability of a claim is only 5%. At most one claim is possible on any given policy during the year. Historical data suggest that claim sizes B measured in hundreds of thousands of dollars on policies for which a claim has occurred can be modeled using a truncated exponential distribution:

$$F_B(b) = \begin{cases} 0 & \text{for } b < 0, \\ 1 - e^{-\lambda b} & \text{for } 0 \le b < m, \\ 1 & \text{for } b \ge m. \end{cases}$$

Here, m is the maximum allowable payout in \$100,000 lots, and λ is the parameter of the distribution. On the policies for which the risk of a claim is 10%, the values of λ and m are 1 and 2.5, respectively, while on the policies for which the risk of a claim is 5%, the values are 2 and 5, respectively.

The insurer would like to ensure that aggregate premiums exceed aggregate payouts at least 95% of the time.

a. Suppose that the insurer charges each policyholder the same premium regardless of risk class. Determine the size of the premium.
b. Suppose that the insurer charges a policyholder with expected loss μ the premium $P = (1 + \theta)\mu$. If θ is the same for all policyholders, determine the premium charged to each type of policyholder.

18. Explain how the formula for the cumulant generating function of a compound distribution generalizes the formula for the cumulant generating function of a deterministic sum of independent, identically distributed random variables. *Hint:* Write out the formula for ψ_C when N has the constant value n.

19. Suppose that $C = W_1 + \cdots + W_N$, where the W_j are independent and identically distributed with $W_j \sim W$ and where each W_j is independent of N. Using the formula for the cumulant generating function of a compound distribution, show that the mean, variance, and skewness of C are given by

$$E[C] = \mu_N \mu_W,$$

$$Var(C) = \mu_N \sigma_W^2 + \sigma_N^2 \mu_W^2,$$

$$\gamma_C = \frac{\gamma_N \sigma_N^3 \mu_W^3 + 3\sigma_N^2 \mu_W \sigma_W^2 + \mu_N \gamma_W \sigma_W^3}{(\mu_N \sigma_W^2 + \sigma_N^2 \mu_W^2)^{3/2}}.$$

20. The number of claims N on a particular policy has the distribution Binomial(4, .25), and the claim size distribution W in hundreds of dollars is given by

$$\Pr(W = 1) = .3, \quad \Pr(W = 2) = .45, \quad \Pr(W = 3) = .25.$$

It is assumed that the sizes of the individual claims are mutually independent and are independent of N. Determine the probability that the aggregate claims on this policy exceed \$500.

21. Suppose that C is a compound Poisson random variable with Poisson parameter λ and claim size random variable W.

a. Show that

$$\psi_C(t) = \lambda(M_W(t) - 1).$$

Deduce that

$$\psi_C^{(j)}(t) = \lambda M_W^{(j)}(t).$$

b. Show that the mean, variance, and skewness of C are given by

$$E[C] = \lambda E[W],$$

$$Var(C) = \lambda E[W^2],$$

$$\gamma_C = \frac{1}{\lambda^{1/2}} \cdot \frac{E[W^3]}{E[W^2]^{3/2}}.$$

What happens to the skewness as $\lambda \to \infty$? What approximation for C does this suggest is reasonable for large λ?

22. Suppose that $Y_1, Y_2, \ldots, Y_k$ are independent compound Poisson distributions with respective arrival intensity parameters λ_j and respective claim amount distributions $W^{(j)}$,

$$Y_1 = W_1^{(1)} + \cdots + W_{N_1}^{(1)},$$

$$Y_2 = W_1^{(2)} + \cdots + W_{N_2}^{(2)},$$

$$\vdots$$

$$Y_k = W_1^{(k)} + \cdots + W_{N_k}^{(k)},$$

where $N_j \sim \text{Poisson}(\lambda_j)$ and $W_i^{(j)} \sim W^{(j)}$. Put $C = Y_1 + \cdots + Y_k$. Using moment or cumulant generating functions, show that C has a compound Poisson distribution with intensity parameter

$$\lambda = \sum_{j=1}^{k} \lambda_j$$

and with claim amount distribution W, where W is a mixture of $W^{(1)}, \ldots, W^{(k)}$ with respective mixing weights $\lambda_1/\lambda, \ldots, \lambda_k/\lambda$; that is, W has distribution function

$$F_W(w) = \sum_{j=1}^{k} \frac{\lambda_j}{\lambda} F_{W_j}(w).$$

Deduce that an independent sum of compound Poisson distributions is compound Poisson.

23. Two portfolios of insurance policies are acquired by an insurance company and combined into a single portfolio. Aggregate claims on each of the original portfolios are independent and follow compound Poisson distributions with intensity parameters $\lambda_1 = 10$, $\lambda_2 = 20$ and claim amount distributions $W^{(1)} \sim \text{Exponential}(1/10)$, $W^{(2)} \sim \text{Exponential}(1/2)$. Let N, W, C denote the claim number, individual claim amount, and aggregate claim for the combined portfolio.

a. What are the distributions of N, W, and C?

b. Calculate the probability that a claim W in the combined portfolio exceeds the amount 5.

c. Determine the expected value, variance, and skewness of the aggregate claims C in the combined portfolio.

d. Estimate $\Pr(C > 150)$ using a normal approximation.

24. A property and casualty insurer has observed that the claim arrival intensity and claim size distribution on a particular block of auto insurance policies vary from season to season. Historical data suggest that the claim arrival intensities in spring, summer, fall, and winter are, respectively,

$$\lambda_1 = 100, \quad \lambda_2 = 75, \quad \lambda_3 = 90, \quad \lambda_4 = 200$$

and the corresponding claim amount distributions are exponential with parameters

$$\alpha_1 = 1, \quad \alpha_2 = 2, \quad \alpha_3 = 1, \quad \alpha_4 = 0.25.$$

Here, claim amounts are measured in thousands of dollars. Suppose that the casualty insurer models aggregate claims in each season using an appropriate compound Poisson distribution and that the claims from season to season are independent.

a. Determine the expected value, variance, and skewness of the insurer's aggregate claims for the year.

b. Using an appropriate approximation, estimate the probability that the insurer's aggregate claims for the year will exceed $1.2 million.

25. A random variable N is said to have a **mixed Poisson distribution** if $N \sim \text{Poisson}(\Lambda)$, where Λ is a random variable. We have already seen that the negative binomial random variable is a mixed Poisson random variable with gamma mixing variable. In this question, you will develop some general properties of mixed Poisson distributions.

a. Using the general formula for calculating the moment generating function of a mixture derived in §9.4 and the moment generating function of the Poisson distribution, show that

$$M_N(t) = E_\Lambda[e^{\Lambda(e^t - 1)}].$$

Deduce that

$$M_N(t) = M_\Lambda(\psi_P(t)),$$

where $P \sim \text{Poisson}(1)$. Conclude that

$$\psi_N(t) = \psi_\Lambda\left(\psi_P(t)\right).$$

b. Show that if $P \sim \text{Poisson}(1)$, then

$$\psi_P^{(k)}(0) = 1 \qquad \text{for all } k > 0.$$

c. Using the formulas derived in parts a and b, show that the mean, variance, and skewness of N are given by

$$E[N] = \mu_\Lambda,$$

$$Var(N) = \mu_\Lambda + \sigma_\Lambda^2,$$

$$\gamma_N = \frac{\gamma_\Lambda \sigma_\Lambda^3 + 3\sigma_\Lambda^2 + \mu_\Lambda}{(\mu_\Lambda + \sigma_\Lambda^2)^{3/2}}.$$

26. Suppose that C is a compound mixed Poisson random variable with claim size random variable W and claim number random variable $N \sim \text{Poisson}(\Lambda)$, where Λ is an unspecified random variable.

   a. Using the cumulant generating function for a mixed Poisson distribution derived in the previous question, show that

   $$\psi_C(t) = \psi_\Lambda(\psi_P(\psi_W(t))),$$

   where $P \sim \text{Poisson}(1)$. Deduce that

   $$\psi_C(t) = \psi_\Lambda(M_W(t) - 1).$$

   b. Show that if Λ has the constant value λ, then

   $$\psi_C(t) = \lambda \cdot (M_W(t) - 1).$$

   Deduce that the formula for the cumulant generating function of a compound mixed Poisson distribution is a generalization of the formula for the cumulant generating function of a compound Poisson distribution.

   c. Using the formula for $\psi_C(t)$ derived in part a, show that the mean, variance, and third central moment of C are given by

   $$E[C] = \mu_\Lambda \mu_W,$$

   $$Var(C) = \sigma_\Lambda^2 \mu_W^2 + \mu_\Lambda E[W^2],$$

   $$E[(C - \mu_C)^3] = \gamma_\Lambda \sigma_\Lambda^3 \mu_W^3 + 3\sigma_\Lambda^2 \mu_W E[W^2] + \mu_\Lambda E[W^3].$$

   d. Show that the variance of C may be written as

   $$Var(C) = Var(\Lambda)E[W]^2 + E[\Lambda]Var(W) + E[\Lambda]E[W]^2.$$

   Explain how this sum decomposes the variability in C into variability due to mixing, variability due to claim size variation, and in the absence of mixing and claim size uncertainty, variability due to claim number variation.

27. A special type of compound mixed Poisson distribution is the **compound negative binomial distribution.** A compound negative binomial random variable is a random sum of the form $C = W_1 + \cdots + W_N$, where $N \sim \text{NegativeBinomial}(r, p)$ and where the W_j are identically distributed, mutually independent, and independent of N.

   a. Show that the moment generating function of C is given by

   $$M_C(t) = \left(\frac{p}{1 - (1 - p)M_W(t)}\right)^r.$$

   b. Show that the mean, variance, and third central moment of C are given by

$$E[C] = \mu_N \mu_W,$$

$$Var(C) = \frac{1}{r} \cdot \left(\mu_N \mu_W\right)^2 + \mu_N E[W^2],$$

$$E[(C - \mu_C)^3] = \frac{2}{r^2} \left(\mu_N \mu_W\right)^3 + \frac{3}{r} \cdot \mu_N^2 \mu_W E[W^2] + \mu_N E[W^3],$$

where $\mu_N = r(1-p)/p$.

28. Consider the compound distribution $C = W_1 + \cdots + W_N$ with $N \sim \text{Geometric}(p)$ and $W_j \sim \text{Exponential}(\alpha)$.

    a. Show that

    $$M_C(t) = p + (1-p)\frac{\alpha p}{\alpha p - t}.$$

    Deduce that C is a mixture of a discrete probability mass and an exponential distribution with parameter αp.

    b. Determine a formula for the distribution function F_C of C.

    c. Determine formulas in terms of p and α for each of the following probabilities: $\Pr(C > \alpha)$, $\Pr(C = 0)$, $\Pr(C \le p\alpha)$.

29. Determine which of the following statements are true and which are false. Where possible, alter the false statements to become true statements.

    a. The Pareto distribution is a continuous mixture of exponential distributions with gamma mixing parameter.

    b. The negative binomial distribution is a continuous mixture of binomial distributions with gamma mixing parameter.

    c. If $(N \mid P = p) \sim \text{Binomial}(m, p)$ and $P \sim \text{Beta}(r, s)$, then $(P \mid N = n) \sim \text{Beta}(n + r, m + s - n)$.

    d. The generalized Pareto distribution is a continuous mixture of beta distributions with gamma mixing parameter. (See exercise 33 in §6.4 for a definition of a generalized Pareto random variable.)

    e. If $(N \mid \Lambda = \lambda) \sim \text{Poisson}(\lambda)$ and $\Lambda \sim \text{Gamma}(r, \alpha)$, then $(\Lambda \mid N = n) \sim \text{Gamma}(r + n, \alpha + 1)$.

    f. The return on a portfolio of stocks is a mixture of the returns on the individual stocks with mixing weights given by the weightings of the securities in the portfolio.

    g. If $(X \mid \Lambda = \lambda) \sim \text{Gamma}(r, \lambda)$ and $\Lambda \sim \text{Gamma}(s, \beta)$, then $(\Lambda \mid X = x) \sim \text{Gamma}(r + s, x + \beta)$.

    h. Mixed distributions are mixtures.

    i. Marginal distributions are mixtures.

    j. The nonnegative random variable W with distribution function given by

    $$F_W(w) = 1 - \frac{2}{5}e^{-w} - \frac{1}{2}e^{-2w}, \qquad \text{for } w \ge 0$$

    is a mixture of exponentials.

30. On a particular insurance policy, at most two claims are possible. There is a 50% chance that no claim will be filed, a 30% chance that one claim will be filed, and a 20% chance that two claims will be filed. Claims (when they occur) are exponentially

distributed with mean 2, are mutually independent, and are independent of the number of claim occurrences. Let X be the aggregate claims received on this policy.

a. Determine a formula for the distribution function of X.

b. Determine the mean and variance of X by considering the formulas for total expectation and variance given in §9.3.

c. Determine the probability that aggregate claims exceed three.

31. Determine formulas for the moment generating function and the moments about zero of a finite mixture of the distributions Gamma(r_i, α_i) with mixing weights k_i.

32. Suppose that $(Y \mid \Lambda = \lambda) \sim$ Exponential(λ) and $\Lambda \sim$ Gamma(s, β). We know from §6.1.3 that the unconditional distribution of Y is Pareto(s, β). In this question, you will derive the formulas for the moments $E[Y^k]$ stated in §6.1.3 using the formula for unconditional expectation:

$$E[Y^k] = E_\Lambda[E[Y^k \mid \Lambda]].$$

a. Using the moment generating function for the exponential distribution, show that

$$E[Y^k \mid \Lambda] = \frac{k!}{\Lambda^k}.$$

b. By considering the gamma density with parameters $s - k$ and β, show that for $k < s$

$$E_\Lambda\left[\frac{k!}{\Lambda^k}\right] = k!\beta^k\frac{\Gamma(s - k)}{\Gamma(s)}.$$

c. Deduce that

$$E[Y^k] = \frac{\beta^k k!}{(s - 1)(s - 2) \cdots (s - k)}.$$

Hint: Use the fact that $\Gamma(x + 1) = x\Gamma(x)$ for all x.

33. Consider the following game involving a balanced die and two spinners—one red and one white—each divided into six equal sections with a single number printed on each of the six sections. The red spinner contains two 4s, three 8s, and one 10. The white spinner contains two 3s, three 10s, and one 12.

The die is rolled and the outcome is observed. If the outcome of the roll is even, the red spinner is spun; otherwise, the white spinner is spun. We win the amount showing on the spinner that was spun.

Determine the mean and variance of our winnings.

10 The Markowitz Investment Portfolio Selection Model

The first nine chapters of this book presented the basic probability theory with which any student of insurance and investments should be familiar. In this final chapter, we discuss an important application of the basic theory: the Nobel Prize winning investment portfolio selection model due to Harry Markowitz.[1] This material is not discussed in other probability texts of this level; however, it is a nice application of the basic theory and it is very accessible.

The Markowitz portfolio selection model has had a profound effect on the investment industry. Indeed, the popularity of index funds (mutual funds that track the performance of an index such as the S&P 500 and do not attempt to "beat the market") can be traced to a surprising consequence of the Markowitz model: that every investor, regardless of risk tolerance, should hold the same portfolio of risky securities. This result has called into question the conventional wisdom that it is possible to beat the market with the "right" investment manager and in so doing has revolutionized the investment industry.

Our presentation of the Markowitz model is organized in the following way. We begin by considering portfolios of two securities. An important example of a portfolio of this type is one consisting of a stock mutual fund and a bond mutual fund. Seen from this perspective, the portfolio selection problem with two securities is equivalent to the problem of asset allocation between stocks and bonds. We then consider portfolios of two risky securities and a risk-free asset, the prototype being a portfolio of a stock mutual fund, a bond mutual fund, and a money-market fund. Finally, we consider portfolio selection when an unlimited number of securities is available for inclusion in the portfolio.

We conclude this chapter by briefly discussing an important consequence of the Markowitz model, namely, the Nobel Prize winning capital asset pricing model due to William Sharpe.[2] The CAPM,[3] as it is referred to, gives a formula for the fair return on a risky security when the overall market is in equilibrium. Like the Markowitz model, the CAPM has had a profound influence on portfolio management practice.

---

[1] Markowitz received the Nobel Prize in Economics for this work in 1990. Markowitz's original paper was published in 1952 in the *Journal of Finance*. For details, consult §10.5.

[2] Sharpe, who was a student of Markowitz, shared the Nobel Prize in Economics with Markowitz in 1990. The capital asset pricing model was actually developed independently by William Sharpe, John Lintner, and Jan Mossin. However, Sharpe's work was the first to be published. Details are given in §10.5.

[3] pronounced "cap M"

10.1 Portfolios of Two Securities

In this section, we consider portfolios consisting of only two securities, S_1 and S_2. These two securities could be a stock mutual fund and a bond mutual fund, in which case the portfolio selection problem amounts to asset allocation, or they could be something else. Our objective is to determine the "best mix" of S_1 and S_2 in the portfolio.

Portfolio Opportunity Set

Let's begin by describing the set of possible portfolios that can be constructed from S_1 and S_2. Suppose that the current value of our portfolio is d dollars and let d_1 and d_2 be the dollar amounts invested in S_1 and S_2, respectively. Let R_1 and R_2 be the simple returns on S_1 and S_2 over a future time period that begins now and ends at a fixed future point in time and let R be the corresponding simple return for the portfolio. Then, if no changes are made to the portfolio mix during the time period under consideration,

$$d(1 + R) = d_1(1 + R_1) + d_2(1 + R_2).$$

Hence, the return on the portfolio over the given time period is

$$R = xR_1 + (1 - x)R_2,$$

where $x = d_1/d$ is the fraction of the portfolio currently invested in S_1. Consequently, by varying x, we can change the return characteristics of the portfolio.

Now if S_1 and S_2 are risky securities, as we will assume throughout this section, then R_1, R_2, and R are all random variables. Suppose that R_1 and R_2 are both normally distributed and their joint distribution has a bivariate normal distribution. This may appear to be a strong assumption. However, data on stock price returns suggest that, as a first approximation, it is not unreasonable. Then, from the properties of the normal distribution (see §6.3.1), it follows that R is normally distributed and that the distributions of R_1, R_2, and R are completely characterized by their respective means and standard deviations. Hence, since R is a linear combination of R_1 and R_2, the set of possible investment portfolios consisting of S_1 and S_2 can be described by a curve in the σ-μ plane.

To see this more clearly, note that from the identity $R = xR_1 + (1 - x)R_2$ and the properties of means and variances, we have

$$\mu_R = x\mu_{R_1} + (1 - x)\mu_{R_2},$$
$$\sigma_R^2 = x^2\sigma_{R_1}^2 + 2x(1 - x)\rho\sigma_{R_1}\sigma_{R_2} + (1 - x)^2\sigma_{R_2}^2,$$

where ρ is the correlation between R_1 and R_2. Eliminating x from these two equations by substituting $x = (\mu_R - \mu_{R_2})/(\mu_{R_1} - \mu_{R_2})$, which we obtain from the equation for μ_R, into the equation for σ_R^2, we obtain

$$\sigma_R^2 = \frac{(\mu_R - \mu_{R_2})^2}{(\mu_{R_1} - \mu_{R_2})^2}\sigma_{R_1}^2 + 2\frac{(\mu_R - \mu_{R_2})(\mu_{R_1} - \mu_R)}{(\mu_{R_1} - \mu_{R_2})^2}\rho\sigma_{R_1}\sigma_{R_2}$$
$$+ \frac{(\mu_{R_1} - \mu_R)^2}{(\mu_{R_1} - \mu_{R_2})^2}\sigma_{R_2}^2,$$

which describes a curve in the σ_R-μ_R plane as claimed.

Notice that σ_R and μ_R change with x, while σ_{R_1}, μ_{R_1}, σ_{R_2}, μ_{R_2}, ρ remain fixed. To emphasize the fact that σ_R and μ_R are variables, let's drop the subscript R from now on. Then, the preceding equation for σ_R^2 can be written as

$$\sigma^2 = A(\mu - \mu_0)^2 + \sigma_0^2,$$

where A, μ_0, σ_0^2 are parameters depending only on $\mathcal{S}_1$ and $\mathcal{S}_2$ with $A > 0$ and $\sigma_0^2 \geq 0$. Indeed,

$$A = \frac{1}{(\mu_{R_1} - \mu_{R_2})^2} \left\{ \sigma_{R_1}^2 - 2\rho\sigma_{R_1}\sigma_{R_2} + \sigma_{R_2}^2 \right\}$$

$$= \frac{1}{(\mu_{R_1} - \mu_{R_2})^2} \left\{ (\sigma_{R_1} - \sigma_{R_2})^2 + 2(1 - \rho)\sigma_{R_1}\sigma_{R_2} \right\}$$

$$> 0$$

(the inequality holding since $-1 \leq \rho \leq 1$), and

$$\sigma_0^2 = \frac{\sigma_{R_1}^2\sigma_{R_2}^2(1 - \rho^2)}{(\sigma_{R_1} - \sigma_{R_2})^2 + 2(1 - \rho)\sigma_{R_1}\sigma_{R_2}}$$

$$\geq 0$$

(again since $-1 \leq \rho \leq 1$). Further,

$$\mu_0 = \frac{\mu_{R_1}\sigma_{R_2}^2 - (\mu_{R_1} + \mu_{R_2})\rho\sigma_{R_1}\sigma_{R_2} + \mu_{R_2}\sigma_{R_1}^2}{(\sigma_{R_1} - \sigma_{R_2})^2 + 2(1 - \rho)\sigma_{R_1}\sigma_{R_2}}.$$

Consequently, the possible portfolios lie on the curve

$$\sigma^2 = A(\mu - \mu_0)^2 + \sigma_0^2, \qquad \sigma \geq 0,$$

which we recognize as being the right half of a hyperbola with vertex at (σ_0, μ_0) (Figure 10.1).

Notice that the hyperbola $\sigma^2 = A(\mu - \mu_0)^2 + \sigma_0^2$ describes a trade-off between risk (as measured by σ) and reward (as measured by μ). Indeed, along the upper branch of

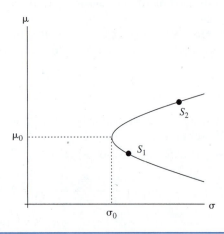

FIGURE 10.1 Set of Possible Portfolios consisting of $\mathcal{S}_1$ and $\mathcal{S}_2$

the hyperbola, it is clear that to obtain a greater reward, we must invest in a portfolio with greater risk; in other words, "no pain, no gain." The portfolios on the lower branch of the hyperbola, while theoretically possible, will never be selected in practice. The reason is that for any selected risk level σ, the portfolio on the upper branch with standard deviation σ will always have higher expected return (i.e., higher reward) than the portfolio on the lower branch with standard deviation σ and, hence, will always be preferred to the portfolio on the lower branch. Consequently, the only portfolios that need be considered further are the ones on the upper branch. These portfolios are referred to as *efficient portfolios*. In general, an **efficient portfolio** is one that provides the highest reward for a given level of risk.

Determining the Optimal Portfolio

Now let's consider which portfolio in the efficient set is best. To do this, we need to consider the investor's tolerance for risk. Since different investors in general have different risk tolerances, we should expect each investor to have a different optimal portfolio. We will soon see that this is indeed the case.

Let's consider one particular investor and let's suppose that this investor is able to assign a number $U(F_R)$ to each possible investment return distribution F_R with the following properties:

1. $U(F_{R_a}) > U(F_{R_b})$ if and only if the investor prefers the investment with return R_a to the investment with return R_b.
2. $U(F_{R_a}) = U(F_{R_b})$ if and only if the investor is indifferent to choosing between the investment with return R_a and the investment with return R_b.

The functional U, which maps distribution functions to the real numbers, is called a *utility* functional. Note that different investors in general have different utility functionals.

There are many different forms of utility functionals. For simplicity, we assume that every investor has a utility functional of the form

$$U(F_R) = \mu - k\sigma^2,$$

where $k > 0$ is a number that measures the investor's level of risk aversion and is unique to each investor. (Here, μ and σ represent the mean and standard deviation of the return distribution F_R.) There are good theoretical reasons for assuming a utility functional of this form. However, in the interest of brevity, we omit the details. Note that in assuming a utility functional of this form, we are implicitly assuming that among portfolios with the same level of risk, greater expected return is preferable, and among portfolios with the same expected return, less risk is preferable.

The portfolio optimization problem for an investor with risk tolerance level k can then be stated as follows:

Maximize: $U = \mu - k\sigma^2$
Subject to: $\sigma^2 = A(\mu - \mu_0)^2 + \sigma_0^2.$

This is a simple constrained optimization problem that can be solved by substituting the condition into the objective function and then using standard optimization techniques from single variable calculus. Alternatively, this optimization problem can be solved using the Lagrange multiplier method from multivariable calculus.

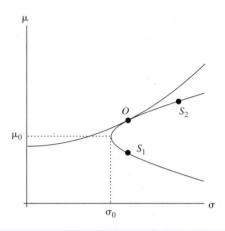

FIGURE 10.2 Portfolio with Greatest Utility

Graphically, the maximum value of U is the number u such that the parabola $\mu - k\sigma^2 = u$ is *tangent* to the hyperbola $\sigma^2 = A(\mu - \mu_0)^2 + \sigma_0^2$. (See Figure 10.2. The optimal portfolio in this figure is denoted by O.) Clearly, the optimal portfolio depends on the value of k, which specifies the investor's level of risk aversion.

Carrying out the details of the optimization, we find that when S_1 and S_2 are both risky securities (i.e., $\sigma_{R_1} \neq 0$ and $\sigma_{R_2} \neq 0$), the risk–reward coordinates of the optimal portfolio O are

$$\sigma^* = \sqrt{\frac{1}{4Ak^2} + \sigma_0^2},$$

$$\mu^* = \mu_0 + \frac{1}{2kA}.$$

Since $\mu = x\mu_{R_1} + (1 - x)\mu_{R_2}$, it follows that the portion of the portfolio that should be invested in S_1 is

$$x = \frac{\mu^* - \mu_{R_2}}{\mu_{R_1} - \mu_{R_2}}.$$

Comment We have assumed that short selling without margin posting is possible (i.e., we have assumed that x can assume any real value, including values outside the interval $[0, 1]$). In the more realistic case, where short selling is restricted, the optimal portfolio may differ from the one just determined.

EXAMPLE 1: The return on a bond fund has expected value 5% and standard deviation 12%, while the return on a stock fund has expected value 10% and standard deviation 20%. The correlation between the returns is 0.60. Suppose that an investor's utility functional is of the form $U = \mu - \frac{1}{100}\sigma^2$. Determine the investor's optimal allocation between stocks and bonds assuming short selling without margin posting is possible.

It is customary in problems of this type to assume that the utility functional is calibrated using percentages. Hence, if R_1, R_2 represent the returns on the bond and stock funds, respectively, then

$$U(F_{R_1}) = 5 - \frac{1}{100}(12^2) = 3.56,$$

$$U(F_{R_2}) = 10 - \frac{1}{100}(20^2) = 6.$$

Note that such a calibration can always be achieved by proper selection of k.

From the formulas that have been developed, the expected return on the optimal portfolio is

$$\mu^* = \mu_0 + \frac{1}{2kA},$$

where $k = 1/100$,

$$
\begin{aligned}
\mu_0 &= \frac{\mu_{R_1}\sigma_{R_2}^2 - (\mu_{R_1} + \mu_{R_2})\rho\sigma_{R_1}\sigma_{R_2} + \mu_{R_2}\sigma_{R_1}^2}{(\sigma_{R_1} - \sigma_{R_2})^2 + 2(1-\rho)\sigma_{R_1}\sigma_{R_2}} \\
&= \frac{(5)(20^2) - (5+10)(0.60)(12)(20) + (10)(12^2)}{(12-20)^2 + 2(0.40)(12)(20)} \\
&= 21.875
\end{aligned}
$$

and

$$
\begin{aligned}
A &= \frac{1}{(\mu_{R_1} - \mu_{R_2})^2}\left\{(\sigma_{R_1} - \sigma_{R_2})^2 + 2(1-\rho)\sigma_{R_1}\sigma_{R_2}\right\} \\
&= \frac{1}{(5-10)^2}\left\{(12-20)^2 + 2(0.40)(12)(20)\right\} \\
&= 10.24.
\end{aligned}
$$

Hence, the portion of the portfolio that should be invested in bonds is

$$
\begin{aligned}
x &= \frac{\mu^* - \mu_{R_2}}{\mu_{R_1} - \mu_{R_2}} \\
&= \frac{26.7578125 - 10}{5 - 10} \\
&= -3.3515625.
\end{aligned}
$$

Thus, for a portfolio of $1000, it is optimal to sell short $3351.56 worth of bonds and invest $4351.56 in stocks. ∎

Special Cases of the Portfolio Opportunity Set

We conclude this section by highlighting the form of the portfolio opportunity set in some special cases. Throughout, we assume that S_1 and S_2 are securities such that

$$\mu_{R_1} < \mu_{R_2} \quad \text{and} \quad \sigma_{R_1} < \sigma_{R_2}.$$

(The situation where $\mu_{R_1} < \mu_{R_2}$ and $\sigma_{R_1} > \sigma_{R_2}$ is not interesting since then S_2 is always preferable to S_1.) We also assume that no short positions are allowed.

Assets Are Perfectly Positively Correlated Suppose that $\rho = 1$ (i.e., R_1 and R_2 are perfectly positively correlated). Then the set of possible portfolios is a straight line, as illustrated in Figure 10.3a.

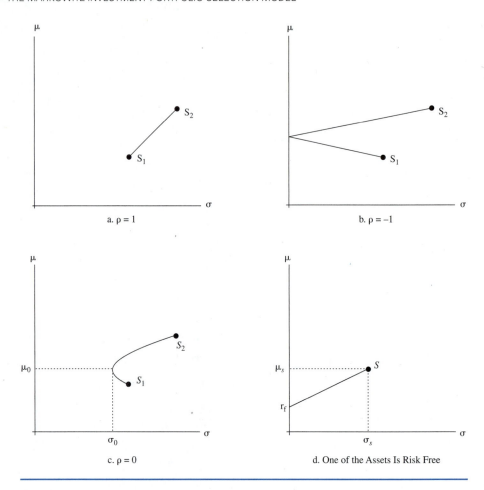

FIGURE 10.3 Special Cases of the Portfolio Opportunity Set

Assets Are Perfectly Negatively Correlated Suppose that $\rho = -1$ (i.e, R_1 and R_2 are perfectly negatively correlated). Then the set of possible portfolios is as illustrated in Figure 10.3b. Note that, in this case, it is possible to construct a perfectly hedged portfolio (i.e., a portfolio with $\sigma = 0$).

Assets Are Uncorrelated Suppose that $\rho = 0$. Then the portfolio opportunity set has the form illustrated in Figure 10.3c. From this picture, it is clear that starting from a portfolio consisting only of the low-risk security S_1, it is possible to decrease risk and increase expected return simultaneously by adding a portion of the high-risk security S_2 to the portfolio. Hence, even investors with a low level of risk tolerance should have a portion of their portfolios invested in the high-risk security S_2. (See also the discussion on the standard deviation of a sum in §8.3.3.)

One of the Assets Is Risk Free Suppose that S_1 is a risk-free asset (i.e., $\sigma_{R_1} = 0$) and put $\mu_{R_1} = r_f$, the risk-free rate of return. Further, let S denote S_2 and write σ_S, μ_S in place of σ_{R_2}, μ_{R_2}. Then the efficient set is given by

$$\mu = r_f + \frac{\mu_S - r_f}{\sigma_S} \cdot \sigma, \qquad \sigma > 0.$$

This is a line in risk–reward space with slope $(\mu_S - r_f)/\sigma_S$ and μ-intercept r_f (see Figure 10.3d).

10.2 Portfolios of Two Risky Securities and a Risk-Free Asset

Suppose now that we are to construct a portfolio from two risky securities and a risk-free asset. This corresponds to the problem of allocating assets among stocks, bonds, and short-term money-market securities. Let R_1, R_2 denote the returns on the risky securities and suppose that $\mu_{R_1} < \mu_{R_2}$ and $\sigma_{R_1} < \sigma_{R_2}$. Further, let r_f denote the risk-free rate.

The Efficient Set

From our discussion in §10.1, we know that the portfolios consisting only of the two risky securities S_1, S_2 must lie on a hyperbola of the type illustrated in Figure 10.4.

We claim that when a risk-free asset is also available, the efficient set consists of the portfolios on the tangent line through $(0, r_f)$ (Figure 10.5). Note that r_f in this figure is the μ-intercept of the tangent line through T.

To see why this is so, consider a portfolio P consisting only of S_1 and S_2 and let T be the tangency portfolio (i.e., the portfolio which is on both the hyperbola and the tangent line). From our discussion in §10.1, we know that every portfolio consisting of the risky portfolio P and the risk-free asset lies on the straight line through P and $(0, r_f)$, and every portfolio consisting of the tangency portfolio T and the risk-free asset lies on the tangent line through T and $(0, r_f)$ (Figure 10.6). Hence, from Figure 10.6, it is clear that every portfolio consisting of P and the risk-free asset is dominated by a portfolio consisting

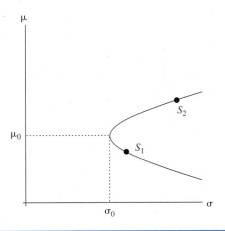

FIGURE 10.4 Portfolio Opportunity Set for Two Securities

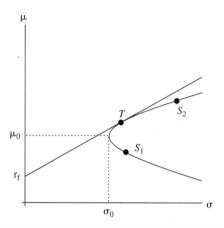

FIGURE 10.5 Efficient Set as a Tangent Line

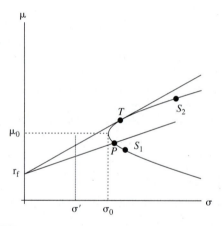

FIGURE 10.6 Portfolios Containing the Tangency Portfolio Dominate All Others

of T and the risk-free asset. Indeed, for any given risk level σ', there is a portfolio on the line through T and $(0, r_f)$ with greater μ than the corresponding portfolio on the line through P and $(0, r_f)$. Hence, given a choice between holding P as the risky part of our portfolio and holding T as the risky part, we should always choose T.

Consequently, the efficient portfolios lie on the line through $(0, r_f)$ and T as claimed. Note, in particular, that the efficient portfolios all have the same risky part T; the only difference among them is the portion allocated to the risk-free asset. This surprising result, which provides a theoretical justification for the use of index mutual funds by every investor, is known as the **mutual fund separation theorem.** In view of this separation theorem, the portfolio selection problem is reduced to determining the fraction of an investor's portfolio that should be invested in the risk-free asset. This is a straightforward problem when the utility functional has the form $U = \mu - k\sigma^2$ (Figure 10.7). The investor's optimal portfolio in this figure is denoted by O. Details are left to the reader.

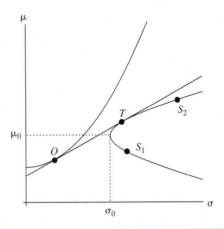

FIGURE 10.7 Optimal Portfolio for a Given Utility Functional

Determining the Tangency Portfolio

The tangency portfolio T has the property of being the portfolio on the hyperbola for which the ratio[4]

$$\frac{\mu - r_f}{\sigma}$$

is maximal. (Convince yourself that this is so.) Hence, one method of determining the coordinates of T is to solve the following optimization problem:

Maximize: $\theta = \frac{\mu - r_f}{\sigma}$

Subject to: $\sigma^2 = A(\mu - \mu_0)^2 + \sigma_0^2.$

We will determine the coordinates of T in a slightly different way, which is more easily adapted when the number of risky securities is greater than two.

Recall that the efficient portfolios are the ones with the least risk (i.e., smallest σ) for a given level of expected return μ. Hence, the efficient set, which we already know is the line through $(0, r_f)$ and T, can be determined by solving the following collection of optimization problems (one for each μ):

Minimize: $Var(R)$

Subject to: $E[R] = \mu.$

Let y_1, y_2, y_3 be the amounts allocated to S_1, S_2, and the risk-free asset, respectively. Then the return on such a portfolio is

$$R = y_1 R_1 + y_2 R_2 + y_3 r_f,$$

and so

$$Var(R) = \sigma_1^2 y_1^2 + 2\sigma_{12} y_1 y_2 + \sigma_2^2 y_2^2$$

---

[4] This ratio is known as the **Sharpe ratio.** It is the slope of the line through (σ, μ) and $(0, r_f)$, and it provides a measure of the excess return to risk for a given security.

and

$$E[R] = \mu_1 y_1 + \mu_2 y_2 + r_f y_3,$$

where $\mu_j = E[R_j]$, $\sigma_j^2 = Var(R_j)$, and $\sigma_{ij} = Cov(R_i, R_j)$. Note that $Var(R)$ does not contain any terms in y_3! Consequently, the optimization problem can be written as

Minimize: $\quad \psi = \sigma_1^2 y_1^2 + 2\sigma_{12} y_1 y_2 + \sigma_2^2 y_2^2$

Subject to: $\quad \mu_1 y_1 + \mu_2 y_2 + r_f y_3 = \mu,$
$\quad\quad\quad\quad y_1 + y_2 + y_3 = 1.$

Note that the conditions in this optimization are equivalent to the conditions

$$(\mu_1 - r_f)y_1 + (\mu_2 - r_f)y_2 = \mu - r_f,$$
$$y_1 + y_2 + y_3 = 1.$$

(Substitute $y_1 + y_2 + y_3 = 1$ into the first condition.) Since the only place that y_3 now occurs is in the condition $y_1 + y_2 + y_3 = 1$, this means that we can solve the general optimization problem by first solving the simpler problem

Minimize: $\quad \psi$

Subject to: $\quad (\mu_1 - r_f)y_1 + (\mu_2 - r_f)y_2 = \mu - r_f$

and then determining y_3 by $y_3 = 1 - y_1 - y_2$. Indeed, ψ will still be minimized because the required y_1, y_2 will be the same in both optimization problems.

The simpler optimization problem can be solved using the Lagrange multiplier method. In general, we will have

$$\nabla \psi = \tau \nabla g \quad \text{and} \quad g = 0,$$

where $g = (\mu_1 - r_f)y_1 + (\mu_2 - r_f)y_2 - (\mu - r_f)$ and τ is the Lagrange multiplier. The letter λ is generally reserved in investment theory for the reward-to-variability ratio $(\mu - r_f)/\sigma^2$ and, hence, will not be used to represent a Lagrange multiplier here. Performing the required differentiation, we obtain

$$\begin{pmatrix} \sigma_1^2 & \sigma_{12} \\ \sigma_{12} & \sigma_2^2 \end{pmatrix} \begin{pmatrix} y_1 \\ y_2 \end{pmatrix} = \left(\frac{\tau}{2}\right) \begin{pmatrix} \mu_1 - r_f \\ \mu_2 - r_f \end{pmatrix},$$

or equivalently,

$$\begin{pmatrix} \sigma_1^2 & \sigma_{12} \\ \sigma_{12} & \sigma_2^2 \end{pmatrix} \begin{pmatrix} z_1 \\ z_2 \end{pmatrix} = \begin{pmatrix} \mu_1 - r_f \\ \mu_2 - r_f \end{pmatrix},$$

where

$$z_1 = \frac{2}{\tau} y_1, \qquad z_2 = \frac{2}{\tau} y_2.$$

Note that the Lagrange multiplier τ will depend in general on μ.

Now the tangency portfolio T lies on the efficient set and has the property that $y_3 = 0$ (i.e., no portion of the tangency portfolio is invested in the risk-free asset). Hence, the values of y_1 and y_2 for the tangency portfolio are given by

$$y_1 = \frac{z_1}{z_1 + z_2}, \qquad y_2 = \frac{z_2}{z_1 + z_2},$$

where (z_1, z_2) is the unique[5] solution of the preceding matrix equation. Indeed, since T lies on the efficient set, we must have $(y_1, y_2) = \frac{\tau}{2}(z_1, z_2)$, and since $y_1 + y_2 = 1 - y_3 = 1$, we must have $\tau/2 = 1/(z_1 + z_2)$. The risk–reward coordinates $(\sigma_{R_T}, \mu_{R_T})$ for the tangency portfolio are then determined using the equations

$$\mu_{R_T} = y_1\mu_1 + y_2\mu_2,$$
$$\sigma_{R_T}^2 = y_1^2\sigma_1^2 + 2y_1y_2\sigma_{12} + y_2^2\sigma_2^2,$$

where y_1, y_2 are the fractions just calculated.

Updating the Tangency Portfolio When the Risk-Free Rate Changes

We have just seen that for a given risk-free rate r_f, the tangency portfolio T has allocations x_1, x_2 in the risky securities S_1, S_2 given by

$$x_1 = \frac{z_1}{z_1 + z_2}, \qquad x_2 = \frac{z_2}{z_1 + z_2},$$

where z_1, z_2 are uniquely determined by the matrix equation

$$\begin{pmatrix} \sigma_1^2 & \sigma_{12} \\ \sigma_{12} & \sigma_2^2 \end{pmatrix}\begin{pmatrix} z_1 \\ z_2 \end{pmatrix} = \begin{pmatrix} \mu_1 \\ \mu_2 \end{pmatrix} - r_f\begin{pmatrix} 1 \\ 1 \end{pmatrix}.$$

From these equations, it is clear that the allocations x_1, x_2 will change as r_f varies. We now describe a procedure that allows us to easily update the allocations of S_1 and S_2 in the tangency portfolio when the risk-free rate changes.

We begin by observing that the matrix equation for (z_1, z_2) given earlier can be written in the form

$$\begin{pmatrix} z_1 \\ z_2 \end{pmatrix} = \begin{pmatrix} c_{11} & c_{12} \\ c_{21} & c_{22} \end{pmatrix}\begin{pmatrix} 1 \\ r_f \end{pmatrix},$$

where $\begin{pmatrix} c_{11} & c_{12} \\ c_{21} & c_{22} \end{pmatrix}$ is a matrix that only depends on $\mu_1, \mu_2, \sigma_1, \sigma_2, \rho$ (i.e., only depends on S_1 and S_2). Indeed, multiplying both sides of the original matrix equation for (z_1, z_2) by $\begin{pmatrix} \sigma_1^2 & \sigma_{12} \\ \sigma_{12} & \sigma_2^2 \end{pmatrix}^{-1}$ (the inverse of the variance–covariance matrix) and simplifying, we obtain a matrix equation of this form.

Now since the c_{ij} only depend on S_1 and S_2 and do not depend on r_f, they need only be determined once. Moreover, we do not actually need to know the values of $\mu_1, \mu_2, \sigma_1, \sigma_2, \rho$ to determine the c_{ij}. Indeed, we need only know the values of (z_1, z_2) and the corresponding[6] r_f for two portfolios on the hyperbola; from this information, we obtain a linear system in the c_{ij} that is easily solved.

Once the values of the c_{ij} are known, it is easy to determine the allocation amounts x_1, x_2 corresponding to a given risk-free rate r_f, and it is easy to update the values of x_1, x_2 when r_f changes.

---

[5] Note that the matrix equation will have a unique solution since the determinant of the variance–covariance matrix must be nonzero when S_1 and S_2 are distinct, which is the case.

[6] Here the phrase "corresponding r_f" means the value of r_f for which the portfolio with z-coordinates (z_1, z_2) is the tangency portfolio.

Another Equation for the Allocations of S_1 and S_2 in the Tangency Portfolio

The allocations x_1, x_2 of S_1, S_2 in the tangency portfolio T corresponding to the risk-free rate r_f actually satisfy the matrix equation

$$\begin{pmatrix} \sigma_1^2 & \sigma_{12} \\ \sigma_{12} & \sigma_2^2 \end{pmatrix} \begin{pmatrix} \lambda x_1 \\ \lambda x_2 \end{pmatrix} = \begin{pmatrix} \mu_1 - r_f \\ \mu_2 - r_f \end{pmatrix},$$

where $\lambda = (\mu_{R_T} - r_f)/\sigma_{R_T}^2$. Note that this is essentially the same matrix equation as before, with $(\lambda x_1, \lambda x_2)$ taking the place of (z_1, z_2). So the only new observation is the form of λ; that is, $\lambda = (\mu_{R_T} - r_f)/\sigma_{R_T}^2$.

To determine the form of λ, let's write the preceding matrix equation as

$$\sum_{j=1}^{2} \lambda x_j \sigma_{ij} = \mu_i - r_f, \quad i = 1, 2.$$

Note that

$$\sum x_j \sigma_{ij} = \sum x_j Cov(R_i, R_j) = Cov\left(R_i, \sum x_j R_j\right) = Cov(R_i, R_T),$$

where R_T is the return on the tangency portfolio T. Hence, the matrix equation is equivalent to

$$\lambda Cov(R_i, R_T) = \mu_i - r_f, \quad i = 1, 2.$$

Multiplying the ith equation by x_i and summing, we have

$$\lambda \sum_{i=1}^{2} x_i Cov(R_i, R_T) = \sum_{i=1}^{2} x_i (\mu_i - r_f).$$

The left side of the latter equation is

$$\lambda Cov\left(\sum x_i R_i, R_T\right) = \lambda Cov(R_T, R_T) = \lambda \sigma_{R_T}^2,$$

and since $x_1 + x_2 = 1$, the right side is

$$\left(\sum x_i \mu_i\right) - r_f = \mu_{R_T} - r_f.$$

Consequently,

$$\lambda = \frac{\mu_{R_T} - r_f}{\sigma_{R_T}^2}$$

as claimed.

Note that, by geometry, we have $\lambda > 0$. This fact is particularly useful when determining the conditions under which it is optimal to be long or short S_1 or S_2.

Conditions for Being Long or Short

From the matrix equation

$$\begin{pmatrix} \sigma_1^2 & \sigma_{12} \\ \sigma_{12} & \sigma_2^2 \end{pmatrix} \begin{pmatrix} z_1 \\ z_2 \end{pmatrix} = \begin{pmatrix} \mu_1 - r_f \\ \mu_2 - r_f \end{pmatrix},$$

it is straightforward to determine whether it is optimal to be long or short S_1 or long or short S_2.[7] Indeed, eliminating z_2 from this equation and using the fact that $\sigma_{12} = \rho\sigma_1\sigma_2$, we obtain

$$\frac{\mu_1 - r_f}{\sigma_1} - \rho\frac{\mu_2 - r_f}{\sigma_2} = (1 - \rho^2)\sigma_1 z_1.$$

Hence,

$$z_1 > 0 \iff \frac{\mu_1 - r_f}{\sigma_1} > \rho\frac{\mu_2 - r_f}{\sigma_2},$$

assuming that $\rho \neq \pm 1$. Since $x_1 = z_1/\lambda$ and $\lambda = (\mu_{R_T} - r_f)/\sigma_{R_T}^2 > 0$, we have

$$x_1 > 0 \iff \frac{\mu_1 - r_f}{\sigma_1} > \rho\frac{\mu_2 - r_f}{\sigma_2}.$$

So the condition for being long S_1 is

$$\frac{\mu_1 - r_f}{\sigma_1} > \rho\frac{\mu_2 - r_f}{\sigma_2}.$$

There is a similar condition for being long S_2.

An important application of this condition arises when considering foreign invest-ments. Indeed, if S_1 represents foreign stocks and S_2 represents domestic stocks, then the foregoing condition tells us when it is optimal to hold foreign stocks in our investment portfolio.

10.3   Portfolio Selection with Many Securities

Suppose now that n risky securities $S_1, \ldots, S_n$ and a risk-free asset are available to construct our portfolio. Let $R_1, \ldots, R_n$ be the returns on the risky securities and let r_f be the risk-free rate. Then arguing as in §10.1 and §10.2, we reach the following conclusions:

1. The set of efficient portfolios is determined by the following family of optimization problems (one for each μ):

---

[7] Being *long* a security means that the fraction x held in the portfolio is such that $x > 0$; being *short* a security means that $x < 0$.

Minimize: $Var(y_1 R_1 + \cdots + y_n R_n + y_{n+1} r_f)$

Subject to: $(\mu_1 - r_f)y_1 + \cdots + (\mu_n - r_f)y_n = \mu - r_f$,

$y_1 + \cdots + y_{n+1} = 1$,

where y_j is the fraction invested in security S_j and y_{n+1} is the fraction invested in the risk-free asset.

2. Applying the Lagrange multiplier method to the system in 1, we find that the optimal risky portfolio (i.e., the optimal portfolio with $y_{n+1} = 0$) for a given risk-free rate r_f has weights $x_1, \ldots, x_n$ in $S_1, \ldots, S_n$ given by

$$x_j = \frac{z_j}{z_1 + \cdots + z_n}, \qquad j = 1, \ldots, n,$$

where $(z_1, \ldots, z_n)$ is the unique solution of the linear system

$$\sum_{j=1}^{n} z_j \sigma_{ij} = \mu_i - r_f, \qquad i = 1, \ldots, n,$$

that is, the system

$$\begin{pmatrix} \sigma_1^2 & \sigma_{12} & \cdots & \sigma_{1n} \\ \sigma_{21} & \sigma_2^2 & \cdots & \sigma_{2n} \\ \vdots & \vdots & & \vdots \\ \sigma_{n1} & \sigma_{n2} & \cdots & \sigma_n^2 \end{pmatrix} \begin{pmatrix} z_1 \\ \vdots \\ z_n \end{pmatrix} = \begin{pmatrix} \mu_1 \\ \vdots \\ \mu_n \end{pmatrix} - r_f \begin{pmatrix} 1 \\ \vdots \\ 1 \end{pmatrix}.$$

3. The collection of portfolios consisting only of $S_1, \ldots, S_n$ forms a convex set whose boundary has the appearance of a hyperbola. (In fact, the boundary is a hyperbola, but we're not claiming this just yet.) The upper half of this "hyperbolic" boundary is simply the collection of optimal risky portfolios determined in 2 and can be traced out by varying r_f. (Consider the geometry and the argument in the case of two risky securities presented in §10.2.)

4. The matrix equation in 2 can be rewritten as

$$\begin{pmatrix} z_1 \\ \vdots \\ z_n \end{pmatrix} = \begin{pmatrix} c_{11} & c_{12} \\ \vdots & \vdots \\ c_{n1} & c_{n2} \end{pmatrix} \begin{pmatrix} 1 \\ r_f \end{pmatrix}.$$

Here, c_{ij} are specific constants depending only on $S_1, \ldots, S_n$. The c_{ij} are completely determined once we know $(z_1, \ldots, z_n)$ and the corresponding r_f for two different boundary portfolios (since in that case we will have $2n$ equations in the $2n$ unknowns c_{ij}). Consequently, the boundary is completely determined by two risky portfolios and hence by our discussion in §10.1, must be a hyperbola as asserted in 3.

5. The equations in 2 for the allocations $x_1, \ldots, x_n$ in the optimal risky portfolio (i.e., the tangency portfolio) can be written as

$$\sum_{j=1}^{n} \lambda x_j \sigma_{ij} = \mu_i - r_f, \qquad i = 1, \ldots, n,$$

where $\lambda = (\mu_{R_T} - r_f)/\sigma_{R_T}^2$. The only new observation here is the form of λ. Note that $(\sigma_{R_T}, \mu_{R_T})$ are the risk–reward coordinates for the tangency portfolio corresponding to the risk-free rate r_f.

6. For any portfolio P on the upper half of the boundary hyperbola, let $Z(P)$ be the portfolio that can be constructed from $\mathcal{S}_1, \ldots, \mathcal{S}_n$ with smallest standard deviation for which $E[R_{Z(P)}] = r$, where r is the μ-intercept of the tangent line passing through P. The portfolio $Z(P)$ is called the **zero beta companion portfolio** to P and has the property that $Cov(R_P, R_{Z(P)}) = 0$.

Fix a portfolio P on the upper half of the boundary hyperbola. Then according to 4, every optimal risky portfolio T can be generated using only the portfolio P and its zero beta companion $Z(P)$. Moreover, if r^* is the risk-free rate, then the optimal risky portfolio T can be constructed from P and $Z(P)$ using the weights

$$w_1 = \left(\frac{E[R_P] - r^*}{\sigma_{R_P}^2} \right) / \lambda,$$

$$w_2 = \left(\frac{E[R_{Z(P)}] - r^*}{\sigma_{R_{Z(P)}}^2} \right) / \lambda,$$

where $\lambda = (\mu_{R_T} - r^*)/\sigma_{R_T}^2$. Indeed, letting P and $Z(P)$ take the roles of $\mathcal{S}_1$ and $\mathcal{S}_2$ in the equation for T given in 2, we have

$$\begin{pmatrix} \sigma_{R_P}^2 & 0 \\ 0 & \sigma_{R_{Z(P)}}^2 \end{pmatrix} \begin{pmatrix} \lambda w_1 \\ \lambda w_2 \end{pmatrix} = \begin{pmatrix} E[R_P] - r^* \\ E[R_{Z(P)}] - r^* \end{pmatrix},$$

where $\lambda = (\mu_{R_T} - r^*)/\sigma_{R_T}^2$, and so the weights w_1, w_2 have the form claimed. Note that $\sigma_{R_P, R_{Z(P)}} = 0$ since P and $Z(P)$ are zero beta companions.

7. From 6, it follows that every efficient portfolio can be constructed using a fixed optimal risky portfolio P, its zero beta companion $Z(P)$, and the risk-free asset. Moreover, as r_f changes, P and $Z(P)$ need not change; only the weights of P and $Z(P)$ need change. This is a more general form of the mutual fund separation theorem discussed in §10.2.

10.4 The Capital Asset Pricing Model

The capital asset pricing model (CAPM) is an important consequence of the Markowitz portfolio selection model which gives a formula for the expected return for an *individual security* when the overall market is in equilibrium.

Statement of the CAPM

Suppose that every investor constructs his or her investment portfolio from the same set of securities using mean-variance optimization (i.e., using the Markowitz model described in §10.3) and that every investor uses the same inputs, (i.e., the same values of μ_j, σ_{ij}) for the securities in the portfolio. Then for any security S in this market, the expected return on S is given by

$$E[R_S] = r_f + \beta_S(E[R_M] - r_f),$$

where r_f is the risk-free rate, R_M is the return on the market portfolio (i.e., the portfolio consisting of all available securities weighted according to their current market values), and β_S is given by

$$\beta_S = \frac{Cov(R_S, R_M)}{\sigma_{R_M}^2} = \rho_{R_S, R_M} \frac{\sigma_{R_S}}{\sigma_{R_M}}.$$

Derivation of the CAPM

Suppose that the possible risky securities are $S_1, \ldots, S_n$ with returns $R_1, \ldots, R_n$. Since every investor uses the Markowitz model with the same input values for μ_i, σ_{ij}, every investor's optimal risky portfolio will be the same and will have weights $x_1, \ldots, x_n$ in $S_1, \ldots, S_n$ given by the equations

$$\sum_{j=1}^{n} \lambda x_j \sigma_{ij} = \mu_i - r_f, \quad i = 1, \ldots, n,$$

where $\lambda = (\mu_{R_T} - r_f)/\sigma_{R_T}^2$, $(\sigma_{R_T}, \mu_{R_T})$ being the coordinates of the optimal risky portfolio. These equations can be written equivalently as

$$\lambda Cov\left(R_i, \sum_{j=1}^{n} x_j R_j\right) = \mu_i - r_f, \quad i = 1, \ldots, n$$

or as

$$E[R_i] = r_f + \lambda Cov\left(R_i, \sum_{j=1}^{n} x_j R_j\right).$$

Now comes the key observation: Since everyone holds the same optimal risky portfolio, the weights of $S_1, \ldots, S_n$ in the overall market portfolio are simply $x_1, \ldots, x_n$. Hence, $R_M = \sum_{j=1}^{n} x_j R_j$ and $\lambda = (E[R_M] - r_f)/\sigma_{R_M}^2$, and the equation for $E[R_i]$ becomes

$$E[R_i] = r_f + \left(\frac{E[R_M] - r_f}{\sigma_{R_M}^2}\right) \cdot Cov(R_i, R_M)$$
$$= r_f + \beta_i(E[R_M] - r_f),$$

where

$$\beta_i = \frac{Cov(R_i, R_M)}{\sigma_{R_M}^2} = \rho_{R_i, R_M} \frac{\sigma_{R_i}}{\sigma_{R_M}}.$$

This is the statement of the CAPM for security S_i. Using the linearity of $E[\cdot]$ and $Cov(\cdot, R_M)$, it follows immediately that

$$E[R_P] = r_f + \beta_P(E[R_M] - r_f),$$

where $\beta_P = Cov(R_P, R_M)/\sigma_{R_M}^2$ for all portfolios P of $S_1, \ldots, S_n$ as well. This completes the derivation of the CAPM.

Some Observations on the CAPM

1. The CAPM specifies the "fair" expected return for a security S given its risk characteristics (σ_{R_S} and ρ_{R_S, R_M}). We can see this more clearly by writing the CAPM in the following form:

$$E[R_S] = r_f + \frac{E[R_M] - r_f}{\sigma_{R_M}}(\rho_{R_S, R_M} \sigma_{R_S}).$$

Note that r_f is the component of the expected return that compensates the investor for postponing current consumption (the time value of money effect) while $((E[R_M] - r_f)/\sigma_{R_M})(\rho_{R_S,R_M}\sigma_{R_S})$ is the component that compensates the investor for taking on risk. From our discussion of the Markowitz model, we recognize $(E[R_M] - r_f)/\sigma_{R_M}$ as being the slope of the efficient set. Hence, for efficient portfolios, $(E[R_M] - r_f)/\sigma_{R_M}$ represents the extra return required for taking on an extra unit of risk (i.e., it is the market price of risk). Consequently, if we interpret $\rho_{R_S,R_M}\sigma_{R_S}$ to be the amount of risk related to market movements, then $((E[R_M] - r_f)/\sigma_{R_M})(\rho_{R_S,R_M}\sigma_{R_S})$ represents the extra return required for taking on the risk associated with S.

2. The quantity $\beta_S = \rho_{R_S,R_M}\sigma_{R_S}/\sigma_{R_M}$ measures the risk in S *relative* to the risk of the market, whereas the quantity σ_{R_S} measures the risk in S without reference to any other portfolio. The CAPM asserts that (in equilibrium) an investor can only expect to receive compensation for the risk due to market movements (so-called "systematic risk"); in particular, the investor cannot expect to be compensated for taking on risk that is specific to an individual security. The reason for this phenomenon is that security-specific risk can be eliminated by holding a diversified portfolio of securities, whereas systematic risk cannot be eliminated because it is common to all securities; hence, the investor must receive compensation in the form of a higher expected return for systematic risk, but will not receive compensation for a specific risk that can be eliminated at essentially no cost.

3. The CAPM asserts that there is a linear relationship between β_S and $E[R_S]$. The line given by the CAPM in the β-μ plane is referred to as the **security market line.** The CAPM asserts that, in equilibrium, all securities and portfolios must lie on this line.

4. The CAPM gives the return on securities expected by the *market* in equilibrium. It may happen that we have access to new information or better estimation technology that results in our obtaining a different estimate for $E[R_S]$ than that predicted by the CAPM.

   Suppose that our estimate for $E[R_S]$ lies above the security market line. Once the new information is available to the market, the price of the security will adjust so that the security again lies on the security market line. Assuming that our estimate of the expected return is correct, this means that the price of the security will rise once the information is absorbed into the market. This suggests that the security is currently undervalued and should be bought. Conversely, if our estimate for $E[R_S]$ lies below the security market line, the security should be sold.

5. The CAPM can also be used to analyze the performance of investment managers. In particular, it provides a method for determining the *risk-adjusted* returns of investment managers. Many investors make the mistake of focusing on absolute returns achieved by managers without considering the risk that was taken to achieve those returns. The CAPM tells us what the manager's return should have been for the level of risk taken and, hence, provides the investor with a benchmark for analyzing a manager's performance. An example will illustrate this point.

   Suppose that during a given period, the risk-free rate is 5%, the market return is 15%, and the return on an actively managed mutual fund is 20%. One might conclude automatically that it was preferable to hold the actively managed fund. However, suppose that the fund's beta was 2. Then, according to the CAPM, we should have expected a return of

$$E[R_S] = r_f + \beta(E[R_M] - r_f)$$
$$= 5\% + 2(15\% - 5\%)$$
$$= 25\%.$$

Hence, the fund performed more poorly than expected, given the level of risk taken.

Now you might say that the actively managed mutual fund still did better than the market. However, you would be deceiving yourself just a bit. For if you had wanted an investment with $\beta = 2$, you could have easily created it by borrowing 100% of your available cash and investing all your funds in the market. For example, if you had $1000 to invest, you could obtain a beta two portfolio by borrowing $1000 and then investing $2000 in the market. The result would have been a return of 25% rather than the 20% actually received in the actively managed mutual fund!

We have only highlighted a few of the important consequences and applications of the CAPM here. As stated earlier, the CAPM has had a profound influence on investment theory and practice, and it is rightly regarded as one of the fundamental results of modern finance. For a more detailed discussion of investment theory in general and the CAPM in particular, the interested reader should consult references listed in the next section.

10.5   Further Reading

The seminal works on portfolio selection and the capital asset pricing model are the following:

J. LINTNER.   "The Valuation of Risk Assets and the Selection of Risky Investments in Stock Portfolios and Capital Budgets," *Review of Economics and Statistics*, February 1965

H. M. MARKOWITZ.   "Portfolio Selection," *Journal of Finance*, March 1952

H. M. MARKOWITZ.   *Portfolio Selection: Efficient Diversification of Investments*, Wiley, New York 1959

J. MOSSIN.   "Equilibrium in a Capital Asset Market," *Econometrica*, October 1966

W. SHARPE.   "Capital Asset Prices: A Theory of Market Equilibrium," *Journal of Finance*, September 1964

A highly readable and entertaining account of the development of modern finance is

P. L. BERNSTEIN.   *Capital Ideas, The Improbable Origins of Modern Wall Street*, The Free Press (Simon and Schuster), New York 1992

Some of the standard textbooks on finance and investments are

Z. BODIE, A. KANE AND A. J. MARCUS.   *Investments*, 4th edition, Irwin/McGraw-Hill, New York 1999

R. A. BREALEY AND S. C. MYERS.   *Principles of Corporate Finance*, 5th edition, Irwin/McGraw-Hill, New York 1996

E. J. ELTON AND M. J. GRUBER.   *Modern Portfolio Theory and Investment Analysis*, 5th edition, Wiley, New York 1995

J. C. HULL. *Options, Futures and Other Derivatives*, 4th edition, Prentice-Hall, New York 2000

R. JARROW AND S. TURNBULL. *Derivative Securities*, 2nd edition, South-Western, Cincinnati, 2000

10.6   Exercises

1. A particular defined contribution pension plan has two investment funds available: a stock fund and a bond fund. Participants in the plan may allocate their own contributions between the stock fund and the bond fund as they see fit. There are no other investment options currently available. Historical data on fund performance are as follows: $\mu_S = 15$, $\sigma_S = 20$, $\mu_B = 5$, $\sigma_B = 10$, $\rho = 0.25$, where the subscript S refers to the stock fund and the subscript B refers to the bond fund.

   a. Determine an equation for the curve in the risk–reward plane which represents the possible individual portfolios. Write your equation in a form where the variance of the global minimum variance portfolio is explicit.

   b. Specify the mean and standard deviation of the portfolio with the least risk.

   c. Sketch the curve of possible portfolios in the risk–reward plane. Indicate the locations of the following portfolios on your graph:

      i. all bonds;

      ii. all stocks;

      iii. portfolio with least risk.

   d. A particular plan participant has utility functional $U = \mu - \frac{1}{100}\sigma^2$. Determine the optimal allocation for this participant and the return on the participant's optimal portfolio.

2. In a particular market, three risky securities S_1, S_2, S_3 are available. The statistics for these securities are as follows:

$$\mu_1 = 8, \mu_2 = 10, \mu_3 = 20,$$
$$\sigma_1 = 1, \sigma_2 = 1.5, \sigma_3 = 2,$$
$$\sigma_{12} = 1, \sigma_{13} = 0, \sigma_{23} = 2.$$

   A risk-free asset returning 4% is also available. Determine the fractions of S_1, S_2, S_3 in the risky portion of an optimal portfolio, assuming unlimited short sales are allowed.

3. The investment universe consists of three risky securities S_1, S_2, S_3, and a risk-free asset. When the risk-free rate is 5%, the optimal risky portfolio has weights

$$x_1 = \frac{74}{93}, x_2 = -\frac{60}{93}, x_3 = \frac{79}{93}$$

   and a reward-to-variability ratio $\lambda = (\mu - r_f)/\sigma^2 = 46.5$. When the risk-free rate is 10%, the optimal risky portfolio has weights

$$x_1 = \frac{12}{19}, x_2 = -\frac{20}{19}, x_3 = \frac{27}{19}$$

and a reward-to-variability ratio $\lambda = (\mu - r_f)/\sigma^2 = 19$. Determine the weights of the optimal risky portfolio when the risk-free rate is 15%.

4. An investor is to construct a portfolio from three risky securities S_1, S_2, S_3. The statistics for these securities are as follows:

$$\mu_1 = 8, \mu_2 = 10, \mu_3 = 20,$$
$$\sigma_1 = 1, \sigma_2 = 1.5, \sigma_3 = 2,$$
$$\sigma_{12} = 1, \sigma_{13} = 0, \sigma_{23} = 2.$$

No risk-free asset is available. The investor determines that the portfolio for which his utility is maximized has expected return 15%. Determine the weights of the three securities in the optimal portfolio.

5. The expected return and standard deviation for the Financial Times Stock Exchange index (an index of stocks listed on the London stock exchange) are $\mu_{FTSE} = 10$ and $\sigma_{FTSE} = 15$, measured in pounds sterling. The expected return and standard deviation for the Standard and Poors 500 index (an index of stocks listed on the New York stock exchange) are $\mu_{SP} = 15$ and $\sigma_{SP} = 25$ measured in pounds sterling. The correlation between the two indexes is .35. Determine conditions on the risk-free rate under which it is optimal for a British investor to hold U.S. securities.

6. An analyst working for your investment firm makes the following projections for a given small cap technology company:

Year	Earnings Per Share	Payout Ratio
1	0.50	10%
2	1.00	20%
3	3.00	30%
4	4.00	40%
5	5.00	50%

Subsequent earnings are projected to grow at 15% per year, and the payout ratio is projected to remain at 50%. The payout ratio is the fraction of earnings paid to shareholders as dividends. All earnings are assumed to be reported at year end.

The long-run expected return on a proprietary technology index developed by your firm is 20% per year. The current risk-free rate is 5%. The sensitivity of the given technology company's stock with respect to movements in the technology index is estimated to be 1.2.

a. Using CAPM, determine the expected return on the given technology company's stock.

b. Determine the theoretical price of the company's stock by discounting each year's dividends to the present time using the rate determined in part a.

7. The expected return and standard deviation of a particular efficient portfolio are 10 and 15, respectively. The expected return and standard deviation of the corresponding zero beta companion portfolio are 5 and 10, respectively. There is no risk-free asset available for lending or borrowing.

Determine the standard deviation of the efficient portfolio with expected return 15 and the expected return of the corresponding zero beta companion.

8. The expected return and standard deviation on the market portfolio are both 15, and the expected return and standard deviation on the global minimum variance portfolio (obtained by considering only risky securities) are both 5. Riskless lending at the rate of 3% is possible, but riskless borrowing is not allowed.

   Let T be the efficient portfolio whose zero beta companion has expected return 3%.

   a. Determine the expected value and standard deviation for the zero beta companion of the market portfolio.
   b. Explain how one can construct the portfolio T using only the market portfolio and the zero beta companion of the market portfolio. Specify explicitly the weights in these two portfolios needed to replicate T.
   c. Determine $E[R_T]$ and β_T.
   d. Determine the security market line for:
      i. efficient portfolios
      ii. individual assets.
   Graph your results in the β-μ plane.

9. An investment management company uses the following procedure when selecting stocks for its portfolios:

   ▪ Earnings, payout ratios, and ultimate growth rates are estimated for each company being considered for investment.
   ▪ Expected returns for each stock are calculated by equating the present value of future dividends to the actual market price and solving for the discount rate.
   ▪ Betas for each stock are estimated using historical returns and price volatilities.
   ▪ An *approximation* to the security market line is constructed by regressing expected returns on *expected* betas.
   ▪ Mispriced securities are identified using the regression line so constructed.

   An analyst working for your firm has provided you with the following data:

Company	Expected Return	Expected Beta
1	14	0.5
2	35	2.0
3	18	1.0
4	40	2.5
5	25	1.5

   a. Determine the least squares regression line for these data. Plot your results. The least square regression line is the line $\mu = a\beta + b$, where a and b are determined such that the quantity

   $$\sum_{i=1}^{5} \left(\hat{\mu}_i - (a\beta_i + b) \right)^2$$

   is minimal. The latter quantity is the sum of the squares of the differences between the "observed" expected returns $\hat{\mu}_i$ and the expected returns predicted by the line $\mu_i = a\beta_i + b$.

b. Which stocks are candidates for inclusion in your portfolio?

10. A pension fund employs three different portfolio managers. Each manager is as-signed a target beta value as a means of characterizing the manager's declared investment style. The returns and betas for each manager during the most recent period are as follows:

Manager	Target Beta	Actual Beta	Return
A	1.5	2.0	25%
B	1.0	0.8	14%
C	0.5	0.7	12%

During the same period, the return on the market was 15% and the return on risk-free investments was 4%.

Analyze the performance of each manager. Recommend which managers should be dismissed and which ones should be retained.

11. A defined benefit pension plan has a current liability of L. This liability is expected to grow by 9.125% in the coming year, with standard deviation 10%; that is, the *liability return* $R_L := \Delta L/L$ has expected value $E[R_L] = 9.125$ and standard deviation $SD(R_L) = 10$.

A pension fund with current assets A is to fund this liability. The only available investments for the pension fund are an equity mutual fund with $\mu_E = 12$ and $\sigma_E = 20$ and a bond mutual fund with $\mu_B = 8$ and $\sigma_B = 15$. The correlations among the equity fund return, the bond fund return, and the pension liability return are as follows: $\rho_{EB} = .375$, $\rho_{EL} = .375$, $\rho_{BL} = .50$.

The pension fund manager would like to allocate the assets between equities and bonds in such a way that the return on *surplus* (i.e., the return on $S = A - L$) is optimal. Since $\Delta S/S$ can be unstable for plans whose assets and liabilities are nearly equal, the pension fund manager measures surplus return as $\Delta S/L$. (One could also measure surplus return using $\Delta S/A$. However, with newly established pension plans, assets may be minimal, whereas the liability for any pension plan is typically significant. Hence, $\Delta S/L$ is likely to be the most stable measure of surplus return.)

Note that

$$\frac{\Delta S}{L} = \frac{\Delta A - \Delta L}{L} = F \cdot \frac{\Delta A}{A} - \frac{\Delta L}{L}$$

where $F = A/L$ is the *funding ratio*. Hence, if R_A, R_L, R_S denote the asset return $\Delta A/A$, the liability return $\Delta L/L$, and the surplus return $\Delta S/L$, respectively, then

$$R_S = F R_A - R_L.$$

a. Determine an equation in μ_{R_S} and σ_{R_S} which describes the set of possible stock–bond portfolios. Write your equation in the form

$$\sigma_{R_S}^2 = a(\mu_{R_S} - \mu_0)^2 + \sigma_0^2.$$

b. Sketch the portfolio opportunity set in the σ_{R_S}-μ_{R_S} plane in each of the following cases:

   i. $F = 2$

   ii. $F = 0.5$.

c. Suppose that the pension fund manager's utility functional is given by

$$U = \mu_{R_S} - \frac{1}{10}\sigma_{R_S}^2.$$

Determine the optimal stock–bond allocation in the cases $F = 2$ and $F = 0.5$.

A

The Gamma Function

The **gamma function** arises frequently in probability and may be defined as follows:

$$\Gamma(r) = \int_0^\infty z^{r-1} e^{-z} dz, \qquad r \neq 0, -1, -2, \ldots .$$

Equivalently, the gamma function may be defined as the "normalizing" constant for the gamma distribution with parameters r and λ.

Recall that the gamma density with parameters r and λ is given by

$$f(x) = \begin{cases} \dfrac{\lambda^r x^{r-1} e^{-\lambda x}}{\Gamma(r)} & \text{for } x \geq 0, \\ 0 & \text{for } x < 0. \end{cases}$$

Hence, $\Gamma(r)$ is the constant that makes f a density function; that is, it makes $\int_{-\infty}^\infty f(x)dx = 1$ a true statement.

Now

$$\int_0^\infty \lambda^r x^{r-1} e^{-\lambda x} dx = \int_0^\infty z^{r-1} e^{-z} dz.$$

(Apply the substitution $z = \lambda x$.) Hence, for $\int_{-\infty}^\infty f(x)dx = 1$ to hold, we must have

$$\Gamma(r) = \int_0^\infty z^{r-1} e^{-z} dz$$

which is precisely the formula given earlier.

The gamma function has many properties, the most important of which is

$$\Gamma(r + 1) = r\Gamma(r) \qquad \text{for all } r \text{ for which this makes sense.}$$

This property follows immediately from the definition of $\Gamma(\cdot)$ using integration by parts. Indeed,

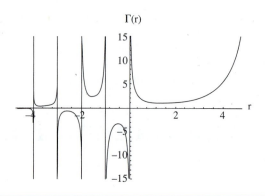

$$\Gamma(r+1) = \int_0^\infty z^r e^{-z} dz$$

$$= z^r(-e^{-z})\Big|_0^\infty - \int_0^\infty (rz^{r-1})(-e^{-z})dz$$

$$= 0 + r\int_0^\infty z^{r-1}e^{-z}dz$$

$$= r\Gamma(r).$$

Applying this formula successively and using the fact that $\Gamma(1) = 1$, which follows directly from the definition of the gamma function, we find that

$$\Gamma(n) = (n-1)! \quad \text{for all positive integers } n.$$

In light of this property, the gamma function can be interpreted as a generalization of the factorial function.

B The Incomplete Gamma Function

The **incomplete gamma function** is the probability distribution function for the gamma distribution with $\lambda = 1$. Hence, the incomplete gamma function with parameter r ($r \neq 0, -1, -2, \ldots$) is given by

$$I_r(t) = \int_0^t \frac{z^{r-1}e^{-z}}{\Gamma(r)} dz, \qquad t \geq 0.$$

Values of this function for particular r and t are given in Table B.1. To calculate $I_r(t)$ for r and t not in this table, one can use linear interpolation and the recurrence formula

$$I_r(t) - I_{r+1}(t) = \frac{t^r e^{-t}}{\Gamma(r+1)}.$$

The latter formula can be derived using integration by parts. Follow the line of reasoning given in Appendix A.

TABLE B.1   Values for the Incomplete Gamma Function $I_r(t)$

t	0.5	1	1.5	2	2.5	3	3.5	4	4.5	5
0.2	0.47291	0.18127	0.05976	0.01752	0.00467	0.00115	0.00026	0.00006	0.00001	0.00000
0.4	0.62891	0.32968	0.15053	0.06155	0.02297	0.00793	0.00256	0.00078	0.00022	0.00006
0.6	0.72668	0.45119	0.24700	0.12190	0.05512	0.02312	0.00907	0.00336	0.00118	0.00039
0.8	0.79410	0.55067	0.34061	0.19121	0.09875	0.04742	0.02136	0.00908	0.00367	0.00141
1.0	0.84270	0.63212	0.42759	0.26424	0.15085	0.08030	0.04016	0.01899	0.00853	0.00366
1.2	0.87866	0.69881	0.50637	0.33737	0.20853	0.12051	0.06556	0.03377	0.01655	0.00775
1.4	0.90574	0.75340	0.57650	0.40817	0.26921	0.16650	0.09713	0.05372	0.02830	0.01425
1.6	0.92636	0.79810	0.63819	0.47507	0.33082	0.21664	0.13410	0.07881	0.04417	0.02368
1.8	0.94222	0.83470	0.69198	0.53716	0.39169	0.26938	0.17548	0.10871	0.06428	0.03641
2.0	0.95450	0.86466	0.73854	0.59399	0.45058	0.32332	0.22022	0.14288	0.08859	0.05265
2.2	0.96406	0.88920	0.77861	0.64543	0.50663	0.37729	0.26728	0.18065	0.11683	0.07250
2.4	0.97154	0.90928	0.81296	0.69156	0.55923	0.43029	0.31565	0.22128	0.14862	0.09587
2.6	0.97741	0.92573	0.84228	0.73262	0.60804	0.48157	0.36443	0.26400	0.18346	0.12258
2.8	0.98204	0.93919	0.86722	0.76892	0.65289	0.53055	0.41285	0.30806	0.22081	0.15232
3.0	0.98569	0.95021	0.88839	0.80085	0.69378	0.57681	0.46025	0.35277	0.26008	0.18474
3.2	0.98859	0.95924	0.90631	0.82880	0.73078	0.62010	0.50611	0.39748	0.30069	0.21939
3.4	0.99088	0.96663	0.92145	0.85316	0.76406	0.66026	0.55000	0.44164	0.34207	0.25582
3.6	0.99271	0.97268	0.93421	0.87431	0.79381	0.69725	0.59164	0.48478	0.38369	0.29356
3.8	0.99416	0.97763	0.94496	0.89262	0.82030	0.73110	0.63082	0.52652	0.42510	0.33216
4.0	0.99532	0.98168	0.95399	0.90842	0.84376	0.76190	0.66741	0.56653	0.46585	0.37116
4.2	0.99625	0.98500	0.96157	0.92202	0.86447	0.78976	0.70135	0.60460	0.50561	0.41017
4.4	0.99699	0.98772	0.96793	0.93370	0.88269	0.81486	0.73266	0.64055	0.54406	0.44882
4.6	0.99758	0.98995	0.97325	0.94371	0.89865	0.83736	0.76139	0.67429	0.58098	0.48677
4.8	0.99805	0.99177	0.97771	0.95227	0.91260	0.85746	0.78760	0.70577	0.61617	0.52374
5.0	0.99843	0.99326	0.98143	0.95957	0.92476	0.87535	0.81143	0.73497	0.64951	0.55951

Table B.1 continued

TABLE B.1 *continued*

| \multicolumn 9 center r |||||||||| |
5.5	6	6.5	7	7.5	8	8.5	9	9.5	t
0.00000	0.00000	0.00000	0.00000	0.00000	0.00000	0.00000	0.00000	0.00000	0.2
0.00002	0.00000	0.00000	0.00000	0.00000	0.00000	0.00000	0.00000	0.00000	0.4
0.00013	0.00003	0.00001	0.00000	0.00000	0.00000	0.00000	0.00000	0.00000	0.6
0.00052	0.00018	0.00006	0.00002	0.00001	0.00000	0.00000	0.00000	0.00000	0.8
0.00150	0.00059	0.00023	0.00008	0.00003	0.00001	0.00000	0.00000	0.00000	1.0
0.00348	0.00150	0.00062	0.00025	0.00010	0.00004	0.00001	0.00000	0.00000	1.2
0.00689	0.00320	0.00144	0.00062	0.00026	0.00011	0.00004	0.00002	0.00001	1.4
0.01219	0.00604	0.00289	0.00134	0.00060	0.00026	0.00011	0.00005	0.00002	1.6
0.01981	0.01038	0.00525	0.00257	0.00122	0.00056	0.00025	0.00011	0.00005	1.8
0.03008	0.01656	0.00881	0.00453	0.00226	0.00110	0.00052	0.00024	0.00011	2.0
0.04328	0.02491	0.01386	0.00746	0.00390	0.00198	0.00098	0.00047	0.00022	2.2
0.05954	0.03567	0.02066	0.01159	0.00631	0.00334	0.00172	0.00086	0.00042	2.4
0.07891	0.04904	0.02948	0.01717	0.00971	0.00533	0.00285	0.00149	0.00076	2.6
0.10132	0.06511	0.04049	0.02441	0.01429	0.00813	0.00450	0.00243	0.00128	2.8
0.12664	0.08392	0.05385	0.03351	0.02025	0.01190	0.00681	0.00380	0.00207	3.0
0.15461	0.10541	0.06962	0.04462	0.02778	0.01683	0.00993	0.00571	0.00321	3.2
0.18496	0.12946	0.08784	0.05785	0.03704	0.02307	0.01401	0.00829	0.00479	3.4
0.21734	0.15588	0.10845	0.07327	0.04814	0.03079	0.01919	0.01167	0.00693	3.6
0.25138	0.18444	0.13135	0.09089	0.06118	0.04011	0.02563	0.01598	0.00974	3.8
0.28670	0.21487	0.15640	0.11067	0.07622	0.05113	0.03345	0.02136	0.01333	4.0
0.32291	0.24686	0.18340	0.13254	0.09325	0.06394	0.04277	0.02793	0.01783	4.2
0.35965	0.28009	0.21212	0.15635	0.11226	0.07858	0.05367	0.03580	0.02334	4.4
0.39656	0.31424	0.24232	0.18197	0.13317	0.09505	0.06622	0.04507	0.02999	4.6
0.43331	0.34899	0.27373	0.20920	0.15588	0.11333	0.08046	0.05582	0.03787	4.8
0.46961	0.38404	0.30607	0.23782	0.18026	0.13337	0.09639	0.06809	0.04705	5.0

Table B.1 continued

TABLE B.1 *continued*

					r					
t	0.5	1	1.5	2	2.5	3	3.5	4	4.5	5
5.2	0.99874	0.99448	0.98455	0.96580	0.93534	0.89121	0.83298	0.76193	0.68092	0.59387
5.4	0.99898	0.99548	0.98714	0.97109	0.94451	0.90524	0.85242	0.78671	0.71033	0.62669
5.6	0.99918	0.99630	0.98931	0.97559	0.95244	0.91761	0.86987	0.80938	0.73775	0.65785
5.8	0.99934	0.99697	0.99111	0.97941	0.95930	0.92849	0.88550	0.83004	0.76319	0.68728
6.0	0.99947	0.99752	0.99262	0.98265	0.96521	0.93803	0.89944	0.84880	0.78669	0.71494
6.2	0.99957	0.99797	0.99387	0.98539	0.97030	0.94638	0.91185	0.86577	0.80831	0.74082
6.4	0.99965	0.99834	0.99491	0.98770	0.97467	0.95368	0.92287	0.88108	0.82813	0.76493
6.6	0.99972	0.99864	0.99578	0.98966	0.97843	0.96003	0.93262	0.89485	0.84624	0.78730
6.8	0.99977	0.99889	0.99650	0.99131	0.98164	0.96556	0.94123	0.90719	0.86272	0.80797
7.0	0.99982	0.99909	0.99709	0.99270	0.98439	0.97036	0.94882	0.91823	0.87767	0.82701
7.2	0.99985	0.99925	0.99759	0.99388	0.98674	0.97453	0.95549	0.92808	0.89121	0.84448
7.4	0.99988	0.99939	0.99800	0.99487	0.98875	0.97813	0.96135	0.93685	0.90342	0.86047
7.6	0.99990	0.99950	0.99835	0.99570	0.99046	0.98124	0.96648	0.94463	0.91441	0.87506
7.8	0.99992	0.99959	0.99863	0.99639	0.99192	0.98393	0.97097	0.95152	0.92428	0.88833
8.0	0.99994	0.99966	0.99887	0.99698	0.99316	0.98625	0.97488	0.95762	0.93312	0.90037
9.0	0.99998	0.99988	0.99956	0.99877	0.99705	0.99377	0.98803	0.97877	0.96483	0.94504
10.0	0.99999	0.99995	0.99983	0.99950	0.99875	0.99723	0.99443	0.98966	0.98209	0.97075
11.0	1.00000	0.99998	0.99993	0.99980	0.99948	0.99879	0.99746	0.99508	0.99112	0.98490
12.0	1.00000	0.99999	0.99998	0.99992	0.99978	0.99948	0.99886	0.99771	0.99570	0.99240
13.0	1.00000	1.00000	0.99999	0.99997	0.99991	0.99978	0.99950	0.99895	0.99796	0.99626
14.0	1.00000	1.00000	1.00000	0.99999	0.99996	0.99991	0.99978	0.99953	0.99905	0.99819
15.0	1.00000	1.00000	1.00000	1.00000	0.99999	0.99996	0.99991	0.99979	0.99956	0.99914

Table B.1 continued

TABLE B.1 *continued*

r = 5.5	6	6.5	7	7.5	8	8.5	9	9.5	t
0.53983	0.41909	0.33906	0.26761	0.20615	0.15508	0.11400	0.08193	0.05762	5.2
0.50519	0.45387	0.37243	0.29833	0.23336	0.17834	0.13323	0.09735	0.06962	5.4
0.57334	0.48814	0.40593	0.32974	0.26171	0.20302	0.15402	0.11432	0.08307	5.6
0.60555	0.52169	0.43932	0.36161	0.29098	0.22897	0.17627	0.13281	0.09800	5.8
0.63636	0.55432	0.47236	0.39370	0.32097	0.25602	0.19986	0.15276	0.11437	6.0
0.66566	0.58589	0.50485	0.42579	0.35147	0.28398	0.22467	0.17409	0.13218	6.2
0.69340	0.61626	0.53662	0.45767	0.38226	0.31268	0.25053	0.19669	0.15135	6.4
0.71955	0.64533	0.56752	0.48916	0.41315	0.34192	0.27730	0.22044	0.17182	6.6
0.74408	0.67302	0.59740	0.52008	0.44394	0.37151	0.30481	0.24523	0.19351	6.8
0.76701	0.69929	0.62616	0.55029	0.47447	0.40129	0.33290	0.27091	0.21631	7.0
0.78836	0.72410	0.65371	0.57964	0.50457	0.43106	0.36139	0.29733	0.24011	7.2
0.80816	0.74744	0.68000	0.60804	0.53408	0.46067	0.39012	0.32435	0.26478	7.4
0.82648	0.76932	0.70497	0.63538	0.56289	0.48996	0.41892	0.35181	0.29020	7.6
0.84336	0.78975	0.72859	0.66159	0.59088	0.51879	0.44765	0.37956	0.31622	7.8
0.85887	0.80876	0.75087	0.68663	0.61795	0.54704	0.47617	0.40745	0.34272	8.0
0.95466	0.88431	0.84248	0.79322	0.73733	0.67610	0.61116	0.54435	0.47756	9.0
0.91842	0.93291	0.90479	0.86986	0.82807	0.77978	0.72577	0.66718	0.60542	10.0
0.97563	0.96248	0.94464	0.92139	0.89220	0.85681	0.81528	0.76801	0.71574	11.0
0.98727	0.97966	0.96887	0.95418	0.93491	0.91050	0.88057	0.84497	0.80385	12.0
0.99351	0.98927	0.98300	0.97411	0.96198	0.94597	0.92554	0.90024	0.86981	13.0
0.99676	0.99447	0.99095	0.98577	0.97843	0.96838	0.95506	0.93794	0.91657	14.0
0.99842	0.99721	0.99529	0.99237	0.98808	0.98200	0.97365	0.96255	0.94820	15.0

C

The Beta Function

The **beta function** arises frequently in probability and may be defined as follows:

$$B(r, s) = \int_0^1 z^{r-1}(1 - z)^{s-1} dz, \qquad r > 0, s > 0.$$

It is the "normalizing" constant for the beta distribution with parameters r and s. Indeed, since the beta density with parameters r and s is given by

$$f(x) = \begin{cases} \dfrac{x^{r-1}(1 - x)^{s-1}}{B(r, s)} & \text{for } 0 < x < 1, \\ 0 & \text{otherwise,} \end{cases}$$

it is clear from the definition of $B(\cdot, \cdot)$ that $B(r, s)$ is the constant that makes f a density.

The beta function is related to the gamma function in the following way:

$$B(r, s) = \frac{\Gamma(r)\Gamma(s)}{\Gamma(r + s)}.$$

This important relationship gives an alternative (and often simpler) way to calculate $B(r, s)$. The proof of this relationship can be found in many textbooks on mathematical analysis.

D

The Incomplete Beta Function

The **incomplete beta function** is the probability distribution function for the beta distribution with parameters r and s. Hence, the incomplete beta function is given by

$$I_x(r, s) = \int_0^x \frac{z^{r-1}(1-z)^{s-1}}{B(r, s)} dz, \qquad 0 \le x \le 1, \qquad (r, s > 0).$$

Note that $I_x(r, s) = \Pr(\text{Beta}(r, s) \le x)$.

In general, the incomplete beta function must be evaluated numerically. However, there are some special cases, where $I_x(r, s)$ has an elementary form:

1. $I_x(r, 1) = x^r$.
2. $I_x(1, s) = 1 - (1 - x)^s$.
3. When s is a positive integer,

$$I_x(r, s) = \sum_{i=0}^{s-1} \left[\frac{\Gamma(r + s)}{\Gamma(r + i + 1)\Gamma(s - i)} \right] x^{r+i}(1 - x)^{s-i-1}.$$

4. When r and s are both positive integers,

$$I_x(r, s) = \sum_{j=r}^{r+s-1} \binom{r + s - 1}{j} x^j (1 - x)^{r+s-1-j}.$$

The latter two formulas follow from the recurrence relation

$$I_x(r, s) = \frac{\Gamma(r + s)x^r(1 - x)^{s-1}}{\Gamma(r + 1)\Gamma(s)} + I_x(r + 1, s - 1),$$

which can be derived using integration by parts, and the relation

$$\Gamma(n) = (n - 1)!$$

for positive integers n. Details are left to the reader.

E The Standard Normal Distribution

The **standard normal distribution function** is the probability distribution function for the normal distribution with mean $\mu = 0$ and standard deviation $\sigma = 1$. Hence, the standard normal distribution function is given by

$$\Phi(x) = \frac{1}{\sqrt{2\pi}} \int_{-\infty}^{x} e^{\frac{-w^2}{2}} dw.$$

Values of this function for selected nonnegative x are given in Table E.1. To calculate $\Phi(x)$ for x not in this table, one can use linear interpolation and the symmetry of the normal density function. In particular, for $x < 0$, one can use the relationship

$$\Phi(x) = 1 - \Phi(-x).$$

TABLE E.1   Values for the Standard Normal Distribution Function $\Phi(x)$

x	.00	.01	.02	.03	.04	.05	.06	.07	.08	.09
.0	.5000	.5040	.5080	.5120	.5160	.5199	.5239	.5279	.5319	.5359
.1	.5398	.5438	.5478	.5517	.5557	.5596	.5636	.5675	.5714	.5753
.2	.5793	.5832	.5871	.5910	.5948	.5987	.6026	.6064	.6103	.6141
.3	.6179	.6217	.6255	.6293	.6331	.6368	.6406	.6443	.6480	.6517
.4	.6554	.6591	.6628	.6664	.6700	.6736	.6772	.6808	.6844	.6879
.5	.6915	.6950	.6985	.7019	.7054	.7088	.7123	.7157	.7190	.7224
.6	.7257	.7291	.7324	.7357	.7389	.7422	.7454	.7486	.7517	.7549
.7	.7580	.7611	.7642	.7673	.7704	.7734	.7764	.7794	.7823	.7852
.8	.7881	.7910	.7939	.7967	.7995	.8023	.8051	.8078	.8106	.8133
.9	.8159	.8186	.8212	.8238	.8264	.8289	.8315	.8340	.8365	.8389
1.0	.8413	.8438	.8461	.8485	.8508	.8531	.8554	.8577	.8599	.8621
1.1	.8643	.8665	.8686	.8708	.8729	.8749	.8770	.8790	.8810	.8830
1.2	.8849	.8869	.8888	.8907	.8925	.8944	.8962	.8980	.8997	.9015
1.3	.9032	.9049	.9066	.9082	.9099	.9115	.9131	.9147	.9162	.9177
1.4	.9192	.9207	.9222	.9236	.9251	.9265	.9279	.9292	.9306	.9319
1.5	.9332	.9345	.9357	.9370	.9382	.9394	.9406	.9418	.9429	.9441
1.6	.9452	.9463	.9474	.9484	.9495	.9505	.9515	.9525	.9535	.9545
1.7	.9554	.9564	.9573	.9582	.9591	.9599	.9608	.9616	.9625	.9633
1.8	.9641	.9649	.9656	.9664	.9671	.9678	.9686	.9693	.9699	.9706
1.9	.9713	.9719	.9726	.9732	.9738	.9744	.9750	.9756	.9761	.9767
2.0	.9772	.9778	.9783	.9788	.9793	.9798	.9803	.9808	.9812	.9817
2.1	.9821	.9826	.9830	.9834	.9838	.9842	.9846	.9850	.9854	.9857
2.2	.9861	.9864	.9868	.9871	.9875	.9878	.9881	.9884	.9887	.9890
2.3	.9893	.9896	.9898	.9901	.9904	.9906	.9909	.9911	.9913	.9916
2.4	.9918	.9920	.9922	.9925	.9927	.9929	.9931	.9932	.9934	.9936
2.5	.9938	.9940	.9941	.9943	.9945	.9946	.9948	.9949	.9951	.9952
2.6	.9953	.9955	.9956	.9957	.9959	.9960	.9961	.9962	.9963	.9964
2.7	.9965	.9966	.9967	.9968	.9969	.9970	.9971	.9972	.9973	.9974
2.8	.9974	.9975	.9976	.9977	.9977	.9978	.9979	.9979	.9980	.9981
2.9	.9981	.9982	.9982	.9983	.9984	.9984	.9985	.9985	.9986	.9986
3.0	.9987	.9987	.9987	.9988	.9988	.9989	.9989	.9989	.9990	.9990
3.1	.9990	.9991	.9991	.9991	.9992	.9992	.9992	.9992	.9993	.9993
3.2	.9993	.9993	.9994	.9994	.9994	.9994	.9994	.9995	.9995	.9995
3.3	.9995	.9995	.9995	.9996	.9996	.9996	.9996	.9996	.9996	.9997
3.4	.9997	.9997	.9997	.9997	.9997	.9997	.9997	.9997	.9997	.9998

F *Mathematica* Commands for Generating the Graphs of Special Distributions

The probability density plots and bar charts presented in this book can be easily generated using the computer software package *Mathematica*. Our definitions of the special distributions are generally consistent with the definitions in the *Mathematica* kernel. However, there are a few notable exceptions. Hence, to avoid unnecessary confusion, we now provide a list of the *Mathematica* keystrokes required to define the probability mass or density function for each of the special distributions discussed in this book.

Distribution	*Mathematica* Keystrokes for Mass Density Function
Binomial(n,p)	`PDF[BinomialDistribution[n,p],x]`
Poisson(λ)	`PDF[PoissonDistribution[`λ`],x]`
NegativeBinomial(r,p)	`PDF[NegativeBinomialDistribution[r,p],x]`
Geometric(p)	`PDF[GeometricDistribution[p],x]`
Exponential(λ)	`PDF[ExponentialDistribution[`λ`],x]`
Gamma(r,λ)	`PDF[GammaDistribution[r,1/`λ`],x]`
Pareto(s,β)	`PDF[ParetoDistribution[`β`,s],`$x + \beta$`]`
Weibull(α,β)	`PDF[WeibullDistribution[`β,α`],x]`
DeMoivre(ω)	`PDF[UniformDistribution[0,`ω`],x]`
Normal(μ,σ)	`PDF[NormalDistribution[`μ,σ`],x]`
Lognormal(μ,σ)	`PDF[LogNormalDistribution[`μ,σ`],x]`
Beta(r,s)	`PDF[BetaDistribution[r,s],x]`

To use these *Mathematica* commands to create density plots and bar charts, one needs to load three special packages from the *Mathematica* kernel at the beginning of a *Mathematica* session. The required keystrokes are as follows:

```
<<Graphics`Graphics`
<<Statistics`ContinuousDistributions`
<<Statistics`DiscreteDistributions`
```

After these packages have been loaded, one can create mass and density plots for any of the special distributions using the keystrokes given earlier. For example, to create a bar chart for Binomial(6,0.25), simply type the following:

```
BarChart[Table[PDF[BinomialDistribution[6,.25],x],{x,0,6}],
    BarLabels -> Table[x,{x,0,6}]]
```

To create a density plot for gamma(3,2) over the interval (0,5), type the following:

```
Plot[PDF[GammaDistribution[3,1/2],x],{x,0,5}]
```

G Elementary Financial Mathematics

On several occasions in this book, we have made reference to some basic facts and terminology from the theory of interest. For the benefit of the reader who may not be familiar with this theory, we provide a brief overview in this appendix.

Interest can be thought of as the cost of borrowing money to increase current consumption beyond what one's resources allow, or alternatively as the compensation for saving money, that is, for delaying consumption until a future date. Interest rates are generally quoted on a per annum basis. However, the *effective* rate of interest will depend on the frequency of interest payments and when they are made during the year. Indeed, if interest payments are made throughout the year, then the effective rate of interest will be higher than the quoted rate since one will also earn interest on the interest payments themselves. This effect is known as **compound interest.**

For example, suppose that interest on a particular certificate of deposit is paid at a rate of 5% per annum and that payments are made in four equal installments with the first payment being made three months from the date of initial deposit. Then on a $100 investment, interest payments of $\left(\frac{1}{4}\right)(.05)(\$100) = \$1.25$ are made 3 months, 6 months, 9 months, and 12 months from the date of the initial investment. If these interest payments are added to the initial investment and continue to earn interest at a rate of 5% per annum, then the value of the certificate of deposit one year after the date of initial investment will be $\$100 (1.0125)^4 \approx \$105.09$. Hence, the effective rate of interest on this investment is 5.09% per year.

More generally, if interest is compounded n times per year at the rate r per annum, then the value of a $1 investment one year from the date of initial investment will be $(1 + \frac{r}{n})^n$, and so the effective rate of interest is $((1 + \frac{r}{n})^n - 1)\,100\%$. The more frequent the compounding, the greater the accumulation and the higher the effective rate for a given rate r. In the limit as $n \to \infty$, the accumulation becomes e^r. This type of compounding is referred to as **continuous compounding.**

Rates that are quoted on a continuously compounded basis are convenient for calculation because they are *additive*. To understand the significance of this property, consider an investment that pays 5% per annum compounded continuously in an economy with a monetary inflation rate of 3% per annum compounded continuously. Then the nominal value of a $1 investment one year from today will be $e^{.05}$. However, the real value of the investment will be less because of inflation. In fact, since the cost of a $1 item will be $e^{.03}$ one year from today, the real value of the $1 investment (i.e., the value in an equivalent economy with no inflation) will be $e^{.05}/e^{.03} = e^{.02}$. This represents a *real* return of 2%

per annum compounded continuously. Note that the interest rate paid (5%) is just the sum of the real rate (2%) and the inflation rate (3%). When rates are not quoted on a continuously compounded basis, this additivity property does not hold. For example, if an investment paying 5% per annum simple interest is subject to a simple inflation rate of 3% per annum, then the real rate of return is $(1.05)/(1.03) - 1 \approx 1.94\%$.

Additivity of rates is a nice property to have, particularly when rates are random variables, as they will be for investments like stocks whose rate of return at the time of purchase is uncertain. For this reason, continuously compounded rates are often used in the investment literature.

Answers to Selected Exercises

§2.6

3a. {HHH, HHT, HTH, HTT, THH, THT, TTH, TTT}

  b. $p_{Y_1}(-3) = \frac{1}{8}$, $p_{Y_1}(-1) = \frac{3}{8}$, $p_{Y_1}(1) = \frac{3}{8}$, $p_{Y_1}(3) = \frac{1}{8}$; $p_{Y_2}(-6) = \frac{1}{8}$, $p_{Y_2}(1) = \frac{3}{8}$, $p_{Y_2}(2) = \frac{3}{8}$, $p_{Y_2}(3) = \frac{1}{8}$.

6a. $p_X(-1) = \frac{1}{3}$, $p_X(\frac{2}{3}) = \frac{2}{3}$

  b. $\frac{1}{9}$

  c. $\frac{1}{3}$

9b. $2X_1 X_2 + X_1(1 - X_2) + (1 - X_1)X_2 - 3(1 - X_1)(1 - X_2)$

12a. {HH, HD, DH, DD}

  b. Identically distributed but not independent

  c. $p_{X_1,X_2}(0, 0) = \frac{1}{6}$, $p_{X_1,X_2}(0, 1) = \frac{1}{3}$, $p_{X_1,X_2}(1, 0) = \frac{1}{3}$, $p_{X_1,X_2}(1, 1) = \frac{1}{6}$

15a. $F_X(x) = 1 - \frac{1}{x^2}, x \geq 1$

  b. 2

  c. $\frac{1}{16}$

18b. Yes

  d. No

21a. $\frac{4}{9} < p \leq 1$

  b. $.5575 < p \leq 1$

24. $E_a[Y] = 1.06$, $E_g[Y] = 0.9835$

§3.6

2. $.2, .6, \frac{2}{3}$

5. $\Pr(E \cup F \cup G) = \Pr(E) + \Pr(F) + \Pr(G) - \Pr(E \cap F) - \Pr(E \cap G) - \Pr(F \cap G) + \Pr(E \cap F \cap G)$

8a. $.7 \leq \Pr(E \cup F) \leq 1, .1 \leq \Pr(E \cap F) \leq .4$

  c. $\Pr(E) + \Pr(F) < 1$

  e. No

11a. .45

c. $\frac{22}{45}$

e. $\frac{2}{7}$

14b. .75

d. $\frac{2}{3}$

17a. .145

c. 65%

20b. $\frac{19}{27}$

23b. Minimize $\Pr(Q^c|P)$

c. Minimize $\Pr(Q|P^c)$

d. .06, .43, Good for screening unqualified applicants

25a. Test 2 better for screening unqualifieds. Test 1 better for ensuring qualifieds not denied admission.

c. Test 2 better for both purposes.

§4.1.13

1a. False

d. True

g. False

j. False

m. True

4. $f_X(x) = \frac{1}{4}$ for $0 < x < 2$, $p_X(0) = \frac{1}{6}$, $p_X(2) = \frac{1}{3}$. $F_X(x) = \frac{1}{2}F_C(x) + \frac{1}{2}F_D(x)$ where $f_C(x) = \frac{1}{2}$ for $0 < x < 2$, $p_D(0) = \frac{1}{3}$, $p_D(2) = \frac{2}{3}$.

7c. $p_{X_1,X_2}(0,0) = \frac{1}{8}$, $p_{X_1,X_2}(0,1) = \frac{1}{4}$, $p_{X_1,X_2}(1,0) = \frac{1}{8}$, $p_{X_1,X_2}(1,1) = \frac{1}{2}$

e. Differences: (1) A jump on F_{X_1,X_2} need not correspond to a probability mass (consider the point $(x_1, x_2) = (1, \frac{1}{2})$). (2) If $x_1 \to \infty$ or $x_2 \to \infty$ with the other variable fixed, it is not necessarily true that $F_{X_1,X_2}(x_1, x_2) \to 1$.

10a. $k = \frac{1}{5}$

b. $k = \frac{27}{14\sqrt{7}-20}$

14. $F_Y(y) = 1 - e^{-y}(1 + y)$, $y > 0$

§4.2.4

1. 16

4. $1386.29

§4.3.3

3. $f_X(x) = \frac{3}{4}e^{-x}$, $x > 0$; $p_X(0) = \frac{1}{4}$.

$$F_X(x) = \begin{cases} 0 & x < 0, \\ 1 - \frac{3}{4}e^{-x} & x \geq 0. \end{cases}$$

6a. 1, 1, 2

  d. 0, 1, 0

§4.5

1c. $E_a[V_k] = \sum_{x=0}^{k}(1+g)^x(1-l)^{k-x}\binom{k}{x}p^x(1-p)^{k-x}$

§5.5

7c. 3, 6

  f. .5

9a. Exact: hypergeometric with $\alpha = 20, \beta = 1980, n = 10$. Approximate: Binomial(10, .01)

  d. NegativeBinomial(100, .10)

  i. Poisson(15)

12c. $2, 2, 1 - 3e^{-2}, 2e^{-2}$

15. .39098125, \$10.97, \$12.04

18. $1 - e^{-1/6}$

24. $(.90)^{103}$

§6.4

5a. Exponential(100) minutes

  d. Exponential(3) months

  g. Normal($67, \sqrt{2}$) inches

  j. Weibull(α, 2.5) years

8b. $3, 3, \frac{5}{2}e^{-1}, 1 - \frac{5}{2}e^{-1}$

11. $.70e^{-4/3} + .20e^{-4/5} + .10e^{-2/5}$

14. .0266 (using continuity correction)

17. .1691

19a. $e^{0.12}$

23a. $1 < r < s$

  c. $r = s = 1$

27. 0.0631 (dollars squared), 0.4806

30a. $\frac{2}{27}$

  c. $S_X(x) = .25(\frac{100}{100+x})^3, x > 0$

35. .07613

38a. Gamma(12, 1.1)

42. Posterior distribution of parameter is Gamma(3, 300)

§7.7

4. $f_Y(y) = \sqrt{\frac{2}{\pi}} y e^{-y^4/2}$, $y > 0$

7. $\int_0^m k x^{k-1} S_X(x) dx$

10. $(\sum_{i=1}^n \lambda_i)/(r + \sum_{i=1}^n \lambda_i)$

14b. \$430.08

17b. \$2334.29

20a. \$9417.65

23. \$836.57

26. $F_T(t) = (1 - e^{-7t/12})(1 - e^{-t/2})(1 - e^{-t})$, $E[T] = 2.97$

§8.7

1a.

$$
F_S(s) = \begin{cases}
0 & s < 0, \\
\frac{s^2}{12} & 0 \le s < 2, \\
\frac{s-1}{3} & 2 \le s < 3, \\
\frac{-s^2 + 10s - 13}{12} & 3 \le s < 5, \\
1 & s \ge 5.
\end{cases}
$$

c. $F_P(p) = \frac{p}{6}\{1 + \log(\frac{6}{p})\}$, $0 < p \le 6$

4. $f_S(s) = s e^{-s}$, $s > 0$

6. .75

8b. Independent, uncorrelated

d. Dependent, uncorrelated

17a. .9878

19a. No

b. Yes

25. Probability of insolvency is about 50% when premium is the expected loss

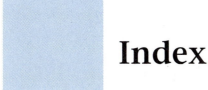

Index

Page numbers followed by an "n" indicate footnotes.

accumulated winnings in law of large numbers, 351–352
actuarial present value, 305
actuarial science, 6–9
addition with random variables, 124–125
additive rates, 434–435
adverse selection problem, 9
ageless property, 207n
AGM (arithmetic-geometric means) inequality, 46–47, 133–134
alcohol study example, 61–64
Alembert, Jean Le Rond d', 3n
American options, 302
annuities, 36, 309–311
arbitrage opportunities, 10
areas under curves, 98–99
arithmetic expectations, 131
  deviation from, 143
  and geometric
    connections with, 45–46
    inequality with, 46–47
    relationship between, 138–139
  in growth, 44–47
  properties of, 139–141
arithmetic-geometric means (AGM) inequality, 46–47, 133–134
arithmetic operations
  beta distribution, 262
  binomial distribution, 191–192
  DeMoivre distribution, 243
  exponential distribution, 223–224
  gamma distribution, 229–230
  geometric distribution, 208
  lognormal distribution, 258
  negative binomial distribution, 204
  normal distribution, 249
  Pareto distribution, 235
  Poisson distribution, 197–198
  on random variables, 124–125
  Weibull distribution, 238

association
  covariance for, 341–342
  degrees of, 342–343
associative laws for sets, 86
asymmetry, informational
  Bayes' theorem in, 72–75
  in insurance, 57
automobile tire lifetimes, Weibull distribution for, 238
axiomatic approaches, 4

balanced dice, 19
bar charts, 22
batteries
  failures, Weibull distribution for, 238–239
  in series and parallel systems, 313–316
Bayes, Thomas, 11n
Bayes' theorem, 58
  distributional form, 123–124
  examples, 73–75
  for insurance, 57, 72–73
Bayesian interpretation of probability, 11–13
beneficiaries, life insurance, 304
Bernoulli random variables, 192
beta distribution
  arithmetic operations with, 262
  definition and interpretation, 260–261
  examples, 263–265
  mean, variance, and higher moments for, 261
  probability calculations for, 261–262
  probability density function for, 261
  relationship with other distributions, 262–263
beta function
  definition, 428
  incomplete, 262, 429
biased coins, 3n

bids, construction, transformations for, 316–317
binomial distribution
  arithmetic operations with, 191–192
  definition and interpretation of, 187–190
  mean, variance, and higher moments for, 190–191
  negative. *See* negative binomial distribution
  probability calculations for, 191
  probability mass function for, 190
  relationship with other distributions, 192, 262–263
birthday problem, 89
births, negative binomial distribution for, 204–205
bivariate density, 106, 116, 249
bivariate distributions, 29, 104–113
bond mutual funds, model for. *See* Markowitz investment portfolio selection model

call options, 302
capital allocation in insurance, 8–9
Capital Asset Pricing Model (CAPM), 396
  derivation of, 412
  observations on, 412–414
  statement of, 411–412
capital markets in financial engineering, 10
caps, insurance with, 368–372
  combination contracts, 300–302
  excess-of-loss contracts, 297–298
  limited excess-of-loss contracts, 298–299
  option contracts, 302–303
  proportional insurance contracts, 300
Cardan, Girolamo, 2
Cauchy distribution, 357–358
  means in, 319
  in normal distributions, 251

Cauchy-Schwarz inequality
  and correlation, 344
  in expectation, 140, 340–341
center of mass, 24, 131
central limit theorem, 352–353
characteristic life, 236–237
charts, 22
Chebyshev's inequality
  definition, 148
  vs. Laplacian distribution, 149
checkout times, gamma distribution for, 233
choosing between payoffs, 25–26
  contingency tables in, 33–34
  independence of random variables in, 31–33
  with not identical distributions, 27–28
  probability distribution in, 26–27
  summary, 34–35
  with unfair coins, 28–31
claims
  frequency of, 372–373
  size of, 121–123
classical probability, 57–58
  Bayes' theorem, 72–75
  conditional probability, 64–68
  formal language of, 58–64
  law of total probability, 68–72
coins
  fair, 3n
  unfair, 28–31
coinsurance, 9, 297–303
combination contracts, 300–302
combinations, 88
combinatorics, 86–89
commutative laws for sets, 86
complements, set, 85
compound distributions, 363, 379–380
  definition, 366
  distribution function for, 380–381
  in insurance claims, 372–373
  mean, variance, and skewness of, 382
  mixed Poisson, 383–384
  moment and cumulant generating
    functions for, 381
  negative binomial, 384
  Poisson, 382–383
compound growth and gains, 24, 42–43
  depressed returns in, 47–48
  geometric and arithmetic expectations
    in, 44–47
  interest, 434
  investments over several periods, 44
  investments over single periods, 43–44
  moment and cumulant generating
    functions with, 160
  summary, 48–49
conditional densities
  formulas for, 114–115
  as scaled cross sections, 116
  for sums, 333
conditional distributions, 114

conditional probability, 58, 64–65
  density function, 114
  examples, 65–68
  mass function, 114
construction bids, transformations for, 316–317
containment of sets, 85
contingency tables, 33–34
continuity correction, 250–251
continuous compounding, 160n, 434
continuous distributions, 94–96
  for modeling lifetimes
    DeMoivre distribution, 241–245
    Weibull distribution, 235–241
  for modeling uncertain sizes
    exponential distribution, 221–226
    gamma distribution, 226–233
    Pareto distribution, 233–235
continuous mixtures, 363–364
continuous quantities, 36
continuous random variables, 96
  density function for, 99–100
  distribution function for, 99
convolutions method, 326, 336–337
correction for continuity, 250–251
correlation
  in expectations of sums and products,
    342–345
  in portfolio opportunity sets, 401–402
covariance
  definition, 145
  in expectations of sums and products,
    341–342
  as measure of association, 341–342
cross-currency options, 10
cross-section survival patterns, 37
cross-sectional study design, 37
cross sections, conditional densities as, 116
cumulant generating functions
  for compound distributions, 381
  definition, 158
  economic interpretations of, 160–161
  examples, 163–167
  graphical properties of, 161–163
  for probability distributions, 155–167
  properties of, 159
cumulative distribution function, 39
cumulative relative frequencies, 37
customer service times, 315

deductibles, insurance with, 9, 368–372
  combination contracts, 300–302
  excess-of-loss contracts, 297–298
  limited excess-of-loss contracts,
    298–299
  option contracts, 302–303
  proportional insurance contracts, 300
defects
  beta distribution for, 263
  Poisson distribution for, 198–199
defined benefit plans, 108
defined contribution plans, 108n

degrees
  of associations, 342–343
  of skewness, 150
delta functions, 178–179
DeMoivre distribution
  arithmetic operations with, 243
  definition and interpretation, 241–242
  examples, 243–245
  mean, variance, and higher moments
    for, 242
  probability calculations for, 242
  probability density function for, 242
  relationship with other distributions,
    243
DeMorgan's laws, 86
densities and density functions, 97–99, 121
  bivariate, 106, 116, 249
  conditional probability, 114
  for continuous random variables,
    99–100
  delta functions, 178–179
  for discrete random variables, 179–181
  examples, 99–100
  generalized, 178–185
  joint, 111–113, 117–118, 326–331
  marginal, 111, 113, 333
  for mixed random variables, 181–185
  probability. See probability density
    functions
dependence of random variables, 113–119
depressed returns in compounding, 47–48
deregulation in financial engineering, 9
descriptive statistics, 130
deviation from expectation, 143–149
dice, balanced, 19
differences for expectation deviation, 143
discontinuities, 100–101
discrete distributions, 94–96, 186–187
  binomial, 187–194
  geometric, 206–209
  negative binomial, 200–206
  Poisson, 195–200
discrete mixtures, 122, 363–364
discrete random variables, 20
  definition, 96
  densities for, 179–181
discrete sets, 20n
disjoint sets, 85–86
dissolved oxygen saturation, beta
    distribution for, 264–265
distributions and distribution functions,
    92–94. See also specific
    distributions by name
  bivariate, 104–113
  for compound distributions, 380–381
  for continuous random variables, 99
  marginal, 34, 110–111, 365
  for maximums, 312
  moment and cumulant generating
    functions for, 155–167
  not identical, 27–28
  of probability

in future lifetimes, 38–39
in payoffs, 21–22, 26–27
of products and quotients, 125, 337–339
of sums, 125, 325–326
    convolutions in, 336–337
    joint density in, 326–331
    law of total probability in, 331–336
of transformed random variables,
    281–289
types of, 94–97
with unfair coins, 28–31
distributive laws for sets, 86
dollar accumulations in compounding,
    47–48
drug and alcohol study example, 61–64

e^x, moments of, 163
effective rate of interest, 434
efficient market hypothesis, 189
efficient portfolios, 399, 403–405
electrical circuits
    gamma distribution for, 232
    normal distribution for, 254
    reliability of, 312–316
    special transformations for, 286–287
    trigonometric transformations for,
        317–318
elements, set, 85
endowment insurance, 308–309
engineering
    financial, 9–11
    normal distribution for, 253–255
    power transformations in, 284
    probability in, 5–6
    series and parallel systems in, 313
    Weibull distribution for, 241
equal likelihood principle, 3
equality of random variables, 102–104
equilibrium in financial engineering, 10
equivalent variables, 27, 102–104
European options, 302
event space, 59
events, 59
    examples, 61–64
    probabilities associated with, 60
    probability measures for, 62
    sample spaces for, 60–62
    Venn diagrams for, 62–64
excess-of-loss contracts, 297–298
    example, 298
    limited, 298–299
exercise prices, 302
expectations, 24, 131
    arithmetic and geometric, 45, 131–133
        connections with, 45–46
        in growth, 44–47
        inequality with, 46–47
        properties of, 139–141
        relationship between, 138–139
    deviation from, 143–149
    exponential distribution for, 135
    in future lifetimes, 41

limited, 280, 294–297
for mixed random variables, 135–137
for payoffs, 23–24
Poisson distribution for, 134
of random variables, 137–139
statistical measures of, 130
of sums and products, 339
    Cauchy-Schwarz inequality for,
        340–341
    correlation in, 342–345
    covariance in, 341–342
    formulas for, 339–340
of transformed random variables,
    289–297
experiments, 20
exponential distribution
    arithmetic operations for, 223–224
    definition and interpretations, 221–222
    examples, 225–226
    for expectation, 135
    mean, variance, and higher moments
        for, 223
    probability calculations for, 223
    probability density function for,
        222–223
    relationship with other distributions,
        224–225
    truncated
        definition, 136
        with moment generating functions,
            165–166
        variance of, 146–147
exponential transformations, 280, 285

face value of whole life insurance, 304
failure rates
    hazard function for, 167
    Weibull distribution for, 238–240
fair coins, 3n
families of distributions, 186
Fermat, Pierre de, 2
financial consequences, actuarial science
    for, 6
financial engineering. *See* investments
financial mathematics, 434–435
finite outcomes in relative frequency, 3
flux in measurement, 5
force of mortality, 167
formal language of classical probability,
    58–64
frequentist interpretation of probability,
    11–12
functions, 21, 58
future lifetimes, 36–38
    distribution of probability in, 38–39
    life expectancy in, 39–41
    summary, 41–42

Galileo, 2
games of chance, 2, 19–21
    distribution of probability in,
        21–22

expected payoffs in, 23–24
summary, 24–25
gamma distribution
    arithmetic operations with, 229–230
    definition and interpretation, 226
    examples, 231–233
    in law of large numbers, 347–348
    mean, variance, and higher moments
        for, 228
    with moment generating functions, 166
    probability calculations for, 229
    probability density function for,
        226–228
    relationship with other distributions,
        230–231
        beta distribution, 262–263
        exponential distribution, 223–224
        negative binomial distribution, 204
gamma function
    definition, 421
    incomplete
        definition, 229
        values of, 423–427
    in negative binomial distribution,
        200–201
    properties of, 421–422
generalized density functions, 178
    delta functions, 178–179
    discrete random variable densities,
        179–181
    mixed random variable densities,
        181–185
generalized Pareto distribution, 277
generating functions, 155–156. *See also*
    moment generating functions
geometric averages, 24
geometric distribution
    arithmetic operations with, 208
    definition and interpretation, 206–207
    examples, 208–209
    mean, variance, and higher moments
        for, 207–208
    probability calculations for, 208
    probability mass function for, 207
    relationship with other distributions,
        208
geometric expectations, 131–133
    and arithmetic
        connections with, 45–46
        inequality with, 46–47
        relationship between, 138–139
    deviation from, 143, 145–146
    in growth, 44–47
    properties of, 139–141
glaucoma example, 107–108
globalization in financial engineering, 10
graphs and graphical properties
    of moment and cumulant generating
        functions, 161–163
    of special distributions, 432–433
groups
    future lifetimes of, 36

groups (*continued*)
  in insurance principle, 7–8
  of observations, 5, 58–59
growth, 42–43
  depressed returns in, 47–48
  geometric and arithmetic expectations
    in, 44–47
  investments over several periods in, 44
  investments over single periods in,
    43–44
  moment and cumulant generating
    functions with, 160
  summary, 48–49

hazard functions, 168
  definition, 167
  relationship to survival functions, 169
higher moments, 149–153
  beta distribution, 261
  binomial distribution, 190–191
  DeMoivre distribution, 242
  exponential distribution, 223
  gamma distribution, 228
  geometric distribution, 207–208
  lognormal distribution, 257
  negative binomial distribution, 203
  normal distribution, 248
  Pareto distribution, 234
  Poisson distribution, 196–197
  Weibull distribution, 237
homogeneous risks, 7–8
hospitalization estimates
  negative binomial distribution for,
    205–206
  Poisson distribution for, 200
human error in measurement, 5

identically distributed variables, 27, 104
imperfect information, 9
incomplete beta function, 262, 429
incomplete gamma function
  definition, 229
  values of, 423–427
independence of random variables, 31–33,
    113–119
independent random variables
  definition, 33
  minimum and maximum distributions
    for, 311–312
independent risks, 7–8
independent sums, 325
index funds, model for. *See* Markowitz
    investment portfolio selection
    model
indicator random variables, 29
inequalities
  arithmetic-geometric means, 46–47,
    133–134
  Cauchy-Schwarz inequality, 140,
    340–341, 344
  Chebyshev's inequality, 148–149
  Jensen's inequality, 292–294

Markov's inequality, 141–142
  for standard deviation of sums, 344–345
inferences, 1
informational asymmetry
  Bayes' theorem in, 72–75
  in insurance, 57
insolvency problem, 9
instantaneous relative frequency density,
    39
insurance
  actuarial science for, 6–9
  Bayes' theorem in, 72–74, 123–124
  beta distribution for, 265
  binomial distributions in, 188, 192–194
  with caps and deductibles, 368–372
    combination contracts, 300–302
    excess-of-loss contracts, 297–298
    limited excess-of-loss contracts,
      298–299
    option contracts, 302–303
    proportional insurance contracts,
      300
  conditional probability in, 64–68
  DeMoivre distribution for, 243–244
  endowment, 308–309
  exponential distribution for, 225–226
  gamma distribution for, 231
  informational asymmetry in, 57, 72–74
  law of large numbers in, 345–349
  law of total probability in, 69–72,
    121–123
  life annuities, 309–311
  lognormal distribution for, 259–260
  mixtures in, 366–373
  negative binomial distributions in, 201
  normal distribution for, 251–253, 255
  Pareto distribution for, 235
  Poisson distribution for, 195, 199
  risk sharing in, 351
  special transformations for, 286
  term life, 307–308
  Weibull distribution for, 240–241
  whole life, 304–307
insurance principle, 6–9
insured persons, 304
interest, 434
Internet service provider calls, exponential
    distribution for, 225
interpretations of probability, 11–13
intersection, set, 86
inverse transformations, 285–286
investments
  binomial distributions in, 188–189, 194
  defined benefit pension plans, 108
  geometric expectation in, 131–133
  growth in, 42–43
    depressed returns in, 47–48
    geometric and arithmetic, 44–47
    investments over several periods in,
      44
    investments over single periods in,
      43–44

moment and cumulant generating
    functions with, 160
  summary, 48–49
lognormal distribution for, 258–259
model for. *See* Markowitz investment
    portfolio selection model
moment and cumulant generating
    functions with, 160
risk-reward analysis for, 345

Jensen's inequality, 292–294
joint density, 111–113, 117–118, 326–331
joint distributions, 29, 111–112
jump discontinuities, 100–101

Kepler, Johannes, 2
Keynes, John Maynard, 2
Kolmogorov, A. N., 4
kth central moments, 152
kth cumulants, 159
kth moments, 152–153, 232–233
kurtosis, 152

lack of memory property, 207
Laplacian distribution, 148–149, 357
law of large numbers, 345–349
  accumulated winnings in, 351–352
  proof of, 349–351
  risk sharing in, 351
law of total probability, 58, 68–69
  distribution of sums, 331–336
  distributional form, 119–123
  examples, 69–72
layers
  in combination contracts, 300–302
  in limited excess-of-loss contracts, 299
l'Hôpital's rule, 353
life annuities, 309–311
life expectancy in future lifetimes, 39–41
life insurance
  endowment insurance, 308–309
  future lifetimes in, 36
  term, 307–308
  whole life, 304–307
lifetimes
  future, 36–38
    distribution of probability in, 38–39
    life expectancy in, 39–41
    summary, 41–42
  mean residual, 296
  modeling
    DeMoivre distribution, 241–245
    Weibull distribution, 235–241
  survival and hazard functions for,
    167–168
light bulb reliability, geometric distribution
    for, 208–209
limited excess-of-loss contracts, 298–299
limited expected value, 280, 294–297
limiters in electrical circuits, special
    transformations for, 286–287
limiting relative frequency, 3

linear transformations, 280, 282–283
loans
    informational asymmetry in, 74–75
    law of total probability in, 70–71
logical approaches, 3
lognormal distribution
    arithmetic operations with, 258
    definition and interpretation, 256–257
    examples, 258–260
    mean, variance, and higher moments
        for, 257
    probability calculations for, 257
    probability density function for, 257
    relationship to other distributions, 258
long conditions in tangency portfolios,
        408–409
long positions for options, 302
longitudinal study design, 37
losses
    average, 23
    in insurance principle, 7–8

marginal density, 111, 113, 333
marginal distributions, 34, 110–111, 365
marine insurance, 2
Markowitz, Harry, 10
Markowitz investment portfolio selection
        model, 396
    and CAPM, 411–414
    for portfolios
        of many securities, 409–411
        of two risky securities and a risk-free
            asset, 403–409
        of two securities, 396–403
mass functions. *See* probability mass
        functions
mass plots, 22, 35
Mathematica commands, 432–433
mathematics, financial, 434–435
mean absolute deviation, 143–144
mean residual lifetimes, 296
means
    beta distribution, 261
    binomial distribution, 190–191
    Cauchy distributions, 319
    compound distributions, 382
    DeMoivre distribution, 242
    with expected values, 131
    exponential distribution, 223
    gamma distribution, 228
    geometric distribution, 207–208
    lognormal distribution, 257
    mixtures, 373–378
    negative binomial distribution, 203
    normal distribution, 248
    Pareto distribution, 234
    Poisson distribution, 196–197
    Weibull distribution, 237

means inequality, arithmetic-geometric,
        46–47, 133–134
measure of association, covariance as,
        341–342
measurements
    normal distribution for, 253–255
    uncertainty in, 5
mechanical systems, reliability of, 312–316
medians, 126, 144
memoryless property, 207
method of convolutions, 326, 336–337
minimum and maximum transformations,
        311–312, 316–317
mixed distributions, 94–96, 100–102
    examples, 102
    mixtures as, 365
mixed random variables, 96
    densities for, 181–185
    expectation for, 135–137
mixtures, 122, 363
    definition and properties, 363–365
    in insurance, 366–373
    as marginal distributions, 365
    mean and variance of, 373–378
    as mixed distributions, 365
    moment generating functions of,
        378–379
    vs. sums, 125–126, 366
modeling
    lifetimes
        DeMoivre distribution, 241–245
        Weibull distribution, 235–241
    portfolio. *See* Markowitz investment
        portfolio selection model
    probability, 13–14
    uncertain sizes
        exponential distribution for, 221–226
        gamma distribution for, 226–233
        Pareto distribution, 233–235
modes, 210
moment generating functions
    for compound distributions, 381
    definition, 156
    economic interpretations of, 160–161
    examples, 163–167
    graphical properties of, 161–163
    of mixtures, 378–379
    for normal distributions, 157–158
    for probability distributions, 155–167
    properties of, 156–157
moments, 152–153
    beta distribution, 261
    binomial distribution, 190–191
    DeMoivre distribution, 242
    e^x, 163
    exponential distribution, 223
    gamma distribution, 228
    geometric distribution, 207–208
    lognormal distribution, 257
    negative binomial distribution, 203
    normal distribution, 248
    Pareto distribution, 234

Poisson distribution, 196–197
    Weibull distribution, 237
Monte Carlo simulation, 243
moral hazard problem, 9
multiple components and processes
        systems, reliability of, 311–317
multiple layer contracts, 301
multiplication rule, 87
multiplication with random variables,
        124–125
multivariate distributions, 29
mutual fund separation theorem, 404
mutual funds, model for. *See* Markowitz
        investment portfolio selection
        model

n-fold convolutions, 337
negative binomial distribution
    arithmetic operations with, 204
    compound, 384
    definition and interpretation, 200–201
    examples, 204–206
    mean, variance and higher moments for,
        203
    probability calculations for, 203–204
    probability mass function for, 202–203
    relationship with other distributions,
        204
negative exponential transformations, 285
negatively correlated assets, portfolio
        opportunity sets with, 402
negatively skewed distributions, 150
no arbitrage, principle of, 10
nonzero probability regions
    with joint density, 328, 330
    for uniform distributions, 342–343
normal distribution
    arithmetic operations with, 249
    definition and interpretations, 245–246
    examples, 251–255
    mean, variance, and higher moments
        for, 248
    moment generating function with,
        157–158
    probability calculations for, 248–249
    probability density function for,
        246–248
    relationship to other distributions,
        249–251
normal power approximations
    derivation of, 355–356
    formulas for, 354–355

objectivist interpretation of probability,
        11–12
observations, groups of, 58–59
one-to-one order preserving
        transformations, 290
opportunity sets, portfolio. *See* portfolios
        and portfolio opportunity sets
optimal portfolios, 399–401
optimality, principle of, 10

option contracts, 302–303
options, 10
overall risks, 8

parallel processors, 315
parallel systems, reliability of, 312–316
parameters, 13
Pareto distribution
  arithmetic operations with, 235
  definition and interpretation, 233–234
  example, 235
  generalized, 277
  mean, variance, and higher moments
    for, 234
  probability calculations for, 235
  probability density function for, 234
  relationship with other distributions,
    235
particle velocity, transformations for, 288
parts manufacturing, binomial distribution
  for, 194
Pascal, Blaise, 2
payoffs
  contingency tables in, 33–34
  distribution of probability in, 21–22,
    26–27
  expected, 23–24
  independence of random variables in,
    31–33
  with not identical distributions, 27–28
  in simple games, 19–25
  summary, 24–25, 34–35
  with unfair coins, 28–31
payouts, insurance, 366
  combination contracts, 300–302
  excess-of-loss contracts, 297–298
  limited excess-of-loss contracts,
    298–299
  option contracts, 302–303
  proportional insurance contracts, 300
  with uncertain claim frequency, 372–373
  with uncertain loss occurrence, 367–372
pension plan example, 108
percentage returns in compounding, 47–48
perfectly correlated assets, portfolio
  opportunity sets with, 401–402
permutations, 88
plots, 22, 35
points of discontinuity, 100–101
Poisson distribution
  arithmetic operations with, 197–198
  compound, 382–383
  definition and interpretation of, 195
  examples, 198–200
  for expectation, 134
  mean, variance, and higher moments of,
    196–197
  probability calculations for, 197
  probability mass function for, 196
  relationship with other distributions,
    198
    exponential distribution, 224–225

negative binomial distribution, 204
  normal distribution, 249–250
Polya distribution, 201
portfolios and portfolio opportunity sets,
  397–399
  efficient, 399, 403–405
  model for. *See* Markowitz investment
    portfolio selection model
  optimal, 399–401
  with perfectly correlated assets,
    401–402
  with risk free assets, 402–403
  tangency, 405–407
    allocations in, 408
    long and short conditions in, 408–409
    risk-free rate changes in, 407
  with uncorrelated assets, 402
positively correlated assets, portfolio
  opportunity sets with, 401–402
positively skewed distributions, 150
posterior probability, 74
power transformations, 280, 283–284,
  287–288
predictability in insurance principle, 7–8
prediction error, 143–144
premiums in insurance, 8–9
present value, 305
pricing in insurance, 8–9
principle of equal likelihood, 3
principle of no arbitrage, 10
principle of optimality, 10
prior probability, 74
probability, 1–2
  classical. *See* classical probability
  distribution of, 26–27
  in engineering and sciences, 5–6
  with events, 60
  interpretations of, 11–13
  measures of, 62
  modeling of, 13–14
  relative frequency in, 3
probability calculations
  beta distribution, 261–262
  binomial distribution, 191
  DeMoivre distribution, 242
  exponential distribution, 223
  gamma distribution, 229
  geometric distribution, 208
  lognormal distribution, 257
  negative binomial distribution,
    203–204
  normal distribution, 248–249
  Pareto distribution, 235
  Poisson distribution, 197
  Weibull distribution, 237
probability density functions, 39, 97–100
  beta distribution, 261
  DeMoivre distribution, 242
  exponential distribution, 222–223
  gamma distribution, 226–228
  lognormal distribution, 257
  normal distribution, 246–248

Pareto distribution, 234
  Weibull distribution, 237
probability mass functions, 21–23, 97
  derivation of, 190
  for geometric distribution, 207
  for negative binomial distribution,
    202–203
  for Poisson distribution, 196
probability rules, set theory for, 89–90
processes, 20
products
  distributions of, 337–339
  expectations of, 339
    Cauchy-Schwarz inequality for,
      340–341
    correlation in, 342–345
    covariance in, 341–342
    formulas for, 339–340
  with random variables, 124–125
proportional insurance contracts, 300
put options, 302

quadratic normal power approximations,
  354
quality control
  beta distribution for, 263
  probability in, 5
quantification of uncertainty, 2–5
quantitative inferences, 1
queuing, 5–6
quota-share contracts, 300
quotients, distributions of, 337–339

random experiments, 91–92
random sampling, binomial distributions
  in, 189
random variables, 20, 91–92
  arithmetic operations on, 124–125
  continuous, discrete, and mixed, 96
  dependence and independence of,
    113–119
  equality and equivalence of, 102–104
  expectation of, 137–139
  independence of, 31–33
  indicator, 29
  transformations of. *See* transformations
    of random variables
random vectors, 29, 104–113
ranges in distributions, 27
rare events, Poisson distribution for, 195
ratios for expectation deviation, 143
Rayleigh density, 241
Rayleigh distribution, 241, 358
regions of nonzero probability
  with joint density, 328, 330
  for uniform distributions, 342–343
relative frequency, 2–3
relative frequency density, 39
reliability
  geometric distribution for, 208–209
  probability in, 5

of systems with multiple components or processes
  minimum and maximums in, 311–312
  series and parallel systems, 312–316
reliability function, 167
reserves in insurance, 9
risk-adjusted rate returns, 10
risk averse persons, 27–28
risk-classification schemes, 57
risk-free assets
  in Markowitz investment portfolio selection model, 403–409
  portfolio opportunity sets with, 402–403
risk-free rate changes in tangency portfolio, 407
risk-neutral investors, 160
risk-reward analysis, 344–345
risk sharing, 351
risk tolerant persons, 28
risks in insurance principle, 7–8
risky securities in Markowitz investment portfolio selection model, 403–409

sample space, 20
  definition, 59
  importance of, 60–62
samples in quality control, 5
scalar-to-scalar transformations, 281–282
  examples, 287–289
  exponential, 285
  inverse, 285–286
  linear, 282–283
  negative exponential, 285
  power, 283–284
  special, 286–287
scale with skewness, 150
scaled cross sections, conditional densities as, 116
science measurements, normal distribution for, 253–255
sciences, probability in, 5–6
securities. See investments
separation theorem, mutual fund, 404
series systems, reliability of, 312–316
sets, 58, 85–86
  discrete, 20n
  for probability rules, 89–90
Sharpe, William, 396
Sharpe ratio, 405n
short conditions in tangency portfolios, 408–409
short positions for options, 302
sign function, 151
signs
  of covariance, 342–343
  with skewness, 150–151
simple games, payoffs in, 19–25
simple growth, 42–43
  depressed returns in, 47–48
  geometric and arithmetic expectations in, 44–47

investments over several periods in, 44
investments over single periods in, 43–44
summary, 48–49
single jump discontinuities, 100–101
single layer contracts, 300
single period investments, growth in, 43–44
skewed distributions, 150
skewness
  compound distributions, 382
  exponential distribution, 223
  gamma distribution, 228
  lognormal distribution, 257
  normal distribution, 248
  Poisson distribution, 196–197
  properties of, 151
special distributions, graphs of, 432–433
special transformations, 286–287
standard deviation
  definition, 144
  properties of, 145
  in skewed distributions, 150
  of sums, inequalities for, 344–345
standard form of X, 246
standard normal distribution
  definition, 245
  values of, 430–431
standard normal random variables, 245
stochastic processes, 20
stocks. See investments
subjectivist interpretation of probability, 11–13
subscriptions, negative binomial distribution for, 205
subsets, 85
sufficiently large groups in insurance principle, 7–8
sums
  central limit theorem for, 352–353
  distributions of, 325–326
    convolutions in, 336–337
    joint density in, 326–331
    law of total probability in, 331–336
  expectations of, 339
    Cauchy-Schwarz inequality for, 340–341
    correlation in, 342–345
    covariance in, 341–342
    formulas for, 339–340
  law of large numbers for, 345–349
    accumulated winnings in, 351–352
    proof of, 349–351
    risk sharing in, 351
  vs. mixtures, 125–126, 366
  normal power approximations for, 354–356
  with random variables, 124–125
  standard deviation of, inequalities for, 344–345
surges of electrical current, gamma distribution for, 232

survival functions, 39, 168
  definition, 167
  for minimums, 312
  properties of, 169–170
  relationship to hazard functions, 169
systems with multiple components or processes, reliability of, 311–317

tangency portfolios, 405–407
  allocations in, 408
  long and short conditions in, 408–409
  risk-free rate changes in, 407
technology in financial engineering, 9
teen drug and alcohol study example, 61–64
telephone hotline service, Poisson distribution for, 198
term life insurance, 307–308
three-dimensional mass plots, 35
time to failure, Weibull distribution for, 238–239
time value of money, 304
tire lifetimes, Weibull distribution for, 238
total expectation, formula for, 374–375
total probability, law of, 58, 68–69
  distribution of sums, 331–336
  distributional form, 119–123
  examples, 69–72
total variance, formula for, 375
transformations of random variables
  exponential, 285
  for insurance with caps, 297–303
  inverse, 285–286
  Jensen's inequality for, 292–294
  for life insurance and annuity contracts, 303–311
  limited expected value for, 294–297
  linear, 282–283
  minimum and maximum, 311–312, 316–317
  negative exponential, 285
  power, 280, 283–284, 287–288
  for reliability of system with multiple components and processes, 311–317
  special, 286–287
  transformed random variables in
    distributions of, 281–289
    expectation of, 289–297
  trigonometric, 317–319
trigonometric transformations, 317–319
truncated exponential distributions
  definition, 136
  with moment generating functions, 165–166
  variance of, 146–147

unbiased coins, 3n
uncertain sizes, modeling
  exponential distribution for, 221–226
  gamma distribution for, 226–233
  Pareto distribution, 233–235

uncertainty in measurement, 5
uncorrelated assets, portfolio opportunity
    sets with, 402
unfair coins, probability distributions with,
    28–31
uniform distributions
    definition, 330n
    regions of nonzero probability for,
        342–343
unions, set, 85
univariate conditional density, 116
universes, set, 85
unpredictable losses, 7–8

variables. *See* random variables
variance
    beta distribution, 261
    binomial distribution, 190–191
    compound distributions, 382
    definition, 144
    DeMoivre distribution, 242
    exponential distribution, 223
    gamma distribution, 228
    geometric distribution, 207–208
    lognormal distribution, 257
    mixtures, 373–378
    negative binomial distribution, 203
    normal distribution, 248
    Pareto distribution, 234
    Poisson distribution, 196–197
    properties of, 145–148
    skewed distributions, 150
    truncated exponential distributions,
        146–147
    Weibull distribution, 237
vector-to-scalar transformations, 282
    definition, 281
    for sums, 325–326
velocity of particles, transformations for,
    288
Venn diagrams, 62–64

voltage measurements, normal distribution
    for, 254

Website visits, Poisson distribution for, 199
Weibull distribution
    arithmetic operations with, 238
    definition and interpretations, 235–236
    examples, 238–241
    for hazard functions, 169
    mean, variance, and higher moments
        for, 237
    probability calculations for, 237
    probability density function for, 237
    relationship with other distributions,
        238
weighting in insurance principle, 8
whole life insurance, 304–307
work time, beta distribution for, 263–264

zero beta companion portfolios, 411